T0324152

Biological Approaches
to Controlling
Pollutants

Advances in Pollution Research
Biological Approaches to Controlling Pollutants

Edited by

Sunil Kumar

Faculty of Biosciences, Institute of Biosciences and Technology,
Shri Ramswaroop Memorial University, Barabanki,
Uttar Pradesh, India

Muhammad Zaffar Hashmi

Department of Chemistry, COMSATS University Islamabad,
Islamabad, Pakistan

ELSEVIER

WP
WOODHEAD
PUBLISHING
An imprint of Elsevier

Woodhead Publishing is an imprint of Elsevier
The Officers' Mess Business Centre, Royston Road, Duxford, CB22 4QH, United Kingdom
50 Hampshire Street, 5th Floor, Cambridge, MA 02139, United States
The Boulevard, Langford Lane, Kidlington, OX5 1GB, United Kingdom

Notices
Knowledge and best practice in this field are constantly changing. As new research and experience broaden
our understanding, changes in research methods, professional practices, or medical treatment may
become necessary.

Practitioners and researchers must always rely on their own experience and knowledge in evaluating
and using any information, methods, compounds, or experiments described herein. In using such
information or methods they should be mindful of their own safety and the safety of others, including
parties for whom they have a professional responsibility.

To the fullest extent of the law, neither the Publisher nor the authors, contributors, or editors, assume
any liability for any injury and/or damage to persons or property as a matter of products liability,
negligence or otherwise, or from any use or operation of any methods, products, instructions, or ideas
contained in the material herein.

Library of Congress Cataloging-in-Publication Data
A catalog record for this book is available from the Library of Congress

British Library Cataloguing-in-Publication Data
A catalogue record for this book is available from the British Library

ISBN: 978-0-12-824316-9

For information on all Woodhead Publishing publications visit our
website at https://www.elsevier.com/books-and-journals

Publisher: Candice Janco
Acquisitions Editor: Marisa LaFleur
Editorial Project Manager: Naomi Robertson
Production Project Manager: Sruthi Satheesh
Cover Designer: Mathew Limbert

Typeset by TNQ Technologies

Contents

CHAPTER 5 **Advances in biodegradation and bioremediation of environmental pesticide contamination**...............79
Shubhra Sharma, Shikha Saxena, Bhawana Mudgil and Siddharth Vats

CHAPTER 9 Advances in bioremediation of industrial wastewater containing metal pollutants 163

Vadivel Karthika, Udayakumar Sekaran,
Gulsar Banu Jainullabudeen and
Arunkumar Nagarathinam

CHAPTER 12   Advances in bioremediation of organometallic pollutants: strategies and future road map..........233

K.S. Vinayaka and Supreet Kadkol

CHAPTER 13   Bioremediation of polycyclic aromatic hydrocarbons from contaminated dumpsite soil in Chennai city, India 241

Sancho Rajan, V. Geethu and Paromita Chakraborty

Contributors

Jafar Ali
Department of Biotechnology, University of Sialkot, Sialkot, Punjab, Pakistan; Research Center for Eco-environmental Sciences, Chinese Academy of Sciences, Haidian, Beijing, PR China

Mahwish Ali
Department of Biological Sciences, National University of Medical Sciences (NUMS), Rawalpindi, Punjab, Pakisan

R. Anandan
Department of Genetics and Plant Breeding, Faculty of Agriculture, Annamalai University, Chidambaram, Tamil Nadu, India

Katerina Atkovska
Faculty of Technology and Metallurgy, Ss. Cyril and Methodius University, Skopje, Republic of North Macedonia

Shehla Batool
Department of Environmental Sciences, Faculty of Biological Sciences, Quaid-i-Azam University, Islamabad, Punjab, Pakistan

Paromita Chakraborty
Department of Civil Engineering, SRM Institute of Science and Technology, Kancheepuram, Tamil Nadu, India; SRM Research Institute, SRM Institute of Science and Technology, Kancheepuram, Tamil Nadu, India

Rati Chandra
Faculty of Biotechnology, Institute of Bio-Sciences and Technology, Shri Ramswaroop Memorial University, Barabanki, Uttar Pradesh, India

Tanushri Chatterji
Department of Microbiology, Babu Banarasi Das College of Dental Sciences, Babu Banarasi Das University, Uttar Pradesh, Lucknow, India

Barbara Clasen
Department of Environmental Science, State University of Rio Grande do Sul, Porto Alegre, RS, Brazil

Akhilesh Dubey
Department of Biological Sciences and Engineering, Netaji Subhas University of Technology, Delhi, India

Abida Farooqi
Department of Environmental Sciences, Faculty of Biological Sciences, Quaid-i-Azam University, Islamabad, Punjab, Pakistan

Mrinmoy Garai
Materials Science Centre, Indian Institute of Technology (IIT), Kharagpur, West Bengal, India

R.K. Gaur
Department of Biotechnology, Deen Dayal Upadhyay University, Gorakhpur, Uttar Pradesh, India

V. Geethu
Department of Civil Engineering, New Horizon College of Engineering, Bangalore, Karnataka, India

Saima Gul
Department of Chemistry, Islamia College Peshawar, Peshawar, Khyber Pakhtunkhwa, Paksitan

Neeraj Gupta
Faculty of Biosciences, Institute of Biosciences and Technology, Shri Ramswaroop Memorial University, Barabanki, Uttar Pradesh, India

Muhammad Zaffar Hashmi
Department of Chemistry, COMSATS University Islamabad, Islamabad, Pakistan

Sajjad Hussain
Faculty of Materials and Chemical Engineering, GIK Institute of Engineering Sciences & Technology, Topi, Khyber Pakhtunkhwa, Pakistan; Faculdade de Engenharias, Arquitetura e Urbanismo e Geografia, Universidade Federal de Mato Grosso do Sul, Cidade Universitária, Campo Grande, MS, Brazil

Pankaj Kumar Jain
Indira Gandhi Centre for Human Ecology, Environmental and Population Studies, Department of Environmental Science University of Rajasthan, Jaipur, Rajasthan, India

Gulsar Banu Jainullabudeen
Central Institute for Cotton Research, Regional Station, Coimbatore, Tamil Nadu, India

Supreet Kadkol
Department of Zoology, Sri Venkataramana Swamy College, Dakshina Kannada, Karnataka, India

Muhammad Kaleem
Department of Plant Sciences, Quaid-i-Azam University, Islamabad, Pakistan

Vadivel Karthika
Department of Crop Management, Kumaraguru Institute of Agriculture, Erode, Tamil Nadu, India

Hammad Khan
Faculty of Materials and Chemical Engineering, GIK Institute of Engineering Sciences & Technology, Topi, Khyber Pakhtunkhwa, Pakistan

Ibrar Khan
Department of Microbiology, Abbottabad University of Science & Technology, Havelian, Khyber Pakhtunkhwa, Pakistan

Abeer Khan
Department of Biotechnology, University of Sialkot, Sialkot, Punjab, Pakistan

Khurram Imran Khan
Faculty of Materials and Chemical Engineering, GIK Institute of Engineering Sciences & Technology, Topi, Khyber Pakhtunkhwa, Pakistan

Sabir Khan
São Paulo State University (UNESP), Institute of Chemistry, Araraquara, São Paulo, Brazil

Anand Kumar
Department of Biotechnology, Faculty of Engineering and Technology, Rama University, Kanpur, Uttar Pradesh, India

Sunil Kumar
Faculty of Biosciences, Institute of Biosciences and Technology, Shri Ramswaroop Memorial University, Barabanki, Uttar Pradesh, India

Stefan Kuvendziev
Faculty of Technology and Metallurgy, Ss. Cyril and Methodius University, Skopje, Republic of North Macedonia

Kiril Lisichkov
Faculty of Technology and Metallurgy, Ss. Cyril and Methodius University, Skopje, Republic of North Macedonia

Aroosa Malik
Department of Environmental Sciences, Faculty of Biological Sciences, Quaid-i-Azam University, Islamabad, Punjab, Pakistan

Sarada Prasanna Mallick
Department of Biotechnology, Koneru Lakshmaiah Education Foundation, Guntur, Andhra Pradesh, India

Mirko Marinkovski
Faculty of Technology and Metallurgy, Ss. Cyril and Methodius University, Skopje, Republic of North Macedonia

Neha Maurya
Faculty of Biotechnology, Institute of Bio-Sciences and Technology, Shri Ramswaroop Memorial University, Barabanki, Uttar Pradesh, India

Hamdije Memedi
Department of Chemistry, Faculty of Natural Sciences and Mathematics,
University of Tetovo, Tetovo, Republic of North Macedonia

Bhawana Mudgil
TGT, Natural Science, Sarvodaya Vidyalaya, Rohini, Delhi, India

Abdul Samad Mumtaz
Department of Plant Sciences, Quaid-i-Azam University, Islamabad, Pakistan

M. Muthukumaran
PG and Research Department of Botany, Ramakrishna Mission Vivekananda
College (Autonomous), Affiliated to the University of Madras, Chennai, Tamil
Nadu, India

Arunkumar Nagarathinam
Department of Microbiology, School of Agriculture and Animal Sciences,
The Gandhigram Rural Institute, Dindigul, Tamil Nadu, India

S. Nalini
Centre for Ocean Research (DST-FIST Sponsored Centre), MoES — Earth
Science & Technology Cell (Marine Biotechnological Studies), Col. Dr. Jeppiaar
Research Park, Sathyabama Institute of Science and Technology, Chennai,
Tamil Nadu, India

Lini Nirmala
Department of Biotechnology, Mar Ivanios College, Thiruvananthapuram, Kerala,
India

R. Parthasarathi
Department of Agricultural Microbiology, Faculty of Agriculture, Annamalai
University, Chidambaram, Tamil Nadu, India

Blagoj Pavlovski
Faculty of Technology and Metallurgy, Ss. Cyril and Methodius University, Skopje,
Republic of North Macedonia

M. Prakash
Department of Genetics and Plant Breeding, Faculty of Agriculture, Annamalai
University, Chidambaram, Tamil Nadu, India

Zainab Rafique
Department of Biotechnology, University of Sialkot, Sialkot, Punjab, Pakistan

Sancho Rajan
Department of Civil Engineering, SRM Institute of Science and Technology,
Kancheepuram, Tamil Nadu, India

V. Ramamoorthy
Department of Plant Pathology, Agricultural College and Research Institute, Tamil Nadu Agricultural University, Madurai, Tamil Nadu, India

Arianit A. Reka
Department of Chemistry, Faculty of Natural Sciences and Mathematics, University of Tetovo, Tetovo, Republic of North Macedonia; NanoAlb, Albanian Unit of Nanoscience and Nanotechnology, Academy of Sciences of Albania, Fan Noli Square, Tirana, Albania

Shikha Saxena
Faculty of Biotechnology, Institute of Bio-Sciences and Technology, Shri Ramswaroop Memorial University, Barabanki, Uttar Pradesh, India

Udayakumar Sekaran
Department of Plant and Environmental Sciences, Clemson University, Clemson, SC, United States

Shreya Sharma
Department of Biological Sciences and Engineering, Netaji Subhas University of Technology, Delhi, India

Shubhra Sharma
Faculty of Biotechnology, Institute of Bio-Sciences and Technology, Shri Ramswaroop Memorial University, Barabanki, Uttar Pradesh, India

Deepti Singh
Faculty of Biosciences, Institute of Biosciences and Technology, Shri Ramswaroop Memorial University, Barabanki, Uttar Pradesh, India

Prama Esther Soloman
Indira Gandhi Centre for Human Ecology, Environmental and Population Studies, Department of Environmental Science University of Rajasthan, Jaipur, Rajasthan, India

Swati Srivastava
Faculty of Biosciences, Institute of Biosciences and Technology, Shri Ramswaroop Memorial University, Barabanki, Uttar Pradesh, India

Shreya Srivastava
Faculty of Biotechnology, Institute of Bio-Sciences and Technology, Shri Ramswaroop Memorial University, Barabanki, Uttar Pradesh, India

Shiburaj Sugathan
Department of Botany, University of Kerala, Thiruvananthapuram, Kerala, India

M. Theradimani
Department of Plant Pathology, Agricultural College and Research Institute, Tamil Nadu Agricultural University, Madurai, Tamil Nadu, India

Sana Ullah
Faculty of Materials and Chemical Engineering, GIK Institute of Engineering Sciences & Technology, Topi, Khyber Pakhtunkhwa, Pakistan

Siddharth Vats
Faculty of Biotechnology, Institute of Bio-Sciences and Technology, Shri Ramswaroop Memorial University, Barabanki, Uttar Pradesh, India

K.S. Vinayaka
Plant Biology Lab., Department of Botany, Sri Venkataramana Swamy College, Dakshina Kannada, Karnataka, India

Hassan Waseem
Department of Biotechnology, University of Sialkot, Sialkot, Punjab, Pakistan

Shriyam Yadav
Faculty of Biotechnology, Institute of Bio-Sciences and Technology, Shri Ramswaroop Memorial University, Barabanki, Uttar Pradesh, India

Acknowledgements

Thanks to the Higher Education Commission of Pakistan's National Research Program for Universities, Projects 7958 and 7964; and the Pakistan Science Foundation, Project PSF/Res/CP/C-CUI/Envr (151). Thanks are also due to the Pakistan Academy of Sciences, Project 3-9/PAS/98, for funding.

Sunil Kumar thanks the Shri Ramswaroop Memorial University, Barabanki (UP), India for continuous support and assistance during the work and scientific writing.

Advances in bioremediation: introduction, applications, and limitations

Anand Kumar[1], Sarada Prasanna Mallick[2], Deepti Singh[3], Neeraj Gupta[3]

[1]*Department of Biotechnology, Faculty of Engineering and Technology, Rama University, Kanpur, Uttar Pradesh, India;* [2]*Department of Biotechnology, Koneru Lakshmaiah Education Foundation, Guntur, Andhra Pradesh, India;* [3]*Faculty of Biosciences, Institute of Biosciences and Technology, Shri Ramswaroop Memorial University, Barabanki, Uttar Pradesh, India*

Chapter outline

1.1 Introduction

Environmental biotechnology is an old field; composting and wastewater treatments are common examples of older environmental biotechnologies. Current studies in ecology and molecular biology present opportunities for extra-efficient biological processes. Notable accomplishments of these studies include the cleanup of polluted water and land areas. Bioremediation is a process in which organic wastes are biologically degraded under controlled conditions to levels below the concentration limits established by regulatory authorities or to innocuous states (Mueller et al., 1996). In other words, bioremediation is the use of living organisms, mainly microorganisms, to degrade environmental pollutants into less toxic forms. It mainly uses bacteria and fungi or plants to degrade or detoxify substances harmful to human health and the environment. The microorganisms and plants may be native to a contaminated area or collected from elsewhere and brought to the contaminated site. Pollutants are

Biological Approaches to Controlling Pollutants. https://doi.org/10.1016/B978-0-12-824316-9.00003-3

transformed by living organisms by biochemical reactions that occur as a part of their metabolic processes. Biodegradation of pollutants is a consequence of the actions of multiple organisms. When microorganisms are added to a contaminated site to supplement and improve degradation, the process is known as bioaugmentation. In the bioremediation process, microorganisms enzymatically attack pollutants and convert them to harmless products. Bioremediation can be effective only when environmental conditions allow microbial growth and activity; its application often involves the manipulation of environmental parameters to allow microbial growth and degradation to proceed at a faster rate.

Bioremediation techniques are typically more economically feasible than traditional methods such as incineration because some pollutants can be treated on site, thus minimizing exposure risks for cleanup personnel or potentially wider exposure as a result of transportation accidents. Because bioremediation is based on natural attenuation of pollutants, it is considered more acceptable than other technologies.

Most bioremediation systems are run under aerobic conditions, but running a system under anaerobic conditions may permit microbial organisms to degrade otherwise recalcitrant molecules (Colberg and Young, 1995). As with other technologies, bioremediation has its limitations.

Some pollutants such as high aromatic hydrocarbons and chlorinated organics are resistant to microbial attack. They are degraded either slowly or not at all; hence, it is not easy to predict the rates of cleanup for a bioremediation exercise; there are no ways to predict whether a contaminant can be degraded.

The traditional method of remediation has been to excavate polluted soil and eliminate it to a landfill and treat the polluted areas of a location. The methods have some disadvantages. An approach superior to these conventional methods is to entirely destroy the pollutants if possible or at least convert them to nondangerous substances. Several tools have been used including incineration and many types of chemical decomposition (e.g., base-catalyzed dechlorination and UV oxidation). They may be very efficient at minimizing the degree of pollutants but have many disadvantages, primarily their technological complexity, the price for small-scale application, and in particular, incineration, which increases the exposure to pollutants for both the workers on location and nearby residents (Vidali, 2001).

Soil pollutants with petroleum hydrocarbons, halogenated organic chemicals, persistent organic pollutants, and toxic metals are a crucial worldwide problem disturbing human and ecological health. Over the last half century, scientific and industrial advancements have led to the growth of many brownfields; principally, these are placed in the center of highly populated cities worldwide. Reestablishing and regenerating cities in a sustainable way for beneficial uses are key priorities for all industrialized nations. Bioremediation is believed to be a safe, cost-effective, competent, and sustainable technology for restoring contaminated sites (Megharaj and Naidu, 2017).

It has been reported that many microorganisms can biodegrade pollutants. However, the pace of biodegradation depends on the physiological condition of the microorganisms, which are susceptible to variable environmental factors. It is identified that immobilization increases microorganism resistance to unfavorable environmental impacts.

Bioremediation is noninvasive, eco-friendly, less expensive than conventional methods, and furthermore, is a permanent solution that can result in the degradation or transformation of environmental pollutants into risk-free or less toxic forms. Soil bioremediation can be carried out at the place of contamination or in a specially prepared place. In situ technology is used when there is no possibility to transfer polluted soil—for example, when pollutants affect an extensive area.

There are three basic methods of in situ bioremediation with microorganisms: natural attenuation, biostimulation, and bioaugmentation (Dzionek et al., 2016).

1.2 Applications of bioremediation

Ecological pollution and its remediation are leading concerns across the globe. An enormous number of contaminants such as fertilizers, pesticides, hazardous hydrocarbons (oily waste), toxic heavy metals, and dyes are the key agents responsible for environmental damage. Bioremediation is an environment-oriented process that degrades highly toxic hazardous substances into less toxic forms.

1.2.1 Solid waste management and sewage treatment

Solid waste management is a globally acknowledged environmental concern. As a consequence of civilization, urbanization, and industrialization, a large amount of waste is produced and dumped into the environment throughout the year. All over the world, waste generation levels are increasing. In 2016, some reports state that the world's urban areas produced 2.0 billion tons of municipal solid waste, equivalent to a footprint of 0.74 kg per person per day (Pandey et al., 2019). Bioremediation is known as an effective approach to minimizing residual contaminants and restoring polluted sites back to their original forms. Bioremediation offers a good opportunity to resolve the issue of solid waste management of unwanted detoxifying components and harmful dumps (Saxena and Bharagava, 2015). The waste management system includes land farming, composting, and soil piles. It is highly efficient in the remediation of organic wastes, hazardous domestic waste, industrial effluents, municipal solid, and sewage wastes. Because of their comparatively less expensive and ecological implications, it ensures attractive and more conventional decontaminating techniques (Muangchinda et al., 2018; Bharagava et al., 2017). Composting is a controlled transformation of decomposable organic wastes matter into stable inorganic by-products with the help of microbes. Composting is one of the safest approaches to solid waste management. It is an aerobic process and converts several complex decayable waste materials into natural products that can be used safely and beneficially as organic fertilizers and soil conditioners. This comprises detoxifying and mineralization, wherein the waste has been consistently reformed into basic natural substances. It has assisted in the prevention of greenhouse impacts by mitigating the production of gases such as methane, whereas carbon dioxide is released by composting, which is minimal compared with alternative methods of waste management (Ayilara et al., 2020). Meanwhile, if environmental pollution is steady, biodegradation is commonly sustained within several stages by utilizing various enzymes or microbial residents. Microorganisms have immensely enhanced the rehabilitation of polluted habitats by mopping up waste in an ecologically safer way including the production of reliable outcomes (Pande et al., 2020). Land filling raises the aerobic degradation approach by assisting the development of microbes that naturally occur (Shinde, 2018). Biopiles are a mixture of landfarming and composting. In biopiles, artificially engineered cells are developed as oxygenated manured piles, and further action is retained by adding compost to the polluted soil. Biopiles are used in the elimination of petroleum hydrocarbons (PHCs) and also control the physical losses of contaminants through evaporation and leaching; therefore, biopiling is a pure form of landfarming (Shinde, 2018). Groundwater quality has emerged as an essential issue of this era because of the growing scarcity of water resources. Deterioration of water quality associated with anthropogenic activities or natural calamities have caused potential environmental effects and health hazards. Sewage effluents are one of the leading sources of consecutive input of these harmful emissions into the aquatic environment. Miscellaneous removal of indisposed wastewaters has a harmful health impact on marine and terrestrial living things. Untreated wastewater mainly consists of organochlorine, nitrogenous and phosphorus compounds, and causative microbial agents such as bacteria, viruses, and protozoa that have toxicological impacts on human health (Goutam et al., 2018; Bharagava et al., 2017; Saxena et al., 2018). A group of researchers has established an innovative sewage treatment scheme that would significantly eliminate conventional contaminates and retrieve advantageous resources that are in sewage treatment plants (STPs) and effluent treatment plants. The primary mission of STP establishment is to transform household wastewater conveniently, and the secondary aim is to restore and recycle the wastewater later in sewage effluent treatment (Raychoudhury and Prajapati, 2020). Ligninolytic enzymes such as lignin peroxidase, manganese peroxidase (MnP), and laccase have been reported in the degradation and detoxification of many contaminants from metropolitan wastewaters. For example, by a ligninolytic enzyme-producing bacterium, *Aeromonas hydrophila*, that causes biodegradation of the crystal violet dye isolated from textile wastewater effluent (Bharagava et al., 2018).

1.2.2 **Removal of toxic metals from polluted water bodies**

The accumulation of toxic metals in the atmosphere has become an emerging issue globally. Due to the large and persistent nature and nonbiodegradable properties of toxic metals, it causes bioaccumulation in the food chain that leads to adverse environmental conditions and health risks from their acute toxic essence in biota (Teles et al., 2018). These toxic metals widely associated with anthropogenic activities, suchlike fossil fuel combusting, uncontrolled usage of agrochemicals, tannery, mining, electroplating, dyeing, and pigment manufacturer industries, fertilizers/pesticides, and discharge of wastewater effluents and several other industrial and agricultural operations, are subsequently discharged in huge amounts into the surrounding environment daily through wastewater effluents. The steadiness of toxic metals has proven itself to be vigorous and has risen in attention in recent years (Osundeko et al., 2014). Therefore, bioremediation is essential to prevent toxic metal militarization or leaching into ecological strata and to encourage their lineages. There have been several reports on the application of biofilms in the removal of toxic metals. Biofilm works as an efficient bioremediation tool along with a biological stabilizing mediator. Biofilms have great tolerance of noxious inorganic components at their lethal concentrations. Microalgae are often used for biological cultivation treatment or sewage treatment. The capacity of algae-based treatments is more efficient in removing radioactive compounds, pathogenic microorganisms, and heavy metals from wastewater. Consequently, microbes have a wide variety of techniques of metal removal that have greater metal biosorption potential (Tarekegn et al., 2020). The biological characteristics of microbes could be rectified by the existence of heavy metals. Different groups of bio-based constituents including bacteria, fungi, yeast, and algae have been explored as biosorbents for accumulating persistent organic contaminants and toxic metals by bioremediation (Gola et al., 2016; Barquilha et al., 2017). Owing to the abundant quality of microorganism and their cost-effectiveness, researchers have investigated many techniques as advantageous in the dismissal of heavy metal ions for contaminated sites, would-be biosorption, biotransformation, or bioaccumulation (Balaji et al., 2016; Jaafari and Yaghmaeian, 2019). A list of microorganisms for the removal of heavy metals is shown in Table 1.1.

1.2.3 **Cleaning of oil spills**

Currently, a major challenge is the cleanup of aquatic resources contaminated by oil. This contamination is caused by regular shipments, tankers, pipelines, wastewater drainage from industries, refineries, disposal, and oil spills. Oil spills release petroleum hydrocarbons into the marine environment, which poses an enormous threat to aquatic microflora. Oil spills may exhibit an immense ecological and commercial effect. Estimates are that more than 250,000 seabirds died from the Exxon-Valdez

Table 1.1 Names of microbial species and toxic chemicals they remove.

Name of the species	Removal elements	References
G. metallireducens	Manganese (Mn)	Gavrilescu (2004)
Bacillus subtilis	Lead (Pb), chromium (Cr), and cadmium (Cd)	Abioye et al. (2018)
Acinetobacter sp. and Arthrobacter sp.	Chromium (Cr)	Puyen et al. (2012)
Desulfovibrio desulfuricans	Copper (Cu) and nickel (Ni)	Samarth et al. (2012)
Bacillus pumilus, Pseudomonas aeruginosa, and Brevibacterium	Cadmium (Cd) and lead (Pb)	Singh et al. (2013)
Aspergillus niger, Rhizopus oryzae, Saccharomyces cerevisiae, and Penicillium chrysogenum	Chromium Cr (VI)	Igiri et al. (2018)
Pseudomonas genus K1 and AK9,	Arsenic (As)	Satyapa et al. (2018)
Agaricus bisporus, Pleurotus ostreatus	Chromium Cr, Copper (Cu), and lead (Pb)	Frutos et al. (2016)

oil spill in 1989 and that the Deepwater Horizon oil spill disaster expense exceeds US$61 billion (Li et al., 2016). Various biological and chemical techniques are available to respond to the oil spills, but among them, bioremediation is certified as a promising approach for the treatment of oil spills. Bioremediation is more environmentally friendly than conventional techniques and more economical with less destructive influences on the environment. Bioremediation for oil spills can be handled in two distinct ways (Doshi et al., 2018). These comprise bioaugmentation that introduces natural or genetically manipulated oil-degrading microorganisms to the polluted area or environment as well as biostimulation that entails additional nourishment to the affected zone to support the current oil-degrading microbes. Emerging bioaugmentation methods are led by manipulated microbes especially for amendment of catalytic properties, a metabolic passage scheme, enlargement of the substrate rate, and developing gene resistance through catabolic operations. The use of biosurfactant is another alternative, convenient technique that improves bioremediation by minimizing the surface energy (Haritash and Kaushik, 2009). Certain polymeric substances may be included to develop immobilization in microbial strains and consequently amplify the degradation rate. Biosurfactant is an attractive approach to degrade hazardous substances and protect the marine environment, whereas several seagoing bacteria and microalgae strains can produce biosurfactants during growth on hydrocarbons. Microbial remediation action possesses a prominent role in the cleanup of an oil spill. Microbial species that are known as an excellent degrader of hydrocarbon substances are classified as *Acinetobacter, Marinobacter, Pseudomonas, Rhodococcus*, and *Roseobacter* (McGenity et al., 2012). Laccase-containing ligninolytic fungi are recognized as excellent degraders for polyaromatic hydrocarbon (García-Delgado et al., 2015). Agricultural waste materials suchlike cotton, kapok, and rice straw are plentiful and optimally used as great oil sorbents for the treatment of oil spills. In some reports, banana peel has shown a high oil sorption capacity for crude and gas oils (El-Din et al., 2018). Several reports have considered the remediation of oil spills through bioelectrochemical systems (BESs). BESs have appeared as an interesting scheme to transform the chemical energy of organic wastes into sustainable electrical energy or hydrogen or valuable bulk chemical products. In BES, a different group of electroactive bacteria have evolved and potentially function as the catalyst. The polymeric oil-based absorbents challenge many drawbacks of expensive, secondary contamination and ecological deterioration. Microbial fuel cells are a novel approach for the utilization of waste for the generation of bioelectricity as a novel way to modify chemical energy into electricity concurrent with contaminant degradation (Srikanth et al., 2018).

1.2.4 Removal of pesticides from agriculture field

As we know, the soil is considered the top layer of our Earth's surface where plants can grow, containing minerals and rocks particles that are mixed with the decayed organic matter. Soil plays several essential roles in providing the basis for biomass as well as food production. One of the most severe problems for the Earth is pollution, which means any change takes place on elements involved in the composition of it because of human activities. Soils are contaminated by heavy metals, plants, humanity, pesticides, herbicides, and continuous farming and are due to several toxic chemicals and industrial wastes. Agricultural expansion in all countries and regions of the world because of the rise in demand for food has resulted in incremental population growth, and a threat occurs from this agricultural expansion on soil expansion. Other issues regarding agricultural expansion as well as soil depletion have emerged, namely the extensive use of agricultural pesticides and fertilizers (Hossain et al., 2015; Damalas and Koutroubas, 2016).

Soil is reported as a nonrenewable natural resource; the time required in the formation of 1 cm of forest soil is estimated to be 200−400 years. From the Second World War onwards, pesticides have been considered useful in increasing the production of agriculture and food preservation quality for some time. Pesticides are helpful in benefiting agriculture by increasing production as well as fighting many human and plant diseases. Supercritical extraction is considered a promising method for the remediation of soil, such as the removal of organic compounds such as PAHs and PCBs. Superficial extraction gets much attention as a promising method for the remediation of soil contaminants. When the dispersion of pesticides takes place in the environment, they become pollutants with ecological effects that require remediation (Damalas and Eleftherohorinos, 2011; Purnomo et al., 2020).

The biological process has been demonstrated to be an excellent method for the remediation of pesticides in comparison with traditional techniques through several useful, important, and advantageous properties such as simplicity of design, low initial operating cost, economics, comfort of operation, and intensive effects on toxic substances. Biosorbents used in pesticide removal have been obtained from several sources such as plant biomass, industrial by-products, and agricultural wastes and have had various degree of success in the application of pesticide effluent treatment (Tran et al., 2020; Purnomo et al., 2020).

1.2.4.1 Remediation methods for pesticides

Bioremediation is a well known, effective, and eco-friendly method for removal of soil contamination. Bioremediation is a complex process for soil decontamination. Several methods, such as phytoremediation and microbial and fungal remediation, are considered main components of bioremediation (Odukkathil and Vasudevan, 2016). Bioremediation reduces the pesticides in the soil by enhancing the biological degradation process via several metabolic reactions of microorganisms (Reddy and Antwi, 2016). Pesticide contamination in the soil is considered a nonpoint source, and several chemical methods are used for remediation of these contaminations (Morillo and Villaverde, 2017). These chemical methods have disadvantages due to adding additional secondary pollutants to the soil. Hence, bioremediation is considered a safe method compared with chemical remediation methods (Wang et al., 2016).

Phytoremediation of soil contamination such as toxic pesticides based on the uptake of pollutants is accomplished using plants (monocots and dicots), vegetation, plant roots, and rhizosphere microorganisms. The bacterial bioremediation is mainly based on the utilization of pesticide molecules and converts them into a nontoxic substance. Some bacterial species secrete extracellular enzymes, and these secreted enzymes are responsible for the degradation of pesticide molecules. The bacterial species having P450 cytochrome genes can effectively participate in the aerobic bioremediation of pesticides (Das et al., 2015).

Electrokinetic soil flushing (EKSF) is a recent and effective method used for the remediation of soil in different ways (Trellu et al., 2016). In this method, a high electric field is applied to polluted soil. When contaminated soil comes in contact with the electric field, the pollutants are transferred into flushing fluid, which can be treated with several methods such as electroosmosis, electrophoresis, etc. The removal of pesticides using EKSF is considered a hot topic of remediation (Vieira et al., 2016). Recently, EKSF has been combined with other remediation techniques like bioremediation, and it shows better results for pesticide removal from agriculture fields.

1.3 **Limitations of bioremediation**

Bioremediation is an eco-friendly and cost-effective method for removal of several pollutants such as pesticides, heavy metals, and dyes. Among these advantages, this method has many disadvantages. Biosorption is known as extraction of pollutants using several biosorbents. The pollutants are adsorbed on the surface of the biosorbent through functional groups present on its surface. This method requires a number of steps that make this method more complex. These steps include preparation and processing of biosorbent. A system biology approach by bioremediation is shown in Fig. 1.1.

Removal of pollutants using living biomass of bacteria, fungi, and algae is also considered an effective method. However, several problems occur in this process such as culture maintenance and use of growth media. Along with these disadvantages, bioremediation has other limitations such as the following:

- Bioremediation is limited to those compounds that are biodegradable. Not all compounds are susceptible to rapid and complete degradation.
- There are some concerns that the products of biodegradation may be more persistent or toxic than the parent compound.
- Biological processes are often highly specific. Important factors required for bioremediation include the presence of metabolically capable microbial populations, suitable environmental growth conditions, and appropriate levels of nutrients and contaminants.
- It is difficult to extrapolate from bench and pilot-scale studies to full-scale field operations.

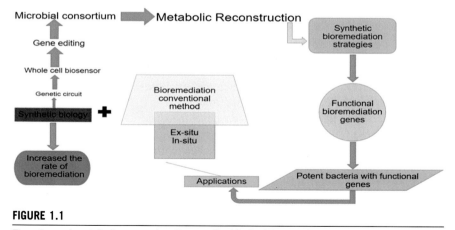

FIGURE 1.1

The plans of metabolic reconstruction applicable for bioremediation.

- Research is needed to develop and engineer bioremediation technologies that are appropriate for sites with complex mixtures of contaminants that are not evenly dispersed in the environment. Contaminants may be present as solids, liquids, and gases.
- Bioremediation often takes longer than other treatment options, such as excavation and removal of soil or incineration.
- Regulatory uncertainty remains regarding acceptable performance criteria for bioremediation. There is no accepted definition of "clean," evaluating performance of bioremediation is difficult, and there are no acceptable endpoints for bioremediation treatment.

1.4 Conclusion

Bioremediation is a very admired and promising technology for the remediation of environments contaminated with petroleum hydrocarbon, Solid waste management includes sewage treatment, removal of toxic metals from polluted water bodies, cleanup of oil spills, and removal of pesticides in agricultural soil.

Petroleum pollution has become a severe environmental problem that causes harmful environmental damage and harmful impacts on human health. The bioremediation is based on the metabolic capabilities of microorganisms and considered the most reliable source to eliminate pollutants, especially petroleum and its recalcitrant compounds. As reported in previous studies, several bioremediation approaches through bioaugmentation and biostimulation have been performed for the removal of petroleum pollution.

A diversity of dangerous pollutants inducing phenols, toxic azo dyes, resins, pharmaceuticals, chlorinated biphenyls, heavy metals, acids/alkalis, polycyclic aromatic hydrocarbons, etc. are being released into water bodies that have severely deteriorated the water and soil ecosystem. The bioremediation technique has been proficiently applied for removing environmental pollutants from water and soil. The numerous methodologies applied in the bioremediation method are ecologically sound and cost-effective (Shivalkar et al., 2021).

The occurrence of heavy metals and their toxicity poses a serious challenge for the treatment of wastewater runoff prior to release into nearby water bodies. Numerous removal techniques have been developed and are functional for the treatment of these wastes to eliminate toxic metal ions. Some technologies such as microbe-assisted phytoremediation, ion exchange, membrane filtration, photocatalytic oxidation and reduction, and adsorption have their own advantages and disadvantages over metal ion sequestrations from environmental matrices. In recent years, developments in adsorption of heavy metals from aqueous solutions have gained tremendous popularity among the scientific community as methods to treat industrial wastewater. Several adsorbents such as clays, LDHs, zeolites, carbon nanotubes and their composites, activated carbons, biomass-derived biosorbents, inorganic nanomaterials, inorganic organic hybrid nanocomposites, and magnetic nanomaterials have been synthesized and investigated for their ability to sequester metal ions from water.

Functionalized magnetic nanoparticles are very promising for applications in catalysis, biolabeling, and bioseparation. In liquid-phase extraction of heavy metals and dyes in particular, such small and magnetically separable particles may be useful, as they combine the advantages of high dispersion, high reactivity, high stability under acidic conditions, and easy separation. In this chapter, we focused mainly on recent developments in the synthesis of active adsorbents and nanoparticles. Further, functionalization and application of magnetic nanoparticles and their nanosorbents for the separation and purification of hazardous metal ions from the environment are discussed in detail in a separate chapter in this book.

The release of wastewater containing toxic materials of heavy metals within the ecosystem is one of the most serious issues for environmental and health challenges in our society. Therefore, there is urgent need for the development of eco-friendly, efficient, novel, and cost-effective methods for the removal of inorganic metals (Pb, Cd, Cr, and Hg) discharged into the environment and to protect the ecosystem. Microbe-based heavy metals have derived bioremediation as a forthcoming alternative to traditional techniques. Heavy metals are nonbiodegradable and may be harmful to microbes. Numerous microorganisms have been developed as detoxifying methods to counter the harmful effects of inorganic metals. This chapter discussed biosorption capacity with respect to the use of fungi, algae, bacteria, genetically immobilized microbial cells, and engineered microbes for the elimination of heavy metals. The application of biofilm has shown synergetic effects, with a manyfold increase in the removal of heavy metals as a sustainable environmental technology in the near future.

References

Abioye, O.P., Oyewole, O.A., Oyeleke, S.B., Adeyemi, M.O., Orukotan, A.A., April 30, 2018. Biosorption of lead, chromium and cadmium in tannery effluent using indigenous microorganisms. Braz. J. Biol. Sci. 5 (9), 25–32.

Ayilara, M.S., Olanrewaju, O.S., Babalola, O.O., Odeyemi, O., January 2020. Waste management through composting: challenges and potentials. Sustainability 12 (11), 4456.

Balaji, S., Kalaivani, T., Sushma, B., Pillai, C.V., Shalini, M., Rajasekaran, C., August 2, 2016. Characterization of sorption sites and differential stress response of microalgae isolates against tannery effluents from Ranipet industrial area—an application towards phycoremediation. Int. J. Phytoremediation 18 (8), 747–753.

Barquilha, C.E., Cossich, E.S., Tavares, C.R., Silva, E.A., May 1, 2017. Biosorption of nickel (II) and copper (II) ions in batch and fixed-bed columns by free and immobilized marine algae *Sargassum* sp. J. Clean. Prod. 150, 58–64.

Bharagava, R.N., Chowdhary, P., Saxena, G., July 6, 2017. Bioremediation: an eco-sustainable green technology: its applications and limitations. In: Environmental Pollutants and Their Bioremediation Approaches, pp. 1–22.

Bharagava, R.N., Mani, S., Mulla, S.I., Saratale, G.D., July 30, 2018. Degradation and decolourization potential of an ligninolytic enzyme producing Aeromonas hydrophila for crystal violet dye and its phytotoxicity evaluation. Ecotoxicol. Environ. Saf. 156, 166–175.

Colberg, P.J., Young, L.Y., 1995. Anaerobic degradation of nonhalogenated homocyclic aromatic compounds coupled with nitrate, iron, or sulfate reduction. In: Microbial Transformation and Degradation of Toxic Organic Chemicals, p. 307330.

Damalas, C.A., Eleftherohorinos, I.G., May 2011. Pesticide exposure, safety issues, and risk assessment indicators. Int. J. Environ. Res. Publ. Health 8 (5), 1402–1419.

Damalas, C.A., Koutroubas, S.D., 2016. Farmers' Exposure to Pesticides: Toxicity Types and Ways of Prevention.

Das, A.C., Das, R., Bhowmick, S., 2015. Non-symbiotic N_2-fixation and phosphate-solubility in Gangetic alluvial soil as influenced by preemergence herbicide residues. Chemosphere 135, 202–207.

Doshi, B., Sillanpää, M., Kalliola, S., May 15, 2018. A review of bio-based materials for oil spill treatment. Water Res. 135, 262–277.

Dzionek, A., Wojcieszyńska, D., Guzik, U., September 2016. Natural carriers in bioremediation: a review. Electron. J. Biotechnol. 19 (5), 28–36.

El-Din, G.A., Amer, A.A., Malsh, G., Hussein, M., September 1, 2018. Study on the use of banana peels for oil spill removal. Alex. Eng. J. 57 (3), 2061–2068.

Frutos, I., García-Delgado, C., Gárate, A., Eymar, E., November 1, 2016. Biosorption of heavy metals by organic carbon from spent mushroom substrates and their raw materials. Int. J. Environ. Sci. Technol. 13 (11), 2713–2720.

García-Delgado, C., Alfaro-Barta, I., Eymar, E., March 21, 2015. Combination of biochar amendment and mycoremediation for polycyclic aromatic hydrocarbons immobilization and biodegradation in creosote-contaminated soil. J. Hazard Mater. 285, 259–266.

Gavrilescu, M., June 2004. Removal of heavy metals from the environment by biosorption. Eng. Life Sci. 4 (3), 219–232.

Gola, D., Dey, P., Bhattacharya, A., Mishra, A., Malik, A., Namburath, M., Ahammad, S.Z., October 1, 2016. Multiple heavy metal removal using an entomopathogenic fungi *Beauveria bassiana*. Bioresour. Technol. 218, 388–396.

Goutam, S.P., Saxena, G., Singh, V., Yadav, A.K., Bharagava, R.N., Thapa, K.B., March 15, 2018. Green synthesis of TiO_2 nanoparticles using leaf extract of *Jatropha curcas* L. for photocatalytic degradation of tannery wastewater. Chem. Eng. J. 336, 386–396.

Haritash, A.K., Kaushik, C.P., September 30, 2009. Biodegradation aspects of polycyclic aromatic hydrocarbons (PAHs): a review. J. Hazard Mater. 169 (1–3), 1–5.

Hossain, M.S., Chowdhury, M.A., Pramanik, M.K., Rahman, M.A., Fakhruddin, A.N., Alam, M.K., June 2015. Determination of selected pesticides in water samples adjacent to agricultural fields and removal of organophosphorus insecticide chlorpyrifos using soil bacterial isolates. Appl. Water Sci. 5 (2), 171–179.

Igiri, B.E., Okoduwa, S.I.R., Idoko, G.O., Akabuogu, E.P., Adeyi, A.O., Ejiogu, I.K., 2018 Sep 27. Toxicity and Bioremediation of Heavy Metals Contaminated Ecosystem from Tannery Wastewater: A Review. J Toxicol 2018, 2568038. https://doi.org/10.1155/2018/2568038. PMID: 30363677; PMCID: PMC6180975.

Jaafari, J., Yaghmaeian, K., February 1, 2019. Optimization of heavy metal biosorption onto freshwater algae (*Chlorella coloniales*) using response surface methodology (RSM). Chemosphere 217, 447−455.

Li, P., Cai, Q., Lin, W., Chen, B., Zhang, B., September 15, 2016. Offshore oil spill response practices and emerging challenges. Mar. Pollut. Bull. 110 (1), 6−27.

McGenity, T.J., Folwell, B.D., McKew, B.A., Sanni, G.O., December 1, 2012. Marine crude-oil biodegradation: a central role for interspecies interactions. Aquat. Biosyst. 8 (1), 10.

Megharaj, M., Naidu, R., September 2017. Soil and brownfield bioremediation. Microb. Biotechnol. 10 (5), 1244−1249.

Morillo, E., Villaverde, J., 2017. Advanced technologies for the remediation of pesticide-contaminated soils. Sci. Total Environ. 576−597.

Muangchinda, C., Rungsihiranrut, A., Prombutara, P., Soonglerdsongpha, S., Pinyakong, O., September 5, 2018. 16S metagenomic analysis reveals adaptability of a mixed-PAH-degrading consortium isolated from crude oil-contaminated seawater to changing environmental conditions. J. Hazard Mater. 357, 119−127.

Mueller, J.G., Cerniglia, C.E., Pritchard, P.H., 1996. Bioremediation of environments contaminated by polycyclic aromatic hydrocarbons. Biotechnol. Res. Ser. 6, 125−194.

Odukkathil, G., Vasudevan, N., 2016. Residues of endosulfan in surface and subsurface agricultural soil and its bioremediation. J. Environ. Manag. 165, 72−80.

Osundeko, O., Dean, A.P., Davies, H., Pittman, J.K., October 1, 2014. Acclimation of microalgae to wastewater environments involves increased oxidative stress tolerance activity. Plant Cell Physiol. 55 (10), 1848−1857.

Pande, V., Pandey, S.C., Sati, D., Pande, V., Samant, M., February 28, 2020. Bioremediation: an emerging effective approach towards environment restoration. Environ. Sustain. 1−3.

Pandey, L.M., Shukla, S.K., July 10, 2019. An insight into waste management in Australia with a focus on landfill technology and liner leak detection. J. Clean. Prod. 225, 1147−1154.

Purnomo, A.S., Sariwati, A., Kamei, I., 2020. Synergistic Interaction of a Consortium of the Brown-Rot Fungus *Fomitopsis pinicola* and the Bacterium *Ralstonia pickettii* for DDT Biodegradation. Heliyon6, p. e04027.

Puyen, Z.M., Villagrasa, E., Maldonado, J., Diestra, E., Esteve, I., Solé, A., December 1, 2012. Biosorption of lead and copper by heavy-metal tolerant *Micrococcus luteus* DE2008. Bioresour. Technol. 126, 233−237.

Raychoudhury, T., Prajapati, S.K., 2020. Bioremediation of pharmaceuticals in water and wastewater. In: Microbial Bioremediation & Biodegradation. Springer, Singapore, pp. 425−446.

Reddy, G.V., Antwi, F.B., 2016. Toxicity of natural insecticides on the larvae of wheat head armyworm, Dargida diffusa (Lepidoptera: noctuidae). Environ. Toxicol. Pharmacol. 42, 156−162.

Samarth, D.P., Chandekar, C.J., Bhadekar, R.K., June 1, 2012. Biosorption of heavy metals from aqueous solution using *Bacillus licheniformis*. Int. J. Pure Appl. Sci. Technol. 10 (2), 12.

Satyapal, G.K., Mishra, S.K., Srivastava, A., Ranjan, R.K., Prakash, K., Haque, R., Kumar, N., March 1, 2018. Possible bioremediation of arsenic toxicity by isolating indigenous bacteria from the middle Gangetic plain of Bihar, India. Biotechno. Rep. 17, 117−125.

Saxena, G., Bharagava, R.N., 2015. Persistent organic pollutants and bacterial communities present during the treatment of tannery wastewater. Environ. Waste Manag. 1 (00), 217−247.

Saxena, G., Bharagava, R.N., Mulla, S.I., Patel, D.K., August 1, 2018. Characterization and identification of recalcitrant organic pollutants (ROPs) in tannery wastewater and its phytotoxicity evaluation for environmental safety. Arch. Environ. Contam. Toxicol. 75 (2), 259−272.

Shinde, O.A., Bansal, A., Banerjee, A., Sarkar, S., 2018 May. Bioremediation of steel plant wastewater and enhanced electricity generation in microbial desalination cell. Water Sci Technol 77 (7-8), 2101−2112. https://doi.org/10.2166/wst.2018.126. PMID: 29722696.

Shivalkar, S., Singh, V., Sahoo, A.K., Samanta, S.K., Gautam, P.K., October 15, 2021. Bioremediation: a potential ecological tool for waste management. Bioremediation Environ. Sustain. 1−21 (Elsevier).

Singh, N., Verma, T., Gaur, R., 2013. Detoxification of hexavalent chromium by an indigenous facultative anaerobic *Bacillus cereus* strain isolated from tannery effluent. Afr. J. Biotechnol. 12 (10).

Srikanth, S., Kumar, M., Puri, S.K., October 1, 2018. Bio-electrochemical system (BES) as an innovative approach for sustainable waste management in petroleum industry. Bioresour. Technol. 265, 506−518.

Tarekegn, M.M., Salilih, F.Z., Ishetu, A.I., January 1, 2020. Microbes used as a tool for bioremediation of heavy metal from the environment. Cogent Food Agric. 6 (1), 1783174.

Teles, Y.V., de Castro, L.M., Junior, É.S., Do Nascimento, A.P., da Silva, H.A., Costa, R.S., do Nascimento Souza, R.D., da Mota, A.J., Pereira, J.O., August 2018. Potential of bacterial isolates from a stream in manaus-amazon to bioremediate chromium-contaminated environments. Water, Air, Soil Pollut. 229 (8), 1-0.

Tran, T.D., Dao, N.T., Sasaki, R., Tu, M.B., Dang, G.H., Nguyen, H.G., Dang, H.M., Vo, C.H., Inigaki, Y., Van Nguyen, N., Sakakibara, Y., May 15, 2020. Accelerated remediation of organochlorine pesticide-contaminated soils with phyto-Fenton approach: a field study. Environ. Geochem. Health 1−2.

Trellu, C., Ganzenko, O., Papirio, S., Pechaud, Y., Oturan, N., Huguenot, D., et al., 2016. Combination of anodic oxidation and biological treatment for the removal of phenanthrene and tween 80 from soil washing solution. Chem. Eng. J306, 588−596.

Vidali, M., July 1, 2001. Bioremediation. an overview. Pure Appl. Chem. 73 (7), 1163−1172.

Vieira, D.S.E., Souza, F., Saez, C., Canizares, P., Lanza, M.R., Martinezhuitle, C.A., et al., 2016. Application of electrokinetic soil flushing to four herbicides: a comparison. Chemosphere 153, 205−211.

Wang, Y., An, X., Shen, W., Chen, L., Jiang, J., Wang, Q., Cai, L., July 2016. Individual and combined toxic effects of herbicide atrazine and three insecticides on the earthworm, *Eisenia fetida*. Ecotoxicology 25 (5), 991−999.

Advances in microbial management of soil

M. Theradimani, V. Ramamoorthy

Department of Plant Pathology, Agricultural College and Research Institute, Tamil Nadu Agricultural University, Madurai, Tamil Nadu, India

Chapter outline

2.1 Introduction

The soil ecosystem is complex and heterogeneous because it comprises inorganic and organic components, water, gases, and various flora and fauna. After cropping, the residual recalcitrant crop wastes from crop plants, trees, and woody plants are dumped in soil. In addition, disposable waste materials containing cellulose are dumped in soil from the paper and pulp, textile, pesticide, and pharma industries. These complex organic materials in soil should be mineralized so that the various nutrients and minerals are freely available to plants for increasing crop productivity. Soil having a high C:N ratio is not suitable for crop cultivation. Thus, the soil

Biological Approaches to Controlling Pollutants. https://doi.org/10.1016/B978-0-12-824316-9.00018-5

containing rich amounts of lignocellulosic material should be mineralized, and the C:N ratio should be reduced to a 10:1 ratio for optimum cultivation of crops and tree plants. For that, a suitable microbial community in the soil ecosystem should be augmented so that mineralization can be carried out to enrich carbon and nitrogen levels and accelerate the development of heterogeneous microbial communities. Microbial degradation of agrowaste is important for a wide spectrum of soil reactions and functions, including organic matter decomposition, nutrient cycling, and substrate stabilization (Huang et al., 2010; Valentín et al., 2010).

Bioremediation refers to the use of microbial agents, such as bacteria, fungi, and algae, to remove or neutralize organic and inorganic wastes in the soil ecosystem. Mycoremediation is a form of bioremediation that uses fungi as biological agents for rectifying the soil ecosystem. Mycoremediation is also designated as fungal bioremediation or mushroom bioremediation. Mycoremediation is of recent origin and gaining importance in bioremediation of crop residues that are rich in lignocellulosic substances. Crop residues are colonized by a wide variety of microorganisms, mainly fungi and bacteria. These organisms cause decay and degradation with negative impacts on standing crops and trees and beneficial effects by recycling crop residues and deadwood and mineralizing the complex organic lignocellulosic matter to CO_2 and water. Isolation and application of a suitable fungus that decomposes the toxic compound efficiently is the prime need of the mycoremediation process. This paper describes in a nutshell the various fungi used in the remediation of agrowaste rich in lignocellulosic material, the mechanisms by which the fungi remediate agrowastes, establishment of the mycoremediation process, and various factors influencing the mycoremediation process.

2.2 Principal fungal species in mycoremediation

2.2.1 White rot fungi

Many Basidiomycetes fungi are the important degraders of lignocellulosic agricultural wastes and thereby recycle carbon and other nutrients in the soil ecosystem. Mainly woody trees are rich in lignocellulosic material, which is quite recalcitrant to degradation and low in nitrogen and other essential nutrients. Wood-decaying fungi degrade lignin and the cellulose constituents of wood. Certain wood-decaying fungi degrade both lignin and cellulose and are called white rot fungi. White rot refers to the bleached appearance of wood due to degradation of lignin, cellulose, and hemicellulose. *Phanerochaete chrysosporium* is the most important white rot fungus. Other well-known white rot fungi are *Trametes versicolor*, *Pycnoporus sanguineus*, *Lentinula edodes* (shiitake mushroom), *Pleurotus* spp. (oyster mushroom), *Ganoderma lucidum*, and *Ganoderma applanatum*, *Phellinus igniarius*, *Irpex lacteus*, and *Phlebia radiata* (Peralta et al., 2017).

Certain white rot fungi can degrade both lignin and cellulosic constituents of wood simultaneously (called simultaneous degrader/simultaneous white rot fungi), whereas others preferentially degrade lignin or remove lignin in advance of cellulose and hemicellulose (called selective delignifiers/selective white rot fungi). *P. chrysosporium* and *I. lacteus* simultaneously degrade all constituents (lignin, cellulose, and hemicellulose) of wood. *Phlebia* spp. and *G. applanatum* are selective delignifiers (Maciel et al., 2010; Dill and Kraepelin, 1986). Selective delignifiers are preferred in certain industrial processes such as biopulping (paper industry), bioethanol production, and formulation of cellulose-enriched animal feed.

Lignicolous edible mushrooms (mushroom that grow and produce fruiting body on wood and high-lignocellulosic waste materials such as straw or any other agricultural waste but not on soil) are well suited for biodegradation of agrowastes. Typical lignicolous edible mushrooms are *Pleurotus* spp. and *L. edodes*. Actually, these lignicolous edible mushrooms are cultivated using straw and wood logs as a substrate, and after cultivation and harvesting of mushrooms, the leftover spent substrates are used as organic manure, and thus mushroom cultivation is used for the recycling of agrowastes. Interestingly, these mushrooms can also be used for cultivation and degradation/composting of various sanitary materials made from cellulose as a major component such as diapers, napkins, sanitary pads (used for cleaning blood and fluid discharged during menstruation). *Pleurotus* spp. are a suitable choice for biodegradation of sanitary materials such as diapers and sanitary pads because these sanitary materials are made up of cellulose as a major component. The nutritive value of mushrooms grown on waste does not differ from that of mushrooms grown using the normal substrate, and they are free from harmful elements (Valdemar et al., 2015; Valdemer, 2014). Experiments clearly show that mushroom fungi can be used for the biodegradation of any used, disposable, and harmful sanitary products made from cellulosic materials as well as for mushroom production. Mushrooms also can be used for bioremediation of agricultural weed wastes (Das and Mukherjee, 2007a,b).

In relation to other fungi, white rot fungi belong to the class Agaricomycetes, which comes under the phylum Basidiomycota and kingdom Fungi. Their life cycle comprises the production of haploid basidiospores that germinate to produce monokaryotic primary mycelium followed by the formation of dikaryotic secondary mycelium by the fusion of the monokaryotic mycelia and later the formation of basidiocarps that produce basidiospores. Most basidiomycetes are heterothallic fungi and have a tetrapolar-type mating system.

2.2.2 Brown rot fungi

Brown rot refers to the brown color of decayed wood because cellulose and hemicellulose are degraded but lignin remains intact in modified form as a brown deposition (Arantes and Goodell, 2014). In advanced stages of fungal colonization, cellulose and hemicellulose are completely removed, leaving modified lignin that appear brown in color. Hence, the type of decay is called brown rot, and the fungi is called brown rot fungi. White fungi are usually saprophytic, whereas brown rot fungi are pathogenic to trees. The most serious issue of brown rot fungi is that cellulose degradation of wood occurs even at the early stage of colonization and can cause serious issues in wood used for construction purposes. Important brown rot fungi are *Gloeophyllum trabeum*, *Serpula lacrymans*, *Coniophora puteana*, *Schizophyllum commune*, *Postia placenta*, and *Fomes fomentarius*.

Like white rot fungi, brown rot fungi also belong to the class agaricomycetes that comes under the phylum basidiomycota and kingdom fungi. The life cycle of brown rot fungi is the same as described for white rot fungi.

2.2.3 Soft rot fungi

Several cellulolytic ascomycetous mold fungi, mainly *Aspergillus*, *Penicillium*, and *Trichoderma* spp., are referred to as soft rot fungi. Among the various *Trichoderma* spp., *Trichoderma reesei* (*Hypocrea jecorina* sexual form) is an important industrial cellulase-producing strain and implicated in the degradation of agrowastes. The cellulase producing *T. reesei* was first isolated from the rotten fabrics of army personnel on Solomon Island where the shoes and clothes of army personnel were heavily damaged by fungi. Among the various fungi isolated from the rotten fabrics, the *T. reesei* strain QM6a was found to produce an enormous amount of cellulase (Reese, 1976). Other cellulase-producing *T. reesei* strains were derived from this strain. *T. reesei* is the best known soft rot fungus that produces cellulases and hemicellulases (Schuster and Schmoll, 2010). Genetic improvement of *T. reesei* strain QM6a by a mutagenesis approach was carried out for the production of higher levels of cellulases to be used in the pulp and paper industry and textile industries (Seiboth et al., 2011). Recently, various genes encoding proteins for regulating the biosynthesis of cellulases in *T. reesei* have been studied for further strain improvement (Hinterdobler et al., 2019; Beier et al., 2020).

Soft rot fungi belong to class ascomycetes that comes under the phylum ascomycotina and kingdom fungi. Ascomycetes fungi typically produce asexual spores called conidia. They mostly propagate by conidia, and rarely, they also produce sexual spores called ascospores. Soft rot fungi are either homothallic (*Aspergillus* spp.) or heterothallic (*Trichoderma* spp.). If they are heterothallic, they have a bipolar mating system.

2.3 Mechanisms in mycoremediation

Because agrowaste consists mainly of cellulose and lignin, the enzymatic action cleaves these complex moieties into the simpler substance and finally recycles into CO_2. The only mechanism by which fungi degrade lignin and cellulose is the production of various extracellular enzymes. Lignin is degraded by various oxidative lignolytic enzymes. Cellulose is degraded by a set of hydrolytic enzymes called carbohydrate active enzymes (CAZy database).

2.3.1 **Lignolytic enzymes**

Lignin is degraded and mineralized into water and carbon dioxide through the action of a set of enzymes called lignin-modifying enzymes. Oxidative lignin breakdown of agrowastes depends on lignin-modifying enzymes produced by fungi, especially white rot fungi. These enzymes include (1) lignin peroxidase, which contains Fe moiety and catalyzes oxidation of lignin in the presence of hydrogen peroxide; (2) manganese peroxidase, which is a manganese-containing protein that oxidizes lignin in the presence of hydrogen peroxide; (3) laccase—a copper containing protein that catalyzes demethylation of lignin. Since lignin-degrading enzymes by white rot fungi are extracellular, these fungi need not take up the toxic metabolites inside their cells. Thus, the fungus can tolerate very high concentrations of insoluble lignin and many of the hazardous environmental pollutants in the soil ecosystem. Moreover, the extracellular system enables fungi to tolerate considerably high concentrations of toxic pollutants (Reddy and Mathew, 2001).

White rot fungi degrade lignin more efficiently than brown rot and soft rot fungi do. White rot fungi are unique in their ability to mineralize lignin using an array of extracellular enzymes, and degradation is an aerobic oxidative process. The best-studied white rot fungus is *P. chrysosporium* because it secretes several isoforms of lignin peroxidase and manganese peroxidase along with an H_2O_2-generating system that are the major components of the lignin-degradation system (Gold and Alic, 1993). These ligninolytic enzymes are important in relation to potential biotechnical applications, such as biopulping, biobleaching, and soil bioremediation (Higson, 1991; Kirk and Farrell, 1987). By the action of lignolytic enzymes, white rot fungi are also used in the degradation of various xenobiotic compounds and dyes (Maciel et al., 2010).

Various white rot fungi produce one or a few lignin-degrading enzymes in different combinations. Based on the type of lignin-degrading enzyme production, white rot fungi are grouped into four types: (1) laccase and lignin peroxidase and manganese peroxidase producers (*T. versicolor*), (2) laccase and at least one of the peroxidase producers (*L. edodes, Pleurotus eryngii*), (3) only laccase producers (*S. commune*), or (4) only peroxidase producers (*P. chrysosporium*). The most frequently observed lignin-degrading enzymes among the white rot Basidiomycetes are laccases and manganese peroxidase. In addition to biodegradation of agrowaste, ligninolytic enzymes of white rot fungi are useful for remediation of inorganic chemicals in various fields such as the chemical, fuel, food, paper, and textile industries (Maciel et al., 2010).

2.3.2 **Lignin degradation occurs during nutrient starvation**

Since white rot fungi produce lignolytic enzymes extracellularly, the degradation of lignin takes place in the external environment, and the fungi do not take the lignin inside their mycelium. This implies that the fungi can tolerate extremely high concentrations of lignin. As mentioned earlier, these extracellular lignolytic enzymes also degrade inorganic toxic elements (pesticides, toxic pollutants) present in the soil ecosystem, and the fungi do not assimilate them in their cells. This indicates that white rot fungi can tolerate extremely high levels of toxic metabolites accumulated in the soil. Interestingly, mineralization/degradation of lignin and other inorganic toxic compounds occurs during late growth phases (stationary phase) and also nutrient-limiting conditions, especially under low nitrogen content. Under laboratory conditions, there is no degradation of lignin when white rot fungus is grown under high nitrogen conditions. White rot fungus is not lignolytic during the early logarithmic growth phase, and hydrogen peroxide, which is needed for the functioning of lignin peroxidase, does not appear in the early logarithmic growth phase. Woody plants and fresh agrowastes are usually low in nitrogen content, which stimulates the lignolytic enzyme systems in white rot fungi for degradation, which also have a competitive advantage in lignin degradation (Aust, 1990). Thus, white rot fungi are the suitable fungi for the degradation of woody wastes.

2.3.3 Cellulolytic enzymes

Degradation of cellulose into glucose involves the sequential action of endocellulases, exocellulases, and beta-glucosidases. Endocellulases hydrolyze and cleave internal glycosidic linkages of cellulose polymer in a random fashion and lead to the production of oligosaccharide. Exocellulases hydrolyze and cleave cellulose polymers from either the reducing or the nonreducing ends, which leads to a rapid release of reducing sugars but little change in polymer length. Both endocellulases and exocellulases act on cellulose to produce cello-oligosaccharides and cellobiose, which are then hydrolyzed and cleaved by beta-glucosidase to glucose.

As mentioned earlier, brown rot fungi degrade cellulose and hemicellulose, leaving the lignin, and they produce enormous amounts of cellulolytic enzymes. White rot fungi also degrade cellulose by producing cellulolytic enzymes. But cellulolytic enzyme production in white rot fungi is comparatively low. However, white rot fungi are good producers of lignolytic enzymes. The most important white rot fungi producing cellulose-degrading enzymes are *P. chrysosporium*, *T. versicolor*, *Agaricus arvensis*, *Pleurotus ostreatus*, and *Phlebia gigantea* (Valásková and Baldrian, 2006). White rot basidiomycetes can completely degrade lignocellulosic materials through some of the hydrolytic enzymes involved in the degradation of the polymers cellulose and hemicellulose (cellulase and xylanase complexes). Various extracellular hydrolytic enzymes described in *P. chrysosporium* include several proteases, amylases, xylanases, and other carbohydrases (Dey et al., 1991; Ishida et al., 2007).

The soft rot Ascomycetes such as *Aspergillus* and *Trichoderma* are well known to produce different cellulolytic enzymes. These soft rot fungi can be easily cultured to produce an enormous amount of cellulase and hemicellulase including xylanases. *T. reesei* is well known for the production of cellulases and hemicellulases (He et al., 2014; Schuster and Schmoll, 2010). *T. reesei* is called the industrial workhorse for cellulase production and used for biodegradation of agrowastes and in the paper and pulp industry (Kuhad et al., 2011; Seiboth et al., 2011). In addition to *T. reesei*, other soft rot fungi produce cellulolytic and xylanolytic enzymes and have great potential in various agro-industries such as food, animal feed, brewing, winemaking, biomass refining, textile, and pulp and paper.

2.4 **Establishing mycoremediation systems**

Biological treatment of waste and disposable toxic agrowaste is preferable to chemical treatment. For biological treatment, fungi are preferred to bacteria because the fungi grow from the point of inoculation to varying extents in soil through filamentous mycelium. Native fungi present in the soils represent heterogeneous potential for remediation. But ecological factors (carbon and nitrogen sources, temperatures, pH, aeration, etc.) can hinder and prolong the remediation process. The remediation potential of native fungi can be enhanced with the addition of suitable carbon and nitrogen sources (biostimulation) or with the addition of mass-multiplied bio-inoculants (bioaugmentation).

Although white rot fungi are well known in mycoremediation, the limitation is that they cannot become established in the soil ecosystem. Their natural habitat is woody substances. When these fungi are inoculated into the soil ecosystem, they encounter competition with soil-inhabiting fungi and bacteria for nutrients, which prevents their establishment.

Isolation and selection of suitable fungal species, suitable substrate for their growth, and application of inoculum in large quantities are the fundamental steps in establishing mycoremediation technology.

The selection of remediating fungal isolate should be indigenous and nonpathogenic to crop plants and other microflora, noninvasive to adjacent areas, efficient in degrading organic and inorganic toxic compounds, competitive, etc. It is well known that white rot fungi are the best fungi for the mycoremediation of organic and inorganic toxic compounds in soil ecosystems because of their extracellular enzyme production. *P. chrysosporium*, *T. versicolor*, and *P. ostreatus* are the most successful soil colonizers and strong competitors (Baldrian, 2008; Tuomela et al., 1998).

To augment the growth of introduced white rot fungi in soil ecosystem, the addition of nitrogen and carbon sources in the form of lignocellulosic substrates such as straw, wood chips, and bark enhances the growth of the white rot fungi and suppresses the growth of the native microbial population because non-ligno-cellulolytic native bacterial and fungal microflora cannot utilize the woody material as a substrate and nutrient for growth and multiplication (Mougin et al., 2009; Piškur et al., 2009).

Additionally, larger quantities of inoculum of white rot fungi should be applied so that faster and successful colonization can occur on the applied lignocellulosic substrates. In addition, young and actively growing fungal inoculum (spawn) produced on woody materials as substrates, application of the inoculum along with fresh woody substrate, and maintenance of optimum moisture content in the soil should be followed. Encapsulated inoculum, which improves the survival and effectiveness of introduced fungi, can be useful (Ford et al., 2007; Schmidt et al., 2005; Piškur et al., 2009).

After application of the fungal inoculum for mycoremediation, the success of the mycoremediation is monitored based on the colonization of the fungus visually by inspecting mycelium growth and fruit bodies and biochemically by measuring ergosterol content (because ergosterol is only found in cell membranes of higher fungi such as Basidiomycotina and Ascomycotina) as well as the activities of ligninolytic enzymes and the content of intermediate chemical products of degraded wood and organic matter. Success in establishing mycoremediation of exogenously added fungal inoculum is monitored with the application of DNA-based molecular methods using PCR and other techniques.

2.5 Factors influencing mycoremediation

Nutritional and environmental conditions that affect the growth and multiplication of white rot fungi influence the mycoremediation process. These factors include carbon and nitrogen sources, aeration, pH, redox potential, temperature, and moisture content (Meysami, 2001). Suitable and specific conditions should be maintained for mycoremediation of agrowaste.

2.5.1 Carbon and nitrogen sources

Fungi are heterotrophs, and prototrophic fungi require carbon and nitrogen sources only for growth and multiplication. In addition, they require mineral elements such as Ca, Mg, K, S, Fe, and P and various trace elements (Odu, 1978) for enhanced growth and multiplication, while nitrogen can be provided in a variety of forms such as nitrate, ammonium salts, and organic compounds like urea. However, white rot fungi prefer low nitrogen, and ligno-cellulolytic activity is maximized under nitrogen-starvation conditions. Wood materials are usually rich in carbon and low in nitrogen, and thus, woody materials are good substrates for carbon and nitrogen. Lignin degradation is induced by nitrogen depletion in the medium. Conversely, lignin degradation is suppressed in nitrogen-rich media, and nitrogen metabolism competes with lignin metabolism in fulfilling the requirements of the same cofactors (Aust, 1990).

2.5.2 pH

The pH of the soil ecosystem ranges between 5 and 9, and most microorganisms have evolved with pH tolerance within this range. Most microorganisms tolerate from pH 5 to 9 but prefer pH 6.5–7.5 (Meysami, 2001). Paper and cardboard waste, household waste, and agrowaste degradation is based on lignin degradation, which is quite sensitive to pH. Hence, maintaining appropriate pH during lignin decomposition is an important factor. The optimum pH for delignification by *P. chrysosporium* is 4.0−4.5. The lignolytic effect of *P. chrysosporium* is decreased when the pH is above 5.5 and below 3.5. The pH requirement for growth and waste degradation varies. The optimum pH for fungal growth was slightly higher than that for lignin degradation. The optimum pH for white rot fungi is acidic and can vary from pH of 4.0 to 7.0 (Baker Lee et al., 1995).

Suitable pH conditions are important for growth and degradation of organic compounds by white rot fungi. For example, culturing *T. versicolor* in a flask containing paper and pulp mill effluent showed 80% reduction in color units after 6 days. However, the same fungus cultivated in a laboratory fermentor with 0.8% glucose and 12 mM ammonium sulfate at pH level 5.0 resulted in 88% reduction in color units within 3 days. This experiment indicates the importance of culture conditions on the efficiency of decolorization of effluent (Bergbauer et al., 1991). *Trichoderma* spp. decreased the color of kraft bleach plant effluent by 85% at pH 4 after 3 days incubation. The maximum total decolorization was noticed at pH 4.0.

2.5.3 Aeration

Lignin is degraded by various lignolytic enzyme by oxidation of lignin that needs the presence of oxygen. Lignin degradation is typically aerobic process. Thus, adequate aeration is needed for biodegradation. Increasing the O_2 content in the culture medium induced lignin degradation due to induction of lignin peroxidases and manganese peroxidases. Lignin is degraded faster in the presence of oxygen than air. For example, *P. chrysosporium* degraded lignin content of pulp when it was grown in 100% oxygen instead of air. The oxygen content has a profound effect on the rate of lignin degradation by *P. chrysosporium*, but not on the growth and multiplication of fungi.

2.5.4 Temperature

Temperature has a considerable effect on the ability of white rot fungi to grow and degrade organic waste and optimum temperature is needed for the growth and multiplication and also ligno-cellulolytic enzyme activity. The most contaminated soil ecosystem will not be at the optimum temperature for bioremediation throughout the year. High temperatures up to 35°C generally increase the growth rate and consequently the metabolic activity whereas low temperature below 15°C reduces the growth and degradation ability of the fungus (Baldrian, 2008). The optimum temperature range for lignin degradation by white rot fungi is from 20 to 40°C (Azadpour et al., 1997). Below and above the optimum temperature range the fungal growth is retarded and consequently the degradation process is affected.

2.5.5 Moisture content

Moisture is also an important factor for the growth and multiplication of fungi. The moisture content of soil also affects bioremediation systems such as the bioavailability of contaminants, the transfer of gases, and species distribution (Meysami, 2001). White rot fungi can grow and degrade within a limited range of favorable soil moisture conditions. The optimum moisture content for biodegradation is 50%–80% (Bossert and Bartha, 1984). Below and above this range drastically reduce the microbial activity.

2.6 Conclusions

Several recalcitrant crop residues and tree timbers after harvest (agrowaste), disposed paper, pulp and cardboard industrial wastes and disposable sanitary wastes containing high cellulose levels are piling up within the soil ecosystem. These agricultural and industrial wastes are dangerous to human health and affect the soil ecosystem, especially soil-beneficial microflora and microfauna. Soil and environmental pollution due to these agro and industrial wastes pose serious threats to human beings.

Remediation of soil dumped with disposable agrowaste and household wastes or pharma-sanitary waste containing high cellulose or ligno cellulosic components are degraded and decomposed by microbial systems. This degradation and decomposition can be either slow or rapid based on the type of microbial community. If the soil ecosystem contains an effective native microbial population for degradation of organic waste, the remediation process takes place naturally. However, introducing a large volume of efficient white rot fungi with suitable substate definitely speeds up the mycoremediation process. Most importantly, the introduced fungi should be nonpathogenic to crop plants and nondetrimental to the native beneficial microbial communities. Thus, selection of a suitable fungus or consortia of various fungi as remediating agents is the important step in the establishment of a mycoremediation plan. White rot fungi and soft rot fungi, especially *T. reesei*, are saprophytic and can be effectively used in the mycoremediation process in the soil ecosystem. Other *Trichoderma* spp. can also be used for decomposing organic matter. *Trichoderma* spp. are also potential biocontrol agents that inhibit many soilborne pathogenic fungi and bacteria through competition for space and nutrients and also by antibiosis through production of various inhibitory secondary metabolites. Since many white rot fungi do not establish in the soil ecosystem, conditioning the soil with adequate woody substrate enhances their efficacy in mycoremediation. Monitoring of the established mycoremediation system should be carried out for success.

References

Arantes, V., Goodell, B., 2014. Current understanding of brown rot fungal biodegradation mechanisms: a review. In: Schultz, T.P., Goodell, B., Nicholas, D.D. (Eds.), Deterioration and Protection of Sustainable Biomaterials. ACS Publication, USA, pp. 3–21.

Aust, S.D., 1990. Degradation of environmental pollutants by *Phanerochaete chrysosporium*. Microb. Ecol. 20 (2), 197–209.

Azadpour, A., Powell, P.D., Matthews, J., 1997. Use of lignin-degrading fungi in bioremediation. Remediation 997, 25–49.

Baker Lee, C.J., Fletcher, M.A., Avila, O.I., Callanan, J., Yunker, S., Munnecke, D.M., 1995. Bioremediation of MGP soils with mixed fungal and bacterial cultures. In: Hinchee, R.E., Fredrickson, J., Alleman, B. (Eds.), Third International in Situ and on-Site Bioremediation Symposium, pp. 123–128.

Baldrian, P., 2008. Wood-inhabiting ligninolytic basidiomycetes in soils: ecology and constraints for applicability in bioremediation. Fung. Ecol. 1, 4–12.

Beier, S., Hinterdobler, W., Bazafkan, H., Schillinger, L., Schmoll, M., 2020. CLR1 and CLR2 are light dependent regulators of xylanase and pectinase genes in *Trichoderma reesei*. Fungal Genet. Biol. 136, 103315. https://doi.org/10.1016/j.fgb.2019.103315.

Bergbauer, M., Eggert, C., Kraepelin, G., 1991. Degradation of chlorinated lignin compounds in a bleach plant effluent by the white-rot fungus *Trametes versicolor*. Appl. Microbiol. Biotechnol. 35, 105–109.

Bossert, I., Bartha, R., 1984. The fate of petroleum in soil ecosystems. In: Atlas, R.M. (Ed.), Petroleum microbiology. Macmillan, New York, pp. 434–476.

Das, K., Mukherjee, A.K., 2007a. Crude petroleum-oil biodegradation efficiency of Bacillus subtilis and *Pseudomonas aeruginosa* strains isolated from a petroleum oil contaminated soil from North-East India. Bioresour. Technol. 98, 1339–1345.

Das, N., Mukherjee, M., 2007b. Cultivation of *Pleurotus ostreatus* on weed plants. Bioresour. Technol. 98 (14), 2723–2726.

Dey, S., Maiti, T.K., Saha, N., Banerjee, R., Bhattacharyya, B.C., 1991. Extracellular protease and amylase activities in ligninase-producing liquid culture of *Phanerochaete chrysosporium*. Process Biochem. 26, 325–329.

Dill, I., Kraepelin, G., 1986. Palo podrido: model for extensive delignification of wood by *Ganoderma applanatum*. Appl. Environ. Microbiol. 52, 1305–1312.

Ford, C.I., Walter, M., Northcott, G.L., Di, H.J., Kameron, K.C., Trower, T., 2007. Fungal inoculum properties: extracellular enzyme expression and pentachlorophenol removal by New Zealand *Trametes* species in contaminated field soils. J. Environ. Qual. 36, 1749–1759.

Gold, M.H., Alic, M., 1993. Molecular biology of the lignin-degrading basidiomycete *Phanerochaete chrysosporium*. Microbiol. Rev. 57, 605–622.

He, J., Wu, A., Chen, D., Yu, B., Mao, X., Zheng, P., et al., 2014. Cost-effective lignocellulolytic enzyme production by *Trichoderma reesei* on a cane molasses medium. Biotechnol. Biofuels 7, 43–51.

Higson, F.K., 1991. Degradation of xenobiotics by white rot fungi. Rev. Environ. Contam. Toxicol. 122, 111–152.

Hinterdobler, W., Schuster, A., Tisch, D., Özkan, E., Bazafkan, H., Schinnerl, J., Brecker, L., Böhmdorfer, S., Schmoll, M., 2019. The role of PKAc1 in gene regulation and trichodimerol production in *Trichoderma reesei*. Fungal Biol. Biotechnol. 10 (6), 12. https://doi.org/10.1186/s40694-019-0075-8.eCollection2019.

Huang, D.L., Zeng, G.M., Feng, C.L., Hu, S., Lai, C., Zhao, M.H., Su, F.F., Tang, L., Liu, H.L., 2010. Changes of microbial population structure related to lignin degradation during lignocellulosic waste composting. Bioresour. Technol. 101, 4062–4067.

Ishida, T., Yaoi, K., Hiyoshi, A., Igarashi, K., Samejima, M., 2007. Substrate recognition by glycoside hydrolase family 74 xyloglucanase from the basidiomycete *Phanerochaete chrysosporium*. FEBS J. 274, 5727—5736.

Kirk, T.K., Farrell, R.L., 1987. Enzymatic "combustion": the microbial degradation of lignin. Annu. Rev. Microbiol. 41, 465—505.

Kuhad, R.C., Gupta, R., Singh, A., 2011. Microbial cellulases and their industrial applications. Enzym. Res. 10. Article ID 280696.

Meysami, P., 2001. Feasibility Study of Fungal Bioremediation of a Flare Pit Soil Using White Rot Fungi. Thesis. The University of Calgary.

Maciel, M.J.M., Silva, A.C., Ribeiro, H.C.T., 2010. Industrial and biotechnological applications of ligninolytic enzymes of the basidiomycota: a review. Electron. J. Biotechnol. 13, 6. https://doi.org/10.2225/vol13-issue6-fulltext-2.

Mougin, C., Boukcim, H., Jolivalt, C., 2009. Soil Bioremediation Strategies Based on the Use of Fungal.

Odu, C.T.I., 1978. The effect of nutrient application and aeration on oil degradation in soils. Environ. Pollut. 15, 235—240.

Peralta, R.M., da Silva, B.P., Côrrea, R.C.G., Kato, C.G., Seixas, F.A.V., Bracht, A., 2017. Enzymes from Basidiomycetes— Peculiar and Efficient Tools for Biotechnology in: Biotechnology of Microbial Enzymes, pp. 119—149.

Piškur, B., Zule, J., Piškur, M., Jurc, D., Pohleven, F., 2009. Fungal wood decay in the presence of fly ash as indicated by gravimetries and by extractability of low molecular weight organic acids. Int. Biodeterior. Biodegrad. 63, 594—599.

Reddy, C.A., Mathew, Z., 2001. Bioremediation potential of white rot fungi. In: Gadd, G.M. (Ed.), Fungi in Bioremediation. Cambridge University Press, Cambridge.

Reese, E.T., 1976. History of the cellulase program at the U.S. Army Natick development center. Biotechnol. Bioeng. Symp. 9—20.

Schmidt, K.R., Chand, S., Gostomski, P.A., Boyd-Wilson, K.S.H., Ford, C., Walter, M., 2005. Fungal inoculum properties and its effect on growth and enzyme activity of *Trametes versicolor* in soil. Biotechnol. Prog. 21, 377—385.

Schuster, A., Schmoll, M., 2010. Biology and biotechnology of Trichoderma. Appl. Microbiol. Biotechnol. 87, 787—799.

Seiboth, B., Ivanova, C., Seidl-Seiboth, V., 2011. Trichoderma reesei: a fungal enzyme producer for cellulosic biofuels. In: Bernardes, M.A.S. (Ed.), Biofuel Production: Recent Developments and Prospects. InTech, Rijeka, Croatia, pp. 309—340.

Tuomela, M., Lyytikäinen, M., Oivanen, P., Hatakka, A., 1998. Mineralization and conversion of pentachlorophenol (PCP) in soil inoculated with the white-rot fungus *Trametes versicolor*. Soil Biol. Biochem. 31, 65—74.

Valásková, V., Baldrian, P., 2006. Degradation of cellulose and hemicelluloses by the brown rot fungus *Piptoporus betulinus*: production of extracellular enzymes and characterization of the major cellulases. Microbiology. 152, 3613—3622.

Valdemar, E.R.M., Vazquez-Morillas, A., Ojeda-Benitez, S., Arango-Escorcia, G., Velasco-Perez, M., Cabrera-Elizalde, S., Quechholac-Pina, X., SoteloNavarro, P.X., 2015. Assessment of gardening wastes as a co-substrate for diapers degradation by the fungus *Pleurotus ostreatus*. Sustain. J. 7 (5), 6033—6045.

Valdemer, E., 2014. Treatment Alternatives for Disposable Diapers: Degrading Baby Diapers, Using Them as a Basis for Cultivating Mushrooms. Mexico. Metropolitan University, Azcapotzalco (UAM-A).

Valentín, L., Kluczek-Turpeinen, B., Willfö, S., Hemming, J., Hatakka, A., Steffen, K., Tuomela, M., 2010. Scots pine (*Pinus sylvestris*) bark composition and degradation by fungi: potential substrate for bioremediation. Bioresour. Technol. 101, 2203–2209.

Further reading

Enzymes. In: Singh, A., Kuhad, R.C., Ward, O.P. (Eds.), Advances in Applied Bioremediation. Springer, Heidelberg, pp 123–149.

Hammel, K.E., Jensen Jr., K.A., Mozuch, M.D., Landucci, L.L., Tien, M., Pease, E.A., 1993. Ligninolysis by a purified lignin peroxidase. J. Biol. Chem. 268, 12274–12281.

Karam, J., Nicell, J.A., 1997. Potential applications of enzymes in waste treatment. J. Appl. Chem. Biotechnol. 69, 141–153.

Piskur, B., Bajc, M., Robek, R., Humar, M., Sinjur, I., Kadunc, A., Oven, P., Rep, G., Al Sayegh Petkovšek, S., Kraigher, H., Jurc, D., Pohleven, F., 2011. Influence of *Pleurotus ostreatus* inoculation on wood degradation and fungal colonization. Bioresour. Technol. 102, 10611–10617.

Sanchez, C., 2010. Cultivation of *Pleurotus ostreatus* and other edible mushrooms. J. Appl. Microbiol. Technol. 85, 1321–1337.

Valdemar, E.R.M., Turpin-Marion, S., Delfin-Alcala, I., Vazquez-Morillas, A., 2011. Disposable diapers biodegradation by the fungus *Pleurotus ostreatus*. Waste Manag. J. 31 (8), 1683–1688.

Adsorption of Cr(VI) ions from aqueous solutions by diatomite and clayey diatomite

Hamdije Memedi[1], Arianit A. Reka[1,2], Stefan Kuvendziev[3], Katerina Atkovska[3], Mrinmoy Garai[4], Mirko Marinkovski[3], Blagoj Pavlovski[3], Kiril Lisichkov[3]

[1]*Department of Chemistry, Faculty of Natural Sciences and Mathematics, University of Tetovo, Tetovo, Republic of North Macedonia;* [2]*NanoAlb, Albanian Unit of Nanoscience and Nanotechnology, Academy of Sciences of Albania, Fan Noli Square, Tirana, Albania;* [3]*Faculty of Technology and Metallurgy, Ss. Cyril and Methodius University, Skopje, Republic of North Macedonia;* [4]*Materials Science Centre, Indian Institute of Technology (IIT), Kharagpur, West Bengal, India*

Chapter outline

3.1 Introduction

Examinations of toxic heavy metals released into the environment without proper treatment are gaining considerable attention today. One of the reasons for taking

Biological Approaches to Controlling Pollutants. https://doi.org/10.1016/B978-0-12-824316-9.00002-1

this issue seriously is that these ions pollute the air, soil, and water, thus having a great effect on human health (Laus et al., 2010). Industrial wastewater, sewage, and waste are the main factors for contamination of the soil and the environment. Most industries discharge toxic wastewater into rivers without proper treatment. Most of the pollutants are heavy metals from industrial wastewater from mining, metal processing, tanneries, medicines, pesticides, organic chemicals, rubber and plastics, and wood products. To avoid health hazards, it is necessary to remove these toxic heavy metals from wastewater prior to discharge. Most of the heavy metals released into wastewater are toxic and carcinogenic and pose a serious threat to human health (Srivastava et al., 1996).

There are different chromium oxides in nature—while the ionic valences of chromium are 0–6 valence, the most common ionic valences are Cr^{6+} and Cr^{3+} (Tchobonglus et al., 2003). Cr(VI) usually exists in aqueous solution such as CrO_4^{2-}, $HCrO^{4-}$, $HCr_2O_7^{2-}$, $Cr_2O_7^{2-}$ (Norseth, 1981). The oxidative form of chromium includes Cr(VI) and Cr(III), of which the Cr(VI) form is toxic to the environment because it is carcinogenic, toxic, and highly soluble. Currently, there are several common methods for treating chromium-containing wastewater: ion exchange, membrane separation, neutralization method, reduction, biological methods, and adsorption method. The adsorption method has unique advantages due to the selectivity of the adsorption material, low cost, regeneration of the adsorbent, and not only solves the expensive problem with chemical methods but also overcomes the disadvantages of the limited adsorption capacity of the biological process (Ngah and Hanafiah, 2008).

In this paper, the adsorption method is used in order to determine the absorption characteristics of natural mineral raw materials in order to eliminate heavy metals or Cr(VI) ions (Afal and Wiener, 2014).

Many materials are studied for adsorption of heavy metal ions such as crust of Eucalyptus (Sarin and Pant, 2006), bran of corn (Singh et al., 2006), shells of peanuts and nuts (Karthikeyan et al., 2005; Atkovska et al., 2018), red mud (Gupta et al., 2001, 2003), silicates (Chiron et al., 2003; Belousov et al., 2019a), zeolites (Jakupi et al., 2016; Wingenfelder et al., 2005; Zendelska et al., 2015a,b, 2018a,b, 2019), activated carbon (Srivastava et al., 1996), *Spirogyra* biosorbents (Gupta et al., 2001), fly ash (Babel and Opiso, 2007; Bayat, 2002; Gupta and Ali, 2004), graphene oxide (Alija et al, 2020) and with inorganic raw materials such as bentonite, aksil, and pemza (Memedi et al., 2016a,b, 2017; Belousov and Krupskaya, 2019; Atkovska et al., 2016; Semenkova et al., 2020).

Diatomite has been used for the adsorption of different heavy metals from water and wastewaters (Guru et al., 2008; Al-Ghouti et al., 2009; Yuan et al., 2010; Caliskan et al., 2011; Abu-Zurayk et al., 2015).

In this paper, diatomite and clayey diatomite are used as adsorbents for the removal of Cr(VI) ions.

Clayey diatomite are very similar in appearance to diatoms. Round pieces of opal mixed with quartz, clay, and limestone particles can be seen under a microscope. The clayey diatomite from the village of Suvodol, Bitola, are typical weakly bound sedimentary rocks of biogenic origin, which is confirmed by numerous pieces of evidence (of phytogenic origin) in the form of very discrete microfossils of the order Algae Diatomei.

The examined clayey diatomite from the village Suvodol is true proof of the continuity of living microorganisms (of phytogenic origin) after the sedimentation of plant products responsible for the formation of coal seams in the former Miocene-Pliocene lake basin. This sedimentary complex, composed of ash and coal, is a de facto biogenic-sedimentary formation (Pavlovski et al., 2011; Cekova et al., 2013; Reka et al., 2016, 2019b).

Diatomite is a natural phenomenon and is a soft, silicate sedimentary rock that breaks down very easily. It consists of fossilized remains of diatoms (Reka et al., 2021). It is used as a means of filtration, adsorption of heavy metals, mild abrasive, toothpaste, insecticide, filler in plastic and rubber, etc. (Reka et al., 2014, 2017).

The aim of this research is to compare the pure natural diatomite and the impure clayey diatomite and their behavior in the process of eliminating Cr(VI) in aqueous solutions.

3.2 **Experimental**

3.2.1 **Materials and methods**

Work raw materials used in this paper are clayey diatomite (Bitola) and pure diatomite (Kozuf). The raw materials are mechanically prepared in a ball mill with granulation from which a fraction of 0.25−0.5 mm prevails.

The determination of the specific surface area of the adsorbents used is done according to the Brunauer−Emmett−Teller (BET) method with ASAP 2010, a Micromeritics instrument.

The following tests were performed for the physical-mechanical characterization of the clayey diatomite and diatomite: density, bulk density, compressive strength, and porosity.

The chemical composition of the raw materials is determined using classical silicate analysis. The raw material is melted in a mixture of carbonates (Na_2CO_3 and K_2CO_3), while the percentage of different oxides is determined by complexometric titration (Kirschenbaum, 1983). The presence of alkaline oxides (Na_2O and K_2O) was determined by flame photometry using the Evans Electroselenium Ltd. 410 instrument.

The mineralogical characterization was performed by the Debye−Scherrer powder method using the Rigaku Ultima IV, 1-dimensional D/teX detector using CuKα radiation ($\lambda = 1.54,178$ nm) in the range of 5 to 60 degrees 2θ, recording speed 1 degree/min.

Room temperature infrared tests were performed with a Perkin-Elmer (2000) spectrophotometer using the KBr pressed tablet method in the range of $450-4000$ cm^{-1}. Namely, 1 mg of the sample is dispersed through 250 mg of KBr after which the thus prepared tablet is recorded. The spectral resolution is 4 cm^{-1}, and the number of scans is 16.

Differential thermal and thermogravimetric (DTA/TGA) testing of the raw material was performed using a Stanton Redcroft apparatus, under the following experimental conditions: temperature range 30–1100°C; heating rate 10°C/min in air, and duration of heat treatment of 98 min; mass of the sample (clayey diatomite 13,577 mg, diatomite 19,085 mg) with a ceramic pot as a carrier of the material.

scanning electron microscopy (SEM) testing of the raw material was performed using the VEGA3 LMU Tescan instrument in combination with X-ray dispersive power (Energy 250 Microanalysis System).

Transmission electron microscopy of the raw materials was performed using a JEOL 2100 transmission electron microscope.

For the preparation of solutions of Cr(VI) with a particular initial concentration, $K_2Cr_2O_7$ standard solution with a concentration of 1000 mg/L was used. The pH of solutions was adjusted using 0.1M HCl or 0.1M HNO_3 and 0.1M NaOH. All reagents and chemicals used in the experiments are of analytical grade.

The quantitative monitoring of the dynamics of studied system regarding the presence of Cr(VI) ions in the model solutions, was performed with the atomic absorption spectrophotometric (AAS) method using AAC Perkin Elmer model AA700 and UV/Vis spectrophotometric analysis instrument type HACH DR/2010.

Determination of the point of zero charge of clayey diatomite and diatomite was conducted by adding 0.2 g adsorbent in Erlenmeyer flasks (300 mL) and added 100 mL of distilled water with pH ($pH_{initial}$) from 2 to 10. pH is adjusted by adding solution of HCl or NaOH as required. Erlenmeyer flasks were continuously mixed on stirrer (140 rpm) for 24 h at room temperature. The suspension was then filtered and the pH (pH_{final}) of every filtrate was determined. The pH of the solutions was measured by pH-meter GMH 3500 Series-Greisinger, Germany.

3.2.2 Adsorption experiment

The working raw mineral materials that are the subject of the research in this paper are tested in a laboratory glass reactor with a volume of 2 dm^3, at room temperature, at a constant mixing regime of 400 rpm with a magnetic stirrer, observing the effect of the following operating parameters: pH of the medium, initial concentration of Cr(VI) ions, absorption time (5–180 min), amount of adsorbent (0.5–5.5 g/L). Aqueous solution of $K_2Cr_2O_7$ was used to prepare the standard solutions.

The removal coefficient is determined by applying the following relation:

$$\%Removal = \frac{C_0 - C_e}{C_0} \cdot 100 \tag{3.1}$$

where C_0 is the initial concentration of adsorbate [mg/L], C_e equilibrium concentration [mg/L].

The adsorption capacity q_t [mg/g] is determined by the relation

$$q_t = \frac{(C_0 - C_t) \cdot V}{m} \tag{3.2}$$

where C_t [mg/L] represents the concentration of Cr(VI) ions, V [l] working volume, and m [g] the amount of adsorbent (Kumar et al., 2011).

3.3 Results and discussion

3.3.1 Physical-mechanical characterization

The analyzed clayey diatomite from the village Suvodol is a sedimentary stone with a grayish or whitish-gray color, very light and soft (1 to 2 according to Moos), fine or superfine, with a granular structure, porous, shell, sticky, etc. The diatomite has a white color, massive homogeneous texture and shell brittleness (Fig. 3.1).

Examinations of the physical properties of the pure diatomite and clayey diatomite indicated the results shown in Table 3.1.

The clayey diatomite shows a specific area of 57.64 m^2/g resulting from the presence of diatomite and clays. The diatomite used does not contain clay minerals and the relatively high specific surface area (63.26 m^2/g) results from the high percentage of micro- and nanopores.

Chemical tests on the clayey diatomite showed that it was an acid rock with approximately 60% SiO_2. The parameters of the amount of SiO_2 and Al_2O_3 present, as well as the mass loss (LOI), obviously indicate the determined mineral phases—opal, illite, quartz, feldspar, chlorite in the treated sample. The total alkaline oxide

A   B

FIGURE 3.1

A piece of clayey diatomite and pure diatomite in natural state.

Table 3.1 Physical-mechanical properties of pure diatomite and clayey diatomite.

Property	Pure diatomite	Clayey diatomite
Density (g/cm^3)	2.08	2.39–2.41
Bulk density (g/cm^3)	0.58	0.64–0.88
Water absorption (%)	59.55	85–95
Total porosity (%)	71.63	68–75

content of 2.53% indicates a feldspar content of 16%. The contents of Fe_2O_3 (6.57%) and MgO (2.16%) are compatible with the specified (Mg, Fe)—chlorite in the clayey diatomite sample. The rather low content of CaO of 2.60% indicates that the specific plagioclase belongs to the type albite-oligoclase (Table 3.2).

Chemical tests have shown that the raw material diatomite contains about 93% SiO_2 and if we take into account that the raw material has about 3% loss during combustion, it can be concluded that it contains only about 3% impurities. The chemical composition of the raw material indicates that it is a high-quality diatomite raw material

3.3.1.1 X-ray analysis

The results of X-ray examinations of the clayey diatomite and diatomite are shown in Fig. 3.2.

X-ray clayey diatomite analysis determined: opal, quartz, illite-hydromica structures, feldspars (plagioclase, K-feldspars) and chlorites (Pavlovski et al., 2011; Cekova et al., 2013).

Based on the diffractogram shown (Fig. 3.2) of a diatomite from Kozuv, it can be concluded that it is a natural raw material in which SiO_2 is mainly in amorphous phase as well as the presence of the main peak of quartz. The diffractogram shows a wide diffuse peak that appears in the range 19–25 degrees 2θ, which is the area for the most intense crystalline modifications of silica, quartz, cristobalite, and tridimit (Reka et al., 2014).

3.3.1.2 Fourier transform infrared analysis

The Fourier transform infrared (FTIR) spectra of clayey diatomite and diatomite are shown in Fig. 3.3.

The absorption bands at 800, 1033 and 1633 cm^{-1} at clayey diatomite are the result of the amorphous SiO_2 present in the clayey. In this test, too, the main band of SiO_2 occurs at 1033 cm^{-1}, shifted in lower frequency values, as a result of the substitution of Si^{+4} ions in the tetrahedral position with trivalent cations. The bands at 466, 525, 642 and 1083 cm^{-1} indicate the presence of quartz in the clayey diatomite. The strips at 1633, 3450 and 3617 cm^{-1} are the result of adsorbed water in the clayey diatomite. The band at 3690 and the deformed band at 936 cm^{-1} indicate the presence of hydroxyl groups in the clayey diatomite.

Based on the infrared spectrum of the diatomite shown in Fig. 3.3, the following can be concluded: the bands at 1620, 1100 and 800 cm^{-1} are due to the amorphous SiO_2 in the diatomite. The band at 469 cm^{-1} is due to the Si—O—Si bond as well as the minimal presence of quartz in the sample. The bands at 1063 and 3620 cm^{-1} are the result of the OH groups in the diatomite (Reka et al., 2015).

3.3.1.3 Thermal and thermogravimetric analysis

The DTA/TGA results are shown in Fig. 3.4.

DTA/TGA of clayey diatomite are compatible with X-ray examinations. A very wide end-peak can be seen on the DTA curve starting at about 80–300°C with a

Table 3.2 Chemical composition of pure diatomite and clayey diatomite.

Oxide	SiO$_2$	Al$_2$O$_3$	Fe$_2$O$_3$	CaO	MgO	K$_2$O	Na$_2$O	SO$_3$	LOI	Total
Pure diatomite	92.97	1.52	0.21	0.43	0.19	0.26	0.08	0.05	3.86	99.73
Clayey diatomite	64.95	11.85	4.51	1.49	1.88	1.40	0.84	1.74	11.20	99.86

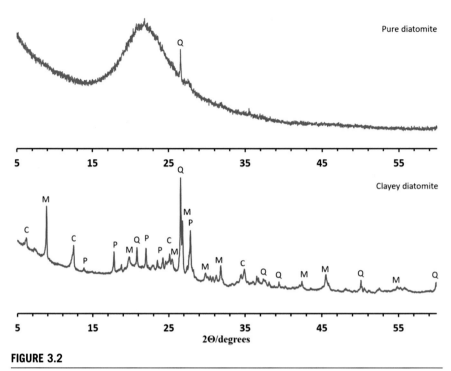

FIGURE 3.2

Pure and clayey diatomite.

minimum of 150°C corresponding to an opal and mineral illite-hydromica structure. Other thermal effects are minor. The TG curve od clayey diatomite shows a total weight loss of 17%, which is compatible with the mineralogical composition— presence of minerals (opal, illite) in an amount of 50%–70%.

TGA results from diatomite show mass loss at two temperature intervals. The first temperature interval is with intense mass loss (7%) when heated from room temperature to 250°C. This is explained by the elimination of absorbed water on the surface and pores of the diatomite. The second temperature range of mass loss is from 250 to 700°C. This mass loss is due to the loss of chemically bound water. The results of the DTA curve show an exothermic peak at 940°C resulting from the crystallization of the amorphous phase in the diatomite.

3.3.1.4 Scanning electron microscopy analysis

Morphological analysis of clayey diatomite and pure diatomite performed by SEM are shown in Fig. 3.5.

Scanning electron microscopy provides very important information about the biogenetic origin of the raw material, which clearly shows fragments of various skeletal forms of microfossils, their morphological characteristics and the pores themselves that are present in the raw material.

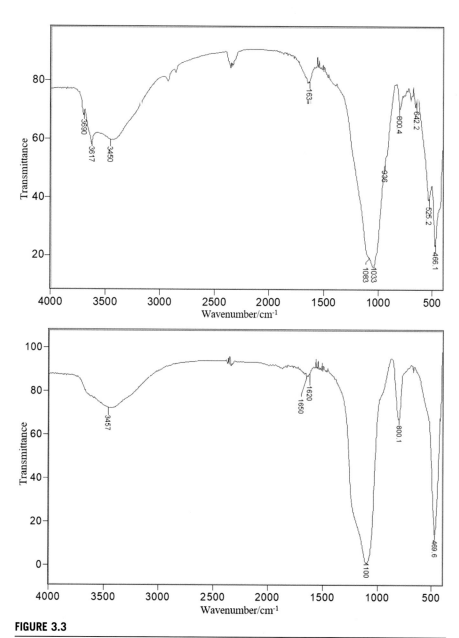

FIGURE 3.3

Fourier transform infrared spectra of raw clayey diatomite (top) and pure diatomite (bottom).

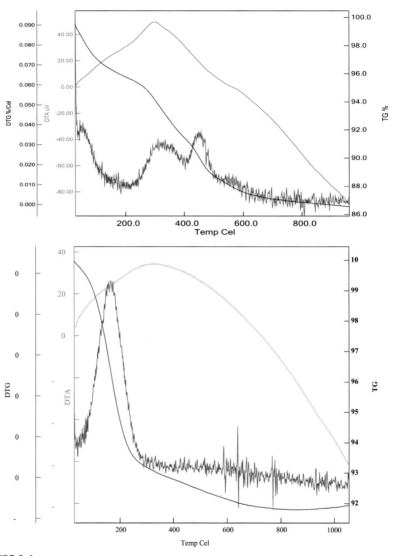

FIGURE 3.4

Differential thermal and thermogravimetric analysis of clayey diatomite and pure diatomite.

Figure A shows the pores present in the clayey diatomite and elements of various microfossils. Figure B shows various skeletal forms of clayey diatomite while Figure C shows a skeleton of a microfossil with well-formed pores.

Based on the presented results from the SEM examinations, the biogenic origin of the raw material diatomite itself can be concluded. The various skeletal shapes shown are the result of algae skeletal remains (diatomite). Nano-sized pores (350–550 nm) are clearly visible in these skeletal forms.

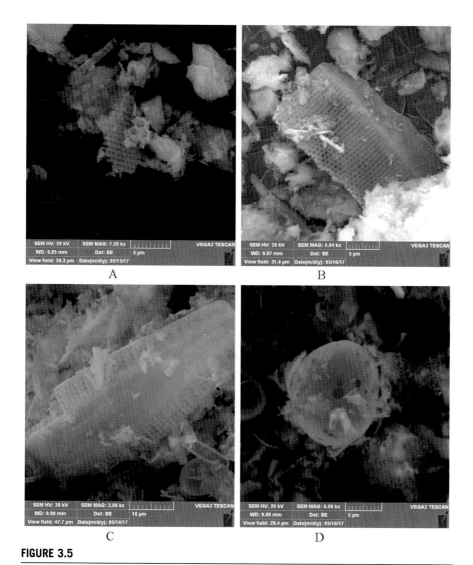

FIGURE 3.5

Scanning electron microscopy photographs of clayey diatomite (A, B) and pure diatomite (C, D).

3.3.1.5 Transmission electron microscopy investigations

Fig. 3.6 shows transmission electron microscopy (TEM) photographs of clayey diatomite (A, B) and pure diatomite (C, D).

With the help of TEM tests, the morphology of the raw material clayey diatomite can be seen, as well as the presence of different microrelics that are present in the raw material. The picture shown shows globular structures of opal nature of

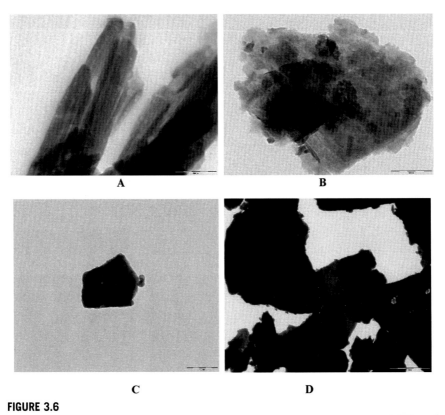

FIGURE 3.6

Transmission electron microscopy images of clayey diatomite (A,B) and pure diatomite (C, D).

biogenetic origin as well as relics with different elongated shapes. TEM shows the clearly visible pores in the pure diatomite as well as the classy phase which is due to the amorphous mass present in the material. TEM of pure diatomite shows remains of algae and the pores in nanometric size (150–200 nm).

3.3.1.6 pH_{PZC} of clayey diatomite and pure diatomite

The pH_{PZC} for clayey diatomite and pure diatomite is shown in Fig. 3.7. The zero point charge of a clayey diatomite is 8.2.

The zero point charge of a clayey diatomite is 8.2 while the zero point charge of pure diatomite is 6.5.

3.3.1.7 Effect of adsorbent dose on adsorption of Cr(VI) on diatomite end clayey diatomite

The efficiency of elimination of Cr(VI) by applying different amount of adsorbent (0.5; 1.5; 2.5; 3.5; 3.5; 4.5 and 5.5 g/L) is analyzed under the following operating

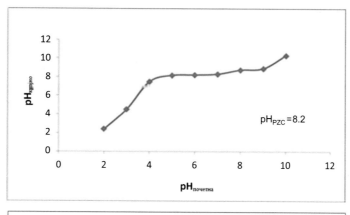

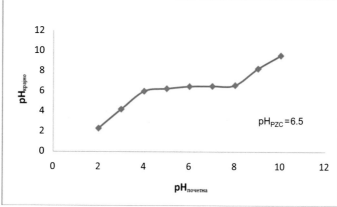

FIGURE 3.7

Zero point charge of clayey diatomite and pure diatomite.

conditions: room temperature, constant stirring mode at 400 rpm, adsorption time of 15 min, initial Cr(VI) concentration 0.5 mg/L. The pH of solutions was 2 (Figs. 3.8 and 3.9).

In diatomite, the percentage of elimination of Cr(VI) ions is as follows: 19.15%; 25.53%; 32.61%; 31%; 30.43%; and 29.79%.

From the studies of the effect of the amount of adsorbent on the removal of Cr(VI) ions, at pH 2, for the amount of adsorbent of 0.5 mg/L, the percentage of removal of Cr (VI) ions is 84.5% while for other values of the amount the percentage of removal is 100%.

3.3.1.8 Effect of contact time

To see the effect of contact time on the elimination coefficient, a series of experiments were performed using adsorbents: diatomite and clayey diatomite. The dependence of the change in adsorbate concentration over time for different initial concentrations is given in Figs. 3.10 and 3.11. These analyzes were performed at

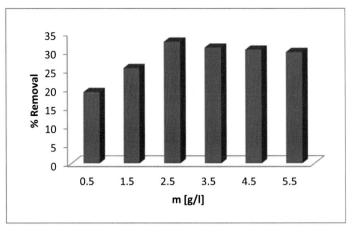

FIGURE 3.8

Effect of adsorbent dose on adsorption of Cr(VI) ions on diatomite ($C_0 = 0.5$ mg/L, pH = 2).

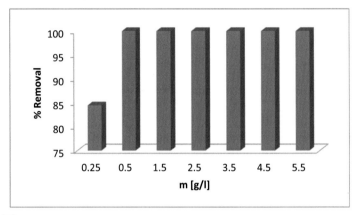

FIGURE 3.9

Effect of adsorbent dose on adsorption of Cr(VI) ions on clayey diatomite ($C_0 = 0.5$ mg/L, pH = 2).

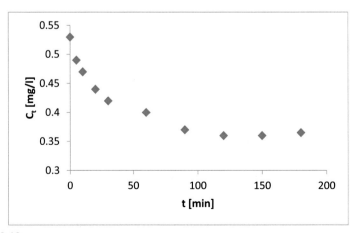

FIGURE 3.10

Dependence of Cr(VI) concentration from time of adsorption for diatomite.

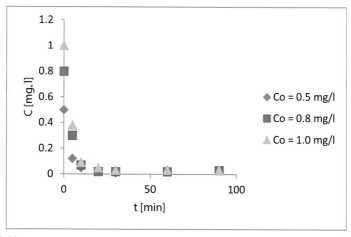

FIGURE 3.11

Dependence of Cr(VI) concentration from time of adsorption for clayey diatomite.

pH of solutions 2, adsorbent mass 2.5 g/L, room temperature, at constant stirring mode of 400 rpm with magnetic stirrer.

From the graphs presented results it can be concluded that the time of the adsorption process has an effect on the degree of elimination. In fact, the adsorption time can be divided into two periods, the period of rapid adsorption and the period when the system begins to enter equilibrium. The results indicate that the system enters equilibrium for diatoms after 95 min, for clayey diatomite after 20 min.

Figs. 3.12 and 3.13 show the results for the amount of adsorption of the adsorbate over time. It can be seen how the mass of adsorbed Cr(VI) ions changes with time for all working raw materials.

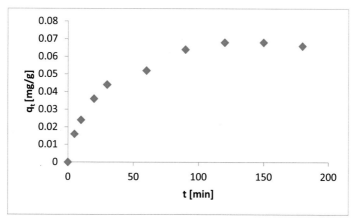

FIGURE 3.12

Dependence of adsorbed amount of Cr(VI) from time of adsorption for diatomite.

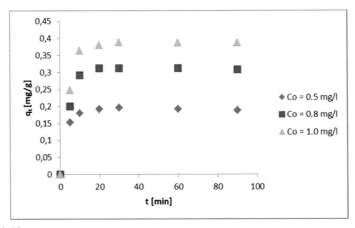

FIGURE 3.13

Dependence of adsorbed amount of Cr(VI) from time of adsorption for clayey diatomite.

From the presented graphs we can see that in the first 20–35 min only for diatoms up to 95 min there is a high degree of adsorption of the adsorbate ions after this period the examined systems enter into equilibrium.

From the presented graphic dependencies for change of the adsorption capacity with time it is evident that at lower pH values due to the phenomenon of protonation on the surface of the adsorbent there is adsorption of 20–95 min in diatomite and in clayey diatomite in the first 10–25 min.

Fig. 3.14 shows the comparison of percentage of removal of Cr(VI) ions for used natural mineral adsorbents. It can be seen from the figure that the removal percentage for diatomite is 28% and clayey diatomite 99.3%.

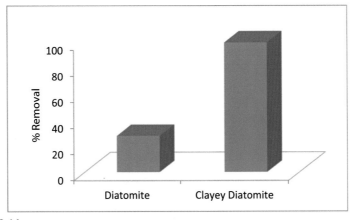

FIGURE 3.14

Function of Cr(VI) removal for adsorbents diatomite and clayey diatomite $C_0 = 0.5$ mg/L; pH = 2.

3.4 Conclusions

Working raw materials used in this paper are clayey diatomite (Bitola) and diatomite (Kozuf). The raw materials are mechanically prepared in a ball mill with granulation from which a fraction of $-0.5 + 0.25$ mm prevails. Aqueous solution of $K_2Cr_2O_7$ is used to prepare the working solutions. To characterize the adsorbents, the following analyses were performed: classical silicate chemical analysis, differential thermal and thermogravimetric analysis (DTA/TGA), XRD analysis, FTIR analysis, SEM, and TEM analysis. The BET model (as, m^2/g) was used to determine the specific area. From the results for the specific surfaces of the materials, it can be concluded that the diatomite used does not contain clay minerals, and the relatively high specific surface area results from the high percentage of micro- and nanopores. The clayey diatomite shows a specific area of 57.64 m^2/g resulting from the presence of diatomite and clays. The AAS method and UV/Vis spectrophotometric analysis (UV/VIS spectrophotometer) were used to determine the concentration before and after the Cr(VI) adsorption process. The obtained results from the determined pHpzc value are pHpzc (clayey diatomite) = 8.2; and pHpzc (pure diatomite) = 6.5. The pH of the medium also has a significant effect on the adsorption process of Cr(VI). We have obtained the best coefficients of elimination at lower pH values. At an initial concentration of adsorbate ions of 0.5 mg/L and a pH equal to 2, the best results were shown by the clayey diatomite, where the percentage of removal is 100%, in diatomite 28%. From the presented results, it can be concluded that the time of the adsorption process has an effect on the degree of elimination. In fact, the adsorption time can be divided into two periods, rapid adsorption and when the system begins to enter equilibrium. The results indicate that for the different types of adsorbents tested, the system enters equilibrium for diatomite after 90 min and for clayey diatomite after 20 min. From the results, it can be concluded that the used natural raw material clayey diatomite, with low cost, is an effective adsorbent for removal of Cr(VI) ions.

References

Afal, A., Wiener, S., 2014. Metal Toxicity. Medscape. Org, Retrieveed, p. 21.

Abu-Zurayk, R.A., Al, B., Ramia, Z., Hamadneh, I., Al-Dujaili, A.H., 2015. Adsorption of Pb(II), Cr(III) and Cr(VI) from aqueous solution by surfactant-modified diatomaceous earth: equilibrium, kinetic and thermodynamic modeling studies. Int. J. Miner. Process. https://doi.org/10.1016/j.minpro.2015.05.00.

Al-Ghouti, M., Khraishehb, M., Ahmad, M., Allenc, S., 2009. Adsorption behaviour of methylene blue onto Jordanian diatomite: a kinetic study. J. Hazard Mater. 165, 589−598.

Alija, A., Gashi, D., Plakaj, R., Omaj, A., Thaci, V., Reka, A., Avdiaj, S., Berisha, A., et al., 2020. A theoretical and experimental study of the adsorptive removal of hexavalent chromium ions using graphene oxide as an adsorbent. Open Chem. 18 (1), 936−942. https://doi.org/10.1515/chem-2020-0148.

Atkovska, K., Bliznakovska, B., Ruseska, G., Bogoevski, S., Boskovski, B., Grozdanov, A., 2016. Adsorption of Fe(II) and Zn(II) ions from landfill leachate by natural bentonite. J. Chem. Technol. Metall. 51 (2), 215−222.

Atkovska, K., Lisichkov, K., Ruseska, G., Dimitrov, A.T., Grozdanov, A., 2018. Removal of heavy metal ions from wastewater using conventional and nanosorbents: a review. J. Chem. Technol. Metall. 53 (2), 202−219.

Babel, S., Opiso, E.M., 2007. Removal of Cr from synthetic wastewater by sorption into volcanic ash soil. Int. J. Environ. Sci. Techol. 4, 99−107.

Bayat, B., 2002. Comparative study of adsorption properties of Turkish fly ashes. J. Hazard Mater. 95, 275−290.

Belousov, P.E., Krupskaya, V.V., 2019. Bentonite clays of Russia and neighboring countries. Georesursy 21, 79−90.

Belousov, P., Semenkova, A., Egorova, T., Romanchuk, A., Zakusin, S., Dorzhieva, O., Tyupina, E., Izosimova, Y., Tolpeshta, I., Chernov, M., Krupskaya, V., 2019a. Cesium sorption and desorption on glauconite, bentonite, zeolite, and diatomite. Minerals 9, 625.

Caliskan, N., Kul, A., Alkan, S., Gokirmak Sogu, E., Alacabey, I., 2011. Adsorption of Zinc(II) on diatomite and manganese-oxide-modified diatomite: a kinetic and equilibrium study. J. Hazard Mater. 193, 27−36.

Cekova, B., Pavlovski, B., Spasev, D., Reka, A., 2013. Structural Examinations of Natural Raw Materials Pumice and Trepel from Republic of Macedonia, Proceedings of the XV Balkan Mineral Processing Congress, Sozopol, Bulgaria, pp. 73−75.

Chiron, N., Guilet, R., Deydier, E., et al., 2003. Adsorption of Cu(II) and Pb(II) onto a grafted silica: isotherms and kinetic models. Water Res. 37 (13), 3079−3086. https://doi.org/10.1016/s0043-1354(03)00156-8.

Gupta, V.K., Ali, I., 2004. Removal of lead and chromium from wastewater using bagasse fly ash − a sugar industry waste. J. Colloid Interface Sci. 271, 321−328.

Gupta, V.K., Gupta, M., Sharma, S., 2003. Process development for the removal of lead and chromium from aqueous solutions using red mud an aluminium industry waste. Water Res. 35, 1125−1134.

Gupta, V.K., Srivastava, A.K., Jain, N., 2001. Biosorption of chromium (VI) from aqueous solutions by green algae Spirogyra species. Water Res. 35, 4079−4085.

Guru, M., Venedik, D., Murathan, A., 2008. Removal of trivalent chromium from water using low-cost natural diatomite. J. Hazard Mater. 160, 318−323.

Jakupi, S., Lisichkov, K., Golomeova, M., Atkovska, K., Marinkovski, M., Kuvendziev, S., Memedi, H., 2016. Separation of Co(II) Ions from water resources by natural zeolite (clinoptilolite). Mater. Environ. Protection V (1), 57−66.

Karthikeyan, T., Rajgopal, S., Miranda, L.R., 2005. Chromium (VI) adsorption from aqueous solution by *Hevea brasiliensis* sawdust activated carbon. J. Hazard Mater. 124, 192−199.

Kirschenbaum, H., 1983. The classical chemical analysis of silicate rocks − the old and the new. Geol. Survey Bull. 1547, 1−55. https://doi.org/10.3133/b1547.

Kumar, D., Gaur, J., 2011. Metal biosorption by two cyanobacterial mats in relation to pH, biomass concentration, pretreatment and reuse. Bioresour. Technol. 102 (3), 2529−2535.

Laus, R., Costa, T.G., Szpoganicz, B., Fávere, V.T., 2010. Adsorption and desorption of Cu (II), Cd (II) and Pb (II) ions using chitosan crosslinked with epichlorohydrin-triphosphate as the adsorbent. J. Hazard Mater. 183 (1−3), 233−241.

Memedi, H., Atkovska, K., Lisichkov, K., Marinkovski, M., Bozinovski, Z., Kuvendziev, S., Reka, A.A., 2016a. Investigation of the Possibility for Application of Natural Inorganic Sorbent (Aksil) for Heavy Metals Removal from Water Resources. Croatian Sci. And Profess. Conf. Water for All, Osijek, Croatia, pp. 185−192.

Memedi, H., Atkovska, K., Lisichkov, K., Marinkovski, M., Bozinovski, Z., Kuvendziev, S., Reka, A.A., Jakupi, S., 2016b. Application of natural inorganic sorbent (Peniza) for removal of Cr(VI) ions from water resources. In: V Int. Conf. Ecology of Urban Areas 2016, Zrenjanin, Serbia, pp. 109−116.

Memedi, H., Atkovska, K., Lisichkov, K., Marinkovski, M., Kuvendziev, S., Bozinovski, Z., Reka, A.A., 2017. Separation of Cr(VI) from aqueous solutions by natural bentonite: equilibrium study. Qual Life 8, 41−47.

Ngah, W.W., Hanafiah, M., 2008. Removal of heavy metal ions from wastewater by chemically modified plant wastes as adsorbents: a review. Bioresour. Technol. 99 (10), 3935−3948.

Norseth, T., 1981. Cancer hazards caused by nickel and chromium exposure. Environ. Health Perspect. 40, 121−130.

Pavlovski, B., Jančev, S., Petreski, L., Reka, A., Bogoevski, S., Boškovski, B., 2011. Trepel — a peculiar sedimentary rock of biogenetic origin from the Suvodol village, Bitola, R. Macedonia. Geol. Maced. 25, 67−72.

Reka, A.A., Anovski, T., Bogoevski, S., Pavlovski, B., Boškovski, B., 2014. Physical-chemical and mineralogical-petrographic examinations of diatomite from deposit near village of Rožden, Republic of Macedonia. Geol. Maced. 28, 121−126.

Reka, A.A., Pavlovski, B., Anovski, T., Bogoevski, S., Bškovski, B., 2015. Phase transformations of amorphous SiO_2 in diatomite at temperature range of 1000−1200°C. Geol. Maced. 29, 87−92.

Reka, A.A., Durmishi, B., Jashari, A., Pavlovski, B., Buxhaku, N., Durmishi, A., 2016. Physical-chemical and mineralogical-petrographic examinations of trepel from Republic of Macedonia. Int. J. Innovat. Stud. Sci. Eng. Technol. 2, 13−17.

Reka, A.A., Pavlovski, B., Ademi, E., Jashari, A., Boev, B., Boev, I., Makreski, P., 2019b. Effect of thermal treatment of trepel at temperature range 800−1200°C. Open Chem. 17, 1235−1243.

Reka, A.A., Pavlovski, B., Fazlija, E., Berisha, A., Pacarizi, M., Daghmehchi, M., Sacalis, C., Jovanovski, G., Makreski, P., Oral, A., et al., 2021. Diatomaceous earth: characterization, thermal modification and application. Open Chem. 19 (1), 451−461. https://doi.org/10.1515/chem-2020-0049.

Reka, A.A., Pavlovski, B., Makreski, P., 2017. New optimized method for low-temperature hydrothermal production of porous ceramics using diatomaceous earth. Ceram. Int. 43, 12572−12578. https://doi.org/10.1016/j.ceramint.2017.06.132.

Sarin, V., Pant, K.K., 2006. Removal of chromium from industrial waste by using eucalyptus bark. Bioresour. Technol. 97, 15−20.

Semenkova, A., Belousov, P., Rzhevskaia, A., et al., 2020. U(VI) sorption onto natural sorbents. J. Radioanal. Nucl. Chem. https://doi.org/10.1007/s10967-020-07318-y.

Singh, K.K., Talat, M., Hasan, S.H., 2006. Removal of lead from aqueous solutions by agricultural waste maize bran. Bioresour. Technol. 97, 2124−2130.

Srivastava, S.K., Gupta, V.K., Mohan, D., 1996. Kinetic parameters for the removal of lead and chromium from wastewater using activated carbon developed from fertilizer waste material. Environ. Model. Assess. 1, 281−290.

Tchobanoglous, G., Burton, F.L., Stensel, H.D., 2003. Wastewater Engineering: Treatment and Reuse, fourth ed. McGraw Hill, p. 1819.

Wingenfelder, U., Nowack, B., Furrer, G., Schulin, R., 2005. Adsorption of Pb and Cd by amine-modified zeolite. Water Res. 39, 3287−3297.

Yuana, P., Liua, D., Fana, B., Yanga, D., Zhuc, R., Gec, F., Zhua, J., Hea, H., 2010. Removal of hexavalent chromium [Cr(VI)] from aqueous solutions by the diatomite-supported/unsupported magnetite nanoparticles. J. Hazard Mater. 173, 614−621.

Zendelska, A., Golomeova, M., Blazev, K., Krstve, B., Golomeov, B., Krstev, A., 2015. Adsorption of copper ions from aqueous solutions on natural zeolite. Environ. Prot. Eng. 41 (4), 17−36. https://doi.org/10.5277/epe150402.

Zendelska, A., Golomeova, M., Golomeov, B., Krstev, B., 2018. Removal of lead ions from acid aqueous solutions and acid mine drainage using zeolite bearing tuff. Arch. Environ. Prot 44 (1), 87−96. https://doi.org/10.24425/118185.

Zendelska, A., Golomeova, M., Golomeov, B., Krstev, B., et al., 2019. Removal of Zinc Ions from acid aqueous solutions and acid mine drainage using zeolite-bearing tuff. Mine Water Environ. 38 (1), 187−196. https://doi.org/10.1007/s10230-018-0560-y.

Zendelska, A., Golomeova, M., Blazev, K., Boev, B., Krstev, B., Golomeov, B., Krstev, A., et al., 2015. Kinetic studies of manganese removal from aqueous solution by adsorption on natural zeolite. Maced. J. Chem. Chem. Eng. 34 (1), 213−220. https://doi.org/10.20450/mjcce.2015.552.

Zendelska, A., Golomeova, M., Jakupi, S., Lisichkov, K., Kuvendziev, S., Marinkovski, M., 2018. Characterization and application of clinoptilolite for removal of heavy metal ions from water resources. Geol. Macedonica 32 (1), 21−32.

Advances in bioremediation of antibiotic pollution in the environment

4

Saima Gul[1], Sajjad Hussain[2,3], Hammad Khan[2], Khurram Imran Khan[2], Sabir Khan[4], Sana Ullah[2], Barbara Clasen[5]

[1]*Department of Chemistry, Islamia College Peshawar, Peshawar, Khyber Pakhtunkhwa, Paksitan;* [2]*Faculty of Materials and Chemical Engineering, GIK Institute of Engineering Sciences & Technology, Topi, Khyber Pakhtunkhwa, Pakistan;* [3]*Faculdade de Engenharias, Arquitetura e Urbanismo e Geografia, Universidade Federal de Mato Grosso do Sul, Cidade Universitária, Campo Grande, MS, Brazil;* [4]*São Paulo State University (UNESP), Institute of Chemistry, Araraquara, São Paulo, Brazil;* [5]*Department of Environmental Science, State University of Rio Grande do Sul, Porto Alegre, RS, Brazil*

Chapter outline

Biological Approaches to Controlling Pollutants. https://doi.org/10.1016/B978-0-12-824316-9.00015-X

4.1 Introduction

Antibiotic obstruction is the ability of bacteria to oppose the impacts of an antibiotic to which they were formerly responsive. This is because of two important factors: genetic mutations and antibiotic-resistant genes (ARGs), which make bacteria antibiotic resistant (Cosgrove, 2006). So far, antibiotics have proven to be the most effective category of drugs to selectively counter microbial infections in humans and animals. Antibiotics are mainly used to treat and forestall irresistible maladies in people, improve development rates, and feed proficiency for livestock. Furthermore, they are also utilized in agribusiness and aquaculture (Cabello, 2006). It has been reported that after the injection of antibiotics, a huge portion is discharged into various ecological compartments (Zhou et al., 2013). Literature reports reveal that somewhere in the range of 30% and 90% of all antibiotics utilized in humans and animals, respectively, are discharged unaltered or as dynamic metabolites into the earth with urine and dung (Jjemba, 2006). Here, it is necessary to mention the nations in different continents with the largest production of these antibiotics. For instance, in Asia as well as worldwide, China is on the top of the list as in 2013, it consumed approximately 92,700 tons of antibiotics (Zhang et al., 2015). Similarly, Italy was the biggest producer of antibiotics among European countries and produced €1020 million of antibiotics, accounting for 34% of the total antibiotics produced in European countries. The World Health Organization (WHO) published data on antibiotic usage in 65 countries which shows that the high-level consumption of antibiotics were from Asia and Southeast Europe (WHO, 2016-2018). The increased dependency on antibiotics in various regions resulted their presence in surface water (Roberts and Thomas, 2006; Zhang et al., 2011), groundwater, drinking water, municipal sewage (Fick et al., 2009; López-Serna et al., 2013; Standley, 2008; Teijon et al., 2010), soil, vegetables (Li, 2014), sediment (Pei et al., 2006), and sludge (Lindberg et al., 2010). Several parameters are involved in the presence of different types of antibiotics and their concentrations in the environment which can be categorized as the utilization, environmental parameters, and environmental behaviors of such antibiotics, e.g., adsorption, biodegradation, and photodegradation (Li and Zhang, 2010; Kümmerer, 2009).

4.2 Sources of antibiotics

A careful survey revealed several sources of antibiotics in the natural environment, such as in veterinaries, pharmaceutical plants, hospital waste disposal (Obayiuwana et al., 2018), dairies, animal excreta, domestics, animal husbandry, municipal waste, and poultry as shown in Fig. 4.1 (Pruden, 2013). However, a large fraction of

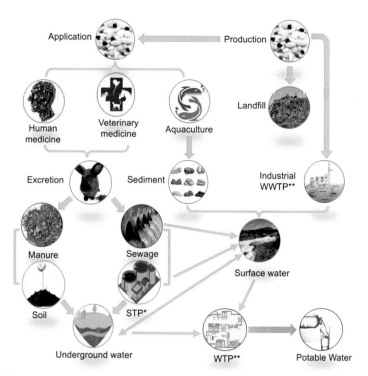

FIGURE 4.1

Different sources of antibiotics in wastewater (*WWTP: wastewater treatment plant, *STP: sewage treatment plant, **WTP: water treatment plant).

antibiotics is consumed by humans and animals, and some of its parts remain unchanged inside their bodies. Up to 90% of the same drugs come out as the result of urination (Jjemba, 2006). Therefore, these excretions are the main source of antibiotics in the environment. Hence, agricultural industries and urban centers are the foremost pathways via which antibiotics enter the aquatic environment. Different activities such as excretion, disposal of waste from households, hospitals, and similarly from industries ultimately end up in sewage. Sewage effluent is then transferred to wastewater treatment plants (WWTPs) or discharged directly to surface water. Sometimes, sewer leakage or overflow results in the mixing of antibiotics in the environment (Sedlak et al., 2004). At WWTPs, antibiotics are removed in either biotic (biodegradation), abiotic (chemical oxidation and advanced oxidation), or biotic-abiotic ways. Moreover, partial biodegradation and mineralization of antibiotics are also implementable because these substances are utilized by bacteria as a source of carbon and energy to grow (Ternes et al., 2004). Current studies proved that antibiotics cannot be removed completely by conventional WWTPs processes and requirement of advanced oxidation processes (AOPs) is essential to completely remove these antibiotics (Kurt et al., 2017). In the agricultural route, the main culprit of antibiotics is animal excreta and plant sprays; however, the consequences of the latter have not yet been reported. Animal excreta that contain antibiotics reach the aquatic environment either by drainage and runoff to surface water or by percolation to groundwater. It is believed that agricultural soil acts as an environmental reservoir for antibiotics because of the ameliorated land application of manure. These compounds are retained and progressively gathered in soils and subsequently released into the aqueous phase (Lee et al., 2007). Therefore, agricultural activities contribute a large fraction of the accumulation of antibiotics in the aquatic environment.

4.2.1 Concentration of antibiotics

Boy-Roura, 2018 focused on alluvial aquifers for the investigation of 53 antibiotics originated from manure application. Around 11 antibiotics were found in groundwater from different groups such as fluoroquinolones, macrolides, quinolones, and sulfonamides. Similarly, in surface water, 5 antibiotics were found from fluoroquinolone and sulfonamide groups. It was confirmed that the incessant antibiotics were ciprofloxacin (CIP) and sulfamethoxazole (Boy-Roura, 2018). New research that analyzed water samples from 91 rivers around the world has proven the presence of antibiotics in nearly two-thirds of all the sites sampled. This research was carried out to look for 14 basic antibiotics in 72 countries across six continents. Metronidazole was found in a concentration exceeding 300 times the safe level, while CIP was found to surpass its safe level in 51 places. Another study conducted by Ansam R. Mahmood et al. (2019) found the presence of fluoroquinolones and B-lactams. The most frequently detected antibiotics were CIP and levofloxacin (LEV) with a concentration of 1.270 and 0.177 µg/L, respectively (Mahmood et al., 2019). Antibiotic traces of erythromycin (320.5 ng/L), CIP (3 ng/L), metronidazole (1195.5 ng/L), clarithromycin (320.5 ng/L), norfloxacin (10 ng/L), tetracycline (23 ng/L), ofloxacin (179 ng/L), trimethoprim (424 ng/L), and sulfamethoxazole (326 ng/L) have been reported in the largest river, The Tagus, and the other rivers Guadarrama, Jarama, Henares, and Manzanares (Valcárcel et al., 2011).

4.2.2 Adverse effects of antibiotics

It is sometimes assumed that risks associated with antibiotics are negligible to humans because of low concentrations in the environment (Jufer et al., 2019). But prolonged exposure results in the accumulation of antibiotics in the human populations via drinking water, food, and other consumer goods, and causes unprecedented consequences. Three mechanisms involved in the interaction between antibiotics and an individual are *ingestion*, *inhalation*, and *exposure* during different processes such as drinking or recreational activities (Manaia, 2017). Although antibiotics are chosen to selectively target the desired bacteria, the residues of these antibiotics also affect the human microbiota. The human gastrointestinal for instance is the colony of approximately 1000 bacterial species and various other strains where a substantial amount of antibiotic residues are ingested daily (Jernberg et al., 2010). Because of the vulnerability of the host-associated microbiome, antibiotic residues strongly interact with the human intestinal microbial community and may change its composition (Blaser, 2016). Although antibiotics are used as a growth-promoter in animals, they also appear to promote the emergence of antibiotic-resistant strains. According to WHO, the sum of total deaths caused by influenza, human immunodeficiency virus, and traffic accidents is less than that caused by antibiotic-resistant strains (Yap, 2013). Hence, the problem of bacterial resistance to antibiotics has attracted more attention in recent years. While the development of antibiotic-resistant strains is frequently related to the utilization of antibiotic agents, obstruction is recognized even in microscopic organisms obtained from uninhabited and meagerly populated places (Chattopadhyay and Grossart, 2010) and completely confined from human mediation (Bhullar, 2012). It bolsters the probability of other sources of antibiotics including animal feed and plant spray. For this reason, researchers believe that growth promoters enhance the evolution of antibiotics-resistant strains at animal farms.

Apart from the impact on human and animal populations, microbial organisms are also affected by antibiotics. A study covering more than 226 antibiotics was conducted to evaluate the associated environmental risks of antibiotics. It deduced that around 45-100 antibiotics out of 226 are highly toxic to algae and daphnids, respectively (Sanderson et al., 2004). Low concentrations of basic anti-infection agents (*streptomycin* and *erythromycin*) have appeared to affect the endurance, and practices of small-scale spineless creatures, for example, *Daphnia magna* (Flaherty and Dodson, 2005) and Artemia under laboratory conditions (Migliore et al., 1997). Similarly, sulfonamides, tetracyclines, and macrolides show threatening effects on *Synechocystis* sp., *Lemna minor*, and *Selenastrum capricornutum* (Pomati et al., 2004). Similar research carried out by Shang et al. (2015) confirmed that aureomycin, oxytetracycline, and tetracycline adversely affect the growth of *Microcystis aeruginosa*.

4.3 Bioremediation

In bioremediation, organic wastes/contaminations are detoxified or converted into less toxic forms under a controlled environment by the application of living microorganisms such as bacteria, fungi, algae, plants, and animals. These microorganisms can be local to the tainted area or may be passed through an isolation process somewhere else and transferred to the polluted location (Vishwakarma et al., 2020). The effectiveness of bioremediation is highly dependent on the enzyme base attack of microorganisms on the pollutant to convert them to less toxic or nontoxic forms. For effective bioremediation, the microorganisms must enzymatically attack the pollutants and convert them to harmless products. Such effectiveness not only depends on the presence of a suitable enzyme in the microorganism but also involves the manipulation of environmental drivers that permit the microbial growth and faster degradation activity of microorganisms. These drivers include optimum pH, temperature, salt concentration, food-to-mass ratio, presence of active microorganisms, electron donor, electron acceptor, and availability of nutrients. However, biodegradation of contaminants is very specific, as many toxic compounds such as heavy metals, coloured compounds, and assimilated metals in the food chain are not quickly absorbed or captured by organisms, so instead of utilizing microorganisms, *plants*, *fungi*, and *algae* are used for the removal, conversion, or reduction of contamination (Vidali, 2001; Yadav et al., 2020).

4.3.1 Bioremediation techniques and strategies

Depending on the degree of contamination, site, and environmental factors, various bioremediation techniques and strategies are available for the decontamination of polluted environment that are mainly classified as discussed here and shown in Fig. 4.2:

a. *In-situ bioremediation*
b. *Ex-situ bioremediation*

4.3.1.1 In situ bioremediation

In situ bioremediation techniques involve at-site application of decontamination methods to treat contaminated soil or groundwater with negligible disruption to soil structure. These bioremediation techniques are economically favorable, as costs

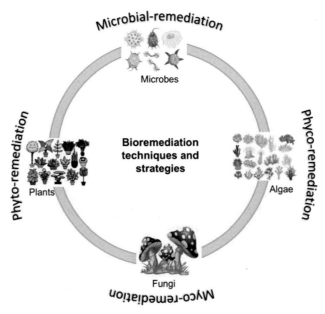

FIGURE 4.2

Bioremediation techniques and strategies.

of excavation processes are avoided; nevertheless, the expense of design and on-site installation of sophisticated equipment for the enhancement of biotic activity in bioremediation is a key concern. In situ bioremediation techniques have been effectively employed to decontaminate chlorinated solvents, dyes, nutrients, heavy metals, and organic waste sites (Rodríguez-Escales et al., 2017; Sarkar et al., 2017). Common examples of in situ applications include *bioverting*, *biosparging*, *bioaugmentation*, and *phytoremediation*.

4.3.1.2 Ex-situ techniques

These techniques involve moving or removing toxins from contaminated locations and transferring them to alternative locations for treatment. The choice of ex situ technique depends on the type, degree, depth, and geology of contamination. Common examples of ex-situ applications include biopiling, windrows, bioreactor, and landfarming. Based on the type of organism, source, and operational condition, a wide classification of conventional and emerging bioremediation practices include *bacterial bioremediation, aerobic & anaerobic methods of bioremediation, phytoremediation, fungal bioremediation*, and *phycoremediation* (Vishwakarma et al., 2020; Yadav et al., 2020).

4.3.1.3 Bacteria–bacterial remediation

One of the most popular, inexpensive, and eco-friendly options for the treatment of antibiotics/industrial wastewater is the exploitation of bacteria. Various strains and consortia of microorganisms are utilized in this treatment approach such as *endophytes, Pseudomonas*, and *Bacillus subtilis*. For instance, Dash and Das (2012) utilized a bacterial consortium consist of *B. subtilis, Bacillus pumilis, Bacillus megatherium, Bacillus licheniformis, phosphate-solubilizing bacteria, Pseudomonas fluorescens, Pseudomonas putida, Nitrosomonas, Nitrobacter, Aspergillus niger*, and *Rhodococcus* to treat various samples from different locations of pharmaceutical industry for the removal of COD, TSS, TDS, and sulfates.

4.3.1.4 Aerobic methods of bioremediation

Microorganisms such as *Pseudomonas, Alcaligenes, Sphingomonas, Rhodococcus*, and *Mycobacterium* have the ability to degrade certain types of contamination such as pesticides and hydrocarbons, both alkanes and polyaromatic compounds, in the presence of oxygen, as these contaminants act as sole energy and carbon source for these microorganisms. Typical aerobic wastewater degradation methods include activated sludge treatment (AS), sequence batch reactors (SBRs), rotating biological contactors (RBCs), trickling filters (TFs), and membrane batch reactors (MBRs). Jafarinejad (2019), for instance, utilized the AS process to treat effluent of three different facilities for 32 pharmaceuticals, hormones, and polymers. Their study shows that 28 of 32 pharmaceuticals were successfully retained by the AS process and proved to be better in performance compared with the TF method. It also expressed the correlation between log k_o/w and drug-removal efficacy and recommended establishing a correlation between hydraulic retention time and active surface area with degradation rate. Similarly, Radjenovic et al. (2007) successfully treated WWTP effluent samples containing various drugs by the application of MBRs and observed the successful removal of certain drugs such as *diclofenac*, *ranitidine*, *ofloxaci*n, and *hydrochlorothiazide* that were not removed by WWTP's AS unit; however, certain drugs were still retained in the effluent of the MBR process, such as *carbamazepine*, as their concentration was not altered before and after treatment.

4.3.1.5 Anaerobic methods of bioremediation

In many situations, aerobic degradation is not feasible, and anaerobic (absence of oxygen) processes are utilized that have the ability to degrade highly loaded organics. Moreover, due to the provision of oxygen starvation, anaerobic processes utilize minimum amount of energy compared to the aerobic process. Fermentation is one of the important types of anaerobic degradation that involve different mechanisms such as acidogenesis, methanogenesis, hydrolysis, or acetogenesis. Various anaerobic processes utilized to treat industrial wastewater include anaerobic contact reactor, upflow anaerobic sludge blanket reactor (UASB), fluidized-bed reactor, and anaerobic biofilm reactor. The choice of process selection depends on various environmental/operational factors as mentioned above. In some situations, only anaerobic treatment is sufficient to degrade the contaminations from wastewater, whereas in others, hybrid aerobic—anaerobic or anaerobic—aerobic methods are employed, as these facilities are not only able to degrade those compounds which are not removed during anaerobic conditions but also offer astonishing productivities and superior process control when the treatment of specific contaminants is desired. Furthermore, the conjunction of bioremediation with AOPs such as the Fenton process present a greener and fascinating area of research. For instance, Akcal Comoglu et al. (2016) studied the treatment efficiency of batch and continuous process for the treatment of effluent from pharmaceutical industry in Istanbul, Turkey, by utilizing mixed mesophilic nongranular sludge. Their study concluded the better performance of the UASB reactor compared with anaerobic film or contact reactor. Similarly, after reviewing many studies, Shi et al. (2017) concluded that a promising technique for the improved removal efficacy of various contamination present in the wastewater, is the utilization of combined anaerobic—aerobic treatment. This is due to the removal of residue after the anaerobic treatment in the form of some organics and nutrients, which can be successfully degraded via the nitrification—denitrification process.

4.3.1.6 Cyanobacteria for bioremediation

Various studies also reported the influence of cyanobacteria on determining water quality that has both +ve and −ve impact on it. The ability to absorb pathogenic elements from the discharged water, low-cost accumulation of the biomass in large quantities to synthesize the surfactants, precipitates, suspended and organic particles, and inhibition to certain bacteria and pathogens, make cyanobacteria a strong candidate in the competition with aerobic and anaerobic treatments. Literature reveals that an increase in cyanobacterial biomass reduces the amount of metal concentrations and nutrients significantly, and the value of purification can be extended upto 97% with the application of micro- and macroalgae and cyanobacteria (González et al., 2008; Basso and Cukierman, 2008; Mehta and Gaur, 2005). Such high treatment efficiency is attributed to the availability of elements necessary for algae that are augmented in the cultivation media in laboratory environments. However, for drinking water treatment, only nontoxic compounds can be utilized, so after the application of cyanobacteria, they must be removed from water, as some cyanobacteria such as *Oscillatoria*, *Microcystis*, *Anabaena*, *Nostoc*, *Aphanizomenon*, and *others* produce neurotoxic and hepatotoxic bye products during transformation of certain compounds, which make the water unusable as food for humans and animals.

4.3.1.7 Antibiotic degradation by fungi (mycoremediation)

Fungi (singular: fungus) are eukaryotic organisms such as molds, yeasts, and the more familiar mushrooms. Fungi are considered a kingdom different from those of plants and animals. There are different organisms in the Fungi kingdom; some are chemoheterotrophic, and some are parasitic or dependent on other living organisms. Some are unicellular, and many are threadlike, and like plants, have cell walls. Generally, the grouping of fungi is established in accord with their mode of reproduction or by using molecular information (Alexopoulos et al., 1996). For bioremediation, different fungi such as *Irpex lacteus*, *Pleurotus ostreatus*, *Lentinula edodes*, and *Leptosphaerulina* have been used to remove diverse antibiotics such as *oxacillin*, *bifonazole*, *oxytetracycline*, *clotrimazole*, *fluoroquinolone*, and *sulfonamides* with zero or limited production of toxic products (Gothwal and Shashidhar, 2015; Copete-Pertuz et al., 2018; Migliore et al., 2012; Kryczyk-Poprawa et al., 2019; Esterhuizen-Londt et al., 2017; Marco-Urrea et al., 2010a,b; Asif et al., 2017; Muszyńska et al., 2019; Yang et al., 2013). Mycoremediation is cost-effective, environmentally friendly, and an overall good alternative compared with other chemical and physical methods for treatment of antibiotic-contaminated water. The detail of some antibiotic degrading fungi is provided in Table 4.1.

Mycoremediation reactors, also termed fungal reactors, have different types based on the fast biodegradation of antibiotics. They can be slurry reactors for suspended growth or trickling-bed reactors or fluidized-bed reactors with immobile systems. Like other bioreactors, the operation of the mycoremediation reactors can be continuous, batch, semibatch, or sequencing modes under aerobic or anaerobic conditions. In order to provide good oxygen and attain good homogenized fluid mixture, generally stirred tank reactor or bubble column is used with mechanical agitation (Crognale et al., 2002). However, the energy cost will be higher because of mechanical agitation, and growth of some organisms can be disturbed by agitation. Bubble column and air-lift reactors are common examples and have cylindrical shapes in which fluid mixing is carried out using compressed air. In these type of reactors, the oxygenation and fluid mixing rates are less efficient than in stirred tank reactors (Chisti and Moo-Young, 1996; Zhong, 2019). In such reactors where fungi can be shaped in the form of pellets after cultivation, air-lift reactors or the

Table 4.1 Detail of important fungi used for degradation of antibiotics.

Antibiotic	Fungi organism	Comments	References
Ciprofloxacin	*Pleurotus ostreatus*	The fungus has demonstrated efficient degradation of 95.07% and a significant decrease in antimicrobial activity of the antibiotic.	Singh et al. (2017)
Ofloxacin	β-1,3-Glucan (*E. coli*)	β-1,3-Glucan enhances ofloxacin tolerance in biofilms.	De Brucker et al. (2015)
Norfloxacin and ciprofloxacin	*Trametes versicolor*	The utilization of *T. versicolor* for degradation of these fluoroquinolones is a viable alternative to physiochemical treatments.	Prieto et al. (2011)
Enrofloxacin	*Gloeophyllum striatum*	Brown rot fungi can generate hydroxyl radicals that can cause deterioration of wood and few xenobiotics. *G. striatum* initiates biodegradation of enrofloxacin and achieves substantial results.	Wetzstein et al. (1997)
Flumequine	Ligninolytic	A rapid transformation of more than 90% flumequine in 3 days was observed by ligninolytic fungi *T. versicolor*.	Čvančarová et al. (2013)
Tetracycline	*Bacillus clausii* T and *Bacillus amyloliquefaciens*	Bio-removal output for tetracycline by coculture of *Bacillus clausii* T and *Bacillus amyloliquefaciens* was measured to be 88.9%. The transformed products indicated a reduced toxicity.	Liu et al. (2020)
Tetracycline	Klebsiella sp. SQY5	Analysis of tetracycline degradation by SQY5 for 4 days revealed its concentration was reduced by 70.38%.	Shao et al. (2019)
Cephalosporin	*Lentinula edodes*	*L. edodes* can swiftly reduce cefuroxime axetil from the system with a removal rate as high as 100% obtainable for 7 day experiment.	Dąbrowska et al. (2018)
Norfloxacin	*Ganoderma lucidum* JAPC1	Encouraging output is witnessed as *G. lucidum* JAPC1 degrades norfloxacin.	Chakraborty & Abraham (2017)
Tetracycline	Cerrena laccase	A degradation of 80% in the first 12 h, and adequate curtailment of antimicrobial activity of tetracycline is observed.	Yang et al. (2017)

fluidized bed are preferred with aeration systems (Crognale et al., 2002). Fungi are capable of degrading a wide application of pharmaceutical pollutants, as their enzymatic systems are constructed on free-radical levels (Pointing, 2001). Antibiotic contaminant degradation seems likely to include any intracellular enzymatic system or extracellular enzymatic system, primarily laccase, lignin peroxidase, versatile peroxidase, and manganese peroxidase (Wesenberg et al., 2003). Overall, studies show a substantial reduction in antibiotic pollutants after treatment, although some constraints found in the literature (Suda et al., 2012; Marco-Urrea et al., 2010a) were shown to be effective in antibiotic removal (Table 4.1). After treatment, the final products were reported to be less harmful or more easily biodegradable than the original complexes, which shows the importance of fungi as helpful remediation agents.

4.3.1.8 Antibiotic degradation by algae (phytoremediation)

Algae (singular alga) is a general term for eukaryotic organisms in which most kinds of algae carry out the photosynthesis process. Many algae organisms are unicellular and called microalgae, for example, *Chlorella* and some others are diatomic and multicellular types. For example, giant kelp, generally called brown algae are a type of macroalgae and can be 50 m long. Most kinds are autotrophic (*photosynthesis* and *chemosynthesis*), aquatic, and have a deficiency of the discrete cell and tissue forms of general plants, for example, *phloem*, *xylem,* and *stomata*. No single definition of algae is generally accepted. Although cyanobacteria, which comes under the definition of prokaryotes, are frequently stated as blue-green algae, most scientists do not consider them under the definition of algae. The *seaweeds* are complex examples of marine algae, and *Charophyta* are complex types of freshwater algae generally known as green algae, stoneworts, and spirogyra, respectively. More detail about algae can be found in the relevant literature (Bergen et al., 2012; Ruggiero, 2015; Guiry, 2012).

When considering phytoremediation systems for the removal of antibiotics, open systems are more cost-effective than closed systems because the algae growth rate is dependent on factors such as algae dilution, mixing, and system depth that are favorable to open systems (Skjånes et al., 2013). Especially, for the growth of microalgae, the open system is preferable because of lower operational and fixed costs. On the other hand, the operation parameters of the algae growth in close system are easy to control and optimize for pure and specific growth of algae organisms (Ugwu et al., 2008; Singh and Singh, 2015). In the antibiotic removal treatment, algae harvesting is a major challenge in phytoremediation, and it is also the cost demanding step for microalgae production (Wiley et al., 2009). However, for microalgae growth, the bio-flocculation process is environmentally friendly, and naturally, flocculate the tiny particles by can be grown using extracellular biopolymers (Akhtar et al., 2004). Phytoremediation is an emerging field for pharmaceuticals, especially antibiotic removal, and limited studies are available for a complete understanding of removal mechanisms (Matamoros et al., 2016). But phytoremediation is cost-effective and successful, versatile pollutant treatment technology that converts antibiotics into less harmful environmentally friendly chemicals (Dixit and Singh, 2015). Guo and Chen (2015) studied the algal-AS combined system for evaluation of the removal efficiency of green algae *Chlorella pyrenoidosa* on cephalosporins. They found that green algae have more than 90% efficiency for the removal of cefradine, cefalexin, ceftazidime, and cefixime. On the other hand, a higher removal rate was obtained based on the combined system of green algae and sludge. It was concluded that compared with the traditional bioremediation, the combined removal system could use the unacclimated AS directly and finally obtain the higher removal efficiency (Guo and Chen, 2015).

Xiong et al. (2017) studied the removal of CIP in a freshwater microalga *Chlamydomonas mexicana*. The impact of CIP on *C. Mexicana* was monitored by observing the total chlorophyll, carotenoid content, malondialdehyde (MDA), and superoxide dismutase activity. *C. Mexicana* demonstrated a relatively less ability to remove CIP (13%) on its own, but when aided by the presence of Sodium Acetate, drastic improvements were witnessed. Its electron donor characteristics enhanced the removal output to 56% for a cultivation period of 11 days (Xiong et al., 2017a). Additionally, another work by J.-Q. Xiong et al. (2016) analyzed the biodegradation of LEV under the environment of freshwater microalga, *Chlorella vulgaris*. Encouraging outcomes were noticed as the microalgae were preexposed to 200 mg/L of LEV for 11 days. A removal efficiency of 16% was observed, and further ascension was witnessed by the application of 1% (w/v) sodium chloride as LEV removal aggravated to more than 80%. Advanced removal of LEV is possible when treated under acclimated *C. vulgaris* and enhanced salinity level. The study indicates favorable results (Xiong et al., 2017b). Yu et al. (2017) studied the removal of the ceftazidime and allied compound structure 7-aminocephalosporanic acid (7-ACA) with algal treatment, and obtained a 92.70% and 96.07% removal rate, respectively. The algal removal mechanism could be understood as three stages: rapid adsorption, slow cell wall-transmission, and final biodegradation. Their study showed that the green algae had a satisfactory growth rate and had the main role in the biodegradation of ceftazidime and 7-ACA antibiotics and attained high removal effectiveness with little environmental impact (Yu et al., 2017).

S. Chen et al. (2020) investigated the behavior of four antibiotics, namely sulfadiazine, sulfamethazine, enrofloxacin, and norfloxacin under algal treatment. The batch cultures of microalgae *C. vulgaris* and the cyanobacterium *Chrysosporum ovalisporum* were utilized for 16 days with an emphasis on alterations in chlorophyll fluorescence parameters (chl a, Fv/Fm, and FPSII) and the responses of the antioxidant system. Both green algae demonstrated an ability to remove antibiotics, with *C. vulgaris* displaying a higher tendency. Sulfonamides (hydrophilic antibiotics) are relatively hard to remove compared with the fluoroquinolones (lipophilic antibiotics), which can be readily removed. The study revealed that appropriate green algae could be availed for the removal of various antibiotics (Chen et al., 2020).

X. Cheng et al. (2020) employed the microalgae *Galdieria sulphuraria* for the removal of antibiotic-resistant bacteria (ARB) and ARGs. A comparison of this system with a run-of-the-mill conventional WWTPs revealed the algal-based system to be more productive and efficient. Adequate reduction in various genes (qnrS, tetW, sul1, and intI1) was observed in the algal system in the bacteriophage phase, which indicated a decline in phage-mediated ARG transfer. The study found that an algal-based treatment system is vastly superior to the conventional treatment in curtailment of ARGs and ARB (Cheng et al., 2020). J. Li et al. (2020) investigated the hazardous effects of roxithromycin (ROX) on MDA, growth, and the chlorophyll content of green algae *Chlorella pyrenoidosa*. Steps involved in the removal of ROX while algae cultivation is also pondered. For the course of the first 14 days, a somewhat sluggish removal of ROX is observed (12.21% ~ 21.37%) with a gradual improvement in removal mechanism, indicating 45.99% ~ 53.30% of ROX removal at 21 d. The study demonstrates biodegradation as a more viable route for ROX removal and reaffirms the importance of microalgae as an emerging source for ROX removal (Li et al., 2019).

4.4 Recent advances in bioremediation of antibiotics

Bioremediation proved to be a very efficient technology for removing numerous contaminants from the environment. Some varieties of organisms have been characterized by degrading or removing antibiotics from the environment. Remediation strategies for antibiotics have shown treatment possibilities, through adsorption mechanisms, bioelectrochemical methods, hydrolysis, redox reactions, and acidic, alkaline, or ionic processes assisted by metal, photolytic, phytolytic, microbial, and enzymatic techniques (Yan et al., 2019; Dangi et al., 2019; Kumar et al., 2019). However, the increase in antibiotic residues in non-degraded or partially degraded form in the environment after certain treatment processes is a major health problem because of the toxic impact of these substances on nontarget organisms (Vasiliadou et al., 2018).

Antibiotic contamination has become a global threat and has increased the risk of antibiotic resistance in microorganisms. Although, antibiotics have a short half-life, but even so, their hydrophobicity leads to their persistence in the environment. Some antibiotics *(sulfonamides, tetracyclines, beta-lactams, aminoglycosides, fluoroquinolones, rifamycins)* have a low molecular weight (<1000 D) and are quickly dissolved in water bodies (Krzeminski et al., 2019). However, WWTPs have less ability to efficiently eliminate the micropollutants such as antibiotics, which end up being continuously released into the environment (He, 2018). These persistent compounds can exert selective pressure on microorganisms and result in the development of resistance (Kim et al., 2018). Understanding the mechanisms of antibiotic resistance is essential for the development of effective treatments for the removal of antibiotics residues (AR), ARB, and ARG, as this knowledge can contribute to the interpretation of the results of technologies used to solve this problem (Wen et al., 2018). However, although the reduction in resistant microorganisms is observed after treatment, reactivation of ARB and

ARG is reported in the literature (Huang, 2012). Thus, the search for technologies that eliminate antibiotic residues, as well as monitoring the presence of ARB and ARG are essential for the treatment of contaminated ecosystems.

Bioremediation of antibiotics through enzymatic degradation has been shown to be an efficient biological technique to minimize the burden of antibiotics in aquatic and terrestrial ecosystems. Studies show that degradation of penicillin by enzymatic strategies can be an excellent technology in addition to being environmentally friendly. For instance, Popovich et al., 2018; Bhattacharya and Yadav, 2018 carried out pre- and postgenomic studies with several *cytochrome P450 enzymes* (P450s or CYPs) in response to exposure to various environmental toxins and their potential in the bioremediation of these chemicals. P450s are a group of enzymes capable of catalyzing numerous reactions that allow the oxidative metabolism of xenobiotics, including antibiotics. Because of their catalytic diversity, microbial P450s have shown great potential in the biodegradation of pollutants (bioremediation). However, because it is a multicomponent system, its application is more complex compared with the enzymes *amylase*, *cellulose*, and *lipase*. According to Bhattacharya and Yadav, 2018, the commercial use of P450s enzymes has major challenges that need to be overcome in order to carry out cost-effective industrial or environmental applications. Bacteria are the most common group of organisms used for bioremediation. Bacterial remediation can play an important role and offers an economical and ecological option for removing antibiotics from the environment. The exploration of bacterial methods, such as enzymatic biosorption and biodegradation processes for the removal of antibiotics, was studied by Al-Gheethi et al. (2015). In this study, the authors emphasize that the choice of organisms used in the process is extremely relevant because of their different needs for growth and survival, as well as affinities with antibiotics. Important advances are being achieved through studies that select microorganisms through molecular analysis and also in research using nanomaterials in bioremediation processes. Molecular studies have contributed to the understanding of the functions and metabolic pathways of microorganisms, and consequently, the development of techniques that can overcome the limitations present in microbial culture methodologies. With the use of genetic engineering, microorganisms can be designed and programmed to perform specific functions in the bioremediation process through the identification of functional genes and enzymes and their role in the degradation of antibiotics. The use of nanomaterials has shown positive results related to the efficiency of microorganisms, increasing the surface area, and decreasing the activation energy (Shah and Shah, 2020).

Because of the versatility of bioremediation techniques, it is evident that the prospect of its applicability in many ways is very likely. One of the most common ways in which these techniques are being implemented in their conjunction with another biological, chemical, or physical method to improve their efficiency. Shah et al., emphasized that a treatment method with two processes, such as bioremediation in conjunction with AOP, has superior performance (Shah and Shah, 2020). In this sense, genetics and biochemistry need to be further explored in the search for efficient organisms for the bioremediation of antibiotics.

4.4.1 Omics approach in bioremediation

It is obvious that bioremediation processes are very effective for the decontamination of wastewater containing different pollutants. In any type of bioremediation, the role of microorganisms is imperative, and therefore their abundance, community, and diversity give intuition to the fate of any bioremediation process, some other parameters also need to be in optimal range for effective decontamination. Various omics approaches such as *genomics, proteomics, transcriptomics, interactomics*, and *metabolomics* could be studied that give insight into the bioremediation process. These techniques are very supportive to correlate DNA sequences with mRNA, protein, and metabolite concentration and give a better image of in-situ bioremediation processes. Similarly, microarray is another strong genomic strategy to explore the microbial community in contaminated soil and can be utilized for tracing biodegradative genes in soil microorganisms. Some postgenomics approaches such as *proteomics, transcriptomics*, and *interactomics* parallel with *metagenomics* are considered to have less correspondence with gene microarrays. It can be utilized for biodegradation with a large amount of genome sequences and metagenomic sequences of environmental samples. These approaches may also represent how particular microorganisms act to modify the environment caused by pollutants.

However, bioremediation has some hurdles such as less nutrient, a small populace community of microbes with inefficient ability of degradation, and recalcitrance of some pollutants. Most of the microbial process depends on biostimulation and bioaugmentation, Biostimulation is related to accumulation of food to contaminated sample can enhance the activities of autochthonous microbes. Microorganisms are universal and naturally destroy the pollutant in sites, however, their accomplishment may increase or decrease in response to the pollutant availability. Although some agricultural waste enriches with nitrogen, phosphorus, and potassium will boost food shortage in sites. Bioaugmentation approach added stains cultures to sites and increased microbial population and their catabolic efficiency toward pollutant. This strategy may work on indigenous or nonindigenous microorganisms and transform the pollutant into nontoxic substances (Herrero and Stuckey, 2015; Poi et al., 2017). It was observed that biostimulation is very effective toward degradation of low molecular weight pollutants and bioaugmentation was more successful for high molecular weight substances. Nguyen et al. (2017) practically utilized bioaugmentation for the removal of antibiotic sulfamethoxazole SMX in wastewater. The isolated a strain *Achromobacter denitrificans* PR1 from AS and was to be more effective for bioaugmentation of SMX. Bioaugmentation of AS with the strain PR1 (batch experiments) led to superior biotransformation rates of SMX (by 60 times) compared to AS, within a complex carbon environment in WWTPs. The bioaugmentation of AS showed a higher removal rate as compared to only AS process, The bioaugmented strain was applied in a membrane bioreactor to enhance the removal of SMX (Nguyen et al., 2017).

4.4.2 Role of nanotechnology in bioremediation

The nanotechnology or simply the nanomaterials demonstrate a modern approach to progress the bioremediation in a forward direction beyond its frontier. This coupled strategy will have conceivable uses with minimum cost and lesser negative impacts on the environment. Nanomaterials could interact with biotic and abiotic systems in different ways and need more efforts to evaluate the synergism of nanomaterials and bioremediation. Recently, researchers incorporated the nanomaterials with the biological processes to enhance the removal of pollutants from environmental samples. For instance, (Cecchin et al., 2017) used the nanoparticles and microorganism or plants to degrade pollutants and El-Rammady et al. rename the process and called these process according to the type of organisms such as *phytonanoremediation*, *microbial nanoremediation*, and *zoo-nanoremediation* (El-Ramady et al., 2017). As living organisms are used to degrade pollutants during bioremediation, the interaction of nanoparticles and living cells is very significant. Similarly, the size of nanoparticles, nanonutrition, and nanotoxicity may affect the living cell and consequently influence the bioremediation process.

Nanomaterials have marvelous properties and enable them to be a competent technology for removal of contaminants from wastewater. Nanobioremediation is a growing area of research with many prospects in the decontamination of pharmaceutical waste. Nanomaterials possess good reactivity with pollutants and may also act as carriers for the immobilization of enzymes and cells. The generic enzymes used for the treatment of pollutant removal are peroxidase, glucose oxidase, horseradish peroxidase, cytochrome reductase (Fe^{+3}), lignin peroxidase, and oxidoreductase. Different types of nanomaterials such as Fe_3O_4/SiO_2, magnetic graphene, and template titania are able to immobilize these enzymes for a variety of pollutants (Vázquez-Núñez et al., 2020). β-Lactamase, a representative enzyme for β-lactam antibiotic degradation, was covalently immobilized on the surface of magnetic nanoparticles modified with amino groups by a simple cross-linking process. β-Lactamase was covalently immobilized on the interface of amino groups modified magnetic nanoparticle and subsequently used for the antibiotic penicillin degradation. The result revealed that 95% of antibiotics were removed by nanomaterials carrying β-Lactamase (Gao et al., 2018). Similarly, 835 mg/g hollow mesoporous carbon sphere nanomaterials were also utilized for immobilization of laccase enzyme. The immobilized enzyme demonstrated superb efficiently toward tetracycline hydrochloride and CIP hydrochloride (Shao et al., 2019).

A novel strategy was applied in the oriented immobilization of laccase from *Echinodontium taxodii* on concanavalin A-activated Fe_3O_4 nanoparticles (GAMNs-Con A) based on *laccase* surface analysis. *Laccase* enzyme from *E. taxodii* on concanavalin immobilized on iron oxide nanoparticles. The nanomaterials exhibited better enzyme loading and with excellent thermal and operational stability and efficiently used for the removal of sulfonamide antibiotics (Shi et al., 2014). Using nanomaterials to achieve functional enzyme mimics (nanozymes) is attractive for both applied and fundamental research. Furthermore, nanomaterials could be modified to obtain enzyme mimics called *nanozymes* and subsequently applied for wastewater treatment. A functional laccase mimic-oriented guanosine monophosphate copper nanozyme was significantly applied in environmental remediation (Liang et al., 2017). *Laccase* from *Trametes versicolor* was immobilized on surface of bentonite nanoporous clay and efficiently utilized for tetracycline degradation. The degradation experiment showed that over 60% antibiotic removal in 3 h (Wen et al., 2019).

4.4.3 Hybrid process for bioremediation

Some researchers have shown interest in the coupling of bioremediation with other processes such as adsorption, membrane filtration, and electrochemical in order to enhance the efficacy of bioremediation process. Bioelectrochemical systems (BES) are a newly emerged technology studying the interrelation between the chemical and electrochemical phenomenon in living organisms that could eventually become an important renewable energy source. The bioelectrochemical processes include microbial electrolysis with metabolisms, microbial fuel cells, and electrochemical redox reactions are supposed to auspicious substitute for the destruction of recalcitrant pollutants. The bioelectrochemical method is a useful electrochemical approach and some consider it as cost-effective because no need to add some reductant in biocathodes and other external potential, bioelectrochemical process demonstrate high antibiotic removal capability as shown in Table 4.2.

The process can also be considered eco-friendly, as it destroyed the antibiotics and generate fewer toxic intermediates that further degraded in microbial fuel cell (Wang et al., 2016). Similarly, the biocathodes has the ability to avoid the generation of hazardous by-products (Liang et al., 2013). The bioelectrochemical also expected to versatile as it can be applied to degradation of antibiotics in solid as well as in liquid phase (Zhang et al., 2016a). Additionally, this process should also be coupled with some other technique such as advanced oxidation, adsorption, and membrane processes. Bioelectrochemical degradation has diverse mode of operation for the remediation of antibiotic waste and the mechanisms can be classified into three types as describe by Yan et al. (2019).

In type I, the microbial fuel cell has a biological anode and abiotic cathode, it may be single compartment (oxygen) or two compartments with potassium ferricyanide that perform the function of electron acceptors in abiotic cathode. Antibiotics work as electron donors and source of carbon in biological anodes. Exoelectrogenic microbes (which transfer electron extracellularly) and degrading antibiotic bacteria, are stick to surface of anodes and provide a biofilm inside the extracellular polymeric compounds that decrease the overpotential of recalcitrant antibiotics and their intermediates. Electrical stimulation with anaerobic biodegradation is prominent step in antibiotic mineralization in microbial fuel cell. This electrical stimulation passing electrons to system and enhance microbial metabolism by direct or indirect electron transfer to bacterial cells. These stimulated bacteria quickly metabolize antibiotics by discharging enzymes as observed in SMX and its intermediates 3A5MI.

Table 4.2 Degradation of different antibiotics by bioelectrochemical process.

Antibiotics	Setup	Initial conc.	Time	Removal efficiency %	References
Cefazolin sodium	Single chamber	50 mg/L	31 h	>70	Zhang et al. (2018)
Metronidazole	Two chamber	10 mg/L	24 h	85.4	Song et al. (2013)
Nitrofurazone	Biocathode	50 mg/L	1 h	70.60	Kong et al. (2017)
Penicillin	Air-cathode single-chamber cell	50 mg/L	24 h	98	Wen et al. (2011)
Ceftriaxone sodium	Single-chamber air-cathode	50 mg/L	24 h	91	Wen et al. (2011)
Cefuroxime	Bio reactor with electrode	0.5 mg/L	12 h	>90	Cheng et al. (2016)
Sulfamethoxazole	Two chamber cell	20 mg/L	48 h	99	Wang et al. (2015)
Tetracycline	Biofilm electrode	200 µL	40 h	89.3	Zhang et al. (2016)
Tetracycline	Photoelectrode	100 mg/L	2 h	70	Jiang et al. (2016)
Tetracycline	Electrochemical cell couple with wetlands	800 µL	2.5 d	>99	Zhang et al. (2016)
Tetracycline	Electrochemical cell with membrance bioreactor	90 mg/L	17 h	99.5	Li et al. (2017)
Oxytetracycline	Two chamber cell	10 mg/L	78 h	99	Yan et al. (2018)
Sulfamethoxazole	Two chamber cells	200 µL	40 h	88.9	Zhang et al. (2016)
Sulfamethoxazole	Biofilm electrode reactor	200 µL	40 h	72.2	Song et al. (2017)
Sulfanilamide	Two chamber cell	30 mg/L	96 h	90	Guo et al. (2016)
Chloramphenicol	Two chamber cell	50 mg/L	12 h	84	Zhang et al. (2017)
Chloramphenicol	Two chamber cell	30 mg/L	48 h	83.7	Guo (2016)
Chloramphenicol	Two chamber fuel cell	32 mg/L	24 h	100	Liang et al. (2013)

In the **second type** of bioelectrochemical system, biocathodes are operated in a microbial electrolysis cell and potential provide through external source because mostly the bioanode has low reduction potentials than antibiotics and anodes used to reduce the power consumption of cell. The antibiotics mainly degraded by direct electrochemical and biological reactions. During the electrochemical process, the cathode donates the electrons to antibiotics and reduced. The microbes act as biocatalyst and boost the reduction reaction of antibiotics by minimizing its overpotential.

In **third type**, different modified materials are frequently used in bioelectrochemical cathode to produce some active radical species that attack and destroyed the antibiotics. The process generally comprises of the following reactions such as the electrons produced by bioanode and transferred to cathode via external circuit. The cathode accepts electrons and reduces dissolved oxygen and form hydrogen peroxide. As the H_2O_2 exposed to UV light it produced hydroxyl radicals that subsequently used for the degradation of antibiotics. Different types of cathode materials have been utilized for this purpose such as CAP degraded using Cu foam cathode and in situ−generated hydroxyl radicals and Fe_0/TiO_2 effectively generated hydroxyl radicals under visible light and applied for fast degradation of tetracycline. The performance of bioelectrochemical process may influence by many parameters such as pH, buffer solution, flow rate, additive and source of inoculation. However, the most important role related to the properties of pollutant, such as electrochemical properties of antibiotics their concentration, applied current density, material of the electrode, carbon source, temperature, and salinity of the medium. Bioelectrochemical systems combined with microbes can also be considered promising alternatives for the degradation of different antibiotics contaminants, capital costs, and effective approach for interpreting observations, comparing systems, and identifying the drawbacks and merits of the system. Furthermore, recent advances have demonstrated that it is possible to engineer microorganisms to develop electron transfer conduits for promoting the degradation of antibiotics in the aspects of waste minimization, energy conversion efficiency due to electron transfer, recovery, and monitoring of various antibiotics such as oxytetracycline (Yan et al., 2018), neomycin sulfate (Catal et al., 2018), oxytetracycline (Yan et al., 2018), and sulfonamide (Li et al., 2020). Moreover, conventional treatment of many antibiotic-containing wastewaters is incapable to accomplish the overall removal of the antibiotics that need additional measurement. But there are some limitations such as limit the surface area of electrode as microbe can clog small pores causing limits the current, still not economically competitive, low energy produced compared with other methods, high initial cost, microbe metabolic looses and have low performance in the degradation of suitable and selective antibiotics and specific electrode.

4.5 Future scope and limitations of bioremediation techniques

The conventional treatment of wastewater treatment aims to reduce or remove the carbon and nitrogen contents of the wastewater, and such treatment is ineffective for the treatment of pharmaceuticals and abused drugs, resulting in the presence of micropollutants in discharge at WWTPs. Although application of chemical and physiochemical methods is effective for the treatment of such micropollutants, high operational costs, and sometimes the formation of harmful by-products, make such applications unsuitable, leaving biological degradation as the only option. However, the efficiency of biodegradation/bioremediation techniques to the influent wastewater is subjected to numerous conditions such as concentration and toxicity of the given compound, stereochemistry, and concentration of competing and noncompeting compounds present in wastewater, microbial strain, retention time, and operational settings during the treatment process (Bali, 2017).

In the study of potentially exploitable microorganisms can generate economic and strategic benefits for modern environmental science. In the environmental area, research needs to be directed to the improvement of bioremediation methods for the clean and efficient removal of an emerging contaminant (Singh et al., 2020; Figueiredo et al., 2016). Several aspects of bioremediation make it an attractive choice in the degradation of pharmaceutical compounds, showing advantages over other treatment methods, as it consists of a natural process; some techniques can be applied in situ with minimal environmental disturbance, often being cost-efficient and safer. In addition, many microorganisms degrade various pollutants permanently, and can also be associated with physical or chemical treatment methods (Silva et al., 2019; Rodríguez-Rodríguez et al., 2010).

In recent years, several studies have indicated that bioremediation is a safe and effective alternative for removing various pharmaceutical compounds from the environment. In future, it would be important to study the removal of these drugs by clay materials in a more complex medium such as effluents, to understand the effect of the matrix on the adsorption capacity of the materials. It would also be interesting to test these materials on a pilot scale (Shah and Shah, 2020). The microbial ecology has progressed in recent years and with an interdisciplinary approach, involving microbiology, molecular biology, engineering, ecology, geology, and chemistry, Moreover, it has provided useful information for the improvement of bioremediation strategies in addition to allowing the evaluation of the impact of the technique on ecosystems (Shah and Shah, 2020; Boopathy, 2000).

4.6 Limitations of bioremediation

Biodegradation can be limited by the inability of indigenous microorganisms to metabolize pollutants or by unfavorable environmental conditions for the survival and activity of degrading microorganisms. Biological processes are often highly specific, limited to those compounds that are easily biodegradable and need suitable environmental conditions and appropriate levels of nutrients and contaminants for growth. Additionally, the production of toxic metabolites, repression of enzymes, presence of preferred substrates, and the lack of inducing substrates can also be considered as limiting factors (Boopathy, 2000; Van Hamme et al., 2003; Esterhuizen-Londt et al., 2016).

Bioremediation must be adapted to specific local conditions, with the need to characterize the environment to be treated and to carry out preliminary studies on a small scale before cleaning the contaminated site (Boopathy, 2000), to clarify some factors to consider ourselves before choosing and applying a bioremediation technique, among which we can mention: **(1)** whether the contaminant is biodegradable; **(2)** whether biodegradation is occurring naturally in the contaminated site; **(3)** whether the environmental conditions are suitable for biodegradation; **(4)** if the pollutant is not completely degraded, what will be the behavior of the residual material (Boopathy, 2000). Some of the non-technical criteria that affect the applicability of bioremediation techniques must also be considered, as the chosen technique must be able to achieve the necessary cleaning; have an advantageous cost concerning other remediation options; the possibility of residual contaminants after bioremediation at acceptable levels; favorable public and regulatory perception; ability to meet time and space limitations (Boopathy, 2000; Esterhuizen-Londt et al., 2016). Other factors constitute limitations to the use of bioremediation, such as economic issues and environmental responsibility. Clients and regulatory agencies usually rigorously evaluate bioremediation, since it is about the application of innovative techniques, there is an imposition of strict standards and a demand for higher performance than for conventional technologies. Thus, investment in bioremediation is slow, and therefore, this activity on a commercial scale still lags behind other industrial sectors. It often takes longer than other treatment procedures, evaluating the performance of bioremediation is difficult, and requires very skilled human power.

4.7 Conclusions

Antibiotics have significant roles in our lives and have served humanity for many years to provide us with a better way of life. However, the residue of these compounds in water bodies poses a serious threat to human and aquatic organisms. Various methodologies have been used to eliminate or reduce the loads of AR, ARB, and ARG from effluents in WWTPs, such as chlorination, UV radiation, ozonation, photocatalysis, photo-Fenton, membrane filtration, and adsorption, among others, as well as combinations of various treatment techniques. These processes have challenges such as production of toxic intermediates that must be further degraded, high operational and handling costs, etc. As an alternative, bioremediation offers a promising way to treat antibiotic wastewater. As discussed earlier, there are various bioremediation processes for antibiotic wastewater decontamination. It must be noted that hybrid or coupled processes such as the coupling of physical, chemical, electrochemical, or photochemical processes with biological processes exhibit significant performance compared with that of individual processes. Numerous studies on alternative means of bioremediation have given way to the study of endophytes for phycoremediation, cyanobacterial remediation, reverse osmosis, and membrane processes that involve the use of biological membranes for the filtration mechanism and can yield competent efficiencies. However, not all studies reviewed, account for the natural qualities such as hardness, amount of silt, salts, etc. of the water samples taken for study and their effects on the treatment process. Moreover, many similar methods have provided inconsistent efficiencies. Each reviewed study has considered a different parameter for efficiency. This creates a need for more studies that can compare and define the superiority of some techniques over others. Many studies have not been carried out using real wastewater samples, and this can offer challenges to real-world applicability. This review emphasizes the merits and demerits of each process which can further aid studies in optimizing the processes for improved efficiencies using advanced tools such as quality by design and advanced modelling techniques to design and operate a treatment system with superior performance. Hence, it can be concluded that bioremediation techniques, with their plasticity and immense potential, can serve as a means of sustainable wastewater management for the pharmaceutical industry to mitigate ecotoxicological impacts.

References

Akcal Comoglu, B., Filik Iscen, C., Ilhan, S., March 2016. The anaerobic treatment of pharmaceutical industry wastewater in an anaerobic batch and upflow packed-bed reactor. Desalin. Water Treat. 57 (14), 6278−6289.

Akhtar, N., Iqbal, J., Iqbal, M., 2004. Enhancement of lead(II) biosorption by microalgal biomass immobilized onto loofa (Luffa cylindrica) sponge. Eng. Life Sci. 4 (2), 171−178.

Alexopoulos, C.J., Mims, C.W., Blackwell, M., 1996. Introductory Mycology, fourth ed. Wiley, New York.

Al-Gheethi, A.A.S., Lalung, J., Noman, E.A., Bala, J.D., Norli, I., December 2015. Removal of heavy metals and antibiotics from treated sewage effluent by bacteria. Clean Technol. Environ. Policy 17 (8), 2101−2123.

Asif, M.B., Hai, F.I., Singh, L., Price, W.E., Nghiem, L.D., 2017. Degradation of pharmaceuticals and personal care products by white-rot fungi—a critical review. Curr. Pollut. Rep. 3 (2), 88−103.

Bali, A., 2017. Handbook of Research on Inventive Bioremediation Techniques. IGI Global.

Basso, M.C., Cukierman, A.L., 2008. Biosorption performance of red and green marine macroalgae for removal of trace cadmium and nickel from wastewater. Int. J. Environ. Pollut. 34 (1/2/3/4), 340.

Bergen, J.Y., Caldwell, O.W., Caldwell, O.W., 2012. Introduction to Botany.

Bhattacharya, S.S., Yadav, J.S., 2018. Microbial P450 enzymes in bioremediation and drug discovery: emerging potentials and challenges. Curr. Protein Pept. Sci. 19 (1), 75–86.

Bhullar, K., et al., 2012. Antibiotic resistance is prevalent in an isolated cave microbiome. PLoS One 7 (4), e34953.

Blaser, M.J., 2016. Antibiotic use and its consequences for the normal microbiome. Science.

Boopathy, R., August 2000. Factors limiting bioremediation technologies. Bioresour. Technol. 74 (1), 63–67.

Boy-Roura, M., et al., 2018. Towards the understanding of antibiotic occurrence and transport in groundwater: findings from the Baix Fluvià alluvial aquifer (NE Catalonia, Spain). Sci. Total Environ. 612, 1387–1406.

Cabello, F.C., 2006. Heavy use of prophylactic antibiotics in aquaculture: a growing problem for human and animal health and for the environment. Environ. Microbiol. 8 (7), 1137–1144.

Catal, T., Yavaser, S., Enisoglu-Atalay, V., Bermek, H., Ozilhan, S., November 2018. Monitoring of neomycin sulfate antibiotic in microbial fuel cells. Bioresour. Technol. 268, 116–120.

Cecchin, I., Reddy, K.R., Thomé, A., Tessaro, E.F., Schnaid, F., April 2017. Nanobioremediation: integration of nanoparticles and bioremediation for sustainable remediation of chlorinated organic contaminants in soils. Int. Biodeterior. Biodegrad. 119, 419–428.

Chakraborty, P., Abraham, J., 2017. Comparative study on degradation of norfloxacin and ciprofloxacin by Ganoderma lucidum JAPC1. Kor. J. Chem. Eng. 34 (4), 1122–1128.

Chattopadhyay, M.K., Grossart, H.P., 2010. Antibiotic resistance: intractable, and here's why. BMJ (Online) 341, c6848 c6848.

Chen, S., et al., 2020. Ecotoxicological effects of sulfonamides and fluoroquinolones and their removal by a green alga (Chlorella vulgaris) and a cyanobacterium (Chrysosporum ovalisporum). Environ. Pollut. 263, 114554.

Cheng, X., et al., 2020. Removal of antibiotic resistance genes in an algal-based wastewater treatment system employing Galdieria sulphuraria: a comparative study. Sci. Total Environ. 711, 134435.

Cheng, Z., Hu, X., Sun, Z., November 2016. Microbial community distribution and dominant bacterial species analysis in the bio-electrochemical system treating low concentration cefuroxime. Chem. Eng. J. 303, 137–144.

Chisti, Y., Moo-Young, M., 1996. Bioprocess intensification through bioreactor engineering. Chem. Eng. Res. Des. 74 (5), 575–583.

Copete-Pertuz, L.S., Plácido, J., Serna-Galvis, E.A., Torres-Palma, R.A., Mora, A., 2018. Elimination of Isoxazolyl-Penicillins antibiotics in waters by the ligninolytic native Colombian strain *Leptosphaerulina* sp. considerations on biodegradation process and antimicrobial activity removal. Sci. Total Environ. 630, 1195–1204.

Cosgrove, S.E., 2006. The relationship between antimicrobial resistance and patient outcomes: mortality, length of hospital stay, and health care costs. Clin. Infect. Dis. 42, S82–S89.

Crognale, S., Federici, F., Petruccioli, M., 2002. Enhanced separation of filamentous fungi by ultrasonic field: possible usage in repeated batch processes. J. Biotechnol. 97 (2), 191–197.

Čvančarová, M., Moeder, M., Filipová, A., Reemtsma, T., Cajthaml, T., 2013. Biotransformation of the antibiotic agent flumequine by ligninolytic fungi and residual antibacterial activity of the transformation mixtures. Environ. Sci. Technol. 47 (24), 14128–14136.

Dąbrowska, M., Muszyńska, B., Starek, M., Żmudzki, P., Opoka, W., 2018. Degradation pathway of cephalosporin antibiotics by in vitro cultures of Lentinula edodes and Imleria badia. Int. Biodeterior. Biodegrad. 127 (June 2017), 104–112.

Dangi, A.K., Sharma, B., Hill, R.T., Shukla, P., Jan. 2019. Bioremediation through microbes: systems biology and metabolic engineering approach. Crit. Rev. Biotechnol. 39 (1), 79–98.

Dash, H.R., Das, S., November 2012. Bioremediation of mercury and the importance of bacterial mer genes. Int. Biodeterior. Biodegrad. 75, 207–213.

De Brucker, K., et al., 2015. Fungal β-1,3-Glucan increases ofloxacin tolerance of *Escherichia coli* in a polymicrobial *E. coli*/Candida albicans biofilm. Antimicrob. Agents Chemother. 59 (6), 3052–3058.

Dixit, S., Singh, D.P., 2015. Phycoremediation: future perspective of green technology. In: Algae and Environmental Sustainability.

El-Ramady, H., et al., 2017. Nanoremediation for Sustainable Crop Production, pp. 335–363.

Esterhuizen-Londt, M., Schwartz, K., Pflugmacher, S., October 2016. Using aquatic fungi for pharmaceutical bioremediation: uptake of acetaminophen by Mucor hiemalis does not result in an enzymatic oxidative stress response. Fungal Biol. 120 (10), 1249–1257.

Esterhuizen-Londt, M., Hendel, A.L., Pflugmacher, S., 2017. Mycoremediation of diclofenac using Mucor hiemalis. Toxicol. Environ. Chem. 99 (5–6), 795–808.

Fick, J., Söderström, H., Lindberg, R.H., Phan, C., Tysklind, M., Larsson, D.G.J., 2009. Contamination of surface, ground, and drinking water from pharmaceutical production. Environ. Toxicol. Chem. 28 (12), 2522.

Figueiredo, M. do V.B., Bonifacio, A., Rodrigues, A.C., de Araujo, F.F., Stamford, N.P., 2016. Beneficial microorganisms: current challenge to increase crop performance. In: Bioformulations: For Sustainable Agriculture. Springer India, New Delhi, pp. 53–70.

Flaherty, C.M., Dodson, S.I., 2005. Effects of pharmaceuticals on Daphnia survival, growth, and reproduction. Chemosphere 61 (2), 200–207.

Gao, X.J., Fan, X.J., Chen, X.P., Ge, Z.Q., October 2018. Immobilized β-lactamase on Fe_3O_4 magnetic nanoparticles for degradation of β-lactam antibiotics in wastewater. Int. J. Environ. Sci. Technol. 15 (10), 2203–2212.

González, C., Marciniak, J., Villaverde, S., García-Encina, P.A., Muñoz, R., October 2008. Microalgae-based processes for the biodegradation of pretreated piggery wastewaters. Appl. Microbiol. Biotechnol. 80 (5), 891–898.

Gothwal, R., Shashidhar, T., 2015. Antibiotic pollution in the environment: a review. Clean Soil Air Water 75 (4), 435–441.

Guiry, M.D., 2012. How many species of algae are there? J. Phycol. 48 (5), 1057–1063.

Guo, R., Chen, J., 2015. Application of alga-activated sludge combined system (AASCS) as a novel treatment to remove cephalosporins. Chem. Eng. J. 260, 550–556.

Guo, W., Song, H., Zhou, L., Sun, J., November 2016. Simultaneous removal of sulfanilamide and bioelectricity generation in two-chambered microbial fuel cells. Desalin. Water Treat. 57 (52), 24982–24989.

Guo, W., June 2016. Removal of chloramphenicol and simultaneous electricity generation by using microbial fuel cell technology. Int. J. Electrochem. Sci. 5128–5139.

He, Y., et al., 2018. Evaluation of attenuation of pharmaceuticals, toxic potency, and antibiotic resistance genes in constructed wetlands treating wastewater effluents. Sci. Total Environ. 631-632, 1572–1581.

Herrero, M., Stuckey, D.C., December 2015. Bioaugmentation and its application in wastewater treatment: a review. Chemosphere 140, 119–128.

Huang, J.J., et al., 2012. Monitoring and evaluation of antibiotic-resistant bacteria at a municipal wastewater treatment plant in China. Environ. Int. 42, 31–36.

Jafarinejad, S., May 2019. Simulation for the performance and economic evaluation of conventional activated sludge process replacing by sequencing batch reactor technology in a petroleum refinery wastewater treatment plant. ChemEngineering 3 (2), 45.

Jernberg, C., Löfmark, S., Edlund, C., Jansson, J.K., 2010. Long-term impacts of antibiotic exposure on the human intestinal microbiota. Microbiology 156 (11), 3216–3223.

Jiang, C., Liu, L., Crittenden, J.C., August 2016. An electrochemical process that uses an Fe0/TiO$_2$ cathode to degrade typical dyes and antibiotics and a bioanode that produces electricity. Front. Environ. Sci. Eng. 10 (4), 15.

Jjemba, P.K., 2006. Excretion and ecotoxicity of pharmaceutical and personal care products in the environment. Ecotoxicol. Environ. Saf. 63 (1), 113–130.

Jufer, H., Reilly, L., Mojica, E.R.E., 2019. Antibiotics pollution in soil and water: potential ecological and human health issues. Encycl. Environ. Heal. 118–131.

Kim, B., Pang, H.-B., Kang, J., Park, J.-H., Ruoslahti, E., Sailor, M.J., December 2018. Immunogene therapy with fusogenic nanoparticles modulates macrophage response to *Staphylococcus aureus*. Nat. Commun. 9 (1), 1969.

Kong, D., et al., October 2017. Response of antimicrobial nitrofurazone-degrading biocathode communities to different cathode potentials. Bioresour. Technol. 241, 951–958.

Kryczyk-Poprawa, A., Żmudzki, P., Maślanka, A., Piotrowska, J., Opoka, W., Muszyńska, B., 2019. Mycoremediation of azole antifungal agents using in vitro cultures of Lentinula edodes. 3 Biotech 9 (6), 207.

Krzeminski, P., et al., January 2019. Performance of secondary wastewater treatment methods for the removal of contaminants of emerging concern implicated in crop uptake and antibiotic resistance spread: a review. Sci. Total Environ. 648, 1052–1081.

Kumar, M., et al., March 2019. Antibiotics bioremediation: perspectives on its ecotoxicity and resistance. Environ. Int. 124, 448–461.

Kümmerer, K., 2009. Antibiotics in the aquatic environment - a review - Part I. Chemosphere 75 (4), 417–434.

Kurt, A., Mert, B.K., Özengin, N., Sivrioğlu, Ö., Yonar, T., 2017. Treatment of antibiotics in wastewater using advanced oxidation processes (AOPs). Physico-Chem. Wastewater Treat. Resour. Recover. https://doi.org/10.5772/67538.

Lee, L.S., Carmosini, N., Sassman, S.A., Dion, H.M., Sepúlveda, M.S., 2007. Agricultural contributions of antimicrobials and hormones on soil and water quality. Adv. Agron. 1–68.

Li, X.W., et al., 2014. Investigation of residual fluoroquinolones in a soil-vegetable system in an intensive vegetable cultivation area in Northern China. Sci. Total Environ. 468-469, 258–264.

Li, H., et al., June 2020. Accumulation of sulfonamide resistance genes and bacterial community function prediction in microbial fuel cell-constructed wetland treating pharmaceutical wastewater. Chemosphere 248, 126014.

Li, B., Zhang, T., 2010. Biodegradation and adsorption of antibiotics in the activated sludge process. Environ. Sci. Technol. 44 (9), 3468–3473.

Li, Y., Liu, L., Yang, F., March 2017. Destruction of tetracycline hydrochloride antibiotics by FeOOH/TiO$_2$ granular activated carbon as expanded cathode in low-cost MBR/MFC coupled system. J. Membr. Sci. 525, 202–209.

Li, J., Min, Z., Li, W., Xu, L., Han, J., Li, P., 2020. Interactive effects of roxithromycin and freshwater microalgae, Chlorella pyrenoidosa: toxicity and removal mechanism. Ecotoxicol. Environ. Saf. 191 (August 2019).

Liang, B., et al., May 2013. Accelerated reduction of chlorinated nitroaromatic antibiotic chloramphenicol by biocathode. Environ. Sci. Technol. 47 (10), 5353−5361.

Liang, H., et al., January 2017. Multicopper laccase mimicking nanozymes with nucleotides as ligands. ACS Appl. Mater. Interfaces 9 (2), 1352−1360.

Lindberg, R.H., Fick, J., Tysklind, M., 2010. Screening of antimycotics in Swedish sewage treatment plants - waters and sludge. Water Res. 44 (2), 649−657.

Liu, C.X., Xu, Q.M., Yu, S.C., Cheng, J.S., Yuan, Y.J., 2020. Bio-removal of tetracycline antibiotics under the consortium with probiotics Bacillus clausii T and Bacillus amyloliquefaciens producing biosurfactants. Sci. Total Environ. 710, 136329.

López-Serna, R., Jurado, A., Vázquez-Suñé, E., Carrera, J., Petrović, M., Barceló, D., 2013. Occurrence of 95 pharmaceuticals and transformation products in urban groundwaters underlying the metropolis of Barcelona, Spain. Environ. Pollut. 174, 305−315.

Mahmood, A.R., Al-Haideri, H.H., Hassan, F.M., 2019. Detection of antibiotics in drinking water treatment plants in Baghdad city, Iraq. Adv. Pub. Health 2019, 1−10.

Manaia, C.M., 2017. Assessing the risk of antibiotic resistance transmission from the environment to humans: non-direct proportionality between abundance and risk. Trends Microbiol. 25 (3), 173−181.

Marco-Urrea, E., Pérez-Trujillo, M., Blánquez, P., Vicent, T., Caminal, G., 2010a. Biodegradation of the analgesic naproxen by *Trametes versicolor* and identification of intermediates using HPLC-DAD-MS and NMR. Bioresour. Technol.

Marco-Urrea, E., Pérez-Trujillo, M., Cruz-Morató, C., Caminal, G., Vicent, T., 2010b. White-rot fungus-mediated degradation of the analgesic ketoprofen and identification of intermediates by HPLC-DAD-MS and NMR. Chemosphere.

Matamoros, V., Uggetti, E., García, J., Bayona, J.M., 2016. Assessment of the mechanisms involved in the removal of emerging contaminants by microalgae from wastewater: a laboratory scale study. J. Hazard Mater. 301, 197−205.

Mehta, S.K., Gaur, J.P., January 2005. Use of algae for removing heavy metal ions from wastewater: progress and prospects. Crit. Rev. Biotechnol. 25 (3), 113−152.

Migliore, L., Civitareale, C., Brambilla, G., Dojmi Di Delupis, G., 1997. Toxicity of several important agricultural antibiotics to Artemia. Water Res. 31 (7), 1801−1806.

Migliore, L., Fiori, M., Spadoni, A., Galli, E., 2012. Biodegradation of oxytetracycline by Pleurotus ostreatus mycelium: a mycoremediation technique. J. Hazard Mater. 215-216, 227−232.

Muszyńska, B., Dąbrowska, M., Starek, M., Żmudzki, P., Lazur, J., Pytko-Polończyk, J., Opoka, W., et al., 2019. Lentinula edodes mycelium as effective agent for piroxicam mycoremediation. Front. Microbiol. 10.

Nguyen, P.Y., Carvalho, G., Reis, A.C., Nunes, O.C., Reis, M.A.M., Oehmen, A., June 2017. Impact of biogenic substrates on sulfamethoxazole biodegradation kinetics by Achromobacter denitrificans strain PR1. Biodegradation 28 (2−3), 205−217.

Obayiuwana, A., Ogunjobi, A., Yang, M., Ibekwe, M., 2018. Antibiotic resistance profiles in wastewaters obtained from pharmaceutical facilities in lagos and Ogun states, Nigeria. Int. J. Environ. Res. Publ. Health 15 (7), 1365.

Pei, R., Kim, S.C., Carlson, K.H., Pruden, A., 2006. Effect of River Landscape on the sediment concentrations of antibiotics and corresponding antibiotic resistance genes (ARG). Water Res. 40 (12), 2427−2435.

Poi, G., Shahsavari, E., Aburto-Medina, A., Ball, A.S., 2017. Bioaugmentation: an effective commercial technology for the removal of phenols from wastewater. Microbiol. Aust. 38 (2), 82.

Pointing, S.B., 2001. Feasibility of bioremediation by white-rot fungi. Appl. Microbiol. Biotechnol. 57 (1–2), 20–33.

Pomati, F., Netting, A.G., Calamari, D., Neilan, B.A., 2004. Effects of erythromycin, tetracycline and ibuprofen on the growth of *Synechocystis* sp. and Lemna minor. Aquat. Toxicol. 64 (4), 387–396.

Popovich, J., et al., January 2018. "Building and breaking the cell wall in four acts: a kinesthetic and tactile role-playing exercise for teaching beta-lactam antibiotic mechanism of action and resistance. J. Microbiol. Biol. Educ. 19 (1).

Prieto, A., Möder, M., Rodil, R., Adrian, L., Marco-Urrea, E., 2011. Degradation of the antibiotics norfloxacin and ciprofloxacin by a white-rot fungus and identification of degradation products. Bioresour. Technol. 102 (23), 10987–10995.

Pruden, A., et al., 2013. Management options for reducing the release of antibiotics and antibiotic resistance genes to the environment. Environ. Health Perspect. 121 (8), 878–885.

Radjenovic, J., Petrovic, M., Barceló, D., 2007. Analysis of pharmaceuticals in wastewater and removal using a membrane bioreactor. Anal. Bioanal. Chem. 387, 1365–1377.

Roberts, P.H., Thomas, K.V., 2006. The occurrence of selected pharmaceuticals in wastewater effluent and surface waters of the lower Tyne catchment. Sci. Total Environ. 356 (1–3), 143–153.

Rodríguez-Escales, P., Fernàndez-Garcia, D., Drechsel, J., Folch, A., Sanchez-Vila, X., May 2017. Improving degradation of emerging organic compounds by applying chaotic advection in Managed Aquifer Recharge in randomly heterogeneous porous media. Water Resour. Res. 53 (5), 4376–4392.

Rodríguez-Rodríguez, C.E., Marco-Urrea, E., Caminal, G., July 2010. Naproxen degradation test to monitor *Trametes versicolor* activity in solid-state bioremediation processes. J. Hazard Mater. 179 (1–3), 1152–1155.

Ruggiero, M.A., et al., 2015. Correction: a higher level classification of all living organisms. PLoS One 10 (6), e0130114.

Sanderson, H., Brain, R.A., Johnson, D.J., Wilson, C.J., Solomon, K.R., 2004. Toxicity classification and evaluation of four pharmaceuticals classes: antibiotics, antineoplastics, cardiovascular, and sex hormones. Toxicology 203 (1–3), 27–40.

Sarkar, P., et al., October 2017. Enrichment and characterization of hydrocarbon-degrading bacteria from petroleum refinery waste as potent bioaugmentation agent for in situ bioremediation. Bioresour. Technol. 242, 15–27.

Sedlak, D.L., Huang, C.-H., Pinkston, K., 2004. Strategies for selecting pharmaceuticals to assess attenuation during indirect potable water reuse. Pharm. Environ. 107–120.

Shah, A., Shah, M., October 2020. Characterisation and bioremediation of wastewater: a review exploring bioremediation as a sustainable technique for pharmaceutical wastewater. Groundw. Sustain. Dev. 11, 100383.

Shang, A.H., et al., 2015. Physiological effects of tetracycline antibiotic pollutants on non-target aquatic Microcystis aeruginosa. J. Environ. Sci. Health Part B Pestic. Food Contam. Agric. Wastes 50 (11), 809–818.

Shao, B., et al., January 2019. Immobilization of laccase on hollow mesoporous carbon nanospheres: noteworthy immobilization, excellent stability and efficacious for antibiotic contaminants removal. J. Hazard Mater. 362, 318–326.

Shao, S., Hu, Y., Cheng, J., Chen, Y., 2019. "Biodegradation mechanism of tetracycline (TEC) by strain *Klebsiella* sp. SQY5 as revealed through products analysis and genomics. Ecotoxicol. Environ. Saf. 185 (September), 109676.

Shi, L., Ma, F., Han, Y., Zhang, X., Yu, H., August 2014. Removal of sulfonamide antibiotics by oriented immobilized laccase on Fe_3O_4 nanoparticles with natural mediators. J. Hazard Mater. 279, 203−211.

Shi, X., Lin, J., Zuo, J., Li, P., Li, X., Guo, X., May 2017. Effects of free ammonia on volatile fatty acid accumulation and process performance in the anaerobic digestion of two typical bio-wastes. J. Environ. Sci. 55, 49−57.

Silva, A., Delerue-Matos, C., Figueiredo, S., Freitas, O., July 2019. The use of algae and fungi for removal of pharmaceuticals by bioremediation and biosorption processes: a review. Water 11 (8), 1555.

Singh, S.P., Singh, P., 2015. Effect of temperature and light on the growth of algae species: a review. Renew. Sustain. Energy Rev. 50, 431−444.

Singh, S.K., Khajuria, R., Kaur, L., 2017. Biodegradation of ciprofloxacin by white rot fungus *Pleurotus ostreatus*. 3 Biotech.

Singh, A., Ummalyma, S.B., Sahoo, D., July 2020. Bioremediation and biomass production of microalgae cultivation in river water contaminated with pharmaceutical effluent. Bioresour. Technol. 307, 123233.

Skjånes, K., Rebours, C., Lindblad, P., 2013. Potential for green microalgae to produce hydrogen, pharmaceuticals and other high value products in a combined process. Crit. Rev. Biotechnol. 33 (2), 172−215.

Song, H., Guo, W., Liu, M., Sun, J., December 2013. Performance of microbial fuel cells on removal of metronidazole. Water Sci. Technol. 68 (12), 2599−2604.

Song, H.-L., Zhang, S., Yang, X.-L., Chen, T.-Q., Zhang, Y.-Y., Febuary 2017. Coupled effects of electrical stimulation and antibiotics on microbial community in three-dimensional biofilm-electrode reactors. Water Air Soil Pollut. 228 (2), 83.

Standley, L.J., et al., 2008. Wastewater-contaminated groundwater as a source of endogenous hormones and pharmaceuticals to surface water ecosystems. Environ. Toxicol. Chem. 27 (12), 2457−2468.

Suda, T., Hata, T., Kawai, S., Okamura, H., Nishida, T., 2012. Treatment of tetracycline antibiotics by laccase in the presence of 1-hydroxybenzotriazole. Bioresour. Technol. 103 (1), 498−501.

Teijon, G., Candela, L., Tamoh, K., Molina-Díaz, A., Fernández-Alba, A.R., 2010. Occurrence of emerging contaminants, priority substances (2008/105/CE) and heavy metals in treated wastewater and groundwater at Depurbaix facility (Barcelona, Spain). Sci. Total Environ. 408 (17), 3584−3595.

Ternes, T.A., Joss, A., Siegrist, H., 2004. Peer reviewed: scrutinizing pharmaceuticals and personal care products in wastewater treatment. Environ. Sci. Technol. 38 (20), 392A−399A.

Ugwu, C.U., Aoyagi, H., Uchiyama, H., 2008. Photobioreactors for mass cultivation of algae. Bioresour. Technol. 99 (10), 4021−4028.

Valcárcel, Y., González Alonso, S., Rodríguez-Gil, J.L., Gil, A., Catalá, M., 2011. Detection of pharmaceutically active compounds in the rivers and tap water of the Madrid Region (Spain) and potential ecotoxicological risk. Chemosphere 84 (10), 1336−1348.

Van Hamme, J.D., Singh, A., Ward, O.P., December 2003. Recent advances in petroleum microbiology. Microbiol. Mol. Biol. Rev. 67 (4), 503−549.

Vasiliadou, I.A., Molina, R., Martinez, F., Melero, J.A., Stathopoulou, P.M., Tsiamis, G., July 2018. Toxicity assessment of pharmaceutical compounds on mixed culture from activated sludge using respirometric technique: the role of microbial community structure. Sci. Total Environ. 630, 809–819.

Vázquez-Núñez, E., Molina-Guerrero, C.E., Peña-Castro, J.M., Fernández-Luqueño, F., de la Rosa-Álvarez, M.G., July 2020. Use of nanotechnology for the bioremediation of contaminants: a review. Processes 8 (7), 826.

Vidali, M., July 2001. Bioremediation. An overview. Pure Appl. Chem. 73 (7), 1163–1172.

Vishwakarma, G.S., Bhattacharjee, G., Gohil, N., Singh, V., 2020. Current status, challenges and future of bioremediation. In: Bioremediation of Pollutants. Elsevier, pp. 403–415.

Wang, L., Wu, Y., Zheng, Y., Liu, L., Zhao, F., 2015. Efficient degradation of sulfamethoxazole and the response of microbial communities in microbial fuel cells. RSC Adv. 5 (69), 56430–56437.

Wang, L., Liu, Y., Ma, J., Zhao, F., January 2016. Rapid degradation of sulphamethoxazole and the further transformation of 3-amino-5-methylisoxazole in a microbial fuel cell. Water Res. 88, 322–328.

Wen, Q., et al., March 2011. Simultaneous processes of electricity generation and ceftriaxone sodium degradation in an air-cathode single chamber microbial fuel cell. J. Power Sources 196 (5), 2567–2572.

Wen, X., et al., May 2019. Immobilized laccase on bentonite-derived mesoporous materials for removal of tetracycline. Chemosphere 222, 865–871.

Wen, Q., Kong, F., Zheng, H., Cao, D., Ren, Y., Yin, J., April 2011. Electricity generation from synthetic penicillin wastewater in an air-cathode single chamber microbial fuel cell. Chem. Eng. J. 168 (2), 572–576.

Wen, X., Wang, Y., Zou, Y., Ma, B., Wu, Y., January 2018. No evidential correlation between veterinary antibiotic degradation ability and resistance genes in microorganisms during the biodegradation of doxycycline. Ecotoxicol. Environ. Saf. 147, 759–766.

Wesenberg, D., Kyriakides, I., Agathos, S.N., 2003. White-rot fungi and their enzymes for the treatment of industrial dye effluents. In: Biotechnology Advances.

Wetzstein, H.G., Schmeer, N., Karl, W., 1997. Degradation of the fluoroquinolone enrofloxacin by the brown rot fungus Gloeophyllum striatum: identification of metabolites. Appl. Environ. Microbiol. 63 (11), 4272–4281.

WHO Report on Surveillance of Antibiotic Consumption, 2016-2018 Early implementation. Geneva. ISBN 978-92-4-151488-0. https://apps.who.int/iris/bitstream/handle/10665/277359/9789241514880-eng.pdf?ua=1.

Wiley, P.E., Brenneman, K.J., Jacobson, A.E., 2009. Improved algal harvesting using suspended air flotation. Water Environ. Res. 81 (7), 702–708.

Xiong, J.-Q., Kurade, M.B., Kim, J.R., Roh, H.-S., Byong-HunJeon, 2017. Ciprofloxacin toxicity and its co-metabolic removal by a freshwater microalga Chlamydomonas mexicana. J. Hazard. Mater. 323, 212–219.

Xiong, J.Q., Kurade, M.B., Kim, J.R., Roh, H.S., Jeon, B.H., 2017a. Ciprofloxacin toxicity and its co-metabolic removal by a freshwater microalga Chlamydomonas mexicana. J. Hazard Mater. 323, 212–219.

Xiong, J.Q., Kurade, M.B., Jeon, B.H., 2017b. Biodegradation of levofloxacin by an acclimated freshwater microalga, *Chlorella vulgaris*. Chem. Eng. J. 313, 1251–1257.

Yadav, A., et al., 2020. Bioremediation of toxic pollutants: features, strategies, and applications. In: Contaminants in Agriculture. Springer International Publishing, Cham, pp. 361–383.

Yan, W., et al., October 2018. The changes of bacterial communities and antibiotic resistance genes in microbial fuel cells during long-term oxytetracycline processing. Water Res. 142, 105−114.

Yan, W.-F., Xiao, Y., Wang, S.-H., Ding, R., Zhao, F., March 2018. [Oxytetracycline wastewater treatment in microbial fuel cells and the analysis of microbial communities]. Huan jing ke xue= Huanjing kexue 39 (3), 1379−1385.

Yan, W., Xiao, Y., Yan, W., Ding, R., Wang, S., Zhao, F., Febuary 2019. The effect of bioelectrochemical systems on antibiotics removal and antibiotic resistance genes: a review. Chem. Eng. J. 358, 1421−1437.

Yang, S., Hai, F.I., Nghiem, L.D., Roddick, F., Price, W.E., 2013. Removal of trace organic contaminants by nitrifying activated sludge and whole-cell and crude enzyme extract of *Trametes versicolor*. Water Sci. Technol. 67 (6), 1216−1223.

Yang, J., Lin, Y., Yang, X., Ng, T.B., Ye, X., Lin, J., 2017. Degradation of tetracycline by immobilized laccase and the proposed transformation pathway. J. Hazard Mater. 322, 525−531.

Yap, M.N.F., 2013. The double life of antibiotics. MO Med. 320 (110), 4.

Yu, Y., Zhou, Y., Wang, Z., Torres, O.L., Guo, R., Chen, J., 2017. Investigation of the removal mechanism of antibiotic ceftazidime by green algae and subsequent microbic impact assessment. Sci. Rep.

Zhang, D., Lin, L., Luo, Z., Yan, C., Zhang, X., 2011. Occurrence of selected antibiotics in Jiulongjiang River in various seasons, South China. J. Environ. Monit. 13 (7), 1953−1960.

Zhang, Q.Q., Ying, G.G., Pan, C.G., Liu, Y.S., Zhao, J.L., 2015. Comprehensive evaluation of antibiotics emission and fate in the river basins of China: source analysis, multimedia modeling, and linkage to bacterial resistance. Environ. Sci. Technol. 59 (11), 6772−6782.

Zhang, S., Song, H.-L., Yang, X.-L., Yang, K.-Y., Wang, X.-Y., December 2016a. Effect of electrical stimulation on the fate of sulfamethoxazole and tetracycline with their corresponding resistance genes in three-dimensional biofilm-electrode reactors. Chemosphere 164, 113−119.

Zhang, S., Song, H.-L., Yang, X.-L., Yang, Y.-L., Yang, K.-Y., Wang, X.-Y., 2016b. Fate of tetracycline and sulfamethoxazole and their corresponding resistance genes in microbial fuel cell coupled constructed wetlands. RSC Adv. 6 (98), 95999−96005.

Zhang, Q., Zhang, Y., Li, D., April 2017. Cometabolic degradation of chloramphenicol via a meta-cleavage pathway in a microbial fuel cell and its microbial community. Bioresour. Technol. 229, 104−110.

Zhang, E., Yu, Q., Zhai, W., Wang, F., Scott, K., Febuary 2018. High tolerance of and removal of cefazolin sodium in single-chamber microbial fuel cells operation. Bioresour. Technol. 249, 76−81.

Zhong, J.J., 2019. Bioreactor engineering. In: Moo-Young, M. (Ed.), Comprehensive Biotechnology. Elsevier, pp. 257−269.

Zhou, L.J., et al., 2013. Occurrence and fate of eleven classes of antibiotics in two typical wastewater treatment plants in South China. Sci. Total Environ. 452−453, 365−376.

Advances in biodegradation and bioremediation of environmental pesticide contamination

Shubhra Sharma[1], Shikha Saxena[1], Bhawana Mudgil[2], Siddharth Vats[1]

[1]*Faculty of Biotechnology, Institute of Bio-Sciences and Technology, Shri Ramswaroop Memorial University, Barabanki, Uttar Pradesh, India;* [2]*TGT, Natural Science, Sarvodaya Vidyalaya, Rohini, Delhi, India*

Chapter outline

Biological Approaches to Controlling Pollutants. https://doi.org/10.1016/B978-0-12-824316-9.00009-4

5.1 Introduction

Pesticides are compounds that can curb, destroy, or mitigate pests. The suffix "-cide" in the word pesticide means to kill, that is, a chemical substance that kills pests. These include a wide variety such as fungicides, herbicides, and insecticides (Vishnoi and Dixit, 2019). Pesticides are categorized on the basis of their nature, toxicity, and chemical composition. The organic group of pesticides contains organochlorine, organophosphate, organometallic, pyrethroids, and carbamates (Kaur et al., 2019). Pesticides cause inimical effects when introduced to organisms. The effect of their toxicity differs according to dosage, time, and other characteristics. However, the existence of pesticides in surroundings determines the dosage and exposure time for which an organism would risk its life depending on its flexibility (Velázquez-Fernández et al., 2012). Mostly, the persistency of pesticides in surroundings is due to their nondegradability and even physicochemical properties. Considerable factors such as heat, moisture, and light can volatize or degrade a lesser amount of pesticides (Chakraborty and Das, 2016). Other technologies, termed green technologies, can be used to degrade pesticides with the help of organisms and are known as biodegradation and bioremediation (Velázquez-Fernández et al., 2012; Varshney, 2019; Jain et al., 2011; Maurya et al., 2014).

Soil and water are the most important resources on planet earth and are now being degraded through various sources: pesticides, heavy metals, and garbage that contains numerous discarded and hazardous materials either from home or industries such as plastic or organic matter (Ahmad et al., 2020; Kaur et al., 2010; Maurya et al., 2013; Negi and Vats, 2013). Heavy metals are introduced in soil and water by means of sewage, irrigation, and the use of pesticides and fertilizers. They are the serious threats to the environment as well as to health due to their toxicity, nonbiodegradability, and bioaccumulation (Niharika et al., 2019; Ojha et al., 2013; Sharma et al., 2014). Contamination of soil and water results in a loss of biodiversity as well as the functionality of soil and water such as the nutrient cycle, and the presence of heavy metals inhibits microbial activity (Roychowdhury et al., 2019). The extensive use of pesticides has created issues not only with the environment and health but also with biodiversity (Maurya et al., 2019; Vats and Miglani, 2011; Vats et al., 2013a, 2013b). Not only does soil quality become degraded, but also these pesticides reach the water table to exploit the aquatic environment, thus harming the diversity of aqueous regions (Ok et al., 2020; Vats and Negi, 2013). Accumulation of these pollutants has been occurring for decades and has also given rise to some new types. Most pollutants can be degraded and metabolized through the natural activities of microorganisms. Microorganisms can initiate several types of metabolic reactions such as dehydrogenation, dichlorination, oxidation-reduction, and hydrolysis (Ramírez-García et al., 2019; Vats et al., 2011, 2012). Remediation has now become a global concern, and various physical, chemical, and biological techniques are being used to address it. Chemical remediation techniques use chemicals to extract chemicals from contaminated media, while physical techniques involve soil vapor extraction and soil washing. Both of these techniques are quite costly and require trained labor. But in biological remediation, both plants and microbes are used for removal of pollutants because (Khan et al., 2000; Vats et al., 2014) biodegradation is a sustainable and eco-friendly process that can remove organic pollutants from the environment more efficiently with the help of microorganisms (Saha et al., 2020). The main objective of biodegradation is to remove pollutants from the ecosystem without creating problems in the biological processes associated with it (Shukla et al., 2020). Compared with other methodologies, biodegradation is considered universally as it provides the best results with cost-effective inputs (Singh et al., 2020). Thus, bioremediation is the application of biosystems involving microbes, plants, and animals to reduce the potential toxicity of any contaminants present in the environment through various processes of degradation and transformation of the undesirable compounds or substances to nontoxic or less harmful compounds (Lone et al., 2008).

5.2 Pesticides: a necessary evil

A report published by the Federation of Indian Chambers of Commerce and Industry has found that in the global market for agrochemical production, India ranks fourth after the United States, Japan, and China (FICCI, 1). A substantial chunk of India's crop output, i.e., 15%—20%, is lost to attacks by diseases, weeds, and pests. The order of agrochemical demand is insecticides (60%) followed by fungicides and herbicides (40%) (FICCI, 2). Table 5.1 provides data about global and national trends in pesticide use in agriculture.

Table 5.1 Global and national (Indian) trends in pesticide use in agriculture.

S. No.	Crop type	Global consumption of pesticides (%)	
		Global	Indian
1.	Fruits and vegetables	26	—
2.	Cereals	18	—
3.	Cotton	6	45
4.	Rice	9	22
5.	Soyabean	10	—
6.	Maize	13	—
7.	Others	18	9
8.	Plantations	—	7
9.	Wheat	—	4
10.	Pulses	—	4
11.	Vegetables	—	9

The existence of pesticides in the surroundings decides the dosage and exposure time for organisms to risk their lives (Moosavi and Seghatoleslami, 2013). Biodegradation means degradation by biological means or biological transformation of pesticides by living microorganisms. The biodegradation or bioremediation of pesticides of living origin is easy. Pesticide degradation is a very important area of research, as pesticides are used in almost all countries and by nearly all farmers in their agricultural practices. From preharvesting to postharvesting stages, pesticides are used. Biotransformation is employed to convert one molecule, generally organic in nature, to simple molecules that are nontoxic or less toxic than previously. Biotransformation is not a complete biodegradation but a few reactions where toxicity is reduced to a safer level. Bioremediation involves the use of microbes to remove toxic compounds from air, water, and soil. Bioremediation is termed as biorestoration, biotreatment, or bioreclamation. We thus can say that biodegradation involves transformation or conversion, whereas bioremediation means removal of the toxic compound.

5.3 Classification of pesticides

On the basis of their action, pesticides are termed as insecticides, herbicides, fungicides, algaecides, bactericides, rodenticides, larvicides, repellents, virucides, ovicides, acaricides, nematicides, molluscicides, moth balls, piscicides, avicides, and lampricides. On the basic of origin, they are classified into biopesticides and chemical pesticides. Biopesticides are microbial, biochemical and plant incorporated protectants. Chemically pesticides are classified into five major categories, namely organophosphates, organochlorines, carbamates, pyrethroids, and substituted urea. Fig. 5.1 and Table 5.2 provide details about pesticide types.

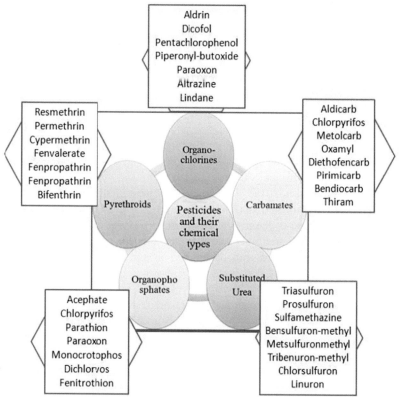

FIGURE 5.1

Classification of chemical-based pesticides.

Table 5.2 Pesticides and their targets.

S. No.	Pesticide	Target
1.	Acaricide	Mites
2.	Algicide	Algae
3.	Avicide	Birds
4.	Defoliant	Plant leaves
5.	Fungicide	Fungi
6.	Herbicide	Weeds
7.	Insecticide	Insects
8.	Molluscicide	Snails and slugs
9.	Nematicide	Nematodes
10.	Piscicide	Fish
11.	Rodenticide	Rodents
12.	Virucide	Viruses

Table 5.3 Various in situ and ex situ methods of bioremediation.

S. No.	In situ bioremediation	Ex situ bioremediation
1.	Bioaugmentation	Land farming
2.	Phytoremediation	Biopiling
3.	Biosparging	Composting
4.	Bioventing	Bioreactors

Pesticides have increased production and have played an important role in the green revolution. Unremediated pesticide content causes health hazards as it mixes with the soil and groundwater (Vats et al., 2019; Singh and Vats, 2019; Sharma et al., 2018; Vats and Bhargava, 2017). The damage that pesticides have done to the soil and environment due to their toxic effects and xenobiotic or recalcitrant natures has also made them one of the major contributors to pollution, especially in water and soil. Bioremediation and biodegradation are the need of the hour and are easy and inexpensive methods that can be applied in situ and ex situ (Gupta et al., 2018) (Table 5.3).

Chemical nature, chemical composition, volatility, structure, physical and chemical characteristics, diffusivity in air and water, soil adsorption coefficient (Kd), half-life, and octanol/water coefficient all have roles to play in in situ bioremediation (Cycon et al., 2017). It cannot be denied that the pesticides used in the fields have become the part of the food chain and have significant negative effects on birds and other animals including humans (Vats et al., 2017; Vats, 2017). The table and figure discuss the various types of pesticides, and one of the major types is persistent organic pollutants (POPs), which includes compounds such as chlordane, dichloro-diphenyltrichloroethane (DDT), toxaphene, heptachlor, and 14 other organochlorine compounds. These compounds have a large octanol versus water partition coefficient, which means they become concentrated in the hepatocytes of living organisms (Tarla et al., 2020). There are various volatile classes of POPs that move from the area of application to other areas and cause air, water, and land pollution. DDT is used by India and some African countries to fight mosquito breeding and to prevent malaria.

5.4 Pesticide stock/banned pesticides

Governments from time to time regulate the use of the pesticides; one major problem that has arisen is that of obsolete pesticides. Obsolete pesticides are banned pesticides that pose a serious threat to the environment and human health. Pesticides that are outdated and not legally allowed to be used are termed obsolete pesticides (Table 5.4) (Tarla et al., 2020).

The sample preparation method must be quick, effective, rugged, cheap, easy, and safe. Successful remediation with clean technologies requires careful

Table 5.4 Obsolete pesticides and their distribution around the globe (Tarla et al., 2020).

S. No.	Obsolete pesticide	Region/Country
1.	210,047 kg and 309,521 L, 4146 kg of POPs	Cameroon
2.	354.7 tons (26 stockpiles)	Kazakhstan
3.	3000 pesticide storage sites and 2000 polluted soils	Ukraine
4.	500,000 tonnes	10% Africa and about 40% in the 12 former Soviet socialist republics

estimation. Liquid/solid-phase extraction, microwave-assisted extraction, solid-phase microextraction, and supercritical fluid extraction are used for sample preparation, and mixing is carried out by sonication. There are various instruments that find application in studying and analyzing pesticide quantities such as liquid chromatography mass spectrometry, polarography with gas chromatography, electron capture, thin-layer chromatography, high-performance liquid chromatography, gas chromatography, spectrophotometry, nitrogen phosphorus, flame photometric detectors, and mass spectrometry (Fang et al., 2007; Tarla et al., 2020; Tandon and Vats, 2016). Third-world countries rely heavily on agriculture, and good productivity completely depends on soil quality. Soil quality becomes deteriorated by the use of pesticides, spills, cleaning of instruments, and postspraying of containers (Vats and Kumar, 2015). Pesticides have a half-life that makes them difficult to degrade, and it has been noted that pesticide residues become bound residues.

5.5 Pesticides and soil ecology

Soil is diverse with different microbes and fauna. Bacteria, protozoa, fungi, nematodes, earthworms, other worms, insects, and arthropods make up healthy soil (Arora and Sahni, 2016). These are the microbes that help in the bioremediation of pesticides or xenobiotics or recalcitrants from the soil. Microbes living close to plants play a major role in bioremediation (Truu et al., 2015). For example, arbuscular mycorrhizal fungi and symbiotic fungi have positive impacts by performing rhizo-degradation and pesticide degradation in the rhizosphere. Mycorrhizal habitation around the roots causes an increase in the root surface area leading to more nutrient uptake. Endophytes protect plants from pesticides such as herbicides, and with rhizospheric microbes, especially bacteria, degradation of pesticides is enhanced (Kumar et al., 2020; Bhargava et al., 2020). Organic pesticides can be degraded by the fauna of soil, such as earthworms, arthropods, and nematodes. Genetically engineered microbes can also be used for the degradation of pesticides (Matsumura, 2012; Kumar et al., 2020). Nowadays, the bioremediation process has been modified accordingly to degrade organic pollutants such as heavy metals. Such properties in microbes are obtained by adding various factors such as electron donors and acceptors, nutrients, and growth factors that can lead to biotransformation. The energy released from the reaction caused by the microbes is taken up by the bioremediation process (Vats et al., 2013, 2014).

5.6 Overview of green technologies

There are some concepts regarding these technologies that should be categorized for better understanding. Therefore, throughout this paper, certain terms are briefly defined. "Bioremediation" is the process through which microorganisms break down and eliminate environmental pollutants, mostly pesticides (Fig. 5.2). In other words, bioremediation may be defined as a way to use biotechnology to detoxify, degrade, and transform the wastes or pesticides present in the environment. The microorganisms that execute the process of bioremediation are known as bioremediators. But through this process, complete elimination is not possible (Rani et al., 2019). Bioremediation can be conducted in two ways, ex situ (Fig. 5.3) and in situ (Varshney et al., 2019) (Table 5.5).

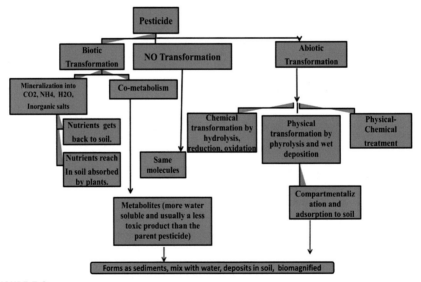

FIGURE 5.2

Bioremediation of pesticides.

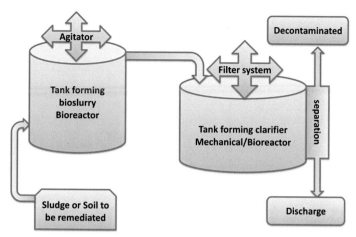

FIGURE 5.3

Ex situ bioremediation process, where contaminated soil is treated with microbes.

Source: https://bioremediationofcontaminatedsoil.wordpress.com/2014/02/07/ex-situ-bioremediation-of-soil/.

Table 5.5 Bioremediation strategies.

S. No.	Technology	Example	Benefit	Limitation
1.	In situ	Bioaugmentation Phytoremediation Biosparging Bioventing	Relatively passive	Extended treatment time
2.	Ex situ	Land farming Biopiling Composting Bioreactors	Extended space and time	Environmental parameters
3.	Bioreactors	Slurry reactors Aqueous reactors	High cost	Toxicity of contaminants

Bioremediation can be subgrouped on the basis of organisms used for treatment:

❖ Microbial remediation: This involves microorganisms (Table 5.6).
❖ Phytoremediation: This involves plants.
❖ Phycoremediation: This involves algae.
❖ Vermiremediation: This involves earthworms.
❖ Zooremediation: This involves the use of animals.

Table 5.6 Microorganisms used for bioremediation.

S. No.	Pollutants	Organisms used
1.	Atrazine	*Pseudomonas* sp. (*ADP*)
2.	Dibenzothiophene (DBT)	*Rhizobium meliloti*
3.	2,4,6-Trinitrotoluene (TNT)	*Methanococcus* sp.
4.	Chlorpyrifos	*Enterobacter strain B-14*
5.	Naphthalene	*Pseudomonas putida*
6.	Polycyclic aromatic hydrocarbon (PAHs)	*Pseudomonas* sp., *Pycnoporus sanguineus, Coriolus versicolor, Pleurotus ostreatus, Daedalea elegans, Fungi*
7.	Hexahydro-1,3-trinitro-1,3,5-triazine (RDX)	*Acetobacterium paludosum, Clostridium acetobutylicum.*
8.	Polychlorinated biphenyl (PCB)	*Rhodococcus erythropolis TA421, Rhizobium* sp.
9.	Phenanthrene	*Agrobacterium, Bacillus, Pseudomonas, Sphingomonas.*
10.	1,4-Dioxane	*Actinomycetes*
11.	Oil	*Arthrobacter*

5.7 Microbial population in bioremediation process or microbial remediation

Isolation of microbes is possible from any environmental condition. Microorganisms are so diverse that they can inhabit every conceivable place on this planet, and most of these places can be considered at the extreme margins of habitability either in physical or chemical conditions (Kaur et al., 2010). Microbes can adapt and grow at extremely low temperatures, i.e., subzero, as well as at extreme heat, in aerobic or anaerobic conditions, with excess water or a scarcity of water, in desert conditions, and in the presence of hazardous compounds or any waste stream (Dassarma et al., 2020). The basic requirement of these enzymes is an energy or carbon source. Microbes can adapt to many biological conditions considered suitable for use in remediating or degrading the hazardous compounds present in the environment (Vishnoi and Dixit, 2019).

5.7.1 On this basis microbes can be divided into several groups

5.7.1.1 Aerobic

These microorganisms require oxygen to survive. Examples of aerobic bacteria used for degradation are *Pseudomonas, Alcaligenes, Sphingomonas, Rhodococcus,* and *Mycobacterium.* These organisms have been found to degrade hydrocarbons and pesticides and compounds such as polyaromatic and alkanes. Most of these microbes use contaminants as their main source of carbon and energy (Mishra et al., 2019).

5.7.1.2 Anaerobic

These microorganisms do not require oxygen for their survival. These bacteria are not used as frequently as aerobic bacteria. Microorganisms within this category are basically used for bioremediation of polychlorinated biphenyls in river sediments, dichlorination of the solvent trichloroethylene, and bioremediation of chloroform (Brusseau et al., 2019).

5.7.1.3 Methylotrophs

This microorganism lies in the category of aerobic bacteria and utilizes methane as a source for carbon and energy. This is used to remediate a wide range of compounds that also includes the chlorinated aliphatics trichloroethylene and 1,2-dichloroethane.

5.7.1.4 Ligninolytic fungi

Fungi such as white rot fungus, *Phanerochaete chrysosporium*, can degrade and remediate a wide range of toxic environmental pollutants. The substrates commonly used are straw, sawdust, or corncobs (Stella et al., 2017).

5.7.1.5 Bioaugmentation

This process requires the addition of indigenous and exogenous microorganisms in the contaminated medium. This can be done under either condition—in situ or ex situ (Sharma et al., 2012).

5.7.1.6 Bioventing

In this method, contaminated soil is treated by adding oxygen to the soil to stimulate microbial activity (Chan and Kjellerup, 2019).

5.7.1.7 Bioreactors

Biodegradation in a container or reactor that can be used to treat liquids or slurries.

5.7.1.8 Land farming

It is a solid-phase treatment system for contaminated soil and can be done in situ or ex situ.

5.7.1.9 Composting

Aerobic, thermophilic treatment process in which contaminated material is mixed with bulking agent.

5.7.1.10 Biofilter

In these microbial stripping columns to treat air emissions is used.

5.7.1.11 Biosparging

This process involves injecting air in pressure below water table so that oxygen concentration can be increased in groundwater so that enhancement can be done in degradation process of contaminants by bacteria (Sharma, 2019).

5.7.1.12 Biopiles

This involves treating surface contamination by petroleum hydrocarbons so as to control physical losses of contaminants by leaching and vitalization.

Now days, bioremediation process had been modified accordingly to degrade organic pollutants also such as heavy metals. Such properties in microbes are obtained by addition of various factors such as electron donors and acceptors, nutrients and growth factors that can lead to biotransformation. The energy released from the reaction caused by the microbes is taken up by bioremediation process (Lone et al., 2008).

5.8 Factors affecting bioremediation

Bioremediation process is optimized and controlled through a complex system consisting of many factors, which include the presence of microbes that can degrade the pollutants and various other factors such as pH, temperature, soil environment, and oxygen presence as well as other electrons. The microorganisms will adapt and customize them according to the conditions such as heat, aerobic and anaerobic conditions, very low temperature or presence of any harmful compound and grow in respect to that (Rani and Dhani, 2014).

Major factors responsible in process of bioremediation are (Table 5.7) (Bewley et al., 1992):

1. Microbial factors
   (Biomass concentration, population diversity, enzyme activities)
2. Environmental factors
   (pH, temperature, moisture content and carbon sources)
3. Substrate
   (Physiochemical characteristics, molecular structure and concentration)
4. Biological aerobic and anaerobic process

Table 5.7 Microbial bioremediation and the factors (http://www.pollutionissues.com/).

S. No.	Factors	Requirement
1.	pH	For bioremediation, the required pH is 6.5–7.5.
2.	Population of microbes	Desired microbes that can degrade contaminant microbes.
3.	Nutrients	N, P, S, and other growth media.
4.	Water	Moisture in the soil must be 50%–70% of the holding capacity.
5.	Temperature	High temperature above 40 degrees can reduce degradation. Up to 40°C is optimum.
6.	Oxygen	2% in gas phase and 0.4 mg/L in liquid phase i.e., in the water in the soil.

5. Growth substrate and cometabolism
6. Physiochemical bioavailability of pollutants
7. Mass transfer limitations

5.9 Advantages of bioremediation

1. This technique is cost-effective compared to other physiochemical treatment methods (Shinde, 2013).
2. This method requires less energy (Vidali, 2001).
3. Low-cost implementation (Shinde, 2013).
4. Bioremediation can be done on sites less expensively than using other conventional technologies (Sharma, 2012).
5. Enhanced regulatory and public acceptance (Sharma, 2012).
6. Soil stabilization and reduced water leaching and transport of organic compounds in the soil (Kensa, 2011).
7. No dangerous chemical compounds are used (Kensa, 2011).

5.10 Disadvantages of bioremediation

1. This process is slower than other remediation technology (Kensa, 2011).
2. It does not remove all the quantities of contaminants from the polluted site (Kensa, 2011).
3. Bioremediation is not useful for treating inorganic contaminants or every organic compound (Shinde, 2013).
4. During in situ bioremediation site must have soil with high permeability (Vidali, 2001).
5. If ex situ process is used, controlling of volatile organic compounds may be difficult (Vidali, 2001).
6. This technology is only applicable to biodegradable compounds (Shinde, 2013).

5.11 Phytoremediation

Phytoremediation is a term formed from Greek word "phyto" which means plant and "remedium" which means to clean and restore (Cunningham et al., 1997). Phytoremediation is the technique that uses green plants and their linked rhizospheric microorganisms to remove pollutants from the environment and can be applied to both, organic and inorganic pollutant that may be found in solid (e.g., soil), liquid (e.g., water) or air (Table 5.8 and Fig. 5.4) (Painuly et al., 2019; Saxena et al., 2019; Bhargava et al., 2019). The criteria of using plants for environmental remediation are that it is cost-effective as well as eco-friendly (Singh et al., 2003).

Table 5.8 Phytoremediation strategies.

Technique	Description
Phytoextraction	Accumulation of pollutants in harvestable biomass by shoots.
Phyto filtration	Sequestration of pollutants from contaminated water by plants.
Phytostabilization	Limiting the mobility and bioavailability of pollutants in soil by plant roots.
Phytovolatilization	Conversion of pollutants to volatile form and their subsequent release to the atmosphere.
Phytodegradation	Degradation of organic xenobiotics by plant enzymes within plant tissues.
Rhizodegradation	Degradation of organic xenobiotics in the rhizosphere by rhizospheric microorganisms.
Phytodesalination	Removal of excess salts from saline soils by halophytes.

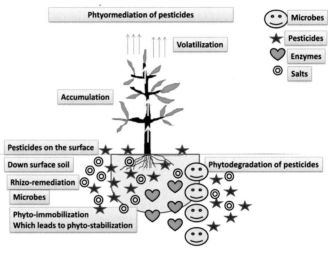

FIGURE 5.4

Phytoremediation of pesticides (Ifon et al., 2019)

5.11.1 Limitations and disadvantages of phytoremediation (Chaudhry et al., 2002)

1. This process may take several years to remediate the contaminants.
2. Hyperaccumulators are usually limited by their slow growth rate and low biomass.
3. It is applicable on sites with low level of metal contamination.
4. This process is influenced by soil and climatic conditions.
5. Toxicity and bioavailability of biodegradation products are not known.
6. It is slower than conventional methods.

5.11.2 Advantages of phytoremediation (Moosavi and Seghatoleslami, 2013)

1. This method is more cost-effective than conventional methods.
2. It is easy to implement and maintain.
3. Through this method natural resources can be conserved.
4. This method is eco-friendly and aesthetically pleasing to the public.
5. This method can be used either in situ or ex situ.

5.12 Phycoremediation

This bioremediation technique involves the action of micro- and macroalgae for the removal or biotransformation of contaminants including nutrients and xenobiotics present in wastewater bodies and CO_2 from air (Upadhyay et al., 2019). Many physical, chemical, and biological methods have been developed for treating wastewater bodies, among which microalgae are observed to be eco-friendly as well as economical (Renuka et al., 2015). *Chlorella vulgaris* is an organism that can be used for remediation technology because it could easily take up the heavy metals from water bodies.

Steps involved in phycoremediation (Phange et al., 2015):

1. Phycostabilization
2. Phycovolatilization
3. Phycofiltration
4. Constructed wetlands
5. Hydraulic barriers

5.12.1 Phycostabilization

It is a procedure of immobilizing organic as well as inorganic pollutants within the surface absorption, erosion or leaching of contaminants to less bioavailable kinds of compounds. This process is suitable for large sites that has low rate of pollution. This process also has one limitation of building up the contaminants over the water surface, which requires monitoring at regular intervals to ensure that conditions are stable (Radziemska and Bes, 2019).

5.12.2 Phycovolatilization

In this process microalgae are used to uptake, translocate as well as modify the pollutants. In this technique metals and inorganic pollutants does not gets accumulated on surface of biomass rather they are transformed to less toxic and volatile form by the mechanism of transpiration and gets released in atmosphere. This method is advantageous over other techniques as it removes metal or metalloids from contaminated sites without need for harvesting and disposal of heavy metals accumulated by microalgae.

5.12.3 **Phycofiltration**

This process is conducted by the help of microalgae and cyanobacteria. This process is called filtration as the contaminants are adsorbed as well as absorbed on the upper surface of microalgae present in aqueous system due to which their minimization of movement of pollutants underwater. The productivities of this process rely on the biochemical characteristics of microalgae.

5.12.4 **Constructed wetlands**

This construction is a manufactured design built to use the natural function of wetland vegetation, soil media, water, and related microbial accumulation of wastewater treatment in a controlled environment (Rajan et al., 2018). These wetlands are suitable for treating different kinds of wastewater such as sludge treatment, hospital wastewaters, industrial wastewaters, effluents, and winery wastewaters. Microalgae provide better conditions for physical purifications, stabilize the surface of beds, and provide a large surface area for microbial growth. They also facilitate transfer of oxygen from the surface by increasing process of aerobic degradation of organic matter (Khan et al., 2020).

5.12.5 **Hydraulic barrier**

This barrier is used to filter out the waste that emerges from different sources through sieve or mesh type of barrier such as algal sieve, matrix or mesh or algal mat. In these filamentous and unicellular algae forms a meshlike surface through which waste is flown out information on thick mat. This kind of technique makes adequate use of time for adsorption, absorption, treatment and degradation of pollutants.

5.12.6 **Factors affecting algae production**

1. pH
2. Temperature
3. Concentration of heavy metals present
4. Metabolic stage of organism
5. Nutrient level
6. Growth rate of algae

5.12.7 **Advantages of phycoremediation (Rajkumar and Takriff, 2016)**

- High yield of microalgae
- Environmental cleanup with less energy and cost due to simple structure of algae
- Can be grown on any water source
- Produces valuable products such as proteins, polysaccharides, etc.
- Does not require herbicide or pesticide
- Biomass yield annually is high

5.12.8 Applications of phycoremediation

- Nutrient removal from municipal wastewater and effluents rich in organic matter (Kiran et al., 2017)
- Treatment of acidic and metal wastewaters (Kiran et al., 2017)
- Nutrient and xenobiotic compounds removal with the aid of algae-based bio sorbents (Olguın et al., 2003)
- Detection of toxic compounds with the aid of algae-based biosensors (Kiran et al., 2017)
- Transformation and degradation of xenobiotics (Olguın et al., 2003)

5.13 Rhizoremediation

In this process, degradation of pollutants occurs by the plants combined with microbial population present in the soil (Table 5.9). The microbes found to be involved in degradation are generally present in rhizospheres of plants (Kuiper et al., 2004). Rhizo means the root, and remediation refers to degradation of organic pollutants (Saravanan et al., 2019). Breakdown of contaminants takes place to convert them to nontoxic compounds (Prabhu et al., 2017). A large quantity of enzymatic and nonenzymatic phenolics or proteins is secreted through plant roots. These secreted products influence the enzyme activity of microbes present in soil for degradation (Etesami and Maheshwari, 2018).

Table 5.9 Microbes that play important roles in rhizoremediation.

Pollutants	Microbes	Catalytic action	Plant associated
Crude oil	*Azospirillum lipoferum*	Promote development of root system.	Wheat
PAHs	*Azospirillum brasilense*	Increase plant tolerance to PAHs.	Tall fescue
PCBs	*Pseudomonas fluorescens* *Pseudomonas putida*	Increase metabolism. Utilization of plant secondary metabolites.	Alfalfa *Arabidopsis*
Trichloroethylene	*P. fluorescens*	Degradation of toluene o-monooxygenase.	Wheat
Lead, zinc	*Azotobacter chroococcum*	Stimulate plant growth.	*Brassica juncea*
Cadmium	*P. fluorescens*	Sequestration of metal from solution.	*None*
Nickel	*Bacillus subtilis*	Facilitate nickel accumulation.	*B. juncea*

5.13.1 Role of rhizospheric microbes in rhizoremediation (Seneviratne et al., 2017)

❖ They encourage ingestion of nutrients plus nonessential metals by roots.
❖ They affect metal dissolvability by changing their synthetic properties.
❖ They discharge organic compounds to increase heavy metal bioavailability.
❖ They cooperate with roots to improve the capability of metal uptake.

5.13.2 Steps taken in process of rhizoremediation

1. Desorption of heavy metal from soil matrix to soil solution.
2. Bioavailability of metals.
3. Bioactivation of metals in the rhizosphere through root−microbe interaction.
4. Complex interactions of transport and chelating activities control the rates of metal uptake and storage by root cells.
5. Metal translocation takes place from symplast of roots to apoplast of xylem and then to stem and leaves, i.e., aerial tissues.
6. Compartmentation and chelation take place in leaves.
7. Detoxification as well as volatilization takes place.

5.13.3 Factors affecting rhizoremediation (Kaur et al., 2019)

1. Contaminant type and concentration
2. Energy sources
3. Microbial population
4. Availability of nutrients
5. Plant root exudates
6. Soil properties and organic matter

5.14 Biodegradation of pesticides

Biodegradation of pesticides is primarily carried out by microbes or their enzymes (Table 5.10). Endosulfan pesticide belonging to the class of organochlorine was degraded by *Alcaligenes faecalis* JBW4, a gram negative, isolated from soil used for planting of coffee (Kong et al., 2013). Fuentes et al. (2011), isolated *Streptomyces* sp. M7 from organochlorine, pesticide-contaminated sediments. This microbe degraded pesticide Lindane. Latifi et al. (2012) isolated *Pseudomonas* sp. IRLM.1, IRLM.2, IRLM.3, IRLM.4 and IRLM.5, all gram negative microbes from soil and El- Helow et al. (2013), isolated *B. subtilis* Y242 a gram positive bacteria from waste water and found that they can degrade pesticide chlorpyrifos belonging to class

Table 5.10 Enzymes and microbes for pesticide degradation (Ortiz-Hernández et al., 2013).

S. No.	Pesticide	Enzyme system	Enzyme	Microbes
1.	Endosulfan	Monooxygenases ese	Monooxygenases	Arthrobacter sp. Mycobacterium sp.
2.	Hexachlorocyclohexane (γ isomers)	Haloalkane dehalogenases LinA	Haloalkane dehalogenases	Sphingobium sp.
3.	Aldrin	Monooxygenases ese	Monooxygenases	Arthrobacter sp.
4.	Chloro-s-triazine herbicides	Haloalkane dehalogenases TrzN	Haloalkane dehalogenases	Nocardioides sp.
5.	Malathion	Monooxygenases ese	Monooxygenases	Arthrobacter sp.
6.	Dicamba	Haloalkane dehalogenases DMO	Haloalkane dehalogenases	Pseudomonas maltophilia
7.	Hexachlorocyclohexane (β and δ isomers)	Haloalkane dehalogenases LIN B	Haloalkane dehalogenases	Sphingomonas sp., Sphingobium sp.
8.	Insecticides phosphotriester	Phosphotriesterase OPH	Phosphotriesterase	Agrobacterium radiobacter, Flavobacterium sp.
9.	Synthetic pyrethroids	Monooxygenases E3	Monooxygenases	Lucilia cuprina
10.	Isoproturon	Monooxygenases Cyp76B1	Monooxygenases	Helianthus tuberosus
11.	2,4 - dichlorophenoxyacetic acid and pyridyl-oxyacetic	Haloalkane dehalogenases TfdA	Haloalkane dehalogenases	Ralstonia eutropha
12.	Insecticides phosphotriester	Monooxygenases E3, Phosphotriesterase OpdA	Monooxygenases, Phosphotriesterase	Lucilia cuprina, Pseudomonas diminuta
13.	Glyphosate	Oxidoreductases gox	Oxidoreductases	Agrobacterium strain T10
14.	Hexachlorobenzene	Monooxygenases P450	Monooxygenases	Pseudomonas putida
15.	Pentachlorobenzene	Monooxygenases P450	Monooxygenases	Pseudomonas putida
16.	Chloro-s-triazine herbicides	Haloalkane dehalogenases AtzA	Haloalkane dehalogenases	Pseudomonas sp. ADP
17.	Linuron	Monooxygenases Cyp76B1	Monooxygenases	Helianthus tuberosus
18.	Herbicide Trifluralin	Dioxygenases (TOD)	Monooxygenases	Pseudomonas putida

organochlorine. Rahman and Alam (2014) isolated *Pseudomonas peli BG1*, *Burkholderia caryophylli BG4*, *Brevundimonas diminuta PD6* all gram-negative bacteria from soil to degrade pesticide Diazinon belonging to the organophosphate class. Similarly, Jariyal et al. (2014) isolated microbes from the sugarcane field treated with phorate belonging to the organophosphate class. Microbes isolated were *Bacillus aerophilus strain IMBL 4.1* (gram positive), *Brevibacterium frigoritolerans strain IMBL 2.1*(gram positive), and *Pseudomonas fulva strain IMBL 5.1* (gram negative). There is a certain specific enzyme system present in microbes that causes the degradation of pesticides. Specific genes are also given credit for microbial ability to degrade pesticides.

5.15 Biodegradation of bound pesticides

Pesticides form bound residues with the soil. On application of pesticides, they interact with the soil. The interaction is influenced by the various factors such as the type of soil, land management, the amount of pesticides used, and the seasons and many other factors. There is physical and chemical aspect to the bound pesticides with the soil. From the time of application, the pesticides undergo change or biotransformation by the action of the microbes or soil biota. Physical and chemical factors also cause physic-chemical transformation. The bioremediation of degradation of the pesticides releases residues. When we talk about the biodegradation of the pesticides then there are various aspects to it. Biodegradation or bioremediation of the pesticides can happen just after the application, post-physico-chemical degradation, free residues in the soil or as residues bound in the soil. Generally free residues run away or may be easily extracted from the soil, while the bound pesticides residues are difficult to extract (Gevao et al., 2000). Bound residues are the chemical molecules/species that are originated from pesticides and used in agricultural practices and cannot be extracted by the methods that do not change the chemical nature of the residues. Strategies that are used to extract and characterize the residues can be supercritical fluid extraction, microwave extraction, high temperature distillation techniques, soxhlet extraction, nuclear magnetic resonance, infrared (IR) (Gevao et al., 2000). It is also to be understood, from all bound residues some are bioavailable, and some are unavailable. There are various factors that affects the availability, nonavailability, and persistence of pesticide residues. According to the Gevao et al. (2000), aging is directly proportional to persistence of the residues. Aging is the time spent and increases in time spent by chemical residues with the soil, immobilization, sorption with the soil particle, or entrapment in the lattice formed by minerals or other organic material, which causes to interact more strongly with the soil and reduces the bioavailability. Increase in time leads to formation of covalent bonds to other residues and parent molecules (Gevao et al., 2000). There are various physic-chemical properties such as polarizability, molecular size, water solubility, ionizability, lipophilicity, or volatility (Gevao et al., 2000). A majority of the interaction falls within the category of adsorption. This interaction takes place in the soil liquid phase. Size, molecule type, molecular structure, shape, polarity, charge, polarizability, charge distribution, chemical functions, interacting species, basic and acid nature of the pesticide. It can be chemical adsorption by electrostatic interaction or physical interaction by van der Waals forces. Pesticide molecules undergo chemical reaction, with their unaltered or residue molecule to form a stable chemical compound and losing their original chemical character. Humus portion of the soil also react with pesticides. This interaction cause reduction in its availability to the microflora, microfauna and microbiota of the soil, less toxic effects, immobilization, and trapping and ultimately preventing it from getting leached or transported. The types of bonding that exists are covalent bonding, sequestration, ligand exchange, charge transfer complexes, ionic bindings, hydrogen bonding, van der Waals forces, and hydrophobic partitioning. There are some agriculture and environmental factors that affects the bound residue formation and reduction in their biodegradation. Concentration of the xenobiotic/pesticide residues is directly proportional to persistence in the soil, and inversely proportional to mineralization and biodegradation. Repeated application is another major factor for decreasing the bioremediation. Every chemical has a half-life that is influenced by type of soil, climate of the soil, nature of the soil, and number of applications in that soil. Repeated applications cause more bound residue formation. On applications of pesticides, it undergoes two major process, dissipation and binding. Dissipation should be enhanced, and binding should be reduced. Accelerated dissipation and decelerated binding reduce the persistence of the pesticides in the soil. And many of the bioremediation process must focus on this concept to remove pesticides from the soil. Aging has its effect on the bound residuals. Time has been the major player for the formation of residues that we cannot extract. Loss of biological activity also takes place with time. This adds up in their resistance of being extracted or degraded. During aging the residues get more adsorbed as well as sequestered by other organic molecules with the help of covalent bonds. Amendment in soil is done with mixing of straw, cow or animal manure. Soil amended with organic matter undergoes show dissipation and formation of bound residues. High temperature also increases the formation of bound residues. The mode of application of pesticides has a role in pesticide binding. Their volatilization and run off can be reduced by tillage. By this, pesticides reach the deeper layers in the soil. Mode of the application of pesticides in the soil have role in bounding rate in the soil. Genetic engineering is also used for degradation of biopesticides with the help of microbes (Table 5.11).

Table 5.11 Genetic engineering and decomposition of pesticides (Ortiz-Hernández et al., 2013).

S. No.	Gene	Organism	Microorganism type
1.	Opd	*Pseudomonas diminuta*	Bacteria
2.	Phn	*Bacillus cereus*	Bacteria
3.	adpB	*Nocardia* sp.	Bacteria
4.	OpdA	*A. radiobacter*	Bacteria
5.	opaA	*Alteromonas* spp.	Bacteria
6.	hocA	*Pseudomonas monteilii*	Bacteria
7.	pepA	*Escherichia coli*	Bacteria
8.	A-opd	*Aspergillus niger*	Fungi
9.	Oph	*Arthrobacter* sp.	Bacteria
10.	P-opd	*Penicillium lilacinum*	Fungi
11.	OpdB	*Lactobacillus brevis.*	Bacteria
12.	ophB	*Burkholderia* sp. JBA3.	Bacteria
13.	ophC2	*Stenotrophomonas* sp. SMSP-1.	Bacteria
14.	Lmh	*Arthrobacter* sp. scl-2.	Bacteria
15.	Mph	*Arthrobacter* sp. L1 (2006).	Bacteria
16.	Mpd	*Ochrobactrum* sp. Yw28, *Rhizobium radiobacter*	Bacteria
17.	opdE	*Enterobacter* sp.	Bacteria
18.	MpdB	*Burkholderia cepacia*	Bacteria
19.	pehA	*Burkholderia caryophylli*	Bacteria

5.16 Conclusion

Through this paper, it is concluded that bioremediation is recognized as a suitable tool for restoring contaminated sites. The combination of phytoremediation, phycoremediation, and rhizoremediation can be used to solve many problems that cannot be solved by these techniques individually. Many studies have shown that these techniques have contributed to the restoration of polluted sites. Different microbes in combination with different plants can be used for remediation purposes. Therefore, further studies are underway for the selection of suitable microbes that can sustain and proliferate over the root systems of plants and will be suitable for bioremediation techniques to yield a useful system. Thus, this system can be improved for wide use in site restoration.

References

Ahamad, A., Madhav, S., Singh, A.K., Kumar, A., Singh, P., 2020. Types of water pollutants: conventional and emerging. In: Sensors in Water Pollutants Monitoring: Role of Material. Springer, Singapore, pp. 21–41.

Arora, S., Sahni, D., 2016. Pesticides effect on soil microbial ecology and enzyme activity-an overview. J. Appl. Nat. Sci. 8 (2), 1126–1132.

Bewley, R.J., 1992. Bioremediation and waste management. In: The Release of Genetically Modified Microorganisms—REGEM 2. Springer, Boston, MA, pp. 33–45.

Bhargava, P., Khan, M., Verma, A., Singh, A., Singh, S., Vats, S., Goel, R., 2019a. Metagenomics as a tool to explore new insights from plant-microbe interface. In: Plant Microbe Interface. Springer, Cham, pp. 271–289.

Bhargava, P., Vats, S., Gupta, N., 2019b. Metagenomics as a tool to explore mycorrhizal fungal communities. In: Mycorrhizosphere and Pedogenesis. Springer, Singapore, pp. 207–219.

Bhargava, P., Gupta, N., Kumar, R., Vats, S., 2020. Plants and microbes: bioresources for sustainable development and biocontrol. In: Plant Microbe Symbiosis. Springer, Cham, pp. 153–176.

Brusseau, M.L., Pepper, I.L., Gerba, C., 2019. Environmental and Pollution Science. Academic Press.

Chakraborty, J., Das, S., 2016. Molecular perspectives and recent advances in microbial remediation of persistent organic pollutants. Environ. Sci. Pollut. Control Ser. 23 (17), 16883–16903.

Chan, A.Y., Kjellerup, B.V., 2019. Bio-and phytoremediation of persistent organic pollutants in storm water containment systems and soil. Microb. Biofilms Bioremed. Wastewater Treat. 225.

Chaudhry, Q., Schröder, P., Werck-Reichhart, D., Grajek, W., Marecik, R., 2002. Prospects and limitations of phytoremediation for the removal of persistent pesticides in the environment. Environ. Sci. Pollut. Control Ser. 9 (1), 4.

Cunningham, S.D., Shann, J.R., Crowley, D.E., Anderson, T.A., 1997. Phytoremediation of Contaminated Water and Soil.

Cycoń, M., Mrozik, A., Piotrowska-Seget, Z., 2017. Bioaugmentation as a strategy for the remediation of pesticide-polluted soil: a review. Chemosphere 172, 52–71.

DasSarma, P., Antunes, A., Simões, M.F., DasSarma, S., 2020. Earth's stratosphere and microbial life. Curr. Issues Mol. Biol. 38, 197–244.

El-Helow, E.R., Badawy, M.E., Mabrouk, M.E., Mohamed, E.A., El-Beshlawy, Y.M., 2013. Biodegradation of chlorpyrifos by a newly isolated Bacillus subtilis strain, Y242. Bioremediat. J. 17 (2), 113–123.

Etesami, H., Maheshwari, D.K., 2018. Use of plant growth promoting rhizobacteria (PGPRs) with multiple plant growth promoting traits in stress agriculture: action mechanisms and future prospects. Ecotoxicol. Environ. Saf. 156, 225–246.

Fang, W., Xin, J.I.A.N.G., Yong-rong, B.I.A.N., Fen-xia, Y.A.O., Hong-jian, G.A.O., Guifen, Y.U., Charles Munch, J., Schroll, R., 2007. Organochlorine pesticides in soils under different land usage in the Taihu Lake region, China. J. Environ. Sci. 19 (5), 584–590.

FICCI, 1. http://ficci.in/events/20563/Add_docs/SectorBrief.pdf.

Fuentes, M.S., Sáez, J.M., Benimeli, C.S., Amoroso, M.J., 2011. Lindane biodegradation by defined consortia of indigenous Streptomyces strains. Water Air Soil Pollut. 222, 217−231.

Gevao, B., Semple, K.T., Jones, K.C., 2000. Bound pesticide residues in soils: a review. Environ. Pollut. 108 (1), 3−14.

Gupta, N., Vats, S., Bhargava, P., 2018. Sustainable agriculture: role of metagenomics and metabolomics in exploring the soil microbiota. In: In Silico Approach for Sustainable Agriculture. Springer, Singapore, pp. 183−199.

Ifon, B.E., Togbé, A.C.F., Tometin, L.A.S., Suanon, F., Yessoufou, A., 2019. Metal-contaminated soil remediation: phytoremediation, chemical leaching and electrochemical remediation. In metals in soil-contamination and remediation. IntechOpen.

Jain, P., Miglani, K., Vats, S., 2011. Aptamers-potential applications. In: Diagnostics and Therapeutics. Everyman's Science: 361, Volume XLV No. 6, Feb−Mar 11.

Kaur, A., Vats, S., Rekhi, S., Bhardwaj, A., Goel, J., Tanwar, R.S., Gaur, K.K., 2010. Physico-chemical analysis of the industrial effluents and their impact on the soil microflora. Proc. Environ. Sci. 2, 595−599.

Jariyal, M., Gupta, V.K., Mandal, K., Jindal, V., Banta, G., Singh, B., 2014. Isolation and characterization of novel phorate-degrading bacterial species from agricultural soil. Environ. Sci. Pollut. Res. 21 (3), 2214−2222.

Kaur, R., Mavi, G.K., Raghav, S., Khan, I., 2019. Pesticides classification and its impact on environment. Int. J. Curr. Microbiol. Appl. Sci 8, 1889−1897.

Kensa, V.M., 2011. Bioremediation-an overview. I Control Pollut. 27 (2), 161−168.

Khan, A.G., Kuek, C., Chaudhry, T.M., Khoo, C.S., Hayes, W.J., 2000. Role of plants, mycorrhizae and phytochelators in heavy metal contaminated land remediation. Chemosphere 41 (1−2), 197−207.

Khan, M.D., Li, D., Tabraiz, S., Shamurad, B., Keith, S., Khan, M.Z., Eileen, H.U., 2020. Integrated air cathode microbial fuel cell-aerobic bioreactor set-up for enhanced bio-electrodegradation of azo dye Acid Blue 29. Sci. Total Environ. https://doi.org/10.1016/j.scitotenv.2020.143752.

Kiran, B., Pathak, K., Kumar, R., Deshmukh, D., 2017. Phycoremediation: an eco-friendly approach to solve water pollution problems. In: Microbial Applications, vol. 1. Springer, Cham, pp. 3−28.

Kong, X., Deng, H., Yan, F., Kim, J., Swisher, J.A., Smit, B., Yaghi, O.M., Reimer, J.A., 2013. Mapping of functional groups in metal-organic frameworks. Science 341 (6148), 882−885. https://doi.org/10.1126/science.1238339.

Kuiper, I., Lagendijk, E.L., Bloemberg, G.V., Lugtenberg, B.J., 2004. Rhizoremediation: a beneficial plant-microbe interaction. Mol. Plant Microbe Interact. 17 (1), 6−15.

Kumar, N., Srivastava, P., Vishwakarma, K., Kumar, R., Kuppala, H., Maheshwari, S.K., Vats, S., 2020a. The rhizobium−plant symbiosis: state of the art. In: Plant Microbe Symbiosis. Springer, Cham, pp. 1−20.

Kumar, N., Balamurugan, A., Shafreen, M.M., Rahim, A., Vats, S., Vishwakarma, K., 2020b. Nanomaterials: emerging trends and future prospects for economical agricultural system. In: Biogenic Nano-Particles and Their Use in Agro-Ecosystems. Springer, Singapore, pp. 281−305.

Latifi, A.M., Khodi, S., Mirzaei, M., Miresmaeili, M., Babavalian, H., 2012. Isolation and characterization of five chlorpyrifos degrading bacteria. Afr. J. Biotechnol. 11 (13), 3140−3146.

Lone, M.I., He, Z.L., Stoffella, P.J., Yang, X.E., 2008. Phytoremediation of heavy metal polluted soils and water: progresses and perspectives. J. Zhejiang Univ. Sci. B 9 (3), 210—220.

Maurya, D.P., Vats, S., Rai, S., Negi, S., 2013. Optimization of enzymatic saccharification of microwave pretreated sugarcane tops through response surface methodology for biofuel. Indian J. Exp. Biol. 51 (11), 992—996.

Maurya, D.P., Singh, D., Vats, S., 2014. In: Ms Aurora, J. (Ed.), Cellulase Production and Utilization. Number of Pages 80, Category- Chemical Technology, Publishing House-LAP LAMBERT. Academic Publishing.

Matsumura, F., 2012. Biodegradation of Pesticides. Springer Science & Business Media.

Maurya, P.K., Malik, D.S., Sharma, A., 2019. Impacts of pesticide application on aquatic environments and fish diversity. Contam. Agric. Environ. Health Risks Remediat. 1, 111.

Mishra, B., Varjani, S., Iragavarapu, G.P., Ngo, H.H., Guo, W., Vishal, B., 2019. Microbial fingerprinting of potential biodegrading organisms. Curr. Pollut. Rep. 5 (4), 181—197.

Moosavi, S.G., Seghatoleslami, M.J., 2013. Phytoremediation: a review. Adv. Agric. Biol. 1 (1), 5—11.

Negi, S., Vats, S., 2013. Pine forest litter based bio-refinery for biofuels and value-added phytochemicals. In: Singh, R.S., Pandey, A., Larroche, C. (Eds.), Advances in Industrial Biotechnology. IK International Publishing House Pvt Ltd., India, pp. 98—116.

Niharika, S., Anita, G., Rajesh, V., Shyam, P., Vinay, M., Kavita, S., 2019. Heavy metals uptake by Alcea Rosea (Holly hock) using phytoremediation technology. Res. J. Chem. Environ. 23 (6), 134—137.

Ojha, A.K., Forster, S., Kumar, S., Vats, S., Negi, S., Fischer, I., 2013. Synthesis of well—dispersed silver nanorods of different aspect ratios and their antimicrobial properties against gram positive and negative bacterial strains. J. Nanobiotechnol. 11 (1), 42.

Ok, Y.S., Rinklebe, J., Hou, D., Tsang, D.C., Tack, F.M., 2020. Soil and Groundwater Remediation Technologies: A Practical Guide. CRC Press.

Olguın, E.J., 2003. Phycoremediation: key issues for cost-effective nutrient removal processes. Biotechnol. Adv. 22 (1—2), 81—91.

Ortiz-Hernández, M.L., Sánchez-Salinas, E., Dantán-González, E., Castrejón-Godínez, M.L., 2013. Pesticide biodegradation: mechanisms, genetics and strategies to enhance the process. Biodegrad. Life Sci. 251—287.

Painuly, A.S., Gupta, R., Vats, S., 2019. Bio-accumulation of arsenic (III) using Nelumbo Nucifera Gaertn. J. Health Pollut. 9 (23), 190902.

Phang, S.M., Chu, W.L., Rabiei, R., 2015. Phycoremediation. In: The Algae World. Springer, Dordrecht, pp. 357—389.

Prabhu, A.A., Chityala, S., Jayachandran, D., Naik, N., Dasu, V.V., 2017. Rhizoremediation of environmental contaminants using microbial communities. In: Plant-Microbe Interactions in Agro-Ecological Perspectives. Springer, Singapore, pp. 433—453.

Radziemska, M., Beś, A., 2019. Assisted phytostabilization of a heavy metals contaminated soil using mineral amendments and *L. perenne*. Mechanization Agric. Conserv. Resour. 65 (4), 132—134.

Rahman, M.A., Alam, M.K., 2014. Degradation of the organophosphorus insecticide diazinon by soil bacterial isolate. Int. J. Biotechnol. 3 (1), 12—23.

Rajan, R.J., Sudarsan, J.S., Nithiyanantham, S., Rajan, R.J., Sudarsan, J.S., Nithiyanantham, S., 2018. Microbial population dynamics in constructed wetlands: review of recent advancements for wastewater treatment. Environ. Eng. Res. 24 (2), 181—190.

Rajkumar, R., Takriff, M.S., 2016. Prospects of algae and their environmental applications in Malaysia: a case study. J. Biorem. Biodegrad. 7, 1–12.

Ramírez-García, R., Gohil, N., Singh, V., 2019. Recent advances, challenges, and opportunities in bioremediation of hazardous materials. In: Phytomanagement of Polluted Sites. Elsevier, pp. 517–568.

Rani, K., Dhania, G., 2014. Bioremediation and biodegradation of pesticide from contaminated soil and water–a noval approach. Int. J. Curr. Microbiol. App. Sci. 3 (10), 23–33.

Rani, R., Patnala, P.K., Pathak, V.V., 2019. Prospects of pesticide contamination and control measures in aquatic systems: a green approach. In: Handbook of Research on the Adverse Effects of Pesticide Pollution in Aquatic Ecosystems. IGI Global, pp. 369–386.

Renuka, N., Sood, A., Prasanna, R., Ahluwalia, A.S., 2015. Phycoremediation of wastewaters: a synergistic approach using microalgae for bioremediation and biomass generation. Int. J. Environ. Sci. Technol. 12 (4), 1443–1460.

Roychowdhury, R., Roy, M., Zaman, S., Mitra, A., 2019. Bioremediation potential of microbes towards heavy metal contamination. Int. J. Res. Anal. Rev. 6 (1), 1088–1094.

Saha, L., Kumari, K., Sinha, S., Bordoloi, N., Tiwari, J., Korstad, J., Bauddh, K., 2020. Removal of organic pollutants from contaminated water bodies by using aquatic macrophytes coupled with bioenergy production and carbon sequestration. In: Emerging Eco-Friendly Green Technologies for Wastewater Treatment. Springer, Singapore, pp. 221–244.

Saravanan, A., Jeevanantham, S., Narayanan, V.A., Kumar, P.S., Yaashikaa, P.R., Muthu, C.M., 2019. Rhizoremediation–a promising tool for the removal of soil contaminants: a review. J. Environ. Chem. Eng. 103543.

Saxena, P., Srivastava, J., Pandey, S., Srivastava, S., Maurya, N., Kaushik, N.C., et al., 2019. Plants for biocontrol and biological control of plant pathogens. In: Plant Biotic Interactions. Springer, Cham, pp. 147–179.

Seneviratne, M., Seneviratne, G., Madawala, H.M.S.P., Vithanage, M., 2017. Role of rhizospheric microbes in heavy metal uptake by plants. In: Agro-Environmental Sustainability. Springer, Cham, pp. 147–163.

Sharma, S., 2012. Bioremediation: features, strategies and applications. Asian J. Pharm. Life Sci. ISSN 2231, 4423.

Sharma, J., 2019. Advantages and limitations of in situ methods of bioremediation. Recent Adv. Biol. Med. 5 (2019), 10941.

Sharma, K.M., Kumar, R., Siddharth, V., Gupta, A. 06/2014. Production, partial purification and characterization of alkaline protease from *Bacillus aryabhattai* K3. Int. J. Adv. Pharm. Biol. Chem. 3 (2), 290–298.

Sharma, D., Javed, S., Arshilekha, P.S., Babbar, P., Shukla, D., Srivastava, P., Vats, S., 2018. Food additives and their effects: a mini review. Int. J. Curr. Res. 10 (06), 69999–70002.

Shinde, S., 2013. Bioremediation. An overview. Recent Res. Sci. Technol. 5 (5).

Shukla, S.K., Tripathi, V.K., Mishra, P.K., 2020. Bioremediation of distillery effluent: present status and future prospects. In: Bioremediation of Industrial Waste for Environmental Safety. Springer, Singapore, pp. 77–97.

Singh, M., Vats, S., 2019. Mathematically designed bioprocess for release of value added products with pharmaceutical applications from wastes generated from spices industries. Int. J. Pharmaceut. Sci. Res. 10 (1), 130–138.

Singh, O.V., Labana, S., Pandey, G., Budhiraja, R., Jain, R.K., 2003. Phytoremediation: an overview of metallic ion decontamination from soil. Appl. Microbiol. Biotechnol. 61 (5–6), 405–412.

Singh, A., Sarma, A.K., Hack, J., 2020. Cost-effective optimization of nature-based solutions for reducing urban floods considering limited space availability. Environ. Process. 7 (1), 297–319.

Stella, T., Covino, S., Čvančarová, M., Filipová, A., Petruccioli, M., D'Annibale, A., Cajthaml, T., 2017. Bioremediation of long-term PCB-contaminated soil by white-rot fungi. J. Hazard. Mater. 324, 701–710.

Tandon, S., Vats, S., 2016. Microbial biosynthesis of cadmium sulfide (Cds) nanoparticles and their characterization. Eur. J. Pharm. Med. Res 3, 545–550.

Tarla, D.N., Erickson, L.E., Hettiarachchi, G.M., Amadi, S.I., Galkaduwa, M., Davis, L.C., Nurzhanova, A., Pidlisnyuk, V., 2020. Phytoremediation and bioremediation of pesticide-contaminated soil. Appl. Sci. 10 (4), 1217.

Truu, J., Truu, M., Espenberg, M., Nõlvak, H., Juhanson, J., 2015. Phytoremediation and plant-assisted bioremediation in soil and treatment wetlands: a review. Open Biotechnol. J. 9 (1).

Upadhyay, A.K., Singh, R., Singh, J.S., Singh, D.P., 2019. Microalgae-assisted phycoremediation and energy crisis solution: challenges and opportunity. In: New and Future Developments in Microbial Biotechnology and Bioengineering. Elsevier, pp. 295–307.

Varshney, K., 2019. Bioremediation of pesticide waste at contaminated sites. J. Emerg. Technol. Innov. Res. 6 (Issue 5), 128–134.

Vats, S., 2017. Methods for extractions of value-added nutraceuticals from lignocellulosic wastes and their health application. In: Ingredients Extraction by Physicochemical Methods in Food. Academic Press, pp. 1–64.

Vats, S., Bhargava, P., 2017. Alternate energy: fuel for "Modi's India" and "smart cities.". Int. J. Curr. Res. 9 (04), 49090–49097.

Vats, S., Kumar, R., 2015. Amylolytic-extremoenzymes: saviour of environments. Eur. J. Biomed. Pharmaceut. Sci. 2 (5), 694–702.

Vats, S., Miglani, K., 2011. Synergistic antimicrobial effect of cow urine and *Azadirachta indica* on infectious microbes. Int. J. Pharmaceut. Sci. Res. 2 (7), 1781.

Vats, S., Negi, S., 2013. Use of artificial neural network (ANN) for the development of bioprocess using *Pinus roxburghii* fallen foliages for the release of polyphenols and reducing sugars. Bioresour. Technol. 140, 392–398.

Vats, S., Kumar, R., Miglani, A.K., 2011. Isolation, characterization and identification of high salinity tolerant, heavy metal contaminant and antibiotics resistant amylolytic-thermophilic pseudomonas Sp. Int. J. Pharmaceut. Sci. Rev. Res. 10 (2), 125–129.

Vats, S., Kumar, R., Negi, S., 2012. Natural food that meet antibiotics resistance challenge: in vitro synergistic antimicrobial activity of *Azadirachta indica*, *Terminalia chebula*, *Piper nigrum* and photoactivated cow urine. Asian J. Pharmaceut. Biol. Res. 2 (2).

Vats, S., Maurya, D.P., Jain, A., Mall, V., Negi, S., 2013a. Mathematical model-based optimization of physico-enzymatic hydrolysis of *Pinus roxburghii* needles for the production of reducing sugars. Indian J. Exp. Biol. 51, 944–953.

Vats, S., Devendra, P.M., Ayushi, agarwal, Mohmmad, S., Sangeeta, N., 2013b. Development of a microbial consortium for the production of blend of enzymes for the hydrolysis of agricultural wastes into sugars. J. Sci. Ind. Res. 72, 585–790.

Vats, S., Rajesh, K., Devendra Prasad, M., (6-1-2014). Alkaline Amylase from Multi Resistant Microbes and its Applications. Number of Pages 100, Category- Microbiology, Publishing House-LAP LAMBERT. Academic Publishing. Ms Elena, A. (Ed.).

Vats, S., Singh, M., Siraj, S., Singh, H., Tandon, S., 2017. Role of nanotechnology in thera-nostics and personalized medicines. J. Health Res. Rev. 4 (1), 1.

Vats, S., Gupta, N., Bhargava, P., 2019. Vulnerability of soil micro biota towards natural and anthropogenic induced changes and loss of pedospheric functionality. In: Mycorrhizo-sphere and Pedogenesis. Springer, Singapore, pp. 191–205.

Velázquez-Fernández, J.B., Martínez-Rizo, A.B., Ramírez-Sandoval, M., Domínguez-Ojeda, D., 2012. Biodegradation and bioremediation of organic pesticides. Pestic. Recent Trends Pestic. Residue Assay 253–272.

Vidali, M., 2001. Bioremediation. an overview. Pure Appl. Chem. 73 (7), 1163–1172.

Vishnoi, N., Dixit, S., 2019. Bioremediation: new prospects for environmental cleaning by fungal enzymes. In: Recent Advancement in White Biotechnology through Fungi. Springer, Cham, pp. 17–52.

Further reading

Bhargava, P., Gupta, N., Vats, S., Goel, R., 2017. Health issues and heavy metals. Austin J. Environ. Toxicol. 3 (1), 3018.

Bioremediation of Contaminated Soils. https://bioremediationofcontaminatedsoil.wordpress.com/2014/02/07/ex-situ-bioremediation-of-soil/.

FCCI, 2. http://ficci.in/spdocument/23103/agrochemical-ficci.pdf.

Luo, X., Zhang, D., Zhou, X., Du, J., Zhang, S., Liu, Y., 2018. Cloning and characterization of a pyrethroid pesticide decomposing esterase gene, Est3385, from *Rhodopseudomonas palustris* PSB-S. Sci. Rep. 8 (1), 1–8.

Advances in biodegradation and bioremediation of arsenic contamination in the environment

Aroosa Malik, Shehla Batool, Abida Farooqi

Department of Environmental Sciences, Faculty of Biological Sciences, Quaid-i-Azam University,
Islamabad, Punjab, Pakistan

Chapter outline

6.1 Introduction

Arsenic (As) is an omnipresent metalloid geogenically present in mineral form and accompanied with various transition metals. It largely exists as arsenates (60%), oxides (10%), arsenides, arsenites, silicates, and elemental arsenic (10%), and the remaining 20% comprising sulfides and sulfosalts (Duarte et al., 2009; Sahoo and Kim, 2013). Arsenic can mobilize in the environment through both natural (geogenic) and anthropogenic (unsewered sanitation) sources. Rock weathering, volcanic eruption, marine aerosols, and thermal waters are natural processes that trigger arsenic release in the environment, whereas agricultural and industrial activities, coal and oil extraction procedures, the use of pesticides, and the mining of metal ores are anthropogenic activities that mobilize arsenic in the environment

Biological Approaches to Controlling Pollutants. https://doi.org/10.1016/B978-0-12-824316-9.00007-0

(Paul et al., 2015). Arsenic is a redox-sensitive element, it readily changes its oxidation state to elemental (As^0), arsine (As^{-3}), arsenite (As^{+3}), or arsenate (As^{+5}), with the last two being the prominent forms, in soil and water (Herath et al., 2016). Furthermore, arsenic can also exist as both organic and inorganic compounds. Arsenic toxicity depends on its chemical form and oxidative state. Inorganic arsenic species are more potent to living organisms than organic species are (Santra et al., 2013). Varying concentrations of arsenic have been recorded in different environmental media as shown in Table 6.1.

Arsenic is carcinogenic in nature and has been dubbed a human carcinogen (Group 1) since the 1980s by the International Agency for Research on Cancer (IARC, 1987). Arsenic contamination in soil and water resources has been reported in numerous countries globally including China, India, Bangladesh, Chile, the United States, Pakistan, Australia, and Taiwan (Lim et al., 2014; Herath et al., 2016; Malik et al., 2020). Arsenic poisoning in humans is manifested as arsenicosis,

Table 6.1 Status of arsenic contamination in various environmental media as reported in several countries.

Environmental media	Arsenic source	Country	Arsenic concentrations	References
Soil (mg/kg)	Paddy soil	Pakistan	12	Javed et al. (2020)
	Loamy soil	Germany	1394	Bauer and Blodau (2006)
	Sediments	India	14.1	Bhattacharya et al. (2009)
	Paddy soil	Japan	38.2	Hasanuzzaman et al. (2014)
	Rice Vegetables			
Water (µg/L)	Groundwater	Pakistan (Lahore)	386	Malik et al. (2020)
	Irrigation water	Pakistan (River Ravi)	65	Javed et al. (2020)
	Groundwater	Taiwan	672	Maity et al. (2011)
	Groundwater	Argentina (La Pampa)	4800	
	Lake water	USA (Tulare Lake)	2600	Cutler et al. (2013)
	Surface water	Vietnam (Red River Delta and Mekong River)	3050	Rahman et al. (2009)

a condition characterized by the appearance of black marks on the skin. If untreated, this can lead to serious ailments such as cancer of vital organs and skin, impairment of foot and legs (disease of blood vessels), gastrointestinal problems, high blood pressure, DNA mutation, and reproductive illnesses (Shankar et al., 2014; Rehman et al., 2020). However, it must be noted that plants exhibit a certain level of tolerance to arsenic. An average arsenic toxicity threshold of 40 mg/kg biomass has been estimated for crop plants and ranges to 200 mg/kg in clayey soil (Paul et al., 2015). Because of their tolerance against arsenic, plants are now considered the most viable option for arsenic remediation. In addition, it has been reported (Satyapal et al., 2018) that various microbes (individually and in association with certain plants) can efficiently remove arsenic from the environment (soil/water). The sections that follow provide a detailed discussion of the biological methods for arsenic remediation, their mechanisms, and advances in the technology.

6.2 Biological methods for arsenic removal

Biological removal methods are used to treat pollutants from environmental media and degrade, remove, or reduce them to less toxic forms. This involves various living organisms such as bacteria, fungi, algae, and plants (Sahkoor et al., 2013; Khoei et al., 2018). The use of a wide range of microorganisms and aquatic and land plants simultaneously not only makes this technique more attractive but also introduces an efficient way to degrade arsenic from the environment (Lim et al., 2014). These biological arsenic removal methods are a cost-friendly and relatively harmless alternative to conventionally used treatment methods for arsenic removal (Paul et al., 2015). However, these methods depend on several factors such as soil type, plant type, microbial strain, depth, quantity, and type (organic, inorganic) of arsenic contamination (Vithanage et al., 2012). The biological methods are primarily divided into two main categories—bioremediation and phytoremediation.

6.2.1 Bioremediation

Bioremediation is a natural process that uses different microorganisms for degradation of environmental contaminants, such as arsenic, lead, and mercury from soil and water (Santra et al., 2013). Microorganisms use arsenic as an energy token in their metabolic activity. Usually, they render it into less bioavailable forms or absorb it within their bodies (Khoei et al., 2018). This technology involves a procedure based on intracellular bioaccumulation, biostabilization, biodegradation, or biovolatilization techniques (Alexander et al., 2016). Different prokaryotic and eukaryotic microorganisms utilize analogous strategies to fight arsenic toxicity. Various studies suggest that numerous indigenous microbial species such as *Lysinibacillus* (Rahman et al., 2014), *Saccharomyces cerevisiae* (Sher et al., 2019), Pseudomonas (Satyapal et al., 2018), Rhodococcus, *Aspergillus niger* (Mukherjee et al., 2010), and Bacillus (Lampis et al., 2015) offer higher capability for arsenic decontamination.

Arsenic entrapment properties of different microbes are attributed to processes such as methylation, redox reaction, adsorption, and complexation from contaminated soil and groundwater (Vithanage et al., 2012; Srivastava and Dwivedi, 2016). Microorganisms living in arsenic-polluted environments develop natural resistance mechanisms by fighting arsenic contamination for survival. They degrade arsenic naturally without human intervention, transforming arsenic from As^{+5} to As^{+3} and As^{+3} to As^{+5} naturally. Their ability to withstand the metal stress condition is well documented in the literature (Chang et al., 2010; Yamamura and Amachi, 2014; Satyapal et al., 2018). For arsenic bioremediation, these naturally resistant strains are isolated based on their tolerance for arsenic pollution and allowed to proliferate in contaminated areas for arsenic remediation (Rahman et al., 2014).

6.2.1.1 Mechanism of arsenic detoxification in microbes

Over time, microorganisms have evolved various arsenic detoxification mechanisms that can be employed for arsenic remediation (Paul et al., 2015). The most common and well-known arsenic degradation mechanism adopted by microorganisms is the oxidation of arsenite to arsenate. The microbes consume arsenic in their metabolic activities as a source for generating energy through arsenite (As^{+3}) oxidation to arsenate (As^{+5}). As^{+5} has stronger adsorption affinity for soil inorganic components, so it results in the immobilization of arsenic in the solid phase (Dhuldhaj et al., 2013; Nguyen, 2015). Further research into the mechanism for arsenic bioremediation has identified arsenic oxidase, a special type of enzyme found in the protoplasm of arsenic-oxidizing microbes, that helps bacteria oxidize As^{+3} to less toxic As^{+5} (Dey et al., 2016).

Reduction of arsenate (As^{+5}) to arsenite (As^{+3}) in aerobic environments is another mechanism for arsenic detoxification (Santra et al., 2013). Arsenate (AsO^{3-}) has structural similarity to phosphate, so the entry of As^{+5} into bacterial cells is made possible by the help of phosphate transporters. The presence of plasmid-encoded detoxifying reductase (arsC enzyme) in their cytoplasm ensures As^{+5} reduction followed by extraction of the reduced form by the assistance of membrane efflux pumps (As^{+3} specific transporter) in their cell walls (Yamamura and Amachi, 2014).

In addition to oxidation/reduction-based mechanisms, another mechanism of arsenic remediation by microbes are based on the transformation of aqueous arsenic into monomethylarsonic acid, dimethylarsine, or trimethylarsine. This process is called biomethylation (Hayat et al., 2017). These compounds are less toxic and volatile in nature, so they released into the atmosphere, this phenomenon is known as biovolatilization (Roy et al., 2015). It usually takes place under nitrogen and phosphate-stress conditions (Vithanage et al., 2012). It involves the activity of S-adenosylmethionine methyltransferase (arsM) enzyme for arsenic remediation especially from the aquatic environment (Lim et al., 2014). Various strains of As and their associated mechanisms are summarized in Table 6.2.

Table 6.2 Different microbial strains bioremediate arsenic (As) with different mechanisms.

Environmental medium	Microbial strains	Mechanism	Arsenic concentrations (mg/g)	References
Bacteria				
Soil	Lysinibacillus sp. (strain B1-CDA)	Intracellular accumulation in bacteria	5.0	Rahman et al. (2014)
	Sulfurospirillum arsenophilum	Arsenate Reduction to Arsenite		Santra et al. (2013)
	Achromobacter sp.	As (V) to MMA and DMA (biomethylation)		Paul et al. (2015)
Water	Klebsiella pneumoniae MNZ6	Oxidize arsenite to arsenate	370 mg/L	Abbas et al. (2014)
	Nocardia sp.	As(V) to TMA (biomethylation)		Paul et al. (2015)
	Pseudomonas	Oxidize arsenite to arsenate		Iqtedar et al. (2019)
	Thermus thermophiles	Oxidize arsenite to arsenate		Paul et al. (2015)
	Pseudomonas stutzeri	Intracellular arsenic accumulation	4.0	Joshi et al. (2008)
	Gram negative	Reduce arsenate to arsenite	125	Iqtedar et al. (2019)
	Bacillus sp. KMO2	Oxidize arsenite to arsenate	4500 mg/L	Dey et al. (2016)
	Aneurinibacillus aneurinilyticus BS-1	Oxidize arsenite to arsenate	550 mg/L	Dey et al. (2016)
	Enterobacter sp. MNZ1	Oxidize arsenite to arsenate	300 mg/L	Abbas et al. (2014)
	Thermus aquaticus	Oxidize arsenite to arsenate		Paul et al. (2015)

Continued

Table 6.2 Different microbial strains bioremediate arsenic (As) with different mechanisms.—*cont'd*

Environmental medium	Microbial strains	Mechanism	Arsenic concentrations (mg/g)	References
Fungi				
	Penicillium sp.	Volatilize the methylated arsenic species	25.8–43.9 µg	Wang and Zhao (2009)
	Aspergillus niger sp.	Intracellular arsenate accumulation in cell	100 mg/L	Mukherjee et al. (2010)
Algae				
	Cladophora sp.	As (V) to MMA and DMA (biomethylation)	6 mg/L	Jasrotia et al. (2015)
Uranium mine waste	*Lemna gibba* (duckweed)	Bioaccumulation (dry mass)	2000 mg/kg	Wang and Zhao (2009)

Abbreviations: DMA, dimethylarsine; MMA, monomethylarsonic acid; TMA, trimethylarsine.

6.2.2 **Advances in bioremediation**

With the advancement of biotechnology, microorganisms have been bioengineered with the desired traits for efficient removal of arsenic from the environment. These traits include overexpression of certain genes encoded for specific proteins and enzymes, the ability to fight arsenic contamination at higher levels, and various other characteristics (Igiri et al., 2018). An example of such a technique is the gram-positive bacterial strain *Corynebacterium glutamicum*. It is genetically modified by the insertion of *ars* operon (*ars*1 and *ars* 2) for biodegradation of arsenic-contaminated soils. These genetic operons encode for metallo-stress regulatory proteins such as arsenite permeases and arsenate reductases (Mateos et al., 2017).

In another study, Chen et al. (2013) engineered the chromosome of *Pseudomonas putida* KT2440 with the arsenite (As^{+3}) (arsM) gene to convert inorganic arsenic to less toxic and more volatile methylated arsine species. Expression of the *ars*M gene evolves *P. putida* with fivefold more resistance to arsenite than the original strain. The activity of this enzyme depends on four cysteine residues at locations 44, 72, 174, and 224 on chromosome CmArsM. Replacement of any of the three may lead to the loss of As (+3) methylation ability. This works both in soil and surface waters (Rosen et al., 2012; Yang and Rosen, 2016).

In a recent advancement, adverse effects of biomethylation have been reported because the end products of biomethylation are genotoxic and can cause DNA impairment (Roy et al., 2015). Furthermore, the reactive nature of methylated arsine compounds made them susceptible to oxidize back to As^{+3}/As^{+5} and mobilize in soil and water. Therefore, it was concluded that biovolatilization can be utilized only as an ex situ method for arsenic remediation under controlled parameters (Wang and Zhao, 2009; Rosen et al., 2012). Furthermore, a study by Yang and Rosen (2016) reports that organic and methylated arsenical compounds can be activated by reduction and sometimes degrade into more lethal inorganic arsenic that can pollute water and soil resources.

In terms of the mechanism of bioremediation, the following advances have been reported:

6.2.2.1 *Bioremediation by biofilms*

The role of biofilms in bioremediation has been recently explored. Dey et al. (2016) reported that the microbial community at times develops protection against contaminants by creating microbial biofilms. Biofilms can remediate arsenic with a high tolerance against its toxic inorganic form even at a quantity that is potent (Igiri et al., 2018).

6.2.2.2 *Arsenic resistance mechanism controlled by genes*

Studies based on the microbial resistance mechanism for arsenic have explored the role of *ars* operons having the genes arsRBC or that are plasmid related. The *ars* genetic system made them capable of utilizing the reduced inorganic (arsenite) and oxidized inorganic (arsenate) forms within their metabolism, ultimately enabling them to resist arsenic toxicity. There are various purposed proteins, encoded with different genes—for example, *ars*B is a carrier protein that assists in arsenic extrusion from the cell. Bacteria that contain both *ars*A and *ars*B genes simultaneously will have significantly more arsenic-resistant capacity (Dey et al., 2016). Table 6.3 reports literature on the *ars* genetic system.

6.2.3 Phytoremediation

Phytoremediation is a biological treatment method in which plants, their associated microrhizoidal communities, and enzymes fight against external pollutants, including arsenic, by removing, degrading, or stabilizing them into harmless forms and reclaiming the contaminated environments (Tangahu et al., 2011; Dhuldhaj et al., 2013). It is nonintrusive, aesthetically soothing, and a socially recognized technology to biodegrade contaminated environments. This type of treatment method can be employed to remediate contaminants from water (surface, municipal, and industrial) and soil (Hasanuzzaman et al., 2014). Based on plant type and contaminant fate, phytoremediation has five main categorical mechanisms: phytoextraction, phytostabilization/phytoimmobilization, phytovolatilization, phytodegradation, and rhizofiltration (Vithanage et al., 2012; Dhuldhaj et al., 2013; Yadav et al., 2018). Studies have indicated three main types of arsenic-resistant plants: tolerant, accumulator, and hyperaccumulator. They can resist and grow in moderate to high levels of arsenic without showing any toxic effects. Hyperaccumulators have an exceptional ability to accumulate metals (Roy et al., 2015).

Table 6.3 Different genes of *ars* genetic system encoded for various functions in arsenic resistance or arsenic resistance (*ars*) genes.

ars genes	Functions	References
*ars*B	Encodes for As^{+3} efflux permease enzyme	Yamamura and Amachi (2014)
ACR3	Functions to extrude As^{+3} from cells	Yang and Rosen (2016)
*ars*A	Encodes for As^{+3} ATPase	Dey et al. (2016)
*ars*D	Encodes for As^{+3} metallochaperone	Roy et al. (2015)
*ars*C	Encodes for As^{+5} reductase enzyme	Paul et al. (2015)
*ars*M	Encodes for As^{+3} methyltransferase enzyme	Chen et al. (2013)
*aox*B	Encodes for As^{+3} oxidase enzyme	Hayat et al. (2017)
*ars*I	Encodes for C—As bond lyase activity	Yang and Rosen (2016)
*ars*H	Encodes for methylarsenite oxidase	Paul et al. (2015)
*ars*R	Encodes for conversion of inorganic methylated As to volatile trimethylarsine (regulatory protein)	Hayat et al. (2017)

6.2.3.1 Mechanism of arsenic detoxification in plants

Plants are delicate and sessile living organisms, but they have very strong defense mechanism that helps them survive different environments through various detoxification strategies. Phytostabilization is one such method, in which root exudates instinctively alter the arsenic electronic state to a less bioavailable and less toxic form (Vithanage et al., 2012, Yadav et al., 2018). Arsenic-tolerant plants use phytostabilization to avert soil erosion and pollutant leaching into groundwater from mine tailings (Paul et al., 2015). Phytoimmobilization is a similar process in which arsenic mobility and bioavailability are reduced to minimum by augmenting the soil conditions that cause metals to precipitate (insoluble compounds) or adsorb on roots, which eventually decreases arsenic mobilization (Dhuldhaj et al., 2013). Phytovolatilization, on the other hand, is a natural process in which plants absorb and subsequently volatilize or transpire arsenic compounds after modifications in the external environment. It can phytoremediate arsenic from almost all mediums including groundwater, sludge, soil, and sediments (Tangahu et al., 2011).

Methods like phytoextraction and phytoaccumulation involve plants' natural abilities to extract, transform, and accumulate arsenic from roots to the aerial parts of plants that is later harvested for safe disposal. They can be employed using tolerant, accumulator, or hyperaccumulator plants. Plants having high biomass can phytoextract and accumulate arsenic contamination (Dhuldhaj et al., 2013). Studies have shown that phytoaccumulation can be achieved by the expression of metal transporters, pumps, alterations in enzymatic activities, changes in the redox states of arsenic, and more importantly, the generation of intermediary moieties (products of metabolism) (Yadav et al., 2018). However, it must be noted that not all plants accumulate arsenic. Arsenic is mostly stored in leaves and seeds (<1 mg/kg) (Nguyen, 2015). Organoarsenic compounds are usually more efficiently translocated from root to shoot and then accumulate in plant tissues (Chen et al., 2013).

Phytotransformation/phytodegradation is another mechanism and involves the uptake of arsenic and its subsequent disintegration through metabolic activities. It is usually achieved by compounds such as enzymes and phytochelatins (Tangahu et al., 2011). It involves the breakdown of complex organic compounds into simpler, stable ones (Shakoor et al., 2013). Another important phytoremediation mechanism is based on a robust root system, i.e., rhizofiltration. In this process, plant roots uptake the arsenic contaminants and growth nutrients through adsorption or precipitation into the plant roots. It can also cause the plant to absorb arsenical compounds from the nearby root zone into the roots, as in constructed wetlands (Shakoor et al., 2013; Yadav et al., 2018). Various plant species and mechanisms are summarized in Table 6.4.

6.2.4 Advances in phytoremediation

The role of aquatic plants is explored in more recent studies employing the phytoremediation of arsenic in water. Normally, two main approaches are adopted in this

Table 6.4 Various plant species with different mechanisms to phytoremediate arsenic (As).

Medium	Plant species	Mechanism	Arsenic concentrations (mg/g)	Accumulation compartment	References
Soil	Pteris vittata	Phytoextraction	157 mg/kg	Shoots	Yang et al. (2017)
Water	Eichhornia crassipes (water hyacinth)	Intracellular Phytoaccumulation	2 mg/L	–	Jasrotia et al. (2015)
Mining Waste	Epilobium dodonaei Vill.	Phytostabilization	4–5 mg/kg	Reduction in shoots	Randelović et al. (2016)
Water	Warnstorfia fluitans	Rhizofiltration	74 µg/L	Aerial parts and roots	Sandhi et al. (2018)
	Pseudomonas	Oxidize arsenite to arsenate			Paul et al. (2015)
	Bacillus sp. KM02	Oxidize arsenite to arsenate	4500 mg/L		Iqtedar et al. (2019) Dey et al. (2016)

regard. The first utilizes free-floating plants. These can absorb arsenic contaminants within their bodies by their detoxifying mechanism. These plants are planted purposefully so that later, after the completion of a task/target, they can be removed from the pond (Lim et al., 2014). The second uses aquatic submerged rooted plants to eliminate arsenical compounds from water, sludge, and bed filters (Yadav et al., 2018). Submerged plants uptake arsenic more efficiently than terrestrial emerged plants, as they directly uptake by their whole body, i.e., it has been observed in the case of *Warnstorfia fluitans*, an aquatic moss plant that directly uptakes through the thallus and accumulates higher arsenic concentrations (Sandhi et al., 2018).

Ferns are the most exceptional plant species that hyperaccumulate arsenic from contaminated soil. They are considered ideal for phytoremediation (Jasrotia et al., 2015). Ferns grow and accumulate arsenic more successfully in arsenic-contaminated soil than on noncontaminated soil (Alexander et al., 2016). Yang and his coworkers in 2017 achieved very promising results when they coplanted *Pteris vittata*, the most hyperaccumulator fern species of arsenic with the castor bean plant *Ricinus communis*. *Pteris vittata* individually can accumulate 157 mg/kg of arsenic in its shoot parts, but when its coplanted results are observed, they are found to be more significant, i.e., 221 mg/kg of accumulated arsenic.

6.3 Conclusion

Arsenic (As), an omnipresent metalloid, is carcinogenic in nature. Arsenic poisoning, as arsenicosis, has also been reported in various parts of the world. To remediate arsenic, chemical, physical, and biological methods are used. Among all the methods, there is an emphasis on biological removal, as they are less harmful to the environment. The biological methods mainly include bioremediation and phytoremediation. They are both cost-friendly and relatively harmless alternatives to conventional chemical methods for arsenic removal. These methods employ microorganisms or plants to degrade, remove, or reduce arsenic into less toxic forms. Microorganisms living in the arsenic-polluted environments develop their own natural resistance mechanism by fighting arsenic contamination for survival. They degrade arsenic naturally without human intervention, transforming arsenic from As^{+5} to As^{+3} and As^{+3} to As^{+5} naturally. These naturally resistant strains are then isolated based on their tolerance for arsenic pollution and can proliferate in contaminated areas for arsenic remediation. Plants and microorganisms adopt various arsenic detoxification mechanisms. With advancements in biotechnology, they are bioengineered with the desired traits for efficient removal of arsenic from the environment. These traits include overexpression of certain genes encoded for specific proteins and enzymes, the ability to fight arsenic contamination at higher levels, and various other characteristics.

References

Abbas, S.Z., Raiz, M., Ramzan, N., Zahid, M.T., Shakoori, F.R., Rafatullah, M., 2014. Isolation and characterization of arsenic-resistant bacteria from wastewater. Braz. J. Microbiol. 45 (4), 1309–1315.

Alexander, T.C., Gulledge, E., Han, F., 2016. Arsenic occurrence, ecotoxicity and its potential remediation. J. Biorem. Biodegrad. 7 (4), 174–176. https://doi.org/10.4172/2155-6199.1000e174.

Bauer, M., Blodau, C., 2006. Mobilization of arsenic by dissolved organic matter from iron oxides, soils and sediments. Sci. Total Environ. 354, 179–190.

Bhattacharya, P., Samal, A.C., Majumdar, J., Santra, S.C., 2009. Transfer of arsenic from groundwater and paddy soil to rice plant (*Oryza sativa* L.): a micro level study in West Bengal, India. World J. Agric. Sci. 5 (4), 425–431.

Chang, J.S., Yoon, I.H., Lee, J.H., Kim, K.R., An, J., Kim, K.W., 2010. Arsenic detoxification potential of aox genes in arsenite-oxidizing bacteria isolated from natural and constructed wetlands in the Republic of Korea. Environ. Geochem. Health 32 (2), 95–105.

Chen, J., Qin, J., Zhu, Y.G., Lorenzo, V., Rosen, B.P., 2013. Engineering the soil bacterium *Pseudomonas putida* for arsenic methylation. Appl. Environ. Microbiol. 79 (14), 4493–4495.

Cutler, W.G., Brewer, R.C., El-Kadi, A., 2013. Bioaccessible arsenic in soils of former sugarcane plantations, Island of Hawaii. Sci. Total Environ. 442, 177–188.

Dey, U., Chatterjee, S., Mondal, N.K., 2016. Isolation and characterization of arsenic-resistant bacteria and possible application in bioremediation. Biotechnol. Reports 10, 1–7.

Dhuldhaj, U., Yadav, I.C., Singh, S., Sharma, N.K., 2013. Microbial interactions in the arsenic cycle: adoptive strategies and applications in environmental management. In: Reviews of Environmental Contamination and Toxicology, vol. 224, pp. 1–38.

Duarte, A.A., Cardoso, S.J.A., Alcada, A.J., 2009. Emerging and innovative techniques for arsenic removal applied to a small water supply system. Sustainability 1, 1288–1304.

Hasanuzzaman, M., Nahar, K., Hakeem, K.R., Ozturk, M., Fujita, M., 2014. Arsenic toxicity in plants and possible remediation. In: Hakeem, K.R., Sabir, M., Ozturk, M., Murmut, A. (Eds.), Soil Remediation and Plants, Edition: 1st. Publisher: Academic press, Elsevier, pp. 433–501.

Hayat, K., Menhas, S., Bundschuh, J., Chaudhary, H.J., 2017. Microbial biotechnology as an emerging industrial wastewater treatment process for arsenic mitigation: a critical review. J. Clean. Prod. 151, 427–438.

Herath, I., Vithanage, M., Bundschuh, J., Maity, J.P., Bhattacharya, P., 2016. Natural arsenic in global groundwaters: distribution and geochemical triggers for mobilization. Curr. Pollut. Rep. 2, 68–89.

IARC, 1987. Summaries & evaluations: Arsenic and arsenic compounds (Group 1). International Agency for Research on Cancer, Lyon, p. 100 (IARC Monographs on the Evaluation of Carcinogenic Risks to Humans, Supplement 7. http://www.inchem.org/documents/iarc/suppl7/arsenic.html.

Igiri, B.E., Okoduwa, S.I.R., Idoko, G.O., Akabuogu, E.P., Adeyi, A.O., Ejiogu, I.K., 2018. Toxicity and bioremediation of heavy metals contaminated ecosystem from tannery wastewater: a review. J. Toxicol. 2018 (9), 2568038.

Iqtedar, M., Aftab, F., Asim, R., Abdullah, R., Kaleem, A., Saleem, F., Aehtisham, A., 2019. Screening and optimization of arsenic degrading bacteria and their potential role in heavy metal bioremediation. Biosci. J. 35 (4), 1237–1244.

Jasrotia, S., Kansal, A., Mehra, A., 2015. Performance of aquatic plant species for phytoremediation of arsenic-contaminated water. Appl. Water Sci. 7 (2), 889–896.

Javed, A., Farooqi, A., Baig, Z., Ellis, T., van Geen, A., 2020. Soil arsenic but not rice arsenic increasing with arsenic in irrigation water in the Punjab plains of Pakistan. Plant Soil 450, 601–611. https://doi.org/10.1007/s11104-020-04518-z.

Joshi, D.N., Patel, J.S., Flora, S.J.S., Kalia, K., 2008. Arsenic accumulation by *Pseudomonas stutzeri* and its response to some thiol chelators. Environ. Health Prev. Med. 13, 257–263.

Khoei, A.J., Joogh, N.J.G., Darvishi, P., Rezaei, K., 2018. Application of physical and biological methods to remove heavy metal, arsenic and pesticides, Malathion and Diazinon from water. Turk. J. Fish. Aquat. Sci. 19 (1), 21–28.

Lampis, S., Santi, C., Ciurli, A., Anderolli, M., Vallini, G., 2015. Promotion of arsenic phytoextraction efficiency in the fern *Pteris vittata* by the inoculation of As-resistant bacteria: a soil bioremediation perspective. Front. Plant Sci. 6, 1–12.

Lim, K.T., Shukor, M.Y., Wasoh, H., 2014. Physical, chemical and biological methods for the removal of arsenic compounds. BioMed Res. Int. 2014, 1–9.

Maity, J.P., Nath, B., Chen, C.Y., Bhattacharya, P., Sracek, O., Bundschuh, J., 2011. Arsenic-enriched groundwaters of India, Bangladesh and Taiwan comparison of hydrochemical characteristics and mobility constraints. J. Environ. Sci. Health 46, 1163–1176.

Malik, A., Parvaiz, A., Mushtaq, N., Hussain, I., Javed, T., Rehman, H., Farooqi, A., 2020. Characterization and role of derived dissolved organic matter on arsenic mobilization in alluvial aquifers of Punjab, Pakistan. Chemosphere 251, 126374.

Mateos, L.M., Villadangos, A.F., de la Rubia, A.G., Mourenza, A., Pascual, L.M., Letek, M., Pedre, B., Messens, J., Gill, J.A., 2017. The arsenic detoxification system in Corynebacteria: basis and application for bioremediation and redox control. Adv. Appl. Microbiol. 99, 103–137.

Mukherjee, A., Das, D., Mondal, S.K., Biswas, R., Das, T.K., Boujedaini, N., Khuda-Bukhsh, A.R., 2010. Tolerance of arsenate-induced stress in *Aspergillus niger*, a possible candidate for bioremediation. Ecotoxicol. Environ. Saf. 73, 172–182.

Nguyen, H., 2015. Bioremediation of arsenic toxicity. In: Narayan, C. (Ed.), Arsenic Toxicity: Prevention and Treatment, Edition: 1st. Publisher: CRC Press, pp. 155–165.

Paul, S., Chakraborty, S., Ali, M.N., Ray, D.P., 2015. Arsenic distribution in environment and its bioremediation: a review. Int. J. Agric. Environ. Biotechnol. 8 (1), 189–204.

Rahman, M.M., Naidu, R., Bhattacharya, P., 2009. Arsenic contamination in groundwater in the Southeast Asia region. Environ. Geochem. Health 31, 9–21.

Rahman, A., Nahar, N., Nawani, N.N., Jass, J., Desale, P., Kapadnis, B.P., Hossain, K., Saha, A.K., Ghosh, S., Olsson, B., Mandal, A., 2014. Isolation and characterization of a Lysinibacillus strain B1-CDA showing potential for bioremediation of arsenics from contaminated water. J. Environ. Sci. Health A 49, 1349–1360.

Randelovic, D., Gajic, G., Mutic, J., Pavlovic, P., Mihailovic, N., Jovanovic, S., 2016. Ecological potential of *Epilobium dodonaei* Vill. for restoration of metalliferous mine wastes. Ecol. Eng. 95, 800–810.

Rehman, M.Y.A., van Herwijnen, M., Krauskopf, J., Farooqi, A., Kleinjans, J.C.S., Malik, R.N., Briede, J.J., 2020. Transcriptome responses in blood reveal distinct biological pathways associated with arsenic exposure through drinking water in rural settings of Punjab, Pakistan. Environ. Int. 135, 105403.

Rosen, B., Marapakala, K., Abdul Salam, A.A., Packianathan, C., Yoshinaga, M., 2012. Pathways of arsenic biotransformations: the arsenic methylation cycle. In: Understanding the Geological and Medical Interface of Arsenic, as 2012 — 4th International Congress: Arsenic in the Environment, pp. 185—188.

Roy, M., Giri, A.K., Dutta, S., Mukherjee, P., 2015. Integrated phytobial remediation for suitable management of arsenic in soil and water. Environ. Int. 75, 180—198.

Sahoo, K.P., Kim, K., 2013. A review of the arsenic concentration in paddy rice from the perspective of geoscience. Geosci. J. 17 (1), 107—122.

Sandhi, A., Landberg, T., Greger, M., 2018. Phytofiltration of arsenic by aquatic moss (*Warnstorfia fluitans*). Environ. Pollut. 237, 1098—1105.

Santra, S.C., Samal, A.C., Bhattacharya, P., Banerjee, S., Biswas, A., Majumdar, J., 2013. Arsenic in foodchain and community health risk: a study in Gangetic West Bengal. Proc. Environ. Sci. 18, 2—13.

Satyapal, G.K., Mishra, S.K., Srivastava, A., Ranjan, R.K., Prakash, K., Haque, R., Kumar, N., 2018. Possible bioremediation of arsenic toxicity by isolating indigenous bacteria from the middle Gangetic plain of Bihar, India. Biotechnol. Rep. 17, 117—125.

Shakoor, M.B., Ali, S., Farid, M., Farooq, M.A., Tauqeer, H.M., Iftikhar, U., Hannan, F., Bharwana, S.A., 2013. Heavy metal pollution, a global problem and its remediation by chemically enhanced phytoremediation: a review. J. Biodivers. Environ. Sci. (JBES) 3 (3), 12—20.

Shankar, S., Shanker, U., Shikha, 2014. Arsenic contamination of groundwater: a review of sources, prevalence, health risks and strategies for mitigation. Sci. World J. 2014, 1—18.

Sher, S., Rehman, A., 2019. Use of heavy metals resistant bacteria—a strategy for arsenic bioremediation. Appl. Microbiol. Biotechnol. 103, 6007—6021.

Srivastava, S., Dwivedi, A.K., 2016. Biological wastes the tool for biosorption of arsenic. J. Biorem. Biodegrad. 7 (1), 323—325. https://doi.org/10.4172/2155-6199.1000323.

Tangahu, B. V., Abdullah, S. R. S., Basri, H., Idris, M., Anuar, N. and Mukhlisin, M. (2011). A review on heavy metals (as, Pb and Hg) uptake by plants through phytoremediation. Int. J. Chem. Eng., 939161:1-31.

Vithanage, M., Dabrowska, B.B., Mukherjee, A.B., Sandhi, A., Bhattacharya, P., 2012. Arsenic uptake by plants and possible phytoremediation applications: a brief overview. Environ. Chem. Lett. 10 (3), 217—224.

Wang, S., Zhao, X., 2009. On the potential of biological treatment for arsenic contaminated soils and groundwater. J. Environ. Manag. 90, 2367—2376.

Yadav, K.K., Gupta, N., Kumar, A., Reece, L.M., Singh, N., Rezania, S., Khan, S.A., 2018. Mechanistic understanding and holistic approach of phytoremediation: a review on application and future prospects. Ecol. Eng. 120, 274—298.

Yamamura, S., Amachi, S., 2014. Microbiology of inorganic arsenic: from metabolism to bioremediation. J. Biosci. Bioeng. 118 (1), 1—9.

Yang, H.-C., Rosen, B.P., 2016. New mechanisms of bacterial arsenic resistance. Biomed. J. 39, 5—13.

Yang, J., Yang, J., Huang, J., 2017. Role of co-planting and chitosan in phytoextraction of as and heavy metals by *Pteris vittata* and castor bean — a field case. Ecol. Eng. 109, 35—40.

Advances in biodegradation and bioremediation of emerging contaminants in the environment

Jafar Ali[1,2], Mahwish Ali[3], Ibrar Khan[4], Abeer Khan[1], Zainab Rafique[1], Hassan Waseem[1]

[1]*Department of Biotechnology, University of Sialkot, Sialkot, Punjab, Pakistan;* [2]*Research Center for Eco-environmental Sciences, Chinese Academy of Sciences, Haidian, Beijing, PR China;* [3]*Department of Biological Sciences, National University of Medical Sciences (NUMS), Rawalpindi, Punjab, Pakisan;* [4]*Department of Microbiology, Abbottabad University of Science & Technology, Havelian, Khyber Pakhtunkhwa, Pakistan*

Chapter outline

7.1 Introduction

Emerging contaminants (ECs) are mainly synthetic organic chemicals that have no regulatory standards because sufficient information about the effects of chronic exposure is lacking (Waseem et al., 2020a). Recently there has been an upsurge in research on the mitigation of ECs due to either increased awareness of their potential risks to humans or their recent detection in environmental matrices through advanced analytical techniques. Removal of ECs from the environment is essential for attaining sustainable development goals. Current wastewater treatment strategies are ineffective at removing ECs, and hence, there is a vital need for the development of cost-effective and efficient treatment systems that can be applied to a range of scales and types of wastes. Although physicochemical methods are commonly used to remediate ECs (Wang et al., 2018), these techniques are becoming obsolete because they generate toxic by-products and have high costs and low sustainability. Bioremediation is the most promising and innovative technology; it utilizes living microorganisms to degrade environmental pollutants or prevent pollution.

Bioremediation-mediated removal of ECs offers several advantages such as high removal efficiency and low costs. Numerous biodegradation approaches have been applied for wastewater treatment during the last few decades. These conventional strategies are unable to completely remove ECs and recalcitrant contaminants. Thus, advancement of current methodologies is a prerequisite for complete removal of ECs and to achieve sustainable goals. In this chapter, we summarize advances in the biodegradation and bioremediation methods for removing ECs from the environment.

7.2 Constructed wetlands

Constructed wetlands (CWs) are human-designed systems for the treatment of wastewater. CWs consist of a shallow basin or beds that are planted with emergent plant species or floating plants. They provide an economical and environmentally friendly treatment compared with other technologies. CWs are mainly divided into three categories depending on the direction of water flow in these ponds. They are classified as vertical flow constructed wetlands (VF-CWs) when water flows from top to bottom within the beds and provides a relatively aerobic environment in the beds. On the other hand, CWs in which water flows from left to right are classified as horizontal flow constructed wetlands (HF-CWs) and provide an anaerobic environment in the beds. CWs with only a shallow basin planted with floating plants are termed free water surface constructed wetlands and have anoxic conditions (Vymazal, 2010).

According to the CW configuration, different types of environments are of primary interest for the biodegradation of emerging pollutants in wastewater. Although limited information is present in the literature about the biodegradation of ECs through CWs, this portion reviews studies on the biodegradation of ECs in CWs.

7.2.1 Pharmaceutical (nonsteroidal antiinflammatory) drugs and personal care products

Nonsteroidal antiinflammatory drugs are a potent source of ECs in waste streams (Żur et al., 2018). Human beings frequently use these drugs, so they have a high frequency of detection in urban wastewater. These analgesics are negatively charged compounds, and they have less sorption capacity for sewage; that is why they can be readily biodegraded in CWs (Carballa et al., 2004). Various studies have presented the use of constructed wetlands for the removal of antibiotics. In many studies, all three types of CWs showed better removal of analgesic drugs than with other remediation techniques. The removal efficiencies of some analgesic drugs evaluated in various CWs are presented in Table 7.1.

On the other hand, personal care products (PCPs) are one of the most critical ECs. They can reside in wastewater and become a source of illness for humankind. Alvia and coworkers studied the biodegradation of three kinds of PCPs in the three types of constructed wetlands (Ávila et al., 2015). These PCPs included tonalide, oxybenzone, and triclosan. They reported the maximum removal of triclosan compound in VF-CWs due to the prevailing aerobic conditions and intermittent flow of the wastewater, which makes the system unsaturated compared with HF-CWs, where there are anaerobic conditions and saturation of the wastewater as shown in Table 7.1. Conversely, tonalide was frequently degraded in HF-CWs compared with the VF-CW system, while oxybenzone concentrations were found below the detection limit in the influents and effluents of CWs. This means that different prevailing conditions in CWs contributed differently toward the degradation of the compounds.

Table 7.1 Summary of percentage removal of analgesic drugs and personal care products in different types of constructed wetlands.

Type of contaminant	Type of wetland			Removal efficiency (%)			References
Pharmaceutical drugs	VF-CW (a)	HF-CW (b)	FWS (c)	(a)	(b)	(c)	
Salicylic acid	✓	✓	—	96	98	—	Carballa et al. (2004)
Ibuprofen	✓	✓	✓	71	99	96	Carballa et al. (2004)
OH-ibuprofen	✓	✓	✓	62	99	33	Buser et al. (1999)
Diclofenac	✓	✓	✓	15	73	96	Bernhard et al. (2006)
Naproxen		✓			50		Matamoros et al. (2017)
Carbamazepine		✓			30		Matamoros et al. (2017)
Personal care products							
Tonalide	✓	✓	✓	27	54	54	Ávila et al. (2015)
Triclosan	✓	✓	✓	61	1	20	Ávila et al. (2015)

7.2.2 **Pesticides**

Pesticides are used in modern agricultural practices to increase crop yields. The excessive use of these pesticides poses a severe threat to aquatic life. Pesticides enter water streams through point or diffused sources. Diffused sources are a significant point of concern and include leaching, runoff with rainwater, and drainage (Zhang and Zhang, 2011). Recently, CWs have been used for the removal of pesticides. They effectively removed pesticides through physical, chemical, and biological and biochemical processes (Imfeld et al., 2009). According to the data available in the literature, it seems that free water surface (FWS) CWs are quite useful in the removal of pesticides from water runoff. There are many studies on the removal of pesticides in CWs; some examples from the last 10 years are presented in Table 7.2.

7.2.3 **Surfactants**

Surfactants are new ECs of concern nowadays. It has been estimated that about 15 million tons of the surfactants that are produced go directly into wastewater streams, as surfactants are easily soluble in water without changing their fate (Ramprasad and Philip, 2016). Surfactants are reported as toxic compounds for humans and mammals, and long-term exposure to surfactants may lead to the development of cancer. Although little literature is available on their treatment through CWs, the scientific community is now devoting their attention to treating these new ECs through CWs. In greywater, a high level of detergents is always present in both ionic and nonionic forms. Sodium dodecyl sulfate is present as an anionic form, propylene glycol is a nonionic surfactant, and trimethylamine is a cationic surfactant, and combined, they make up the major constituents of surfactants in wastewater. Few studies have been done until now on CWs in the treatment of surfactants; some of these are presented in Table 7.3.

7.2.4 **Hormones**

CWs prove to be efficient in the removal of naturally occurring hormones in wastewater. Estradiol 17α, Estradiol 17β, estrone and testosterone are mostly present in wastewater. The estrogen concentration may vary from few micrograms to milligrams in wastewater. However, they harm the reproductive system of fish, even at their lowest concentration (Khanal et al., 2006). The literature indicates the removal of these hormones through CWs. It seems that hormones are removed by the mechanism of sorption by organic matter in the waste stream and secondly due to the interaction with biofilms. Retention time in CWs plays an essential role in the removal of hormones as long retention time give more time for the interaction between hormones and organic matter, which ultimately increase the capacity of CWs to remove these hormones. Few studies are given in the literature for the removal of hormones through CWs are given in Table 7.4.

Table 7.2 Summary of percentage removal of pesticides in different types of constructed wetlands.

Type of pesticide	Type of constructed wetland	Removal efficiency (%)	References
Dicamba, dimethoate, trifloxystrobin, metamitron, tebuconazole	FWS	—	Elsaesser et al. (2011)
AMPA, azoxystrobin, cymoxanil, cyprodinil, carbendazim, dimethomorph, diuron, flufenoxuron, gluphosinate, glyphosate, isoxaben, kresoxim methyl, metalaxyl, pyrimethanil, simazine, terbuthylazine, tetraconazole	FWS/HF	—	Maillard et al. (2011)
Atrazine, diazinon, permethrin	FWS	—	Moore et al. (2013)
Isoproturon, metazachlor, S-metolachlor, chlorotoluron, iprodione, azoxystrobin, tebuconazole, napropamide, epoxiconazole, prosulfocarb, pendimethalin, diflufenican, aclonifen	FWS	—	Tournebize et al. (2013)
Indoxacarb, tebuconazole, thiacloprid, trifloxystrobin	FWS	—	Elsaesser et al. (2013)
Metolachlor, alachlor, acetochlor	VF-CW	—	Elsayed et al. (2014)
λ-Cyhalothrin, imidacloprid	FWS	—	Mahabali and Spanoghe (2014)
Terbuthylazine	HF-CW	73	Gikas et al. (2018)
Chlorpyrifos	VF-CW	86	Tang et al. (2019)

FWS, *free water surface;* HF, *horizontal flow;* HF-CW, *horizontal flow constructed wetland;* VF-CW, *vertical flow constructed wetland.*

7.2.5 **Antibiotic-resistant genes**

Antibiotic resistance is becoming a severe aspect of environmental risks, as pathogenic microorganisms are becoming more resistant nowadays. Antibiotic resistance not only is associated with hospitals but also is present in wastewater, surface water, and other aquatic bodies (Waseem et al., 2017, 2019). Domestic wastewater treatment plants serve as a potential reservoirs for antibiotic-resistant genes (ARGs) and act as a significant route for introducing them from human excreta into the ecosystem (Novo and Manaia, 2010; Servais and Passerat, 2009). Minimal information is available for the prevalence of antibiotic resistance among pathogenic and fecal indicators in CWs. In recent years, various studies have reported the use of CWs in treating micropollutants, especially ARGs (Cheng et al., 2016; Liu et al., 2013).

Table 7.3 Summary of percentage removal of surfactants in different types of constructed wetlands.

Type of surfactant	Type of wetland			Removal efficiency (%)			References
	VF-CW(a)	HF-CW (b)	FWS (c)	(a)	(b)	(c)	
Sodium dodecyl sulfate	✓	✓	—	89	85	—	Ramprasad and Philip (2016)
Propylene glycol	✓	✓	—	95	90	—	Ramprasad and Philip (2016)
Trimethylamine	✓	✓	—	98	95	—	Ramprasad and Philip (2016)
Linear alkylbenzene sulfonate	✓	✓		95	95	—	Thomas et al. (2017)

FWS, *free water surface;* HF-CW, *horizontal flow constructed wetland;* VF-CW, *vertical flow constructed wetland.*

Table 7.4 Summary of percentage removal of hormones in different types of constructed wetlands.

Type of hormone	Type of wetland			Removal efficiency (%)			References
	VF-CW (a)	HF-CW (b)	FWS (c)	(a)	(b)	(c)	
Estradiol 17β	—	—	✓			36	Gray and Sedlak (2005)
Estradiol 17α	—	—	✓			41	Gray and Sedlak (2005)
Estrogens	✓	✓	✓	90	90	83–93	(Masi et al., 2005) (Shappell et al., 2007)

Liu et al. (2013) established a study and found that the three *tet* genes were 50% removed through CWs. Various other genes, which include *intl1*, *sul1*, *sul2*, and *qnrA*, also showed a decrease after treatment. Hence, CWs could be a complement option in the treatment of these ARGs. These resistant genes are removed by the process of filtration through the beds. Among all three types of CWs, HF-CWs have proven to be the most efficient in the removal of ARGs, especially sulfonamide-resistant genes, because of the prevailing anaerobic conditions in these systems (Chen et al., 2016) In one study by Chen et al. (2015), 99% removal of ARGs was demonstrated through subsurface flow constructed wetlands. However, very poor reduction of macrolide-resistant genes was observed in CWs (Vacca et al., 2005). Different studies reported different reasons and phenomena for the removal mechanism of ARGs in CWs. In one phenomenon, the removal of ARGs depended on reducing the antibiotic rate in CWs (Vacca et al., 2005). On the other hand, another study suggested that the type of substrate used in the beds of subsurface flow CWs can affect the removal efficiency of ARGs (Liu et al., 2013). However, thus far, research on the elimination of resistant genes by CWs remains inadequate, and more effective studies are needed.

The performance of the three types of wetlands are remarkable for the removal of new ECs. Different wetlands show different removal efficiencies among all ECs. The possible utility of constructed wetlands for mitigating ECs is higher than for other, high-cost technologies. They can be considered in future studies to further evaluate the exact mechanism of removal of ECs through CWs.

7.3 Membrane bioreactors

It is broadly realized that numerous regions on the planet have scant water resources. In these regions, groundwater springs are seen to be in critical condition because of excessive use. That is because in such areas, reuse of wastewater is a typical practice, and skillful specialists have attempted numerous approaches to empower its reuse. Among the newly rising advancements is the utilization of smaller-scale and ultra-filtration membranes as profoundly proficient economically feasible systems for acquiring recycled water of high quality.

Membrane bioreactors (MBRs) consolidate a membrane system with a biological reaction, offering a one-of-a-kind chance to limit the physical space of a biocatalyst, which can be an enzyme, a microorganism, or a plant/animal cell. Because of this broad scope of biological responses, the construction of such an MBR system can be somewhat extraordinary and dependent on the biological processes occurring (Galinha et al., 2018). MBRs have been utilized for both urban and industrial wastewater treatment and recovery (Friha et al., 2014). The MBR innovation has the following advantages over ASP: high-quality effluent, higher volumetric loading rates, shorter hydraulic retention times (HRTs), longer solid retention times (SRTs), less sludge formation, and the potential for synchronous nitrification/denitrification in long SRTs (Mutamim et al., 2013). The incorporation of membranes in the system removes the requirement for supplementary clarifiers. The disposal of these additional clarifiers and activity of the MBR at a shorter HRT brings about essentially diminished plant area necessities. However, the utilization of MBR innovation has disadvantages, including higher energy costs, the need to control membrane fouling issues, and potentially significant expenses of occasional membrane replacement.

The most cited investigation report demonstrates an annual development rate of 13.2% and predicted a global market of $627 million in 2015. MBRs have been employed in over 200 countries. Especially striking is the situation of China and some European countries with usage rates of over 50% and 20%, respectively

(Delgado et al., 2011). There are two basic kinds of membrane bioreactors on the basis of their operations:

(1) Anaerobic membrane bioreactors (AnMBR) have a crucial role in future arrangements for sustainable wastewater treatment and resource recovery because they have no energy escalated oxygen transfer prerequisites and can produce biomethane for renewable energy (Kappell et al., 2018).

(2) Aerobic MBRs, on the other hand, represent a significant choice for wastewater reuse, as they are extremely compact and productive systems for separating suspended and colloidal matter to achieve the highest effluent quality standards for sanitization and clarification. The principal limitation for their broad application is their high energy demand—somewhere in the range of 0.45 to 0.65 kWh m-3 for the highest optimum activity from a demonstration plant as indicated by ongoing investigations (Delgado et al., 2011).

MBRs have shown to be a very adaptable and competent system for wastewater treatment for a massive range of operating conditions. A total of 99 pharmaceuticals and personal care products (PPCPs) were examined in influent, final effluent, and biosolids samples from a wastewater treatment plant employed an MBR. There were huge concentrations of influents such as acetaminophen, caffeine, metformin, 2-hydroxy-ibuprofen, paraxanthine, ibuprofen, and naproxen (10^4-10^5 ng/L). Final effluents contained clarithromycin, metformin, atenolol, carbamazepine, and trimethoprim (>5 00 ng/L) as the highest concentrations. In contrast, triclosan, ciprofloxacin, norfloxacin, triclocarban, metformin, caffeine, ofloxacin, and paraxanthine were found at high concentrations in biosolids ($>10^3$ ng/g dry weight). PPCP removals varied from −34% to >99%, and 23 PPCPs had $\geq$90% removal (Kim et al., 2014).

Five pesticide formulations had active ingredients—azinphos-methyl, chlorpyrifos, diazinon, malathion, and phorate—registered for use in Canada, and as organophosphate insecticides were subjected to treatment by MBR technology. The target active ingredients were introduced to the MBR at ppm level concentrations. The biodegradation of these compounds was analyzed daily using specific ion monitoring gas chromatography-mass spectrometry followed by extraction of the analytes with the use of solid-phase extraction. A considerable amount surmounting to amounts measuring 83%−98% of the target analytes when removed with steady-state concentrations was achieved within 5 days of their introduction (Ghoshdastidar et al., 2012).

ARGs are representative contaminants because of public health concerns about the infectious spread with resistance to common antibiotics. Mixed and unsatisfactory results of ARG removal have been recorded by the demonstration of conventional activated sludge processes. The purpose of this study was to study the impact of an AnMBR on ARG removal when treating municipal primary clarifier effluent at 20°C. AnMBR treatment results produced 3.3 to 3.6 log reduction of ARG and the horizontal gene transfer determinate, *intI*1, copies infiltrate. The total biomass was lessened apparently by membrane treatment, as shown by a decrease in

16S rRNA gene concentration. Microbial community analysis via Illumina sequencing unveiled the putative pathogens' abundance higher up in the membrane filtrate compared with primary effluent, despite that the overall bacterial 16S rRNA gene concentrations had lower filtrate (Kappell et al., 2018).

MBRs are utilized for the treatment of surfactants. Surfactants are commonly present in chemical, machinery, petroleum, metallurgy, and other fields, and linear alkylbenzene sulfonate (LAS) is the central component. This primary component of surfactant is harmful to the ecological environment and causes severe destruction. The removal rates of chemical oxygen demand (COD) and LAS were somewhat stable, at the beginning about as high as 85.49%−93.31% to as low as about 80%. The rate of removal of LAS also declined, when the LAS concentration reached over 175 mg/L, and the COD declined to about 83%, and it kept declining to about 60% when the dosage reached 200 mg/L, showing that the defense of microbes against LAS toxicity also decreased. *Dechloromonas*, *Gemmata*, *Pseudomonas*, and *Zoogloea* can easily make LAS degeneration happen in the system (Ran et al., 2018).

7.4 **Electromicrobiology**

The well-established biological treatment processes usually fail to remove these ECs and may lead to their uncontrolled discharge into the environment (Roccaro, 2018). The resulting accumulations of numerous xenobiotic compounds in soil and water are of great concern because of their carcinogenicity, high toxicity, and ability to bioaccumulate in living organisms (Maculewicz et al., 2020). Development of innovative technologies is highly desired to control the pollution of ECs into the environment. Microbial electrochemical technologies provide sustainable wastewater treatment and energy production (Ali et al., 2018). Microbial electrochemical technology is the use of an electrochemical process using microbes as the catalyst. The microbial electrochemical systems are based on microbe-electrode interactions. Microbial electrochemical systems can be classified into microbial fuel cells (MFCs), microbial desalination cells and microbial electrolysis cells. Generally, the working principle is the same in all MET-based systems (Ali et al., 2019).

A typical MFC consists of an anode chamber and a cathode chamber divided by a proton exchange membrane (PEM) in which anaerobic sludge is used to inoculate the anodic chamber. Electroactive bacteria grow and formulate an electroactive biofilm after the anodic inoculation. During anaerobic respiration of electroactive bacteria, electrons, and protons are released, the electrons flow from anode to cathode via an external circuit, while protons move through the PEM. In the cathode chamber, terminal electron acceptor receives the electrons, and power is generated along with wastewater treatment (Ali et al., 2019). Various configurations of MFCs/MET have been used for degradation of different xenobiotic compounds or ECs such as antiinflammatory drugs, estrogens, antibiotics dyes, pesticides, herbicides, PAHs, and heavy metals with high mineralization rates (Ali et al., 2020a). The removal of such pollutants is more problematic than that of organic wastes. Antibiotics have been frequently detected in the effluents of wastewater treatment plants and receiving water bodies, which shows the inability of conventional treatment systems to metabolize the trace levels of such contaminants (Waseem et al., 2020c).

Long-term exposure to these contaminants could lead to the evolution of antibiotic-resistant bacteria and genes, which pose serious risks for human health (Waseem et al., 2020b). Therefore, effective treatment is needed for the removal of recalcitrant compounds and antibiotics. Hybrid microbial electrochemical technologies have been proposed for enhanced bioremediation of ECs. These hybrid technologies, including constructed wetland, soil-MFC, sediment-MFC, and plant-MFC, can remove the varying degree of environmental pollutants (Ali et al., 2020b). However, various studies have reported significant removal of ECs from wastewater at lab scale (Ghangrekar et al., 2020). For the broad-scale application of MFC-based technologies, their efficiency and durability must be enhanced. Moreover, degradation kinetics can be improved by the augmentation of MFC with external power sources, which is not a cost-effective way.

Advanced oxidation processes have emerged as a promising technology for the treatment of ECs with complete mineralization rates. Higher treatment efficiencies for organic and persistent inorganic compounds have been reported in Fenton based reactions (Cheng et al., 2016). In Situ generated hydrogen peroxide (H_2O_2) during Fenton reactions have increased the importance of advanced oxidation processes for the treatment of ECs. Recently, several studies have proven the concept that cathodic H_2O_2 could be synthesized on the surfaces of carbon materials via electrons generated in MFCs without the use for external power supply (Olvera-Vargas et al., 2018). Thus, MFCs could provide the conditions for advance oxidation reactions and treatment of ECs and recalcitrant contaminants. Further studies must target the enhancement of extracellular electron transfer and reaction kinetics in the cathodic chamber of MFCs.

7.5 Nanotechnology for bioremediation

Nanobioremediation is an emerging technique that characterizes an innovative strategy to forge bioremediation ahead beyond its limitations. It is vastly explored for the reduction of pollutants from the environment by utilizing biosynthetic nanoparticles from biological sources such as microbes which include fungi, bacteria, and plants to eradicate contaminants, for example, organic, inorganic pollutants and heavy metals from the groundwater, wastewater, and soil implying the advantage of nanotechnology (Ali et al., 2017). Nanoparticles or nanomaterials have several advantages in bioremediation which includes the increase in the surface area, providing greater points of contact for numerous materials, thus escalating reactivity and boosting up phytoremediation effectiveness. For bioremediation, phytoremediation technique can be utilized in conjunction with plants. To improve phytoremediation considerably, various strategies have been applied and may include infusion of CeO_2

and ZnO nanoparticles in soil. It has shown to enhance the growth of root and shoot in cucumber (*Cucumis sativus*) plants (Ramírez-García et al., 2018; Zhao et al., 2014).

For nanoparticle synthesis, microbes are showing up as efficacious nano-factories (Jafar Ali et al., 2019). Even at elevated metal ion concentration microbes can exist and grow. One of the significant functions of microorganisms in the remediation of metals is the depletion of metal ions. To counter the toxicity of foreign metal ions or metals existing in the microenvironment, they are sometimes exposed to severe environmental conditions which leads them to evolve particular defense mechanisms. By changing the oxidoreduction state of the metal ions and/or the precipitation of the metals whether intracellularly or extracellularly, the toxicity of metal ions is diminished or eradicated accordingly laying the foundation of the synthesis of nanoparticles (Cameotra and Dhanjal, 2010).

As nanoparticles constitute a larger surface area and contain phylogenetic relation with target compounds, they thus present considerable adsorptive property which proves to be a great advantage of using them. To eliminate heavy metals, metal-based nano-adsorbents are substantially utilized in wastewater treatment as yet. The adsorbents like manganese oxides (Gupta et al., 2011), magnesium oxides (Gao et al., 2008) copper oxides (Goswami et al., 2012), ferric oxides (Feng et al., 2012), cerium oxide (Cao et al., 2010), and silver nanoparticles (Fabrega et al., 2011) are few examples of them. As an adsorbent for the elimination of Cr (VI) from aqueous solutions, nanofibers like polypyrrole-polyaniline (PPy-PANI) are also regarded effective (Theron et al., 2008). For removal of PAHs, which has been vigorously absorbed by soils and are hard to eliminate, Amphiphilic polyurethane nanoparticles have been produced (Tungittiplakorn et al., 2004). Biogenic gold nanomaterials incorporated into a radiation-resistant bacterium known as *Deinococcus radiodurans* R1 (Au-DR) through the process of biomineralization and efficiently eliminated radioactive iodine from an aqueous solution through absorption by AuDR (Choi et al., 2017; Ramírez-García et al., 2018).

Carbon-based nanotube sheets are another type of nanotechnology used for bioremediation; they are utilized as an adsorbent for heavy metals like Co^{2+}, Cu^{2+}, Pb^{2+}, Zn^{2+}, and Cd^{2+} (Tofighy and Mohammadi, 2011). Owing to their small size, nanoparticles can penetrate deeper and hence can cause remediation of wastewater which is almost impossible by prevailing established technologies. Pertaining to their reduced cost, high adsorption property, easy separation, and increased stability, Iron oxides are substantially utilized for industrial wastewater treatment (Ramírez-García et al., 2018). Recently, due to their ultrafine structure and eminent proficiency, superparamagnetic iron oxide nanoparticles (SPION) are utilized for the separation of pollutants from wastewater. In this method, the carriers constitute a polymeric shell possessing functional groups and a magnetic core (FeO, Fe3O4, and Fe2O3) that yields a robust magnetic response (Mody et al., 2010). Zerovalent iron nanoparticles (nZVIs) are widely acknowledged to eliminate various groundwater pollutants such as heavy metals, pesticides, and chlorinated compounds. Even so, it is affected by agglomeration, higher mobility, and settlement problems by the nontarget compounds. However, it has been observed that the general removal competence of entrapped nZVIs for pollutants was equivalent to bare nZVIs (Bezbaruah et al., 2009; Gaur et al., 2014).

The nanoparticles, in conjunction with zerovalent ions, can potentially degrade organic pollutants, for example, chlorpyrifos, atrazine, and molinate. The nanoscale zerovalent iron was produced and substantiated for the eradication of toxic arsenic species termed as As(III) and As(V) in the anoxic aquifer (Kanel et al., 2006). It is also essential to have the dyestuff be degraded as they cause a significant hazard to wastewater, as dyeing materials are released in several tons into the environment increasing the toxicity in water for example by absorbing the oxygen and thus posing hazardous effects on the environment (Ramírez-García et al., 2018) The research has also shown that *N. mucronata* is an accumulator for Pbs and an efficient accumulator for Zn, Cu, and Ni.

For detoxification and bioremediation of soil, aquifer, and wastewater, nanoparticles are attained from those kinds of bacteria, fungi, and plants that are effective in yielding desired results. The increased stability of enzymes is a further advantage of utilizing nanoparticles. It proffers a method to eradicate their susceptibility to mechanical shearing and 3-D structural loss by protease attack, which can be averted with the help of encapsulation of enzymes inside the nanoparticles and can be reused multiple times (Lee et al., 2007; Ramírez-García et al., 2018).

For devising or developing nanobioremediation strategies, research innovation in nanobiotechnology offers certain new realms. The integration of nanobioremediation constitutes the potency to mitigate the aggregated price for the reduction of environmental pollution at wide-ranging levels of implementation. In in situ site nanobioremediation, the decrease in the toxicant level could even reduce to nearly zero. Ex situ nanobioremediation is still in the developing phase. However, it has exhibited beneficial outcomes in the detoxification of organic and inorganic contaminants and microbes. As nanobioremediation is in a dynamic stage of development, it is regarded as an imminent technique for effective environmental cleanup. Though probable hazards linked to environmental nanotoxicity have yet to be determined, these technological advancements are still likely to be a promising and advantageous substitute for generally prevailing methods. Nanoremediation is an undeniable potential technology for environmental sustainability as soon as ambiguities in research are resolved.

References

Ali, J., Ali, N., Jamil, S.U.U., Waseem, H., Khan, K., Pan, G., 2017. Insight into eco-friendly fabrication of silver nanoparticles by *Pseudomonas aeruginosa* and its potential impacts. J. Environ. Chem. Eng. 5 https://doi.org/10.1016/j.jece.2017.06.038.

Ali, J., Ali, N., Wang, L., Waseem, H., Pan, G., 2019. Revisiting the mechanistic pathways for bacterial mediated synthesis of noble metal nanoparticles. J. Microbiol. Methods 159, 18−25. https://doi.org/10.1016/j.mimet.2019.02.010.

Ali, J., Sohail, A., Wang, L., Rizwan Haider, M., Mulk, S., Pan, G., 2018. Electro-microbiology as a promising approach towards renewable energy and environmental sustainability. Energies 11, 1822. https://doi.org/10.3390/en11071822.

Ali, J., Wang, L., Waseem, H., Djellabi, R., Oladoja, N.A., Pan, G., 2020a. FeS@rGO nano-composites as electrocatalysts for enhanced chromium removal and clean energy generation by microbial fuel cell. Chem. Eng. J. 384, 123335. https://doi.org/10.1016/j.cej.2019.123335.

Ali, J., Wang, L., Waseem, H., Sharif, H.M.A., Djellabi, R., Zhang, C., Pan, G., 2019. Bio-electrochemical recovery of silver from wastewater with sustainable power generation and its reuse for biofouling mitigation. J. Clean. Prod. 235, 1425−1437. https://doi.org/10.1016/j.jclepro.2019.07.065.

Ali, J., Wang, L., Waseem, H., Song, B., Djellabi, R., Pan, G., 2020b. Turning harmful algal biomass to electricity by microbial fuel cell: a sustainable approach for waste management. Environ. Pollut. 266, 115373. https://doi.org/10.1016/j.envpol.2020.115373.

Ávila, C., Bayona, J.M., Martín, I., Salas, J.J., García, J., 2015. Emerging organic contaminant removal in a full-scale hybrid constructed wetland system for wastewater treatment and reuse. Ecol. Eng. 80, 108−116. https://doi.org/10.1016/j.ecoleng.2014.07.056.

Bernhard, M., Müller, J., Knepper, T.P., 2006. Biodegradation of persistent polar pollutants in wastewater: comparison of an optimised lab-scale membrane bioreactor and activated sludge treatment. Water Res. 40, 3419−3428. https://doi.org/10.1016/j.watres.2006.07.011.

Bezbaruah, A.N., Krajangpan, S., Chisholm, B.J., Khan, E., Elorza Bermudez, J.J., 2009. Entrapment of iron nanoparticles in calcium alginate beads for groundwater remediation applications. J. Hazard Mater. 166, 1339−1343. https://doi.org/10.1016/j.jhazmat.2008.12.054.

Buser, H.R., Poiger, T., Muller, M.D., 1999. Occurrence and environmental behavior of the chiral pharmaceutical drug ibuprofen in surface waters and in wastewater. Environ. Sci. Technol. 33, 2529−2535. https://doi.org/10.1021/es981014w.

Cameotra, S.S., Dhanjal, S., 2010. Environmental nanotechnology: nanoparticles for bioremediation of toxic pollutants. In: Bioremediation Technology. Springer, Netherlands, pp. 348−374. https://doi.org/10.1007/978-90-481-3678-0_13.

Cao, C.Y., Cui, Z.M., Chen, C.Q., Song, W.G., Cai, W., 2010. Ceria hollow nanospheres produced by a template-free microwave-assisted hydrothermal method for heavy metal ion removal and catalysis. J. Phys. Chem. C 114, 9865−9870. https://doi.org/10.1021/jp101553x.

Carballa, M., Omil, F., Lema, J.M., Llompart, M., García-Jares, C., Rodríguez, I., Gómez, M., Ternes, T., 2004. Behavior of pharmaceuticals, cosmetics and hormones in a sewage treatment plant. Water Res. 38, 2918−2926. https://doi.org/10.1016/j.watres.2004.03.029.

Chen, Jun, Liu, You Sheng, Su, Hao Chang, Ying, Guang Guo, Liu, Feng, Liu, Shuang Shuang, He, Liang Ying, Chen, Zhi Feng, Yang, Yong Qiang, Chen, Fan Rong, 2015. Removal of antibiotics and antibiotic resistance genes in rural wastewater by an integrated constructed wetland. Environmental Science and Pollution Research 22 (3), 1794−1803.

Chen, J., Wei, X.D., Liu, Y.S., Ying, G.G., Liu, S.S., He, L.Y., Su, H.C., Hu, L.X., Chen, F.R., Yang, Y.Q., 2016. Removal of antibiotics and antibiotic resistance genes from domestic sewage by constructed wetlands: optimization of wetland substrates and hydraulic loading. Sci. Total Environ. 565, 240−248. https://doi.org/10.1016/j.scitotenv.2016.04.176.

Cheng, M., Zeng, G., Huang, D., Lai, C., Xu, P., Zhang, C., Liu, Y., 2016. Hydroxyl radicals based advanced oxidation processes (AOPs) for remediation of soils contaminated with organic compounds: a review. Chem. Eng. J. 284, 582−598. https://doi.org/10.1016/j.cej.2015.09.001.

Choi, M.H., Jeong, S.W., Shim, H.E., Yun, S.J., Mushtaq, S., Choi, D.S., Jang, B.S., Yang, J.E., Choi, Y.J., Jeon, J., 2017. Efficient bioremediation of radioactive iodine using biogenic gold nanomaterial-containing radiation-resistant bacterium, *Deinococcus radiodurans* R1. Chem. Commun. 53, 3937−3940. https://doi.org/10.1039/c7cc00720e.

Delgado, S., Villarroel, R., Gonzalez, E., Morales, M., 2011. Aerobic membrane bioreactor for wastewater treatment − performance under substrate-limited conditions. In: Biomass - Detection, Production and Usage. InTech. https://doi.org/10.5772/17409.

Elsaesser, D., Blankenberg, A.G.B., Geist, A., Mæhlum, T., Schulz, R., 2011. Assessing the influence of vegetation on reduction of pesticide concentration in experimental surface flow constructed wetlands: application of the toxic units approach. Ecol. Eng. 37, 955−962. https://doi.org/10.1016/j.ecoleng.2011.02.003.

Elsaesser, D., Stang, C., Bakanov, N., Schulz, R., 2013. The Landau stream mesocosm facility: pesticide mitigation in vegetated flow-through streams. Bull. Environ. Contam. Toxicol. 90, 640−645. https://doi.org/10.1007/s00128-013-0968-9.

Elsayed, O.F., Maillard, E., Vuilleumier, S., Nijenhuis, I., Richnow, H.H., Imfeld, G., 2014. Using compound-specific isotope analysis to assess the degradation of chloroacetanilide herbicides in lab-scale wetlands. Chemosphere 99, 89−95. https://doi.org/10.1016/j.chemosphere.2013.10.027.

Fabrega, J., Luoma, S.N., Tyler, C.R., Galloway, T.S., Lead, J.R., 2011. Silver nanoparticles: behaviour and effects in the aquatic environment. Environ. Int. 37 (2), 517−531. https://doi.org/10.1016/j.envint.2010.10.012.

Feng, L., Cao, M., Ma, X., Zhu, Y., Hu, C., 2012. Superparamagnetic high-surface-area Fe_3O_4 nanoparticles as adsorbents for arsenic removal. J. Hazard Mater. 217−218, 439−446. https://doi.org/10.1016/j.jhazmat.2012.03.073.

Friha, I., Karray, F., Feki, F., Jlaiel, L., Sayadi, S., 2014. Treatment of cosmetic industry wastewater by submerged membrane bioreactor with consideration of microbial community dynamics. Int. Biodeterior. Biodegrad. 88, 125−133. https://doi.org/10.1016/j.ibiod.2013.12.015.

Galinha, C.F., Sanches, S., Crespo, J.G., 2018. Membrane bioreactors. In: Fundamental Modeling of Membrane Systems: Membrane and Process Performance. Elsevier, pp. 209−249. https://doi.org/10.1016/B978-0-12-813483-2.00006-X.

Gao, C., Zhang, W., Li, H., Lang, L., Xu, Z., 2008. Controllable fabrication of mesoporous MgO with various morphologies and their absorption performance for toxic pollutants in water. Cryst. Growth Des. 8, 3785−3790. https://doi.org/10.1021/cg8004147.

Gaur, N., Flora, G., Yadav, M., Tiwari, A., 2014. A review with recent advancements on bioremediation-based abolition of heavy metals. Environ. Sci. Process. Impacts 16 (2), 180−193. https://doi.org/10.1039/c3em00491k.

Ghangrekar, M.M., Sathe, S.M., Chakraborty, I., 2020. In situ bioremediation techniques for the removal of emerging contaminants and heavy metals using hybrid microbial electrochemical technologies. In: Emerging Technologies in Environmental Bioremediation. Elsevier, pp. 233−255. https://doi.org/10.1016/b978-0-12-819860-5.00009-2.

Ghoshdastidar, A.J., Saunders, J.E., Brown, K.H., Tong, A.Z., 2012. Membrane bioreactor treatment of commonly used organophosphate pesticides. J. Environ. Sci. Heal. B Pestic. Food Contam. Agric. Wastes 47, 742−750. https://doi.org/10.1080/03601234.2012.669334.

Gikas, G.D., Pérez-Villanueva, M., Tsioras, M., Alexoudis, C., Pérez-Rojas, G., Masís-Mora, M., Lizano-Fallas, V., Rodríguez-Rodríguez, C.E., Vryzas, Z., Tsihrintzis, V.A., 2018. Low-cost approaches for the removal of terbuthylazine from agricultural wastewater: constructed wetlands and biopurification system. Chem. Eng. J. 335, 647−656. https://doi.org/10.1016/j.cej.2017.11.031.

Goswami, A., Raul, P.K., Purkait, M.K., 2012. Arsenic adsorption using copper (II) oxide nanoparticles. Chem. Eng. Res. Des. 90, 1387−1396. https://doi.org/10.1016/j.cherd.2011.12.006.

Gray, J.L., Sedlak, D.L., 2005. The fate of estrogenic hormones in an engineered treatment wetland with Dense macrophytes. Water Environ. Res. 77, 24−31. https://doi.org/10.2175/106143005x41582.

Gupta, K., Bhattacharya, S., Chattopadhyay, D., Mukhopadhyay, A., Biswas, H., Dutta, J., Ray, N.R., Ghosh, U.C., 2011. Ceria associated manganese oxide nanoparticles: synthesis, characterization and arsenic(V) sorption behavior. Chem. Eng. J. 172, 219−229. https://doi.org/10.1016/j.cej.2011.05.092.

Imfeld, G., Braeckevelt, M., Kuschk, P., Richnow, H.H., 2009. Monitoring and assessing processes of organic chemicals removal in constructed wetlands. Chemosphere 74 (3), 349−362. https://doi.org/10.1016/j.chemosphere.2008.09.062.

Kanel, S.R., Greneche, J.M., Choi, H., 2006. Arsenic(V) removal from groundwater using nano scale zero-valent iron as a colloidal reactive barrier material. Environ. Sci. Technol. 40, 2045−2050. https://doi.org/10.1021/es0520924.

Kappell, A.D., Kimbell, L.K., Seib, M.D., Carey, D.E., Choi, M.J., Kalayil, T., Fujimoto, M., Zitomer, D.H., McNamara, P.J., 2018. Removal of antibiotic resistance genes in an anaerobic membrane bioreactor treating primary clarifier effluent at 20°C. Environ. Sci. Water Res. Technol. 4, 1783−1793. https://doi.org/10.1039/c8ew00270c.

Khanal, S.K., Xie, B., Thompson, M.L., Sung, S., Ong, S.K., Van Leeuwen, J., 2006. Fate, transport and biodegradation of natural estrogens in the environment and engineered systems. Environ. Sci. Technol. 40 (21), 6537−6546. https://doi.org/10.1021/es0607739.

Kim, M., Guerra, P., Shah, A., Parsa, M., Alaee, M., Smyth, S.A., 2014. Removal of pharmaceuticals and personal care products in a membrane bioreactor wastewater treatment plant. Water Sci. Technol. 69, 2221−2229. https://doi.org/10.2166/wst.2014.145.

Lee, J.H., Hwang, E.T., Kim, B.C., Lee, S.M., Sang, B.I., Choi, Y.S., Kim, J., Gu, M.B., 2007. Stable and continuous long-term enzymatic reaction using an enzyme-nanofiber composite. Appl. Microbiol. Biotechnol. 75, 1301−1307. https://doi.org/10.1007/s00253-007-0955-3.

Liu, L., Liu, C., Zheng, J., Huang, X., Wang, Z., Liu, Y., Zhu, G., 2013. Elimination of veterinary antibiotics and antibiotic resistance genes from swine wastewater in the vertical flow constructed wetlands. Chemosphere 91, 1088−1093. https://doi.org/10.1016/J.CHEMOSPHERE.2013.01.007.

Maculewicz, J., Świacka, K., Kowalska, D., Stepnowski, P., Stolte, S., Dołżonek, J., 2020. In vitro methods for predicting the bioconcentration of xenobiotics in aquatic organisms. Sci. Total Environ. 739, 140261 https://doi.org/10.1016/j.scitotenv.2020.140261.

Mahabali, S., Spanoghe, P., 2014. Mitigation of two insecticides by wetland plants: feasibility study for the treatment of agricultural runoff in Suriname (South America). Water. Air. Soil Pollut 225, 1−12. https://doi.org/10.1007/s11270-013-1771-2.

Maillard, E., Payraudeau, S., Faivre, E., Grégoire, C., Gangloff, S., Imfeld, G., 2011. Removal of pesticide mixtures in a stormwater wetland collecting runoff from a vineyard catchment. Sci. Total Environ. 409, 2317−2324. https://doi.org/10.1016/j.scitotenv.2011.01.057.

Masi, F., Conte, G., Lepri, L., Martellini, T., Del Bubba, M., 2005. Endocrine Disrupting Chemicals (EDCs) and Pathogens Removal in an Hybrid CW System for a Tourist Facility Wastewater Treatment and Reuse.

Matamoros, V., Rodríguez, Y., Bayona, J.M., 2017. Mitigation of emerging contaminants by full-scale horizontal flow constructed wetlands fed with secondary treated wastewater. Ecol. Eng. 99, 222−227. https://doi.org/10.1016/j.ecoleng.2016.11.054.

Mody, V., Siwale, R., Singh, A., Mody, H., 2010. Introduction to metallic nanoparticles. J. Pharm. Bioallied Sci. 2, 282. https://doi.org/10.4103/0975-7406.72127.

Moore, M.T., Tyler, H.L., Locke, M.A., 2013. Aqueous pesticide mitigation efficiency of *Typha latifolia* (L.), *Leersia oryzoides* (L.) Sw., and *Sparganium americanum* Nutt. Chemosphere 92, 1307−1313. https://doi.org/10.1016/j.chemosphere.2013.04.099.

Mutamim, N.S.A., Noor, Z.Z., Hassan, M.A.A., Yuniarto, A., Olsson, G., 2013. Membrane bioreactor: applications and limitations in treating high strength industrial wastewater. Chem. Eng. J. 225, 109−119. https://doi.org/10.1016/j.cej.2013.02.131.

Novo, A., Manaia, C.M., 2010. Factors influencing antibiotic resistance burden in municipal wastewater treatment plants. Appl. Microbiol. Biotechnol. 87, 1157−1166. https://doi.org/10.1007/s00253-010-2583-6.

Olvera-Vargas, H., Trellu, C., Oturan, N., Oturan, M.A., 2018. Bio-electro-Fenton: a new combined process − principles and applications. In: Handbook of Environmental Chemistry. Springer Verlag, pp. 29−56. https://doi.org/10.1007/698_2017_53.

Ramírez-García, R., Gohil, N., Singh, V., 2018. Recent advances, challenges, and opportunities in bioremediation of hazardous materials. In: Phytomanagement of Polluted Sites: Market Opportunities in Sustainable Phytoremediation. Elsevier, pp. 517−568. https://doi.org/10.1016/B978-0-12-813912-7.00021-1.

Ramprasad, C., Philip, L., 2016. Surfactants and personal care products removal in pilot scale horizontal and vertical flow constructed wetlands while treating greywater. Chem. Eng. J. 284, 458−468. https://doi.org/10.1016/j.cej.2015.08.092.

Ran, Z., Zhu, J., Li, K., Zhou, L., Xiao, P., Wang, B., 2018. Study on the membrane bioreactor for treating surfactant wastewater. J. Water Clim. Chang. 9, 240−248. https://doi.org/10.2166/wcc.2018.046.

Roccaro, P., 2018. Treatment processes for municipal wastewater reclamation: the challenges of emerging contaminants and direct potable reuse. Curr. Opin. Environ. Sci. Heal. 2, 46−54. https://doi.org/10.1016/j.coesh.2018.02.003.

Servais, P., Passerat, J., 2009. Antimicrobial resistance of fecal bacteria in waters of the Seine river watershed (France). Sci. Total Environ. 408, 365−372. https://doi.org/10.1016/j.scitotenv.2009.09.042.

Shappell, N.W., Billey, L.O., Forbes, D., Matheny, T.A., Poach, M.E., Reddy, G.B., Hunt, P.G., 2007. Estrogenic activity and steroid hormones in swine wastewater through a lagoon constructed-wetland system. Environ. Sci. Technol. 41, 444−450. https://doi.org/10.1021/es061268e.

Tang, X.Y., Yang, Y., McBride, M.B., Tao, R., Dai, Y.N., Zhang, X.M., 2019. Removal of chlorpyrifos in recirculating vertical flow constructed wetlands with five wetland plant species. Chemosphere 216, 195−202. https://doi.org/10.1016/j.chemosphere.2018.10.150.

Theron, J., Walker, J.A., Cloete, T.E., 2008. Nanotechnology and water treatment: applications and emerging opportunities. Crit. Rev. Microbiol. 34 (1), 43−69. https://doi.org/10.1080/10408410701710442.

Thomas, R., Gough, R., Freeman, C., 2017. Linear alkylbenzene sulfonate (LAS) removal in constructed wetlands: the role of plants in the treatment of a typical pharmaceutical and personal care product. Ecol. Eng. 106, 415−422. https://doi.org/10.1016/j.ecoleng.2017.06.015.

Tofighy, M.A., Mohammadi, T., 2011. Adsorption of divalent heavy metal ions from water using carbon nanotube sheets. J. Hazard Mater. 185, 140−147. https://doi.org/10.1016/j.jhazmat.2010.09.008.

Tournebize, J., Passeport, E., Chaumont, C., Fesneau, C., Guenne, A., Vincent, B., 2013. Pesticide de-contamination of surface waters as a wetland ecosystem service in agricultural landscapes. Ecol. Eng. 56, 51−59. https://doi.org/10.1016/j.ecoleng.2012.06.001.

Tungittiplakorn, W., Lion, L.W., Cohen, C., Kim, J.Y., 2004. Engineered polymeric nanoparticles for soil remediation. Environ. Sci. Technol. 38, 1605−1610. https://doi.org/10.1021/es0348997.

Vacca, G., Wand, H., Nikolausz, M., Kuschk, P., Kästner, M., 2005. Effect of plants and filter materials on bacteria removal in pilot-scale constructed wetlands. Water Res. 39, 1361−1373. https://doi.org/10.1016/j.watres.2005.01.005.

Vymazal, J., 2010. Constructed wetlands for wastewater treatment. Water 2, 530−549. https://doi.org/10.3390/w2030530.

Wang, L., Miao, X., Ali, J., Lyu, T., Pan, G., 2018. Quantification of oxygen nanobubbles in particulate matters and potential applications in remediation of anaerobic environment. ACS Omega 3, 10624−10630. https://doi.org/10.1021/acsomega.8b00784.

Waseem, H., Ali, J., Syed, J.H., Jones, K.C., 2020a. Establishing the relationship between molecular biomarkers and biotransformation rates: extension of knowledge for dechlorination of polychlorinated dibenzo-p-dioxins and furans (PCDD/Fs). Environ. Pollut. 263, 114676 https://doi.org/10.1016/j.envpol.2020.114676.

Waseem, H., Jameel, S., Ali, J., Jamal, A., Ali, M.I., 2020b. Recent Advances in Treatment Technologies for Antibiotics and Antimicrobial Resistance Genes. Springer, Cham, pp. 395−413. https://doi.org/10.1007/978-3-030-40422-2_18.

Waseem, H., Jameel, S., Ali, J., Saleem Ur Rehman, H., Tauseef, I., Farooq, U., Jamal, A., Ali, M., Waseem, H., Jameel, S., Ali, J., Saleem Ur Rehman, H., Tauseef, I., Farooq, U., Jamal, A., Ali, M.I., 2019. Contributions and challenges of high throughput qPCR for determining antimicrobial resistance in the environment: a critical review. Molecules 24, 163. https://doi.org/10.3390/molecules24010163.

Waseem, H., Saleem ur Rehman, H., Ali, J., Iqbal, M.J., Ali, M.I., 2020c. Global trends in ARGs measured by HT-qPCR platforms. In: Antibiotics and Antimicrobial Resistance Genes in the Environment. Elsevier, pp. 206−222. https://doi.org/10.1016/b978-0-12-818882-8.00014-0.

Waseem, H., Williams, M.R., Stedtfeld, R.D., Hashsham, S.A., 2017. Antimicrobial resistance in the environment. Water Environ. Res. 89, 921−941. https://doi.org/10.2175/106143017X15023776270179.

Zhang, X., Zhang, M., 2011. Modeling effectiveness of agricultural BMPs to reduce sediment load and organophosphate pesticides in surface runoff. Sci. Total Environ. 409, 1949−1958. https://doi.org/10.1016/j.scitotenv.2011.02.012.

Zhao, L., Peralta-Videa, J.R., Rico, C.M., Hernandez-Viezcas, J.A., Sun, Y., Niu, G., Servin, A., Nunez, J.E., Duarte-Gardea, M., Gardea-Torresdey, J.L., 2014. CeO_2 and ZnO nanoparticles change the nutritional qualities of cucumber (*Cucumis sativus*). J. Agric. Food Chem. 62, 2752—2759. https://doi.org/10.1021/jf405476u.

Żur, J., Piński, A., Marchlewicz, A., Hupert-Kocurek, K., Wojcieszyńska, D., Guzik, U., 2018. Organic micropollutants paracetamol and ibuprofen—toxicity, biodegradation, and genetic background of their utilization by bacteria. Environ. Sci. Pollut. Res. 25 (22), 21498—21524. https://doi.org/10.1007/s11356-018-2517-x.

Further reading

Fan, F.L., Qin, Z., Bai, J., Rong, W.D., Fan, F.Y., Tian, W., Wu, X.L., Wang, Y., Zhao, L., 2012. Rapid removal of uranium from aqueous solutions using magnetic $Fe_3O_4@SiO_2$ composite particles. J. Environ. Radioact. 106, 40—46. https://doi.org/10.1016/j.jenvrad.2011.11.003.

Advances in dye contamination: health hazards, biodegradation, and bioremediation

Siddharth Vats[1], Shreya Srivastava[1], Neha Maurya[1], Shikha Saxena[1], Bhawana Mudgil[2], Shriyam Yadav[1], Rati Chandra[1]

[1]*Faculty of Biotechnology, Institute of Bio-Sciences and Technology, Shri Ramswaroop Memorial University, Barabanki, Uttar Pradesh, India;* [2]*TGT, Natural Science, Sarvodaya Vidyalaya, Rohini, Delhi, India*

Chapter outline

Biological Approaches to Controlling Pollutants. https://doi.org/10.1016/B978-0-12-824316-9.00020-3

8.1 Introduction

Dyes and dyestuffs have been used in industries related to textiles, plastics, manufacturing, paints, inks, fabrics, cosmetics, and other materials. Waste (solid/liquid/fumes) coming from these industries poses serious health hazards to humans and animals. In ancient times, there were many basic strategies used to avoid dye contamination. Various techniques including cultural activities, microbial activities, and the application of enzymes are processes that can be carried out for the degradation of dyes. To remove their natural colors, many fabrics are bleached with hydrogen peroxide, and optical brightening agents are added if the fabric is to be sold white and not dyed. Aqueous dyeing involves the application of colors to textile with synthetic dyes at elevated pressure and temperatures in some steps. During the dyeing process, chemical aids such as acids, alkali, electrolytes, carriers, promoting agents, emulsifying oils, and softening agents are applied to textiles to achieve a uniform depth of color with color fastness properties for use of the fabric. Different fastness properties may be required depending on the type of fabric. Finishing aims to improve the quality of fabric treatment with chemical compounds. Examples of fabric treatments applied in the finishing process are waterproofing, softening, soil resistance, stain release, and fungal protection. A batch process is carried out for dyeing. The dye is gradually transferred from the dye bath to the material being dyed over a relatively long period, the dyeing occurs in the presence of dilute chemicals in closed equipment such as a kettle, beam, or jet.

8.2 Health hazards of dyes to humans

Wastewater from the textile industry contains a variety of polluting substances including dyes. Wastewater from the textile industry is a complex mixture of polluting substances ranging from organochlorine-based pesticides to heavy metals associated with dyes and the dyeing process. Textile industries use stain-resistant, antistatic, antibacterial, permanent press, and nonshrink textile products that create a health hazard by the chemicals used in these industries. Allergic/dermatologic conditions such as dermatitis arise from using clothing colored by dyes. Clothing comes into contact with skin, and toxic chemicals are absorbed into skin when the body is warm and skin pores have opened to allow perspiration. The other common hazard of dyes is respiratory problems due to the inhalation of dye particles. When a person inhales the dyes, their body reacts dramatically. It is called respiratory sensitization, with symptoms including watery eyes, sneezing, and wheezing; the toxicants arising from dyeing and related finishing process act as irritants. Textile industries produce a large volume of liquid wastes. During the process of dyeing, not all dyes are applied to fabrics or fixed on them, and a portion of dyes remains unfixed to fabrics and washed out or gets released slowly when it comes into contact with sweat. Many dyes and chemicals are nonbiodegradable as well as carcinogenic and pose threats to health and the environment. Cotton fibers are dyed using azo dyes, which make up one of the largest groups of synthetic colorants used in industry. These dyes alter the physical and chemical properties of soil and cause harm to flora and fauna in the environment. The toxic nature of dye causes death to soil microorganisms, which affects agricultural productivity (Kaur et al., 2010). Dyes are also used in food. Food dyes are one of the most used and dangerous additives. Various studies have proven the side effects of applying dyes in food. Swanson and Kinsbourne (1980) proved the role of dyes in impairing the performances of hyperactive children. Synthetic dyes are recalcitrant compounds that not only cause environmental damage but also pose major carcinogenic risks. Tsuda et al. (2001) studied the genotoxicity of tar dyes (synthetic red dyes) that find applications as food additives for color in countries such as Japan. Kobylewski, and Jacobson (2012) determined the negative effects of food dyes, namely Red 3 (it has carcinogenic effects on animals), Red 40, Yellow 5, and Yellow 6 result from being contaminated with carcinogenic chemicals such as benzidine and others. Blue 1, Red 40, Yellow 5, and Yellow 6 caused hypersensitivity reactions. Yellow 5 shows genotoxicity in microbes and rodents. Toxicity tests on two dyes (Citrus red 2 and Orange B) suggest safety concerns, but Citrus red 2 is used at low levels and only on some Florida oranges, and dye Orange B has not been used for several years. Ragunathan et al. (2013) studied chromosome aberrations in human blood lymphocytes of persons exposed to a mixture of chemicals for a long time. Benzidine is used as a reactant in dyes and directly exposes workers to carcinogens (Sakthisharmila et al., 2018). Mutagenicity testing chemicals in the Ames assay pose a danger of being exposed to chemicals that are mutagenic in the bacterium and carcinogenic in laboratory animals and present a risk of cancer to humans (Zeiger et al., 2019). Factory workers use harmful dyes without protective equipment and clothing, and dyes laced with banned ingredients can cause cancer. Examples of harmful chemical in dyes are dioxin, formaldehyde, and zinc, and when these substances come into contact with human skin, they cause negative effects ranging from disturbances in hormones in children and cancer in people who drink contaminated water as well as factory workers (Songur et al., 2010).

8.2.1 Health hazards of dyes to nature

Environmental pollution is one of the most urgent problems worldwide. In industry, wastewater in composition is produced, and colored water released during the dyeing of fabrics is the most problematic, and even a trace of dye can be highly visible (Hossain et al., 2018). Dyes are stable organic pollutants in the environment, and most artificial compounds are xenobiotic. With the damage and loss that has been witnessed in the environment, environmentalists are developing treatments that are environmentally conscious and economically viable (Mani et al., 2019). The current technologies that control pollution biodegradation of synthetic dyes by microbes are an effective and promising approach. The uses of single techniques, efficiency-wise, are not completely able to decolorize wastewater (Shen et al., 2021). Forty thousand different synthetic dyes and pigments are used in industry, and 450,000 tons of dyestuffs are produced. The fibers to which they can be applied comprise various classes. Dyes such as azo dyes (50% of total) are used to dye natural and synthetic fibers as reactive dyes. Acid and basic dyes are used to dye fibers, while inks, plastic fat, and mineral oil are colored by solvent dyes (Ramzan et al., 2019). To reduce the toxic effects of dyes and their content in effluents, textile industries are put under immense pressure to reduce the use of harmful substances and be responsible for mutagenic and carcinogenic effects. Various physicochemical and biological technologies have been developed for the removal of synthetic dyes from wastewater to decrease their environmental impact. These include physical methods such as nanofiltration, reverse osmosis, and sorption techniques; chemical methods such as flocculation combined with filtration, electrokinetic coagulation, irradiation; and biological methods such as aerobic and anaerobic microbial degradation and the use of enzymes (Hussain et al., 2020). The application of biological materials allow the recovery and disposal of concentrations and accumulations of pollutants from aqueous/wastewater effluents to environmentally acceptable levels (Hussain et al., 2020). Enzymes can be used in remediation treatments to target specific pollutants (Vats and Kumar, 2015). Dyes absorb and reflect sunlight in water. This diminishes the photosynthetic activity of algae and seriously influences the food chain (Saini, 2017). Triple primary cancers involving kidney, urinary bladder, and liver of dye workers have been reported (Nakano et al., 2018). Most processes performed in textile mills produce atmospheric emission. Gaseous emissions have been identified as the second-greatest pollution problem for the textile industry. Speculation concerning the amounts and types of air pollutants emitted from textile operations has been widespread, but generally, air emission data for textile manufacturing operations are not readily available (Mia et al., 2019). In addition, increased demand for textile products and the proportional increase in their production as well as the use of synthetic dyes have together contributed to dye wastewater becoming a substantial source of severe pollution problems currently. Dyes can remain in the environment for an extended period because of their high thermal and photo stability in resisting biodegradation (Mia et al., 2019).

8.2.2 Health hazards of dyes to flora and fauna

Natural dyes are derived from fruits, animals, plants, insects, and minerals. Some sources of natural dye, such as logwood and bloodroot, can be toxic. Logwood produces a range of colors, but its active ingredients, hematoxylin and hematein, are harmful when the enter the body by inhalation and skin absorption (Cooksey, 2020). Bloodroot can be harmful when it is inhaled because it causes inflammation and irritation. The mordants used for their application can be toxic. Mordants are the substances used to make natural dyes stick to fabrics and include iron, chrome, aluminum, and copper (Choudhury, 2018). Dyes are the most important uses of plants. Natural dyes derived from flora and fauna are believed to be safe and nontoxic in nature. Natural dyes are in demand not only in the textile industry but in leather, pharmaceuticals, and food. Plants possess many medicinal properties. In ayurvedic medicine, flowers are the most effective fermentation agent. Various parts of plants have been used to extract dyes and include flowers, vegetables, and leaves, and various mordants are used to fix dyes into fabrics (Choudhury, 2018). Natural dyes can be broken into two categories, adjective and substantive. Medieval dyers mordanted their fabrics and yarns before dyeing them. A protein-based fiber found in Europe and dating back to 2000 BCE is wool. Primitive humans used plant dyestuffs to color their skin and animal skins during religious festivals and during wars (Choudhury, 2018). Dyes might have been discovered accidently, but their use has become a part of human custom, and it is difficult to imagine life without them. Primitive dyeing techniques included sticking plants on fabric or crushing pigments into cloth. Many natural stains and dyestuffs are obtained mainly from plants and are dominant sources of natural dyes, producing various colors such as yellow, blue, red, and brown (Choudhury, 2018). All parts of the plant are used, such as leaves, fruit, wood, and flowers. They harmonize colors for subtle and restful effects. Modern and biotechnological techniques are necessary for improving the quantity and quality of dye production (Richardson, 2017). Other sources of plant dyes rich in naphthoquinones, such as lawsone from *Lawsonia inermis* L., lapachol from alkanet, and juglone from walnut exhibit antifungal and antibacterial activities (Adeel et al., 2019). Negative effects of dyes on human life are dysfunction of nervous system, brain, liver, etc. Dyes are divided into two types, namely cationic and anionic, and carry negative and positive charges. Basic dyes are the class of cationic dyes, whereas acid, reactive, and direct dyes are anionic. Many textile industries use activated carbon for dye wastewater treatment (El Harfi and El Harfi, 2017). Dye adsorption capacity increases with decreases in particle size. The characteristics of basic dyes are brilliance and intensity of color. The clarity of hue with basic dyes cannot be matched by other dye classes. Basic dyes are not soluble in water and form a sticky mass and are difficult to distribute in solution. Basic dyes are precipitated under conditions mixed with anionic dyes such as acid dyes. Basic dyes have poor to moderate washing fastness on wool (El Harfi and El Harfi, 2017). Basic dyes have several chemical classes:

(a) Diphenylmethane (presence of C=NH group), e.g., CI Basic yellow 1.
(b) Thiazine (nitrogen and sulfur atoms forming a ring with benzene carbons), e.g., CI Basic violets 3 and 14.
(c) Azine (two nitrogen atoms forming a ring with benzene carbons, e.g., CI Basic red 5.
(d) Acridine (two benzene rings linked by a nitrogen atom and −CH = group) derivatives, e.g., CI Basic orange 14.
(e) Azo groups, e.g., CI Basic brown 1.

The significance affinity is the hydrophilic substituent affinity of dye for acrylic fibers. The ranges of cationic dyes for acrylic fibers are manufactured by companies such as Ciba, DyStar, Sevron, and Clariant (Shindy, 2017).

Dyes can alter the physical and chemical properties of biodegradation and bioremediation, cause the deterioration of water bodies, and cause harm to flora and fauna in the environment. It has also been observed that the toxic nature of dyes causes death to the soil and microorganisms that in turn affect agricultural productivity. Most dyes are toxic and highly resistant to degradation because of their complex chemical structures. Thus, biodegradation is a promising approach for the remediation of synthetic dye wastewater because of its cost-effectiveness, efficiency, and environmentally friendly nature.

Dye sources can be either natural or synthetic.

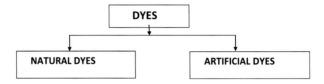

8.3 Natural dyes

Natural dyes are the dyes or colorants derived from plants, invertebrates or minerals. The majority of natural dyes are vegetable dyes from plant sources, e.g., root, berries, bark, leaves etc. There are two types of natural dyes. Adjective or additive dyes such as madder. This type of natural dye has been in use for more than 2000 years. Substantive dyes bond with fiber without the use of a mordant. Common mordants are weak organic acids such as acetic or tannic acid, Basically, the natural dyes in the market are exotic, historic or classic. In the proper combination with mordants, they yield well (El Harfi and El Harfi, 2017).

Natural dyes are categorized as in the basis of plants and animals:

- Plants- Madder (Madder root)
- Animals- Tyrian purple (sea snails)

Natural dyes are used in small quantities by craftspeople. Some commercial use of natural dyes is a response to concerns about synthetic dyes and environmental pollution. These dyes are also play a role in the economic development, and natural dyes are a renewable resource. The dyestuff is harvested or collected, and commercially available natural dye extracts make the dyeing process less involved. The dyes are harvested or collected, soaked in water for several hours. Here we taking an example of Indigo, Indigo is extracted from the stems and leaves of plants of the *Indigofera* species from India, Central America, and Africa, whereas Indigo, originally from India, is used for cotton, wool, and silk (El Harfi and El Harfi, 2017).

8.3.1 **Madder**

It is a type of natural dye obtained from the rose madder. It is highly expensive and a member of the *Rubia* family.

This type of dye has been used since ancient times.

8.3.2 **Tyrian purple**

Tyrian purple is a reddish-purple natural dye produced by the several species of predatory sea snails in the family Muricidae.

8.4 **Synthetic dyes**

It is said that people see all the colors in their daily life are the synthetic dyes. Synthetic dyes are used everywhere in everything from clothes to food. This is because they are cheaper to produce and provide a dark contrast. They are easy to apply on clothes, paper, and fabrics (El Harfi and El Harfi, 2017). Examples are azo dyes, basic dyes, etc.

Synthetic dyes are also known as artificial dyes. They are industrial-based in various developments and are cheaper. Synthetic dyes, from a chemical point of view, are of three types:

- azo dyes
- triphenylmethane dyes
- anthraquinone dyes

8.4.1 **Azo dyes**

Azo dyes are the largest class in the group of synthetic dyes.

Their general formula is $R-N=N-R'$.

R1 and R2 are usually aryl.

They are commercially important for the family of azo compounds, i.e., compounds containing linkage.

Azo dyes are widely used to treat textiles, leather articles, etc.

8.4.2 **Triphenylmethane dyes**

The second group consists of triphenylmethane dyes.

This group consists of a basic body.

Its phenyl rings carry at least one activated substituent in the amino groups—for example, phenolphthalein.

They are very popular in food coloring and cosmetics.

8.4.3 **Anthraquinone dyes**

They are also termed anthracenedione or deoxoanthracene.

It is an aromatic organic compound.

Anthraquinone refers to the isomers 9,10-anthraquinone.

It is used in bleaching pulp for papermaking.

It is a yellow highly crystalline solid that is poorly soluble in water but soluble in hot organic solvents.

It is completely soluble in ethanol.

8.5 **Bioremediation**

It is a branch of biotechnology which deals with the use of living organisms such as microbes to remove contaminants, pollutants and toxins from soil and water. It is the act of treating waste or pollutants by the use of microorganisms (as bacteria) which is used for the breakdown of the undesirable substances. This is not a natural process (Elbanna et al., 2017; Krishnamoorthy et al., 2018). This is the process applied after the dye contamination. It is a fast process which takes a fixed duration. Basically, bioremediation is a process used to treat contaminated media, including water, soil and subsurface material by altering environmental conditions to stimulate growth of microorganisms and degrade the target pollutants. The bioremediation process involved oxidation-reduction reactions involving an electron acceptor.

In many cases bioremediation is less expensive and more sustainable than other.

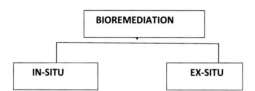

8.5.1 **Bioremediation of dyes**

Bioremediation is carried out by using microorganisms to remove the pollution from the environment, a key research area in the environmental engineering (Vats and Negi, 2013). Microorganism adapts selves to the toxic wastes and develops into new resistant strains naturally, which transform various chemicals into less harmful forms. The biodegradation of recalcitrant compounds by the microbial system occurs by the action of enzymes causing biotransformation (Ihsanullah et al., 2020). Biodegradation of organic substances can be brought by enzymatic mechanism with laccase, tyrosinase, lignin peroxidase and hexane oxidase (Sarkar et al., 2017). The biotechnologists remove the dyes from streams with regard to tackling azo dye pollution in eco-efficient manner. These dyes are xenobiotic in nature and recalcitrant to biodegradation, enzymatic treatment method may be useful for degradation of such dyes from textile effluent. Biodegradation by enzymes hold several advantages such as being inexpensive and producing less sludge. Because of complete mineralization, the intermediate products formed are nontoxic. Microbial effectiveness of decolorization depends on adaptability and activity of microorganisms (Sarkar et al., 2017). Biotechnologists have isolated and cultured number of microbial species including algae, yeasts and actinomycetes. that have been tested for decolorization and mineralization of various dyes. The isolation potent species and degradation is biological aspects of effluent treatment (Sarkar et al., 2017). Compared with fungi, bacterial biomass increases significantly during wastewater treatment and in the decolorization of synthetic dyes (Ihsanullah et al., 2020). Bioremediation is microbial cleanup approach. Strain develops naturally, which transform various toxic chemicals to less harmful forms. Fungi are found in the environment, ecological niches or in association with plants using organic matter. Fungi have the ability to adapt their metabolism to varied nitrogen and carbon sources, which are important for their survival (Bhargava et al., 2020). Metabolic active fungi achieve the degradation of dyes, by the production of large set of extracellular and intraenzymes. For the biodegradation of azo dyes, the focus is on fungal cultures belonging to rot fungi for dyes mineralization. The experimental work the anaerobic decolorization of dyes was conducted using monocultures. The permeation of dyes through biological membrane into microbial cells was citied the principal rate-limiting factor for decolorization. Conventional biological systems are not efficient for decolorization and bioremediation of dyes (Krishnamoorthy et al., 2018).

8.6 **Health hazards**

Health hazards that are toxic in nature, as in dye contamination, are a global issue because dyes are used in food, and when it becomes contaminated, it affects the health of the people who have consumed it. This process includes the organic decompositions as it directly affects the human health and causes various diseases. The pollutant characteristics, environmental conditions, soil and vegetation type and is the source to create a complex set of conditions influencing pollutant lifestyles (Elbanna et al., 2017). Where soils are the reservoirs for environmental pollutants with deposition and persistence being dependent on basic factors.

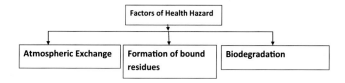

8.7 Biodegradation

Biodegradation of dyes is a process in which the large dye molecules are broken down chemically into smaller molecules. Dye molecules persist in the environment because many of them are reactive toward light, acids, bases and oxygen. Dye biodegradation has been extensively exported as a varied biological treatment method. There are different pathways in which the biodegradation mechanism works (El Harfi and El Harfi, 2017). The isolation and basic identification of the dye intermediates is important in the exportation of the biodegradation mechanism and degradation pathways. The biodegradation pathways of dyes mainly depend on the molecular structures of dyes, microorganisms, related enzymes and other factors. Biodegradation is a process that is highly dependent on the energy involved in breaking contaminants down into various by-products through the action of enzymes. The process of biodegradation can be accomplished microbiologically, such as through process optimization for the biosynthesis of mono- and bimetallic alloy nanoparticle catalysts for the degradation of dyes in individual and ternary mixtures. The contamination of soil and water by dye-containing effluents is of environmental and according to the new research the current pollution control technologies, biodegradation of synthetic and natural dyes by different microbes is emerging as an effective and promising approach (Krishnamoorthy et al., 2018).

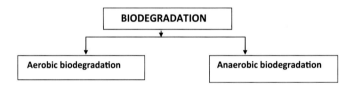

8.8 Aerobic biodegradation

In this process, organic decomposition occurs when oxygen is present, whereas organic pollutants are degraded under the aerobic conditions by aerobic bacteria.

8.9 Anaerobic biodegradation

In this process organic decompositions are done when oxygen is not present.

There are four key components which are involved in the process of anaerobic biodegradation such as Hydrolysis, Acidogenesis, Acetogenesis and Metagenesis.

8.10 **Biodegradation of dyes**

- Decolorization and degradation of dye under aerobic conditions

Dye degradation under aerobic conditions is mainly catalyzed by specific enzymes and most of them are reductase.

- Decolorization and degradation of dyes under anaerobic conditions

Anaerobic digestion of dye wastewater is a promising technology. The decolorization and biodegradation of different dyes under the anaerobic conditions is a relatively simple and a nonspecific reduction process involving azo reductase in various microorganisms (Xiao et al., 2018).

- Decolorization and degradation of dyes under anaerobic—aerobic conditions

With the involvement of both anaerobic and aerobic conditions, the aromatic amines form a different group that becomes properly biodegraded under aerobic conditions.

8.11 **Methods for biodegradation of dyes**

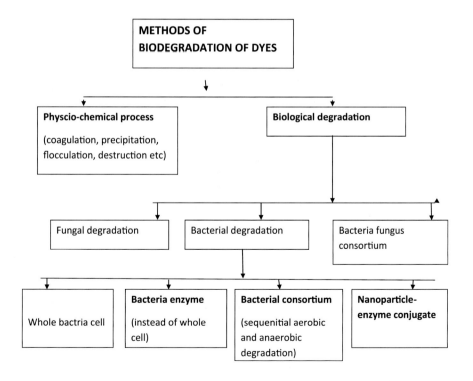

8.12 Past strategies

In ancient times there are many basic strategies used to avoid the dyes contamination. Various cultural and microbial activities play important roles in atmospheric conditions and temperature. In ancient times, people took a culture of microorganisms to prepare solutions used at prestaging to avoid contamination. As dyes become thicker, contamination lasts longer, and the culture or basic strategies used to avoid it become highly expensive. In ancient times, people believed that biological contaminants were easy to detect, such as bacteria, yeasts, and protozoa, while following a proper routine in using good aseptic techniques. Dyes are formed by the use of dyes in chemicals, so basic polymer reactions play a significant role. Removal of the inorganic compounds as they are highly toxic in nature. In this type of process people used the colony of microorganisms as in their strategies. Different efforts made, such as fertilization and growth and basic treatments in the presence of a hypodermis or multiple epidermes, have contributed to taxonomic and phylogenetic studies.

8.13 Microbes used in biodegradation of dyes

Various chemicals discharged from various industries become continuing environmental contaminants. Lot of harmful chemical substances such as dyes, pigments, aromatic molecular and structural compounds are widely applied for the manufacturing of textiles, printing, pharmaceuticals, food, toys, paper, plastic and cosmetics in various large industries (Mohana et al., 2008). In most of the countries industries related to the processing of textiles were found and established and which keeps on increasing their numbers day by day. The use of coloring material in these textile manufacturing industries increases the utilization of the harmful synthetic dyes. These industries produce 30 million tons of textiles annually and cause the utilization of 700,000 tons of dyes such as acidic, basic, reactive, disperse, azo, diazo, and anthraquinone dyes, which leads to substantial environmental contamination and pollution problems (Zollinger, 1987). There is demand for the utilization of synthetic dyes in other areas as well. Reactive dyes are widely used in all industries rather than synthetic dyes. The reactive dye which is a type of azo dye commonly used in industries with distinct reactive groups. The originated reactive dyes from the dyeing factories leads to various difficult causes such as increase in chemical oxygen demand and biological oxygen demand (BOD), modification in the pH of water of rivers, lakes and other water bodies and also causes trouble to the flora and fauna of the ecosystem and to the human beings (Shyamala Gowri et al., 2014). It causes serious problems in plant, animal and human beings. Standard wastewater treatment is very costly and produces high amount of dangerous byproducts, dirt production and also problems related to their disposal, hence treatment of wastewater biologically is relatively done at very low cost for the

Table 8.1 Dye removal treatment methods for textile effluents.

Physicochemical methods	Chemical methods	Biological methods

• Filtration
• Flocculation
• Adsorption
• Reverse osmosis
	• Advanced oxidation processes:
	  1. Fenton
	  2. Photocatalysis
	• Ozonation
		• Enzymes
		• Microbiological treatment
		  1. Activated sludge
		  2. Culture
		  3. Bacterial consortium

removal of dyes from the wastewater coming out from the industries. The victorious removal and degradation of these harmful dyes from the effluent only depends on the selection and grouping of the microorganisms such as bacteria, fungi, algae, etc., which causes the conversion of toxic substances into nontoxic substances (Shyamal Gowri et al., 2014). The success of the dye treatment can be determined by the activity and pliability of the microorganisms. Adsorption of dye on microbial biomass or by the biodegradable activity of cells and enzyme of dyes are the two major methods by which the textile dyes may be decolorized by the microbes. The biomass is useful when the wastewater effluent is more toxic and does not favor the microorganisms to grow and maintain their cells. Following microbes such as bacteria, algae, fungi are used as an adsorbent and the dye does not breakdown into fragments by adsorption. Rather than the process of biosorption, in the process of biodegradation, the dye's original structure becomes disrupted to cause decomposition properly or entirely. Hence, biodegradation is more feasible and empirical option for the degradation of dyes (Khan et al., 2013) (Table 8.1).

8.14 Biodegradation of dyes by bacteria

In 1970, the work on biodegradation of dyes by the use of bacterial strain Bacillus subtilis were started (Horitsu et al., 1977), then other number of bacterial strain were used for the degradation of dyes such as *Aeromonas hydrophila* (Idaka and Ogawa, 1998), *Lysinibacillus fusiformis* JTP-23, *Bacillus cereus* (Wuhrmann et al., 1980), *E. Coli*, and *Pseudomonas* sp. (Kulla, 1981). *Bacillus subtilis, Rhabdobacter* sp., *Micrococcus, Dermacoccus* sp., *Lactobacillus*, and *Aeromonas* are additional bacterial strains reported for the biodegradation of azo dyes (Stolz, 2001; Pearce et al., 2003; Olukanni et al., 2006; Vijaykumar et al., 2006; Banat et al., 1996; Lin and Leu, 2008). The azo dyes do not easily get metabolized in aerobic conditions, during the biodegradation of azo dyes intermediates formed which directly causes the disturbance in metabolic pathways and the dyes were not lapidify, whereas under anaerobic conditions, azo dyes get reduced by the bacterial soluble and unspecific enzyme, cytoplasmic reductase also known as azo reductase

(Shyamala Gowri et al., 2014). Aromatic amines were produced as a result of azo reductase enzyme activity and these aromatic amines may cause toxicity, mutation and cancer to animals (Mcmulan et al., 2001). Distinct members of the bacterial consortium require distinct circumstances for favorable reaction and the azo-based cleft needs the enzyme azo reductase which requires anaerobic condition to perform their activity. Different members of the (Haug et al., 1991). Some of the bacterial strains of aerobic use acts as a root of carbon and nitrogen (Coughlin et al., 2002); other strains only causes the reduction of azo groups by the special enzyme azo reductase which tolerates oxygen.

8.15 Decolorization of azo dyes by bacteria

The first step is the reductive breakdown of the −N=N- bond of the biodegradation of azo dyes by the bacteria. By distinct groups of bacterial strains, the decolorization process of azo dyes under different oxygen level conditions such as complete absence of oxygen (anaerobic), methanogenic, low levels of oxygen (anoxic), and in the presence of oxygen (aerobic). It has been reported that the decolorization of dyes involves many halophiles (SalahUddin et al., 2007). For the decolorization, the modest halotolerant strain of bacteria *(Bacillus* sp.*)* was isolated for azo dyes to a level as high as 64.89%. Due to the high metabolic variety shown by halophiles, this decolorization rate of azo dyes may be seen as due to their extreme natural tolerance (Ventosa et al., 1998). After incubation in the stationary phase, the *Brevibacterium sp.* bacterial strain (VN-15) causes exotic reduction in the toxicity of the solution containing dyes. For the achievement of this phase, first of all, a link must be setup between the electron transport organization intracellularly and also the molecules of azo dye with high molecular mass. For the setup of this type of link, it is important that the constituents of the electron transport organization must translocate to the outer layer of the bacterial cells (only when it is gram-negative bacteria), so that with the surface of azo dye or the redox mediator present at the surface of the cell, they can establish a direct contact (Myers and Myers, 1992). The biodegradation of Navitan fast blue, which is also dye by the *Pseudomonas aeruginosa* (gram-negative) bacterial strain was investigated by Nachiyar and Rajkumar (2003). Ammonium salts and glucose were needed by the organisms for the codegradation of the dye. A fount of organic nitrogen did not assist the decolorization of dyes, whereas a fount of inorganic nitrogen worked in the direction of widening of both the decolorization and the growth of dyes. Senan and Abraham (2004) reported that the bacterial strain of gram-negative bacteria (*Pseudomonas putida*) was evolved for the aerobic biodegradation of textile azo dyes in combination or individual molecules of azo dyes at basic pH (9−10) and salinity (0.9−3.8 g/L) at the temperature of 28.2°C. The degradation ability of bacterial strains in distinct media and at variable dye concentration were also reported and studied. Crude supernatant contains enzymes

Table 8.2 Biodegradation of dyes by bacteria.

Strain: Bacteria (Phugare et al., 2011; Saratale et al., 2011; Kuhad et al., 2004; Pandey et al., 2007; Enayatzamir et al., 2009)	
Organisms	**Dye degraded by bacterial strains**
Rhizobium radiobacter	Reactive red 141
Pseudomonas luteola	Reactive azo dyes
Citrobacter sp., CK3	Reactive red 180
Enterococcus faecalis YZ66	Reactive orange 2
Enterobacter agglomerans	Methyl Red
Bacillus subtilis	Acid blue 113
Geotrichum sp.	Reactive black 5 Reactive red 158 Reactive yellow 27
Citrobacter sp.	Crystal violet Gentian violet Malachite green Brilliant green Basic fuchsin Methyl red Congo red
Serratia sp.	Reactive red 195

which are reused for the degradation of dyes. The range of decolorization of dyes by the strain Bacillus cereus was found up to 95% in ideal conditions and by *Bacillus megaterium* strain was found up to 98% (Shah et al., 2013). By the recent inspection it is believed that the decolorization by the bacteria (*Bacillus cereus* and *Bacillus megaterium*) of azo dyes under in vitro conditions. In previous studies of the degradation of anthraquinone dye, a number of indigenous bacterial strains were isolated such as *Bacillus* sp., *Rhodococcus* sp., and *Klebsiella* sp. (Wang et al., 2013) (Table 8.2).

8.16 Biodegradation of dyes by fungi

Much research has been done on white-rot fungi for degrading and lapidifying synthetic dyes. These white-rot fungi produce certain enzymes such as extracellular oxidoreductases that help in the breakdown of lignin and aromatic complexes. Its enzyme system consists of lignin peroxidases (LiPs), laccase, and manganese peroxidase (MnP), which are structurally indifferent and nonstereoselective. Laccase are extensively studied for the elimination of dyestuffs and phenols from the wastewater coming out rom industries which is produce by *Phanerochaete chrysosporium* and *Neurospora crassa* (Singh et al., 2015). Strains of *Pleurotus* and *Phlebia* such as

Trametes versicolor and *Aspergillus ochraceus* were also recognized (Saratale et al., 2011). Motivating results were obtained for industrial wastewater treatment only by the use of gel capturing and stickiness to a matrix (Saratale et al., 2011). Despite the favorable outcome, the fact of not naturally found or absence of white-rot fungi in wastewater makes the unpredictable enzyme production. In addition, the other disadvantages related to white-rot fungi are such as it requires long time to grow and its depending nature for nutrient limitations. Limitation associated with decolorization of dyes due to its long hydraulic holding time for the total decolorization (Khan et al., 2013). *P. chrysosporium,* a strain of white-rot fungi, can clamp onto the Ca-alginate beads used to decolorize distinct recalcitrant dyes such as Direct violet 51, Ponceau xylidine, and Reactive black 5 in consecutive batch cultures (Enayatzamir et al., 2009). The majority of organo-pollutants are get metabolized by fungi and is closely associated with ligninolytic metabolism. Decolorization of dye is connected to the procedure of enzyme extracellular oxidases, specifically manganese peroxidases (Singh et al., 2012). Comparative examinations of the three fungi were conducted on the basis of the time course for decolorization of dyes out of which *Penicillium chrysogenum* and *Aspergillus niger* shows a high order of activity with the result of 100% decolorization of dyes. Kumar Praveen and Sumangala (2012), has reported high-level activity associated with decolorization by *Cladosporium* sp. The decolorization of dyes was mainly affected by some of the physicochemical parameters such as pH of the culture medium, components and its concentration within a culture medium and temperature of the culture medium. The type of fungal strain also affects the decolorization of dyes. Forss and Welander (2009) has recorded that fungi due to its ability of secretion of extracellular enzymes which are capable of breaking of some of the structures for surely but not completely that bacteria are unable to breakdown or degrade. Degradation of various dyes by microorganisms has been reported such as Congo Red which is get degraded by fungal strain *Gliocladium virens* (Singh and Singh, 2010), and other harmful dyes were also degraded, such as Acid Red and Bromophenol blue by the fungal strain *Trichoderma harzianum* (Singh and Singh, 2010). It is reported that by the use of various distinct strains of fungus, plant wastes substances were also get degraded Singh and Singh (2010). Equivalent results were obtained from the degradation of Congo Red dye by fungal strains as well as the degradation of bromophenol blue dye by another fungal strain, *T. harzianum*, in a culture which is semisolid in nature (Singh and Singh 2010). Some species of *Aspergillus* fungal strain were responsible for the degradation of Gentian violet, Malachite green, Cotton blue, Methylene blue, Methyl red, and Sudan **black** (Muthezhilan et al., 2008) in a liquid medium. It is investigated that few of the azo dyes were also degraded by fungal strain *P. chrysosporium* such as Congo red dye, Tropaeolin O and Orange 2 (Singh and Singh, 2010). It is investigated that the remote fungal strain *A. niger* and *A. oryzae* and their mixed strain group are an essential source for the consumption and breakdown of various toxic dyes (Singh and Singh, 2010). It is found that

Table 8.3 Degradation of dyes by fungal strains.

Strain: Fungi (Sen et al., 2016; McMullan et al., 2001; Ali, 2010; Bumpus, 2004; Rani et al., 2014; Couto, 2009)	
Organisms	**Dyes degraded by fungal strain**
Acremonium kiliense	Malachite green
Aspergillus niger	Reactive red 120
	Direct red 81
Funalia trogii	Reactive blue 19
	Reactive blue 49
	Acid violet 43
	Reactive orange 16
	Acid black 52
Trametes villosa	Drimarene brilliant blue
Pleurotus ostreatus	Remazol brilliant blue R
Trametes versicolor	Amaranth
	Congo red
Aspergillus ochraceus	Reactive blue 25
Pycnoporus sanguineus	Drimarene brilliant blue
Shewanella sp. NTO4	Crystal violet
Phanerochaete chrysosporium	Orange 2

A. niger shows its greater activity toward the decolorization of dye within 16 days of incubation (Manikandan, 2012). White-rot fungi are capable of secreting enzymes which are unspecific and extracellular oxidative such as laccase which are responsible for the degradation of lignin (Banat et al., 1996). It is investigated on a large scale that one of the fungal strain Pleurotus eryngii F032 has its greater use in the biodegradation of naphthalene (Hadibarata et al., 2013). With the help of their powerful process and mechanism of action laccase is able to breakdown the phenolic and nonphenolic complexes, whereas the production of LiPsP from the fungal strains will intensify the decolorization of dye (Hadibarata et al., 2012; Hazeroual et al., 2006). The decolorization of dyes is the subsidiary metabolic activity which is connected to the fungus lignolytic breakdown activity. The breakdown of some xenobiotic is known to happen under the state which is nonlignolytic and it would mostly be through the activity of the fungal secreted laccase enzyme (Dhawale et al., 1992). Some broad-spectrum dyes that are structurally distinct, such as azo, triphenylmethane, anthraquinone, and polymeric dyes, are not only blanched or decolorized but also degraded and lapidified by Basidiomycetes fungi, and these fungi also break down the number of toxic organic complexes. Basidiomycetes fungi have an unspecific enzyme system associated with the degradation of pollutants that also reacts to the composition of pollutants (Machado et al., 2005; Wesenberg et al., 2003) (Table 8.3).

8.17 **Phytoremediation of dyes**

Phytoremediation attempts to make use of plants and the microorganism associated with the root system of plants for the protection of environment by elimination of pollutants from the environment in form of organic wastes and inorganic wastes. Phytoremediation is also able to treat the pollutants from different sources which comes from dye waste. The basic reason behind the phytoremediation of dye is the adaptation of plants at genetic level. There are different type of pollutants in the environment among them effluents and textile dye are predominant pollutants (Bharathiraja et al., 2018). Various plants namely *Scirpus grossus*, aquatic plant *Spirodela polyrrhiza*, water hyacinth *Eichhornia crassipes*, *Tecoma stans* var. *angustata*, have been discussed for their capacity of degradation of dyes. *Gaillardia grandiflora* and *Petunia grandiflora* consortium has been recognized for its part in the degradation of dyes.

Plants are able to eliminate pollutants by acting as filters. Roots of the plant mostly uptake contaminants present at the site and thus save the environment from the toxicity caused by these contaminants (Mahar et al. 2016). Thus, contaminants coming from different sources must be prepared to be absorbed/taken up by roots, and the phenomenon helps in the removal of various dyes combined with the bioavailability of the dyes. The essential nutrients and minerals absorbed by the root system from the soil along with these nutrient contaminants are also taken up as a result of adaptation.

Complex toxic organic pollutants are degraded into nontoxic, simpler organic complexes (carbon, hydrogen, and oxygen) in nature. A mixed supply of textile wastewater and canal water showed improved growth and improved yield in mustard (*Brassica campestris* L.) (Yaseen et al., 2017). Recently, many researchers have studied the unfamiliar area of textile dye removal by phytoremediation.

A major amount of dye wastewater comes from the textile industries, and this has drastically increased the number of suspended particles (solid), and BOD has created unfavorable pH in wastewater (Yaseen and Scholz, 2018). Different varieties of ferns, weeds, agricultural wastes, and grasses have been used for the removal of pollutants. Native *Phragmites australis* populations have been studied widely for textile effluent remediation and especially for Acid orange 7 dye removal. Yaseen et al. (2017) studied the removal of dye Basic red 46 under seminatural and controlled conditions by the plant *Lemna minor* L. through a pond system in a warmer region with a low concentration of textile dyes in wastewater. Degradation, dissipation, accumulation, and immobilization are the processes used by plants for wastewater treatment.

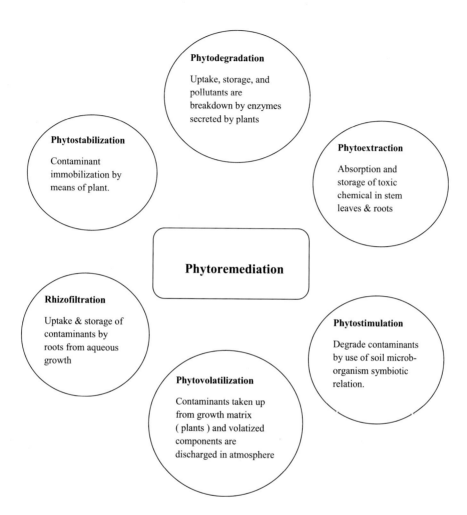

Török et al. (2015) reported the tolerance of plant *L. minor* on two synthetic dyes, Malachite green and Crystal violet. Stress response studies of the plant *L. minor* stated that uptake of 80% of Crystal violet happened because of phytoextraction and 90% of Malachite green was due to phytodegradation. Nisha and Emilia (2016) used a callus culture of *T. stans* var. *angustata* (plant) and reported the effect of biodegradation on the dye Brilliant green which is teratogenic and carcinogenic in nature. It was observed that removal of contaminants and pollutants from wastewater was due to the ability of the enzyme peroxidase to neutralize the ill effects of dyes in the presence of hydrogen peroxide. Some research work and studies have shown that the association of plants and microorganisms (e.g., bacteria and fungi) within the plant microbiome improves the process of phytoremediation compared with the use of either plants or microbes alone because of fungal—clay interactions, mycogenic mineral formation, element cycling, and bioweathering (Deng and Cao, 2017). Phytoremediation processes are widely classified on the basis of applicability, type of contaminant, and fundamental processes.

8.18 Conclusion

Dyes are an integral part of life, and dye contamination poses a threat to health. Biodegradation and bioremediation are important strategies to help in the eco-friendly degradation of dyes. Industries that use dyes produce dyes containing effluents that are detrimental to the health of the environment and all its components. Air, water, soil, and the biosphere are all affected by dye toxicity. Synthetic dyes are now a global concern due to their side effects. Their persistence needs remediation. Biodegradation and bioremediation are promising and effective approaches.

References

Adeel, S., Rehman, F.U., Rafi, S., Zia, K.M., Zuber, M., 2019. Environmentally friendly plant-based natural dyes: extraction methodology and applications. In: Plant and Human Health, vol. 2. Springer, Cham, pp. 383—415.

Ali, H., 2010. Biodegradation of synthetic dyes—a review. Water, Air, Soil Pollut. 213 (1), 251—273.

Banat, I.M., Nigam, P., Singh, D., Marchant, R., 1996. Microbial decolorization of textile-dye containing effluents: a review. Bioresour. Technol. 58 (3), 217—227.

Bharathiraja, B., Jayamuthunagai, J., Praveenkumar, R., Iyyappan, J., 2018. Phytoremediation techniques for the removal of dye in wastewater. In: Bioremediation: Applications for Environmental Protection and Management. Springer, Singapore, pp. 243—252.

Bhargava, P., Gupta, N., Kumar, R., Vats, S., 2020. Plants and microbes: bioresources for sustainable development and biocontrol. In: Plant Microbe Symbiosis. Springer, Cham, pp. 153—176.

Bumpus, J.A., 2004. Biodegradation of azo dyes by fungi. Mycology 21, 457—470.

Choudhury, A.K.R., 2018. Eco-friendly dyes and dyeing. Adv. Mater. Technol. Environ. 2, 145—176.

Cooksey, C.J., 2020. Hematoxylin in the 21st century. Biotech. Histochem. 1—8.

Coughlin, M.F., Kinkle, B.K., Bishop, P.L., 2002. Degradation of acid orange 7 in an aerobic biofilm. Chemosphere 46, 11—19.

Couto, S.R., 2009. Dye removal by immobilised fungi. Biotechnol. Adv. 27 (3), 227—235.

Deng, Z., Cao, L., 2017. Fungal endophytes and their interactions with plants in phytoremediation: a review. Chemosphere 168, 1100—1106.

Dhawale, S.W., Dhawale, S.S., Dean Ross, D., 1992. Degradation of phenanthrene by Phanerochaete chrysosporium occurs under ligninotytic as well as nonligninolytic conditions. Appl. Environ. Microbiol. 58, 3000—3006.

El Harfi, S., El Harfi, A., 2017. Classifications, properties and applications of textile dyes: a review. Appl. J. Environ. Eng. Sci. 3 (3), 00000-3.

Elbanna, K., Sarhan, O.M., Khider, M., Elmogy, M., Abulreesh, H.H., Shaaban, M.R., 2017. Microbiological, histological, and biochemical evidence for the adverse effects of food azo dyes on rats. J. Food Drug Anal. 25 (3), 667−680.

Emilia Abraham, T., Rani, N.D., 2016. A Potential Tissue Culture Approach for the Phytoremediation of Dyes in Aquaculture Industry.

Enayatzamir, K., Alikhani, H.A., Couto, S.R., 2009. Simultaneous production of laccase and decolouration of the diazo dye reactive black 5 in a fixed-bed bioreactor. J. Hazard Mater. 164 (1), 296−300.

Forss, J., Welander, U., 2009. Decolourization of reactive azo dyes with microorganisms growing on soft wood chips. Int. Biodeterior. Biodegrad. 63 (6), 752−758.

Hadibarata, T., Yusoff, A.R.M., Kristanti, R.A., 2012. Decolorization and metabolism of anthraquinone-type by laccase of white-rot fungi Polyporus sp. S133. Water Air Soil Pollut. 223, 933−941.

Hadibarata, T., Teh, Z.C., Rubiyatno, Z., M. M, F.A., Khudhair, A.B., Yusoff, A.R.M., et al., 2013. Identification of naphthalene metabolism by white rot fungus *Pleurotus eryngii*. Bioproc. Biosyst. Eng. https://doi.org/10.1007/s00449-013-0884-8.

Haug, W., Schmidst, A., Nortemann, B., Hempel, D.C., Stolz, A., Knackmuss, H.J., 1991. Mineralization of the sulphonated azo dye mordant yellow 3 by a 6aminonaphthalene 2 sulphonate degrading bacterium consortium. Appl. Environ. Microbiol. 57, 3144−3149.

Hazeroual, Y., Kim, B.S., Kim, C.S., Blaghen, M., Lee, K.M., 2006. Biosorption of bromophenol blue from aqueous solution by *Rhizopus stolonifer* biomass. Water Air Soil Pollut. 177, 135−146.

Horitsu, H., Takada, M., Idaka, E., Tomoyeda, M., Ogawa, T., 1997. Degradation of aminoazo benzene by *Bacillus subtilis*. Eur. J. Appl. Microbiol. 4, 217−224.

Hossain, L., Sarker, S.K., Khan, M.S., 2018. Evaluation of present and future wastewater impacts of textile dyeing industries in Bangladesh. Environ. Dev. 26, 23−33.

Hussain, C.M. (Ed.), 2020. The Handbook of Environmental Remediation: Classic and Modern Techniques. Royal Society of Chemistry.

Idaka, E., Ogawa, Y., 1998. Degradation of azo compounds by *Aeromonas hydroplhila* Var.2413. J. Soc. Dyers Colorists 94, 91−94.

Ihsanullah, I., Jamal, A., Ilyas, M., Zubair, M., Khan, G., Atieh, M.A., 2020. Bioremediation of dyes: current status and prospects. J. Water Process Eng. 38, 101680.

Kaur, A., Vats, S., Rekhi, S., Bhardwaj, A., Goel, J., Tanwar, R.S., Gaur, K.K., 2010. Physico-chemical analysis of the industrial effluents and their impact on the soil microflora. Procedia Environ. Sci. 2, 595−599.

Khan, R., Bhawana, P., Fulekar, M.H., 2013. Microbial decolorization and degradation of synthetic dyes: a review. Rev. Environ. Sci. Biotechnol. 75−97.

Kobylewski, S., Jacobson, M.F., 2012. Toxicology of food dyes. Int. J. Occup. Environ. Health 18 (3), 220−246.

Krishnamoorthy, R., Jose, P.A., Ranjith, M., Anandham, R., Suganya, K., Prabhakaran, J., et al., 2018. Decolourisation and degradation of azo dyes by mixed fungal culture consisted of *Dichotomomyces cejpii* MRCH 1-2 and *Phoma tropica* MRCH 1-3. J. Environ. Chem. Eng. 6 (1), 588−595.

Kuhad, R.C., Sood, N., Tripathi, K.K., Singh, A., Ward, O.P., 2004. Developments in microbial methods for the treatment of dye effluents. Adv. Appl. Microbiol. 56, 185−213.

Kulla, H.G., Leisinger, T., Cook, A.W., Hutter, R., Nuesch, J., 1981. Aerobic Bacterial Degradation of Azo Dyes in Microbial Degradation of Xenobiotics and Recalcitrant Compounds. FEMS Symposium. Academic Press, London, pp. 387−399.

Kumar Praveen, G.N., Sumangala, K.B., 2011. Fungal degradation of azo dye- red 3BN and optimization of physico-chemical parameters. Int. J. Environ. Sci. 1 (No 6).

Lin, Y.H., Leu, J.Y., 2008. Kinetics of reactive azo-dye decolorization by *Pseudomonas luteola* in a biological activated carbon process. Biochem. Eng. J. 39, 457−467.

Machado, K.M.G., Matheus, D.R., Bononi, V.L.R., 2005. Liginolytic enzymes production and remazol brilliant blue R decolorization by tropical brazillian basidomycetes fungi. Braz. J. Microbiol. 36, 246−252.

Mahar, A., Wang, P., Ali, A., Awasthi, M.K., Lahori, A.H., Wang, Q., et al., 2016. Challenges and opportunities in the phytoremediation of heavy metals contaminated soils: a review. Ecotoxicol. Environ. Saf. 126, 111−121.

Mani, S., Chowdhary, P., Bharagava, R.N., 2019. Textile wastewater dyes: toxicity profile and treatment approaches. In: Emerging and Eco-Friendly Approaches for Waste Management. Springer, Singapore, pp. 219−244.

Manikandan, N., Kuzhali, S.S., Kumuthakalavalli, R., 2012. Decolorisation of textile dye effluent using fungal microflora isolated from spent mushroom substrate (SMS). J. Microbiol. Biotech. Res. 2 (1), 57−62.

McMullan, G., Meehan, C., Conneely, A., Kirby, N., Robinson, T., Nigam, P., Banat, I.M., Marchant, R., Smyth, W.F., 2001. Microbial decolorization and degradation of the textile dyes. Appl. Microbiol. Bitechnol. 56 (1 -2), 81−87.

Mia, R., Selim, M., Shamim, A., Chowdhury, M., Sultana, S., 2019. Review on various types of pollution problem in textile dyeing & printing industries of Bangladesh and recommandation for mitigation. J. Textile Eng. Fashion Technol. 5 (4), 220−226.

Mohana, S., Shrivastava, S., Divehi, J., Medawar, D., 2008. Response surface methodolgy for optimization of medium for decolorization of textile dye Direct black 22 by a novel bacterial consortium. Bioresour. Technol. 99, 562−569.

Muthezhilan, R., Yogananth, N., Vidhya, S., Jayalakshmi, S., 2008. Dye degrading mycoflora from industrial effluents. Res. J. Microbiol. 3 (3), 204−208.

Myers, C.R., Myers, J.M., 1992. Localization of cytochromes to the outer membrane of anaerobically grown *Shewanella putrefaciens* MR-1. J. Bacteriol. 174 (11), 3429−3438.

Nachiyar, C.V., Rajkumar, G.S., 2003. Degradation of a tannery and textile dye, Navitan fast blue S5R by *Pseudomonas aeruginosa*. World J. Microbiol. Biotechnol. 19 (6), 609−614.

Nakano, M., Omae, K., Takebayashi, T., Tanaka, S., Koda, S., 2018. An epidemic of bladder cancer: ten cases of bladder cancer in male Japanese workers exposed to ortho-toluidine. J. Occup. Health 60 (4), 307−311.

Olukanni, O.D., Osuntoki, A.A., Gbenle, G.O., 2006. Textile effluent biodegradation potentials of textile effluent adapted and non adapted bacteria. Afr. J. Biotechnol. 5, 1980−1984.

Pandey, A., Singh, P., Iyengar, L., 2007. Bacterial decolorization and degradation of azo dyes. Int. Biodeterior. Biodegrad. 59 (2), 73−84.

Pearce, C.I., Lloyd, J.T., Guthrie, J.T., 2003. The removal of color from textile waste water using whole bacteria cells: a review. Dyes Pigments 58, 179−196.

Phugare, S.S., Kalyani, D.C., Patil, A.V., Jadhav, J.P., 2011. Textile dye degradation by bacterial consortium and subsequent toxicological analysis of dye and dye metabolites using cytotoxicity, genotoxicity and oxidative stress studies. J. Hazard Mater. 186 (1), 713−723.

Raghunathan, V.K., Devey, M., Hawkins, S., Hails, L., Davis, S.A., Mann, S., et al., 2013. Influence of particle size and reactive oxygen species on cobalt chrome nanoparticle-mediated genotoxicity. Biomaterials 34 (14), 3559−3570.

Ramzan, H., 2019. eISSN 2521-0130 Ramzan et al. LGU. J. Life Sci. 2019 LGU Journal of LGU Society of LIFE SCIENCES life sciences strategies to control the pollution caused by textile industry effluents containing dyes. LGUJLS 3 (3), 163−174.

Rani, B., Kumar, V., Singh, J., Bisht, S., Teotia, P., Sharma, S., Kela, R., 2014. Bioremediation of dyes by fungi isolated from contaminated dye effluent sites for bio-usability. Braz. J. Microbiol. 45 (3), 1055−1063.

Richardson, R., 2017. Britain's Wild Flowers: A Treasury of Traditions, Superstitions, Remedies and Literature. Pavilion Books.

Saini, R.D., 2017. Textile organic dyes: polluting effects and elimination methods from textile waste water. Int. J. Chem. Eng. Res. 9 (1), 121−136.

Sakthisharmila, P.N.P.P., Palanisamy, P.N., Manikandan, P., 2018. Removal of benzidine based textile dye using different metal hydroxides generated in situ electrochemical treatment-A comparative study. J. Clean. Prod. 172, 2206−2215.

Salah Uddin, M., Jiti, Z., Yuan, Q., Jianbo, G., Wang, P., et al., 2007. Biodecolourization of Azo-dye acid red B under high salinity condition. Bull. Environ. Contamin. Toxicol. 79 (4), 440−444.

Saratale, R.G., Saratale, G.D., Chang, J.S., Govindwar, S.P., 2011. Bacterial decolorization and degradation of azo dyes: a review. J. Taiwan Inst. Chem. Eng. 42 (1), 138−157.

Sarkar, S., Banerjee, A., Halder, U., Biswas, R., Bandopadhyay, R., 2017. Degradation of synthetic azo dyes of textile industry: a sustainable approach using microbial enzymes. Water Conserv. Sci. Eng. 2 (4), 121−131.

Sen, S.K., Raut, S., Bandyopadhyay, P., Raut, S., 2016. Fungal decolouration and degradation of azo dyes: a review. Fungal Biol. Rev. 30 (3), 112−133.

Senan, R.C., Abraham, T.E., 2004. Bioremediation of textile azo dyes by aerobic bacterial consortium aerobic degradation of selected azo dyes by bacterial consortium. Biodegradation 15 (4), 275−280.

Shah, M.P., Patel, K.A., Nair, S.S., Darji, A.M., Maharaul, S., 2013. Optimization of environmental parameters on decolorization of remazol black B using mixed culture. Am. J. Microbiol. Res. 3, 53−56.

Shen, J.H., Jiang, Z.W., Liao, D.Q., Horng, J.J., 2021. Enhanced synergistic photocatalytic activity of TiO_2/oxidant for azo dye degradation under simulated solar irradiation: a determination of product formation regularity by quantifying hydroxyl radical-reacted efficiency. J. Water Process Eng. 40, 101893.

Shindy, A., 2017. Problems and solutions in colors, dyes and pigments chemistry: a review. Chem. Int. 3 (2), 97−105.

Shyamala Gowri, R., Vijayaraghavan, R., Meenambigai, P., 2014. Microbial degradation of reactive dyes- A review, 3 (Number 3), 421−436.

Singh, L., Singh, V.P., 2010. Biodegradation of textile dyes, bromophenol blue and congored by fungus *Aspergillus flavus*. Environ. We Int. J. Sci. Tech 5, 235−242.

Singh, L., Singh, V.P., 2015. Textile dyes degradation: a microbial approach for biodegradation of pollutants. In: Microbial Degradation of Synthetic Dyes in Waste Waters, Environmental Science and Engineering. Springer International Publishing Switzerland.

Singh, A.K., Singh, R., Soam, A., Shahi, S.K., 2012. Dye orange 3R by Aspergillus strain (MMF3) and their culture optimization. Curr. Discov. 1 (1), 7−12.

Songur, A., Ozen, O.A., Sarsilmaz, M., 2010. The toxic effects of formaldehyde on the nervous system. Rev. Environ. Contam. Toxicol. 105−118.

Stolz, A., 2001. Basic and applied aspects in the microbial degradation of azo dyes. Appl. Microbiol. Biotechnol. 56, 69−80.

Swanson, J.M., Kinsbourne, M., 1980. Food dyes impair performance of hyperactive children on a laboratory learning test. Science 207 (4438), 1485−1487.

Török, A., Buta, E., Indolean, C., Tonk, S., Silaghi-Dumitrescu, L., Majdik, C., 2015. Biological removal of triphenylmethane dyes from aqueous solution by Lemna minor. Acta Chim. Slov. 62 (2), 452−461.

Tsuda, S., Murakami, M., Matsusaka, N., Kano, K., Taniguchi, K., Sasaki, Y.F., 2001. DNA damage induced by red food dyes orally administered to pregnant and male mice. Toxicol. Sci. 61 (1), 92−99.

Vats, S., Kumar, R., 2015. Amylolytic-extremoenzymes: saviour of environments. Eur. J. Biomed. Pharmaceut. Sci. 2 (5), 694−702.

Vats, S., Negi, S., 2013. Use of artificial neural network (ANN) for the development of bioprocess using *Pinus roxburghii* fallen foliages for the release of polyphenols and reducing sugars. Bioresour. Technol. 140, 392−398.

Ventosa, A., Nieto, J.J., Oren, A., 1998. Biology of moderately halophilic aerobic bacteria. Microbiol. Mol. Biol. Rev. 62 (2), 504−544.

Vijaykumar, M.H., Veeranagouda, Y., Neelkanteshwar, K., Karegoudar, T.B., 2006. Degradation of 1:2 metal complex dye acid blue 193 by a newly isolated fungus *Cladosporium cladosporioide*. World J. Microbiol. Biotechnol. 22, 157−162.

Wang, Z.W., Liang, J.S., Liang, Y., 2013. Decolorization of Reactive Black 5 by a newly isolated bacterium Bacillus sp. YZU1. Int. Biodeterior. Biodegr. 76, 41−48.

Wesenberg, D., Kyziakydes, I., Agathos, S.N., 2003. White rot fungi and their enzymes for the treatment of industrial dye effluents. Biotechnol. Adv. 22, 162−187.

Wuhrman, K., Mechsner, K.I., Kappeler, T.H., 1980. Investigation on rate determining factors in the microbial reduction of azodyes. Eur. J. Appl. Microbiol. 9, 325−338.

Xiao, X., Li, T.T., Lu, X.R., Feng, X.L., Han, X., Li, W.W., et al., 2018. A simple method for assaying anaerobic biodegradation of dyes. Bioresour. Technol. 251, 204−209.

Yaseen, D.A., Scholz, M., 2018. Treatment of synthetic textile wastewater containing dye mixtures with microcosms. Environ. Sci. Pollut. Control Ser. 25 (2), 1980−1997.

Yaseen, M., Aziz, M.Z., Komal, A., Naveed, M., 2017. Management of textile wastewater for improving growth and yield of field mustard (*Brassica campestris* L.). Int. J. Phytoremediation 19 (9), 798−804.

Zeiger, E., 2019. The test that changed the world: the Ames test and the regulation of chemicals. Mutat. Res. Genet. Toxicol. Environ. Mutagen 841, 43−48.

Zollinger, H., 1987. Color Chemistry- Synthesis, Properties and Application of Organic Dyes and Pigments. VCH Publishers, New York, pp. 92−100.

Advances in bioremediation of industrial wastewater containing metal pollutants

9

Vadivel Karthika[1], Udayakumar Sekaran[2], Gulsar Banu Jainullabudeen[3], Arunkumar Nagarathinam[4]

[1]*Department of Crop Management, Kumaraguru Institute of Agriculture, Erode, Tamil Nadu, India;* [2]*Department of Plant and Environmental Sciences, Clemson University, Clemson, SC, United States;* [3]*Central Institute for Cotton Research, Regional Station, Coimbatore, Tamil Nadu, India;* [4]*Department of Microbiology, School of Agriculture and Animal Sciences, The Gandhigram Rural Institute, Dindigul, Tamil Nadu, India*

Chapter outline

Biological Approaches to Controlling Pollutants. https://doi.org/10.1016/B978-0-12-824316-9.00001-X

9.1 Introduction

Heavy metal pollution is a major environmental problem that reduces crop production and food quality due to excessive application of agricultural inputs such as fertilizers and pesticides (Su, 2014). Most pesticides are organic compounds, a few are organic—inorganic compounds or pure minerals, and some contain Hg, As, Cu, Zn, and other heavy metals (Arao et al., 2010). Unlike organic contaminants, metals are not degradable and thus remain in the environment for long periods. At high concentrations, these metal pollutants can negatively affect plant metabolism (Ferraz et al., 2012). There is a need for innovative treatment technologies to remove heavy metal ions from the environment, particularly from industrial wastewater. Various microbes have been proposed as efficient and economical alternatives for the removal of heavy metals from soil and water through the process of bioremediation (Ahirwar et al., 2016). Bioremediation is an option to transform toxic heavy metals into a less harmful state using microbes or their enzymes and is an eco-friendly, cost-effective technique for revitalizing wastewater-polluted environments (Vidali, 2001).

9.2 Sources of heavy metal contaminants

Heavy metal contamination is the exposure to chromium, lead, cadmium, mercury, and arsenic metals that enter the food chain and cause serious, lethal health concerns for humans (Järup, 2003). Heavy metal contamination occurs through a number of human activities (Table 9.1), such as mining, smelting, electroplating, pesticide use, sludge dumping, and phosphate fertilizer use as well as biosolids in agriculture (Ali et al., 2013). Contamination of agricultural fields with heavy metals such as chromium (Cr) is toxic to human beings and plants and has been a major environmental concern for several decades (Tiwari et al., 2013). The release of heavy metals into the environment is due to electroplating, leather tanning, metal finishing, corrosion control, and pigment manufacturing (Liu et al., 2011).

9.3 Role of microbes in bioremediation process

Physicochemical methods for removing heavy metals from contaminated soils is very difficult, expensive, and not feasible for large-scale applications (Danh et al., 2009). Hence, bioremediation implies a process that is more feasible and economical for the removal of heavy metals without additional contamination (Khudhaier et al., 2020). Bioremediation activity through microbes is stimulated by supplementing nutrients (nitrogen and phosphorus), an electron acceptor (oxygen), and substrates (methane, phenol, and toluene) or by introducing microorganisms with desired catalytic capabilities (Ma et al., 2007; Baldwin et al., 2008). Some common microorganism used in the process of remediation are *Achromobacter, Alcaligenes, Arthrobacter, Bacillus, Acinetobacter, Corynebacterium, Flavobacterium, Micrococcus, Mycobacterium, Nocardia, Pseudomonas, Vibrio, Rhodococcus,* and

Table 9.1 Anthropogenic sources of specific heavy metals in the environment.

S. No	Heavy metal	Sources	Reference
1.	Arsenic (As)	Building materials and electronics industry	Christodoulidou et al. (2012) Yang et al. (2015b)
2.	Barium (Ba)	Paper making, pesticides, and chemical industry	Giri and Singh (2014) Gutierrez et al. (2012)
3.	Chromium (Cr)	Automobile manufacturing, vehicle exhaust, chemical industry, and electroplating	Chan et al. (2011) Avudainayagam et al. (2003) Johnson et al. (2006)
4.	Vanadium (V)	Smelting, chemical industry, and battery manufacturing	Giri and Singh (2014)
5.	Cobalt (Co)	Building materials, smelting, chemical industry, and battery manufacturing	Shan et al. (2013), Petrucci et al. (2014)
6.	Lead (Pb)	Welding, mining, building materials, smelting, chemical industry, battery manufacturing, fertilizers, and pesticides	Huang et al. (2007)
7.	Zinc (Zn)	Vehicle components, electroplating, building materials, and electronics Industry	Huston et al. (2012) Mendiguchía et al. (2007) Robert-Sainte et al. (2009)
8.	Iron (Fe)	Smelting, chemical industry, mining, pharmaceuticals, and pesticides	Zarazúa et al. (2013)
9.	Manganese (Mn)	Food service industry, smelting, mining, and refrigeration	Brankov et al. (2012)
10.	Molybdenum (Mo)	Smelting, electronics industry, pharmaceuticals, and pesticides	Shan et al. (2013)
11.	Nickel (Ni)	Electroplating, fertilizers and pesticides, mining, vehicle exhaust, and smelting	Lee et al. (2016) Huston et al. (2012)
12.	Selenium (Se)	Smelting, glass, ceramics, electronics, and food	Silva et al. (2013)
13.	Tin (Sn)	Smelting and electroplating	Karbasdehi et al. (2016)
14.	Strontium (Sr)	Smelting and electronics industry	Uchida and Tagami (2017)

Sphingomonas species (Gupta et al., 2001; Kim et al., 2007; Jayashree, 2012). The main species involved in effective wastewater treatment include lactic acid bacteria—*Lactobacillus plantarum, L. casei,* and *Streptococcus lactis,* and photosynthetic bacteria *Rhodopseudomonas palustris* and *Rhodobacter sphaeroides* (Narmatha and Kavitha, 2012).

9.4 Mechanism of microbial detoxification of heavy metals

Microorganisms adopt different mechanisms to interact and survive in the presence of inorganic metals. Mechanisms used by microbes to survive metal toxicity are biotransformation, extrusion, enzyme use, exopolysaccharide production, and metallothionein synthesis (Fig. 9.1) (Dixit et al., 2015; Igiri et al., 2018). In response to metals in the environment, microorganisms have developed ingenious mechanisms of metal resistance and detoxification. The mechanisms involve several procedures together with electrostatic interaction, ion exchange, precipitation, redox process, and surface complexation (Yang et al., 2015a). The major mechanical means by which microorganisms resist heavy metals are metal oxidation, methylation, enzymatic decrease, metal–organic complexion, metal decrease, metal–ligand degradation, metal efflux pumps, demethylation, intracellular and extracellular metal sequestration, exclusion by permeability barrier, and production of metal chelators such as metallothioneins and biosurfactants (Ramasamy et al., 2006).

Microorganisms can decontaminate metals by valence conversion, volatilization, or extracellular chemical precipitation (Ramasamy et al., 2006). Microorganisms have a negative charge on their cell surface because of the presence of anionic structures that empower the microbes to bind to metal cations (Gavrilescu, 2004). The negatively charged sites of microbes involved in the adsorption of metal are the hydroxyl, alcohol, phosphoryl, amine, carboxyl, ester, sulfhydryl, sulfonate, thioether, and thiol groups (Gavrilescu, 2004). The uptake of heavy metals by microbial cells through biosorption mechanisms can be classified into metabolism-independent biosorption, which mostly occurs on the exterior of the cell, and metabolism-dependent bioaccumulation, which comprises sequestration, redox reaction, and species-transformation methods (Vijayaraghavan and Yun, 2008;

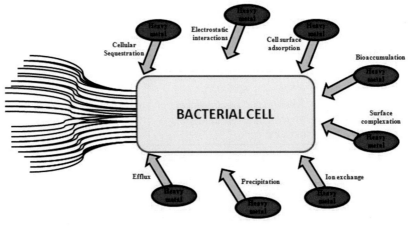

FIGURE 9.1

Mechanism of microbial detoxification of heavy metals.

Godlewska and Zyłkiewicz, 2006). Biosorption can be carried out by dead biomass or living cells as passive uptake through surface complexation onto the cell wall and surface layers. Bioaccumulation depends on a variety of chemical, physical, and biological mechanisms, and these factors are intracellular and extracellular processes, where biosorption plays a limited and ill-defined role (Fomina and Gadd, 2014). Elaborate research on these various mechanisms has been conducted previously by many researchers and is documented in detail as follows.

9.4.1 Intracellular sequestration

Intracellular sequestration is the complexation of metal ions by various compounds in the cell cytoplasm. The concentration of metals within microbial cells can result from interaction with surface ligands followed by slow transport into the cell. The ability of bacterial cells to accumulate metals intracellularly has been exploited in practice, predominantly in effluent treatment. The cadmium-tolerant *Pseudomonas putida* strain possesses the ability of intracellular sequestration of copper, cadmium, and zinc ions with the help of cysteine-rich low molecular weight proteins (Higham et al., 1986). In addition, intracellular sequestration of cadmium ions by glutathione has been revealed in *Rhizobium leguminosarum* cells (Lima et al., 2006).

The rigid cell wall of fungi is made up of chitin, mineral ions, lipids, nitrogen-containing polysaccharide, polyphosphates, and proteins. They can decontaminate metal ions by energetic uptake, extracellular and intracellular precipitation, and valence conversion, with several fungi accumulating metals to their mycelium and spores. The exterior of the cell wall of fungi behaves like a ligand used for labeling metal ions and brings about the elimination of inorganic metals (Xie et al., 2016; Gupta et al., 2015; Selvam et al., 2013; Jha ct al., 2011). Peptidoglycan, polysaccharide, and lipid are components of the cell wall rich in metal-binding ligands (e.g., $-OH$, $-COOH$, $-HPO42-$, $SO42-$ $-RCOO-$, R_2OSO_3-, $-NH2$, and $-SH$). The amine present in the cell can be more active in metal uptake among these functional groups, as it binds to anionic metal species via electrostatic interaction and cationic metal species through surface complexation.

9.4.2 Extracellular sequestration

Extracellular sequestration is the accumulation of metal ions by cellular components in the periplasm or complexation of metal ions as insoluble compounds. Copper-resistant *Pseudomonas syringae* strains produce copper-inducible proteins CopA, CopB (periplasmic proteins), and Cop C (outer membrane protein) that bind copper ions and microbial colonies (Cha and Cooksey., 1991). Bacteria can eject metal ions from the cytoplasm to sequester the metal within the periplasm. Zinc ions can cross from the cytoplasm by the efflux system where they are accumulated in the periplasm of the *Synechocystis* PCC 6803 strain (Telwell et al., 1998).

Iron-reducing bacteria such as *Geobacter* spp. and sulfur-reducing bacteria such as *Desulfuromonas* spp. can reduce harmful metals to less toxic or nontoxic forms.

G. metallireducens, a strict anaerobe, can reduce manganese (Mn) from lethal Mn (IV) to Mn (II) and uranium (U) from poisonous U (VI) to U (IV) (Gavrilescu, 2004). *G. sulfurreducens* and *G. metallireducens* have the ability to decrease chromium (Cr) from the very lethal Cr (VI) to less toxic Cr (III) (Bruschi and Florence, 2006). Sulfate-reducing bacteria generate large amounts of hydrogen sulfide, which causes precipitation of metal cations (Luptakova and Kusnierova, 2005; White and Knowles, 2000). The *Klebsiella planticola* strain generates hydrogen sulfide from thiosulfate under anaerobic conditions and precipitates cadmium ions as insoluble sulfides (Sharma et al., 2000).

9.4.3 Extracellular barrier preventing metal entry into microbial cell

The microbial plasma membrane, cell wall, or capsule could prevent metal ions from entering the cell. Bacteria can absorb metal ions by ionizable groups of the cell wall (amino, carboxyl, phosphate, and hydroxyl groups). El-Helow et al. (2000), Pardo et al. (2003), Taniguchi et al. (2000), and Green-Ruiz (2006) observed a high level of passive biosorption of heavy metal ions for nonviable cells of *Pseudomonas putida*, *Brevibacterium* sp., and *Bacillus* sp. *Pseudomonas aeruginosa* biofilm cells show higher resistance to ions of copper, lead, and zinc than that of planktonic cells, whereas cells located at the periphery of the biofilm are killed. Extracellular polymers of biofilm accumulate metal ions and then protect bacterial cells inside the biofilm (Teitzel and Parsek, 2003).

9.4.4 Methylation of metals

Methylation increases metal toxicity because of increased lipophilicity and thus increases permeation across cell membranes. Microbial methylation plays a significant function in metal remediation. Methylated compounds are regularly explosive; for instance, Hg (II) can be biomethylated by bacteria such as *Bacillus* spp., *Escherichia* spp., *Clostridium* spp., *and Pseudomonas* spp. to gaseous methyl mercury. Biomethylation of selenium (Se) to volatile dimethyl selenide and arsenic (As) to gaseous arsines, as well as lead (Pb) to dimethyl lead, has been witnessed in polluted topsoil (Ramasamy et al., 2006).

9.4.5 Reduction of heavy metal ions by microbial cells

Microbial cells can convert metal ions from one oxidation state to another, hence reducing their harmfulness (Jyoti and Harsh, 2014). Bacteria use metals and metalloids as electron donors or acceptors for energy generation. Metals in oxidized form could serve as terminal acceptors of electrons during the anaerobic respiration of bacteria. Reduction of metal ions through an enzymatic activity could result in the formation of a less toxic form of mercury and chromium (Barkay et al., 2003; Viti et al., 2003).

9.4.6 **Bioremediation capacity of microorganisms on heavy metals**

The uptake of heavy metals by microorganisms occurs via bioaccumulation, which is an active process, or through adsorption, which is a passive process. Several microorganisms such as bacteria, fungi, and algae have been used to clean up heavy metal—contaminated environments (Kim et al., 2015; Neha et al., 2013). The application of metal-resistant strains in single, consortium, and immobilized form for the remediation of heavy metals has yielded effective results, while the immobilized form could have more chemisorption sites to absorb heavy metals.

9.4.7 **Bacteria remediation capacity of heavy metals**

Microbial biomass has different biosorption abilities that vary significantly among microbes. The biosorption ability of each microbial cell depends on its pretreatment and on experimental conditions. The microbial cell must adapt to alteration of the physical, chemical, and bioreactor configuration to enhance biosorption (Fomina and Gadd, 2014). Bacteria are important biosorbents due to their ubiquity, size, and ability to grow under controlled conditions as well as their resilience to environmental conditions (Srivastava et al., 2015; Wang and Chen, 2009).

De (2008) used mercury-resistant bacteria such as *Alcaligenes faecalis, Bacillus pumilus, Pseudomonas aeruginosa,* and *Brevibacterium iodonium* for the removal of cadmium (Cd) and lead (Pb). In the study, *P. aeruginosa* and *A. faecalis* removed 70% and 75% cadmium (Cd) with a reduction of 1000 to 17.4 mg/L of cadmium (Cd) by *P. aeruginosa* and to 19.2 mg/L by *A. faecalis* in about 72 h. *B. iodonium* and *B. pumilus* remove greater than 87% and 88% of lead (Pb) with a reduction of 1000 to 1.8 mg/L in 96 h. In another study, (Singh et al., 2013) used indigenous facultative anaerobic *Bacillus cereus* to detoxify hexavalent chromium. *B. cereus* has an excellent capacity of 72% Cr (VI) removal at 1000 μg/mL chromate concentration. The bacteria could reduce Cr (VI) under a wide range of temperatures (25—40°C) and pH (6—10) with an optimum at 37°C and initial pH 8.0.

Several heavy metals have been tested using bacteria species such as *Flavobacterium, Pseudomonas, Enterobacter, Bacillus,* and *Micrococcus* sp. Their great biosorption ability is due to high surface-to-volume ratios and potentially active chemosorption sites (teichoic acid) on the cell wall (Mosa et al., 2016). Bacteria are more stable and survive better when they are in mixed culture (Sannasi et al., 2006). Therefore, consortia of cultures are metabolically superior for biosorption of metals and are more appropriate for field application (Kader et al., 2007).

Abioye et al. (2018) investigated the biosorption of lead (Pb), chromium (Cr), and cadmium (Cd) in tannery effluent using *Bacillus subtilis, B. megaterium, Aspergillus niger,* and *Penicillium* sp. *B. megaterium* recorded the highest lead (Pb) reduction (2.13—0.03 mg/L), followed by *B. subtilis* (2.13—0.04 mg/L). Kim et al.(2015), designed a batch system using zeolite-immobilized *Desulfovibrio desulfuricans* for the removal of chromium (Cr^{6+}), copper (Cu), and nickel (Ni) with a removal efficiency of 99.8%, 98.2%, and 90.1%, respectively. Ashruta (2014) reported efficient removal of chromium, zinc, cadmium, lead, copper, and cobalt by bacterial consortia of 75%—85% within less than 2 h of contact duration.

9.4.8 Fungi remediation capacity of heavy metals

Fungi are widely used as biosorbents for the removal of toxic metals with excellent capacities for metal uptake and recovery (Fu et al., 2012; Akar et al., 2005; Dursun et al., 2003). Most studies have shown that active and lifeless fungal cells play a significant role in the adhesion of inorganic chemicals (Karakagh et al., 2012; Vankar and Bajpai, 2008). Srivastava and Thakur (2006) reported the efficiency of *Aspergillus* sp. in removing 85% of chromium at pH 6 in a bioreactor system from synthetic medium compared with 65% removal from the tannery effluent. *Coprinopsis atramentaria* is studied for its ability to bioaccumulate 76% of Cd^{2+} at a concentration of 1 mg/L of Cd^{2+}, and 94.7% of Pb^{2+}, at a concentration of 800 mg/L of Pb^{2+}. Therefore, it has been documented as an effective accumulator of heavy metal ions for mycoremediation (Lakkireddy and Kues, 2017). Park (2005) reported that dead fungal biomass of *A. niger*, *Rhizopus oryzae*, *Saccharomyces cerevisiae*, and *Penicillium chrysogenum* could be used to convert toxic Cr (VI) to less toxic or nontoxic Cr (III). Luna et al. (2016) also observed that *Candida sphaerica* produces biosurfactants with a removal efficiency of 95%, 90%, and 79% for Fe (iron), zinc (Zn), and lead (Pb), respectively. These surfactants could form complexes with metal ions and interact directly with heavy metals before detachment from the soil. *Candida* spp. accumulate a substantial quantity of nickel Ni (57%−71%) and copper Cu (52%−68%), but the process was affected by initial metal ion concentration and pH (optimum 3−5) (Donmez and Aksu, 2001).

Biosurfactants have received increased interest in recent years owing to their low toxicity, biodegradable nature, and diversity. Mulligan et al. (2001) assessed the viability of using surfactin, rhamnolipid, and sophorolipid for the removal of heavy metals (Cu and Zn). A single washing with 0.5% rhamnolipid removed 65% of copper (Cu) and 18% of the zinc (Zn), whereas 4% sophorolipid removed 25% of the copper (Cu) and 60% of zinc (Zn). Several strains of yeast such as *Hansenula polymorpha*, *S. cerevisiae*, *Yarrowia lipolytica*, *Rhodotorula pilimana*, *Pichia guilliermondii*, and *Rhodotorula mucilage* have been used to bio-convert Cr (VI) to Cr (III) (Chatterjee et al., 2012; Ksheminska et al., 2006; Ksheminska et al., 2008).

9.4.9 Heavy metal removal using biofilm

There are several reports on the application of biofilms for the removal of heavy metals. Biofilm acts as a proficient bioremediation tool as well as biological stabilization agent. Biofilms have a very high tolerance to toxic inorganic elements even at a lethal concentration. It was revealed in a study conducted on *Rhodotorula mucilaginosa* that metal removal efficiency was from 4.79% to 10.25% for planktonic cells and from 91.71% to 95.39% for biofilm cells (Grujic et al., 2017). Biofilms mechanisms of bioremediation could either be via biosorbent or by exopolymeric substances present in biofilms that contain molecules with surfactant or emulsifier properties (Masry et al., 2004).

9.4.10 **Algae remediation capacity of heavy metals**

Algae are autotrophic and hence require low nutrients and produce enormous biomass compared with other microbial biosorbents. These biosorbents have also been used for heavy metal removal with a high sorption capacity (Abbas et al., 2014). Algae biomass is used for bioremediation of heavy metal—polluted effluent via adsorption or by integration into the cells. Phytoremediation is the use of various types of algae and cyanobacteria for the remediation of heavy metals by either removal or degradation of toxicant (Chabukdhara et al., 2017). Algae have various chemical moieties on their surface such as hydroxyl, carboxyl, phosphate, and amide, that act as metal-binding sites (Abbas et al., 2014; J. He and Chen, 2014). Goher (2016) used dead cells of *Chlorella vulgaris* to remove cadmium (Cd^{2+}), copper (Cu^{2+}), and lead (Pb^{2+}) ions from an aqueous solution under various conditions of pH, biosorbent dosage, and contact time. The results suggested that the biomass of *C. vulgaris* is an extremely efficient biosorbent for the removal of cadmium (Cd^{2+}), copper (Cu^{2+}), and lead (Pb^{2+}) at 95.5%, 97.7%, and 99.4%, respectively, from a mixed solution of 50 mg/dm of each metal ion.

9.4.11 **Immobilized biosorption of heavy metals**

The use of encapsulated biomass enhances biosorption performance and increases its physical and chemical stability. Immobilizations of microbial biomass in polymeric matrixes confer rigidity and heat resistivity with optimum porosity for practical applications. *Agrobacterium* biomass was encapsulated in alginate with iron oxide nanoparticles and showed an adsorption capacity of 197.02 mg/g for Pb and was effective for five consecutive cycles (Tiwari et al., 2017).

9.5 **Conclusion**

Industrial pollution has continued to be a major factor causing degradation of the environment around us, affecting the water we use, the air we breathe, and the soil we live on. The exponential increase in industrialization not only consumes large areas of agricultural lands but also simultaneously causes serious environmental degradation of soil. Water originating from various industries finds its place in agriculture. The challenge is to properly incorporate the disposal of the wastes in a controlled management program so that the applied industrial solid wastes do not contribute pollution to soil, soil microbes, or the environment. To stop pollution and prevent metal toxicity, there is a clear need for an overall waste treatment strategy with the goal of eliminating priority pollutants at their source. This can be achieved by indigenous microorganisms found in various industrial effluents that can be used as an indicator of pollution and to resist, process, metabolize, and detoxify chromate-polluted wastewater.

References

Abbas, H.S., Ismail, M.I., Mostafa, M.T., Sulaymon, H.A., 2014. Biosorption of heavy metals: a review. J. Chem. Sci. Technol. 3, 74–102.

Abioye, O.P., Oyewole, O.A., Oyeleke, S.B., Adeyemi, M.O., Orukotan, A.A., 2018. Biosorption of lead, chromium and cadmium in tannery effluent using indigenous microorganisms. Brazilian J. Biol. Sci. 5 (9), 25–32.

Ahirwar, N.K., Gupta, G., Singh, R., Singh, V., 2016. Isolation, identification and characterization of heavy metal resistant bacteria from industrial affected soil in central India. Int. J. Pure Appl. Biosci. 4 (6), 88–93.

Akar, T., Tunali, S., Kiran, I., 2005. Botrytis cinerea as a new fungal biosorbent for removal of Pb(II) from aqueous solutions. Biochem. Eng. J. 25 (3), 227–235.

Ali, H., Khan, E., Sajad, M.A., 2013. Phytoremediation of heavy metals—concepts and applications. Chemosphere 91, 869–881.

Arao, T., Ishikawa, S., Murakam, I.M., Abe, K., Maejima, Y., Makino, T., 2010. Heavy metal contamination of agricultural soil and counter measures in Japan. Paddy Water Environ. 8 (3), 247–257.

Ashruta, G.A., Nanoty, V., Bhalekar, U., 2014. Biosorption of heavy metals from aqueous solution using bacterial EPS. Int. J. Life Sci. 2 (3), 373–377.

Avudainayagam, S., Megharaj, M., Owens, G., Kookana, R.S., Chittleborough, D., Naidu, R., 2003. Chemistry of chromium in soils with emphasis on tannery waste sites. Rev. Environ. Contam. Toxicol. 178, 53–91.

Baldwin, B.R., Peacock, A.D., Park, M., Ogles, D.M., Istok, J.D., McKinley, J.P., 2008. Multilevel samplers as microcosms to assess microbial response to bio stimulation. Ground Water 46, 295–304.

Barkay, T., Miller, S.M., Summers, A.O., 2003. Bacterial mercury resistance from atoms to ecosystems. Fed. Eur. Microbiol. Soc. Microbiol. Rev. 27 (2–3), 355–384.

Brankov, J., Evi, D.M., Milanovi, A., 2012. The assessment of the surface water quality using the water pollution index: a case study of the Timok River (the Danube River Basin), Serbia. Arch. Environ. Protect. 38, 49–61.

Bruschi, M., Florence, G., 2006. New bioremediation technologies to remove heavy metals and radionuclides using Fe(III)-, sulfate-and sulfur-reducing bacteria. In: Singh, S.N., Tripathi, R.D. (Eds.), Environmental Bioremediation Technologies, pp. 35–55.

Cha, J.S., Cooksey, D.A., 1991. Copper resistance in Pseudomonas syringae mediated by periplasmic and outer membrane proteins. Proc. Natl. Acad. Sci. U.S.A. 88 (20), 8915–8919.

Chabukdhara, M., Gupta, S.K., Gogoi, M., 2017. Phycoremediation of heavy metals coupled with generation of bioenergy. Algal Biofuels 163–188.

Chan, Y.C., Hawas, O., Hawker, D., Vowles, P., Cohen, D.D., Stelcer, E., 2011. Using multiple type composition data and wind data in PMF analysis to apportion and locate sources of air pollutants. Atmos. Environ. 45, 439–449.

Chatterjee, S., Chatterjee, C.N., Dutta, S., 2012. Bioreduction of chromium (VI) to chromium (III) by a novel yeast strain Rhodotorula mucilaginosa (MTCC9315). Afr. J. Biotechnol. 1, 14920–14929.

Christodoulidou, M., Charalambous, C., Aletrari, M., Kanari, P.N., Petronda, A., Ward, N.I., 2012. Arsenic concentrations in groundwaters of Cyprus. J. Hydrol. 468 (469), 94–100.

Danh, L.T., Truong, P., Mammucari, R., Tran, T., Foster, N., 2009. Vetiver grass, Vetiveria Zizanioides: a choice plant for phytoremediation of heavy metals and organic wastes. Int. J. Phytoremediation 11, 664–691.

De, J., Ramaiah, N., Vardanyan, L., 2008. Detoxification of toxic heavy metals by marine bacteria highly resistant to mercury. Mar. Biotechnol. 10 (4), 471−477.

Dixit, Wasiullah, R., Malaviya, D., 2015. Bioremediation of heavy metals from soil and aquatic environment: an overview of principles and criteria of fundamental processes. Sustainability 7 (2), 2189−2212.

Donmez, G., Aksu, Z., 2001. Bioaccumulation of copper(ii) and nickel(ii) by the non-adapted and adapted growing Candida SP. Water Res. 35 (6), 1425−1434.

Dursun, A.Y., Uslu, G., Cuci, Y., Aksu, Z., 2003. Bioaccumulation of copper(II), lead(II) and chromium(VI) by growing Aspergillus Niger. Process Biochem. 38 (12), 1647−1651.

El-Helow, E.R., Sabry, S.A., Amer, R.M., 2000. Cadmium biosorption by a cadmium resistant strain of Bacillus thuringiensis: regulation and optimization of cell surface affinity for metal cations. Biometals 13 (4), 273−280.

Ferraz, P., Fidalgo, F., Almeida, A., Teixeira, J., 2012. Phytostabilization of nickel by the zinc and cadmium hyperaccumulator *Solanum nigrum* L. Are metallothioneins involved? Plant Physiol. Biochem. 57, 254−260.

Fomina, M., Gadd, G.M., 2014. Biosorption: current perspectives on concept, defnition and application. Bioresour. Technol. 160, 3−14.

Fu, Q.Y., Li, S., Zhu, Y.H., 2012. Biosorption of copper (II) from aqueous solution by mycelial pellets of *Rhizopus oryzae*. Afr. J. Biotechnol. 11 (6), 1403−1411.

Gavrilescu, M., 2004. Removal of heavy metals from the environment by biosorption. Eng. Life Sci. 4 (3), 219−232.

Giri, S., Singh, A.K., 2014. Risk assessment, statistical source identification & seasonal fluctuation of dissolved metals in the Subarnarekha River, India. J. Hazard Mater. 265, 305.

Godlewska-Zyłkiewicz, B., 2006. Microorganisms in inorganic chemical analysis. Anal. Bioanal. Chem. 384 (1), 114−123.

Goher, M.E., El-Monem, A.M.A., Abdel-Satar, A.M., Ali, M.H., Hussian, A.E.M., Napiorkowska-Krzebietke, A., 2016. Biosorption of some toxic metals from aqueous solution using nonliving algal cells of *Chlorella vulgaris*. J. Elementol. 21 (3), 703−714.

Green-Ruiz, C., 2006. Mercury (II) removal from aqueous solutions by nonviable Bacillus sp. from a tropical estuary. Bioresour. Technol. 97 (15), 1907−1911.

Grujic, S., Vasic, S., Radojevic, I., Comic, L., Ostojic, A., 2017. Comparison of the *Rhodotorula mucilaginosa* biofilm and planktonic culture on heavy metal susceptibility and removal potential. Water Air Soil Pollut. 228 (2), 73.

Gupta, V.K., Nayak, A., Agarwal, S., 2015. Bioadsorbents for remediation of heavy metals: current status and their future prospects. Environ. Eng. Res. 20 (1), 1−18.

Gupta, V.K., Shrivastava, A.K., Jain, N., 2001. Biosorption of chromium (VI) from aqueous solutions by green algae Spirigyra species. Water Res. 35 (17), 4079−4085.

Gutierrez, M., Gomez, V.M.R., Herrera, M.T.A., Daniel, N.L., 2012. Exploratory analysis of sediment geochemistry to determine the source and dispersion of Ba, Fe, Mn, Pb and Cu and in Chihuahua, Northern Mexico. J. Geogr. Geol. 4, 4.

He, J., Chen, J.P., 2014. A comprehensive review on biosorption of heavy metals by algal biomass: materials, performances, chemistry, and modeling simulation tools. Bioresour. Technol. 160, 67−78.

Higham, D.P., Sadler, P.J., Scawen, M.D., 1986. Cadmium-binding proteins in *Pseudomonas putida*: Pseudothioneins. Environ. Health Perspect. 65 (5−11).

Huang, S.S., Liao, Q.L., Hua, M., Wu, X.M., Bi, K.S., Yan, C.Y., 2007. Survey of heavy metal pollution and assessment of agricultural soil in Yangzhong District, Jiangsu Province, China. Chemosphere 67, 2148−2155.

Huston, R., Chan, Y.C., Chapman, H., Gardner, T., Shaw, G., 2012. Source apportionment of heavy metals and ionic contaminants in rainwater tanks in a subtropical urban area in Australia. Water Res. 46, 1121−1132.

Igiri, B.E., Okoduwa, S.I., Idoko, G.O., Akabuogu, E.P., Adeyi, A.O., Ejiogu, I.K., 2018. Toxicity and bioremediation of heavy metals contaminated ecosystem from tannery wastewater: a review. J. Toxicol. https://doi.org/10.1155/2018/2568038.

Järup, L., 2003. Hazards of heavy metal contamination. Br. Med. Bull. 68 (1), 167−182.

Jayashree, R., Nithya, S.E., Rajesh, P.P., Krishnaraju, M., 2012. Biodegradation capability of bacterial species isolated from oil contaminated soil. J. Academia Indust. Res. 1 (3), 127−135.

Jha, S., Dikshit, S., Pandy, G., 2011. Comparative study of agitation rate and stationary phase for the removal of cu^{2+} by *A. lentulus*. Int. J. Pharm. Biol. Sci. 2, 208−211.

Johnson, J., Schewel, L., Graedel, T.E., 2006. The contemporary anthropogenic chromium cycle. Environ. Sci. Technol. 40, 7060−7069.

Jyoti, B., Harsh, K.S.N., 2014. Utilizing *Aspergillus niger* for bioremediation of tannery effluent. Octa J. Environ. Res. 2 (1), 77−81.

Kader, J., Sannasi, P., Othman, O., Ismail, B.S., Salmijaj, S., 2007. Removal of Cr (VI) from aqueous solutions by growing and non-growing populations of environmental bacterial consortia. Global J. Environ. Resour. 1, 12−17.

Karakagh, R.M., Chorom, M., Motamedi, H., Kalkhajeh, Y.K., Oustan, S., 2012. Biosorption of Cd and Ni by inactivated bacteria isolated from agricultural soil treated with sewage sludge. Ecohydrol. Hydrobiol. 12 (3), 191−198.

Karbasdehi, V.N., Dobaradaran, S., Nabipour, I., Arfaeinia, H., Mirahmadi, R., Keshtkar, M., 2016. Data on metal contents (As, Ag, Sr, Sn, Sb, and Mo) in sediments and shells of *Trachycardium lacunosum*in the northern part of the Persian Gulf. Data Brief 8, 966−971.

Khudhaier, S.R., Al-Lami, A.M.A., Abbas, R.F., 2020. A review article-technology of bioremediation. Int. J. Res. Appl. Sci. Biotechnol. 7 (5), 349−353.

Kim, I.H., Choi, J.-H., Joo, J.O., Kim, Y.-K., Choi, J.-W., Oh, B.-K., 2015. Development of a microbe-zeolite carrier for the effective elimination of heavy metals from seawater. J. Microbiol. Biotechnol. 25 (9), 1542−1546.

Kim, S.U., Cheong, Y.H., Seo, D.C., Hu, J.S., Heo, J.S., Cho, J.S., 2007. Characterization of heavy metal tolerance and biosorption capacity of bacterium strains CPB4 (Bacillus Sp.). Water Sci. Technol. 55 (1), 105−111.

Ksheminska, H.P., Honchar, T.M., Gayda, G.Z., Gonchar, M.V., 2006. Extra-cellular chromate-reducing activity of the yeast cultures. Cent. Eur. J. Biol. 1 (1), 137−149.

Ksheminska, H., Fedorovych, D., Honchar, T., Ivash, M., Gonchar, M., 2008. Yeast tolerance to chromium depends on extracellular chromate reduction and Cr(III) chelation. Food Technol. Biotechnol. 46 (4), 419−426.

Lakkireddy, K., Kues, U., 2017. Bulk isolation of basidiospores from wild mushrooms by electrostatic attraction with low risk of microbial contaminations. AMB Express 7 (1), 28.

Lee, D.H., Kim, J.H., Mendoza, J.A., Lee, C.H., Kang, J.H., 2016. Characterization and source identification of pollutants in runoff from a mixed land use watershed using ordination analyses. Environ. Sci. Pollut. Res. Int. 23, 9774.

Lima, A.I.G., Corticeiro, S.C., de Almeida Paula Figueira, E.M., 2006. Glutathione-mediated cadmium sequestration in *Rhizobium leguminosarum*. Enzym. Microb. Technol. 39 (4), 763−769.

Liu, J., Duan, C., Zhang, X., Zhu, Y., Lu, X., 2011. Potential of *Leersia hexandra* swartz for phytoextraction of Cr from soil. J. Hazard Mater. 188, 85−91.

Luna, J.M., Rufno, R.D., Sarubbo, L.A., 2016. Biosurfactant from *Candida sphaerica* UCP0995 exhibiting heavy metal remediation properties. Process Saf. Environ. Protect. 102, 558–566.

Luptakova, A., Kusnierova, M., 2005. Bioremediation of acid mine drainage contaminated by SRB. Hydrometallurgy 77 (1–2), 97–102.

Ma, X., Novak, P.J., Ferguson, J., Sadowsky, M., Lapara, T.M., Semmens, M.J., 2007. The impact of H_2 addition on dechlorinating microbial communities. Ann. Finance 11, 45–55.

Masry, M H El, Bestawy, E El, Adl, N I El, 2004. Bioremediation of vegetable oil and grease from polluted wastewater using a sand bioflm system. World J. Microbiol. Biotechnol. 20 (6), 551–557.

Mendiguchía, C., Moreno, C., Garcíavargas, M., 2007. Evaluation of natural and anthropogenic influences on the Guadalquivir River (Spain) by dissolved heavy metals and nutrients. Chemosphere 69, 1509–1517.

Mosa, K.A., Saadoun, I., Kumar, K., Helmy, M., Dhankher, O.P., 2016. Potential biotechnological strategies for the cleanup of heavy metals and metalloids. Front. Plant Sci. 7.

Mulligan, C.N., Yong, R.N., Gibbs, B.F., 2001. Remediation technologies for metal-contaminated soils and groundwater: an evaluation. Eng. Geol. 60 (1–4), 193–207.

Narmadha, D., Kavitha, M.S., 2012. Treatment of domestic wastewater using natural flocculants. Int. J. Life Sci. Biotechnol. Pharma Res. 1 (3), 206–213.

Neha, S., Tuhina, V., Rajeeva, G., 2013. Detoxification of hexavalent chromium by an indigenous facultative anaerobic *Bacillus cereus* strain isolated from tannery effluent. Afr. J. Biotechnol. 12, 10.

Pardo, R., Herguedas, M., Barrado, E., Vega, M., 2003. Biosorption of cadmium, copper, lead and zinc by inactive biomass of *Pseudomonas putida*. Anal. Bioanal. Chem. 376 (1), 26–32.

Park, D., Yun, Y.-S., Jo, J.H., Park, J.M., 2005. Mechanism of hexavalent chromium removal by dead fungal biomass of *Aspergillus niger*. Water Res. 39 (4), 533–540.

Petrucci, G., Gromaire, M.C., Shorshani, M.F., Chebbo, G., 2014. Nonpoint source pollution of urban stormwater runoff: a methodology for source analysis. Environ. Sci. Pollut. Res. Int. 21, 10225.

Ramasamy, K., Kamaludeen, S., Parwin, B., 2006. Bioremediation of metals microbial processes and techniques. In: Singh, S.N., Tripathi, R.D. (Eds.), Environmental Bioremediation Technologies. Springer Publication, New York, NY, USA, pp. 173–187.

Robert-Sainte, P., Gromaire, M.C., Gouvello, B.D., Saad, M., Chebbo, G., 2009. Annual metallic flows in roof runoff from different materials: test-bed scale in Paris conurbation. Environ. Sci. Technol. 43, 5612–5618.

Sannasi, P., Kader, J., Ismail, B.S., Salmijah, S., 2006. Sorption of Cr(VI), Cu(II) and Pb(II) by growing and non-growing cells of a bacterial consortium. Bioresour. Technol. 97 (5), 740–747.

Selvam, K., Arungandhi, B., Vishnupriya, B., Shanmugapriya, T., Yamuna, M., 2013. Biosorption of chromium (vi) from industrial efuent by wild and mutant type strain of saccharomyces cerevisiae and its immobilized form. Biosci. Discov. 4 (1), 72–77.

Shan, Y., Tysklind, M., Hao, F., Ouyang, W., Chen, S., Lin, C., 2013. Identification of sources of heavy metals in agricultural soils using multivariate analysis and GIS. J. Soils Sediments 13, 720–729.

Sharma, P.K., Balkwill, D.L., Frenkel, A., Vairavamurthy, M.A., 2000. A new *Klebsiella planticola* strain (Cd-1) grows anaerobically at high cadmium concentrations and precipitates cadmium sulfide. Appl. Environ. Microbiol. 66 (7), 3083–3087.

Silva, L.F.O., Vallejuelo, F.O.D., Martinez-Arkarazo, I., Castro, K., Oliveira, M.L.S., Sampaio, C.H., 2013. Study of environmental pollution and mineralogical characterization of sediment rivers from Brazilian coal mining acid drainage. Sci. Total Environ. 447, 169−178.

Singh, N., Tuhina, V., Rajeeva, G., 2013. Detoxification of hexavalent chromium by an indigenous facultative anaerobic *Bacillus cereus* strain isolated from tannery effluent. Afr. J. Biotechnol. 12 (10), 1091−1103.

Srivastava, S., Takur, I.S., 2006. Isolation and process parameter optimization of Aspergillus sp. for removal of chromium from tannery effluent. Bioresour. Technol. 97 (10), 1167−1173.

Srivastava, S., Agrawal, S.B., Mondal, M.K., 2015. A review on progress of heavy metal removal using adsorbents of microbial and plant origin. Environ. Sci. Pollut. Control Ser. 22 (20), 15386−15415.

Su, C., 2014. A review on heavy metal contamination in the soil worldwide: situation, impact and remediation techniques. Environ. Skeptics Critics 3 (2), 24.

Taniguchi, J., Hemmi, H., Tanahashi, K., Amano, N., Nakayama, T., Nishino, T., 2000. Zinc biosorption by a zinc-resistant bacterium, Brevibacterium sp. strain HZM-1. Appl. Microbiol. Biotechnol. 54 (4), 581−588.

Teitzel, G.M., Parsek, M.R., 2003. Heavy metal resistance of bioflm and planktonic *Pseudomonas aeruginosa*. Appl. Environ. Microbiol. 69 (4), 2313−2320.

Telwell, C., Robinson, N.J., Turner-Cavet, J.S., 1998. An SmtBlike repressor from synechocystis PCC 6803 regulates a zinc exporter. Proc. Natl. Acad. Sci. U.S.A. 95 (18), 10728−10733.

Tiwari, K.K., Singh, N.K., Rai, U.N., 2013. Chromium phytotoxicity in radish (*Raphanus sativus*): effects on metabolism and nutrient uptake. Bull. Environ. Contam. Toxicol. 91, 339−344.

Tiwari, S., Hasan, A., Pandey, L.M., 2017. A novel bio-sorbent comprising encapsulated *Agrobacterium fabrum* (SLAJ731) and iron oxide nanoparticles for removal of crude oil co-contaminant, lead Pb(II). J. Environ. Chem. Eng. 5 (1), 442−452.

Uchida, S., Tagami, K., 2017. Comparison of coastal area sediment−seawater distribution coefficients (Kd) of stable and radioactive Sr and Cs. Appl. Geochem. 85, 148−153.

Vankar, P.S., Bajpai, D., 2008. Phyto-remediation of chrome-VI of tannery efuent by Trichoderma species. Desalination 222 (1−3), 255−262.

Vidali, M., 2001. Bioremediation. An overview. Pure Appl. Chem. 73 (7), 1163−1172.

Vijayaraghavan, K., Yun, Y.-S., 2008. Bacterial biosorbents and biosorption. Biotechnol. Adv. 26 (3), 266−291.

Viti, C., Pace, A., Giovannetti, L., 2003. Characterization of Cr(VI)-resistant bacteria isolated from chromium-contaminated soil by tannery activity. Curr. Microbiol. 46 (1), 1−5.

Wang, J.L., Chen, C., 2009. Biosorbents for heavy metals removal and their future. Biotechnol. Adv. 27 (2), 195−226.

White, V.E., Knowles, C.J., 2000. Effect of metal complexion on the bioavailability of nitriloacetic acid to *Chelatobacter heintzil* ATCC 2900. Achiv. Microbiol. 173 (5−6), 373−382.

Xie, Y., Fan, J., Zhu, W., 2016. Effect of heavy metals pollution on soil microbial diversity and bermudagrass genetic variation. Front. Plant Sci. 7, 775.

Yang, T., Chen, M., Wang, J., 2015a. Genetic and chemical modification of cells for selective separation and analysis of heavy metals of biological or environmental significance. Trac. Trends Anal. Chem. 66, 90–102.

Yang, Y.Y., Liu, L.Y., Guo, L.L., Lv, Y.L., Zhang, G.M., Lei, J., 2015b. Seasonal concentrations, contamination levels, and health risk assessment of arsenic and heavy metals in the suspended particulate matter from an urban household environment in a metropolitan city, Beijing, China. Environ. Monit. Assess. 187, 409.

Zarazúa, G., Ávila-Pérez, P., Tejeda, S., Valdivia-Barrientos, M., Zepeda-Gómez, C., Macedo-Miranda, G., 2013. Assessment of heavy metal Cr, Mn, Fe, Cu, Zn and Pb in water sombrerillo (*Hydrocotyle ranunculoides*) high river course Lerma, Mexico. Rev. Int. Contam. Ambient. 29, 17–24.

Advances in microbial and enzymatic degradation of lindane at contaminated sites

10

R. Parthasarathi[1], M. Prakash[2], R. Anandan[2], S. Nalini[3]

[1]Department of Agricultural Microbiology, Faculty of Agriculture, Annamalai University, Chidambaram, Tamil Nadu, India; [2]Department of Genetics and Plant Breeding, Faculty of Agriculture, Annamalai University, Chidambaram, Tamil Nadu, India; [3]Centre for Ocean Research (DST-FIST Sponsored Centre), MoES — Earth Science & Technology Cell (Marine Biotechnological Studies), Col. Dr. Jeppiaar Research Park, Sathyabama Institute of Science and Technology, Chennai, Tamil Nadu, India

Chapter outline

10.1 Introduction

Lindane is a chemical pesticide within the organochlorine group of insecticides, and its use was banned in the 1970s by most developed countries. Lindane is synthesized by the chlorination of benzene in the presence of UV light. This organochlorine insecticide is obtainable in two formulations, viz., technical grade hexachlorocyclohexane (HCH) and lindane. Technical-grade HCH is a mixture of different isomers: alpha-HCH (60%−70%), beta-HCH (5%−12%), gamma-HCH (10%−15%), delta-HCH (6%−10%), and epsilon −HCH (3%−4%). Lindane is the gamma isomer (>99% pure) of HCH (Ashizawa et al., 2005).

Lindane is used in crops, in warehouses, as a measure of public safety to control insect-borne diseases, and as a seed treatment agent along with fungicides (Donald et al., 1997). Lindane is also used to control lice and scabies in humans in products such as lotions, creams, and shampoos. The use of lindane has led to environmental pollution of global dimensions (Li and Macdonald, 2005).

In India, research studies have revealed the startling concentration of lindane in the Bay of Bengal, the groundwater of the Andhra Pradesh (Shukla et al., 2006), and in uniform diet samples in the agriculturally productive Punjab state (Battu et al., 2005).

Smith (1991) reported symptoms of acute toxicity in humans that include dizziness, headache, seizures, diarrhea, and irritation of the skin. Chronic toxicity effects were witnessed from long-term exposure to lindane. In human beings, chronic toxicity manifests in the gastrointestinal tract and cardiovascular and musculoskeletal systems. Lindane has been alleged to possess carcinogenic, persistent, bioaccumulative, and endocrine-disrupting properties (UNEP, 2005a, b).

10.2 Lindane and India

Lindane has been explicitly banned in 40 countries and restricted from import in 65 countries. However, it is being permitted for special uses by exemption (Orme and Kegley, 2006). Presently, India is well known as the largest consumer and only producer of lindane worldwide. Lindane-polluted soils are widespread in the country due to its continuous use and indiscriminate industrial production (Abhilash and Singh, 2010). In India, lindane is manufactured (1300 t per annum) by two major companies: (1) Kanoria Chemicals Ltd. (1000 t/a capacity) and (2) India Pesticides Ltd (IPL, 300 t/a capacity) (Abhilash and Singh, 2008a, b). Kanoria Chemicals has now stopped the production of lindane (Abhilash and Singh, 2009), and the only operating plant worldwide is IPL (Lucknow). Lindane is used in major food crops in Kerala; it is currently approved only for limited use, but it has shown increasing consumption over the years. The annual compound growth rate of lindane consumption is as high as 107.54% (Indira Devi, 2010).

10.3 Lindane degradation

Lindane can persist for a prolonged time in the soil matrix due to its interactions with soil organic matter and its poor solubility in water (7 mg/L), thus indicating its threat to public health. Lindane can be degraded by physicochemical and biological methods. Chemical degradation can be performed by photolytic (Balakrishnan et al., 2005), electrochemical (Gözmen et al., 2003), and ozonation processes (Boschin et al., 2007), and it has a very low rate of reaction constant with ozone (Ikehata and Gamal El-Din, 2005). The physical methods of degradation include incineration and thermal desorption, but these methods release highly toxic gases (Nagpal and Paknikar, 2006).

10.3.1 **Microbial diversity in lindane degradation**

Microbial diversity offers environmentally friendly options for the transformation of pesticides into less harmful substances. Different microorganisms use halogenated compounds as their growth substrate. The main reaction during microbial degradation of halogenated compounds is dehalogenation of organic halogen, i.e., halogen atoms are replaced by hydrogen or a hydroxyl group, which decreases the risk of toxic intermediate formation. The low aqueous solubility and chlorinated nature of lindane is responsible for its environmental persistence and resistance to degradation (Phillips et al., 2005). Despite its persistence, HCH biodegradation has been reported in various environments, including aerobic and anaerobic soils. The γ—HCH-degrading microorganisms include fungi, cyanobacteria, and anaerobic and aerobic bacteria. The majority of the members belong to the genera *Sphingomonas*, *Rhodanobacter*, and *Pandoraea* (Phillips et al., 2005). The γ-HCH biodegradation is largely an anaerobic process, and variable levels of anaerobic degradation of α, β, γ, and δ-HCH have been detected. Isolates capable of degrading one or more of the other four HCH isomers under anaerobic conditions include *Clostridium rectum* (Ohisa et al., 1980), *Clostridium sphenoides* (Heritage and MacRae, 1977), *Clostridium butyricum*, *Clostridium pasteurianum* (Heritage and MacRae, 1977), *Citrobacter freundii* (Jagnow et al., 1977), *Desulfovibrio gigas*, *Desulfovibrio africanus*, *Desulfococcus multivorans* (Boyle et al., 1999), and *Dehalobacter* sp. (van Doesburg et al., 2005).

Pesce and Wunderlin (2004) reported that bacteria and bacterial consortia under anaerobic conditions can degrade lindane. Previously, many authors have reported that under anaerobic conditions with mixed bacterial cultures in flooded soils, lake sediments, and sludges, the degradation of a chlorinated cycloaliphatic compound, γ -hexachlorocyclohexane (γ,-HCH, lindane) was highly successful (Fianko et al., 2013). In addition, there are reports of pure bacterial strains such as *Clostridium* sp. that can degrade lindane (Ohisa et al., 1980).

Senoo and Wada (1989) first reported an aerobic bacterial strain, *Pseudomonas paucimobilis* SS86, isolated from Japan that degraded HCH. Further, *Sphingomonas paucimobilis* UT26 is a nalidixic acid resistant strain of *P. paucimobilis* SS86 which degraded α-, γ- and δ-HCH aerobically (Senoo and Wada, 1989). Aerobic degradation of lindane was found to occur best with the two bacterial strains UT26, 3identified as *S. paucimobilis*, and a new isolate named RP5557T (Renaud et al., 1999). The pathway for aerobic HCH degradation in *S. paucimobilis* strain UT26 involves *lin* genes that encode the HCH biodegradation pathway (Miyazaki et al., 2006). The well-known bacterium capable of degrading lindane under anaerobic conditions, *P. paucimobilis* as *S. paucimobilis*, is associated with plant roots. Bacteria such as *Escherichia coli*, *Rhodanobacter lindaniclasticus*, and *Pandorea* species degrade lindane from 10% to 90%.

Two bacterial isolates (LIN-1 and LIN-3) from enrichment culture can grow on γ- HCH as a sole source of carbon and energy (Okeke et al., 2002). The growth characteristics and degradation capacity of *Arthrobacter citreus* BI-100 was noted in mineral salt medium with γ-HCH (100 mg/L) as the sole source of carbon (Datta et al., 2000). The ex situ and in situ degradation of lindane with different concentrations of (10 and 100 ppm) by *Azotobacter chroococcum* was studied by Anupama and Paul (2009), and they observed that lindane was completely degraded by the end of the period.

Gupta et al. (2000) reported degradation of γ-HCH by *Alcaligenes faecalis* isolated from agricultural fields. Manickam et al. (2006) isolated and characterized the *Microbacterium* sp. strain ITRC1. Subsequently, Manickam and his coworkers isolated a *Xanthomonas* sp. from contaminated soil that utilized γ-HCH as a sole carbon energy source by successive dechlorination. A root-colonizing HCH-degrading *Sphingomonas* isolated by a two-step enrichment process opens up the feasibility of rhizoremediation. *Pseudomonas aeruginosa* that degraded technical HCH under an aerobic batch process was investigated by Lodha et al. (2007). At between 20 and 50 mg/L technical HCH, the degradation was found to be lower, whereas at 1–10 mg/L, degradation efficiency was found to be 99%.

Bioremediation of lindane-contaminated soil by *Streptomyces* sp. M7 and its effects on the growth of *Zea mays* was reported by Benimeli et al. (2006). *Sphingobium quisquiliarum* P25 (T), degrading a yellow-pigmented HCH bacterial strain, was isolated from an HCH dump site situated in the northern part of India (Bala et al., 2009). There was a similar report by Dadhwal et al. (2009) regarding the isolation of another yellow-pigmented bacterium, *Sphingobium chinhatense*.

Zheng et al. (2011) reported suitability of bacterial strain *Sphingobium indicum* B90A for cold regions to decontaminate γ-HCH from soil/water. Comparative studies on the remediation potential of four rhizospheric bacterial species, viz., *Kocuria rhizophila*, *Microbacterium resistens*, *Staphylococcus equorum*, and *Staphylococcus cohnii* on lindane removal were reported by Abhilash et al. (2011).

10.3.1.1 Algal degradation

Algal biomass as a remediation agent is attractive because it often does not produce toxic substances, and being autotrophic, it produces a large biomass and has low nutrient requirements. Kuritz and Wolk (1995) reported that a cyanobacteria, *Anabaena* sp. strain PCC7120, and *Nostoc ellipsosporum* transformed lindane first to γ pentachlorocyclohexene and then to a mixture of chlorobenzenes; this process was cometabolic and depended on the presence of nitrate.

Environmental cyanobacterial species isolated from the two Egyptian Lakes (Qarun and Mariut) were exposed to 5 and 10 ppm lindane for 7 days and the growth inhibition or stimulation percentage of lindane removal efficiency (RE) was calculated by El-Bestawy et al. (2007). Lindane showed different degrees of toxicity or stimulation for the selected cyanobacteria. Stimulation of growth was noted between 0.0- and 13.16-fold higher than controls, while inhibition was between 0.0% and 100%. The study also revealed that Mariut species were more resistant to lindane toxicity than Qarun species were. Resistance to lindane among Qarun species was in the following order: *Oscillatoria* sp. 12, *Oscillatoria* sp. 13, *Synechococcus* sp., *Nodularia* sp., *Nostoc* sp., *Cyanothece* sp., and *Synechococcus* sp. Among the Mariut species, the order was *Microcystis aeruginosa* MA1, *Anabaena cylindrical*, *M. aeruginosa* MA15, *Anabaena spiroides,* and *Aphanizomenon flos-aquae.* Lindane was removed by all the species as either individuals or mixtures at both concentrations. The lindane RE percentage of Qarun species varied between 71.6% and 99.6%, whereas that of Mariut species was between 45.23% and 100.0%. Mixed-culture RE percentages were found to be between 91.6% and 100% at 5 ppm, while at 10 ppm, it was between 90.4% and 100%. Later, adaptation of microalgae to lindane for lindane bioremediation was reported by González et al. (2012) by adopting three concentrations of lindane exposure (4, 15, and 40 mg/). In these exposures, lindane-resistant cells showed great capacity to remove lindane until 99% of the lindane had been removed.

10.3.1.2 Actinomycetes degradation

The lindane removal ability of isolated actinomycetes, individually and as a mixed culture under controlled laboratory conditions, was evaluated. Mixed cultures were found to be more suitable for bioremediation than pure cultures because of their biodiversity, which could enhance environmental survival and thereby increase the catabolic pathways available for contaminant biodegradation (Murthy and Manonmani, 2007). Fuentes et al. (2010) reported Argentina isolates, namely *Streptomyces* sp. M7 (M7) and *Streptomyces coelicolor* A3 (ScA3), and four actino-mycete isolates (A2, A5, A8 and A11) that were cultivated as pure and mixed cultures in minimal medium with lindane (1.66 mg/L), and they found that native *Streptomycetes* showed the ability to grow as a microbial consortium and remove lindane.

10.3.1.3 Fungal degradation

Bumpus and Aust (1987) reported on biodegradation of lindane by the white rot fungi *Phanerochaete chrysosporium*. Mougin et al. (1997) reported on improved mineralization in soils, added with lindane by *Phanerochaete* sp. and the fungus was found to change the lindane degradation pathway by increasing the conversion of volatile intermediates to CO_2. Comparative studies of the white-rot fungus *Trametes hirsutus* with *P. chrysosporium* to degrade lindane in liquid culture was investigated by Singh and Kuhad (1999). It was shown that *T. hirsutus* degraded lindane faster than *P. chrysosporium*, but the mechanism of degradation for both the fungi looked to be identical (Singh and Kuhad, 2000).

Rigas et al. (2009) studied the performance of *Cyathus bulleri* and *Phanerochaete sordida*, two white rot fungi for γ-HCH degradation, and reported that *C. bulleri* degraded lindane more efficiently than *P. sordida* did. Certain fungi like *Pleurotus ostreatus* and subtropical white rot fungus DSPM95 also have the ability to degrade lindane (Tekere et al., 2002). Quintero et al. (2007) assessed degradation of HCH isomers present in a spiked soil by the white rot *Bjerkandera adusta* in a slurry system. At optimal conditions, maximal degradations of 94.5%, 78.5%, and 66.1% were attained after 30 d for γ-, α- and δ-HCH isomers. Biodegradation of lindane by The Phycomycetes fungus *Conidiobolus* 03-1-56, a non−white rot fungus, was reported by Nagpal et al. (2008). The fungus completely degraded lindane on the 5th day in the culture medium. Two *Fusarium* species (*F. poae* and *F. solani*) isolated from pesticide-contaminated soil showed better degradability of lindane and used as a sole carbon source when equated with the growth performance of other fungal isolates from the matching contaminated soil (Sagar and Singh, 2011).

10.3.2 Genes and enzymes for lindane degradation

Sphingobium japonicum UT26, converts γ-HCH to β-ketoadipate through the action of six enzymes, viz., Lin A (dehydrochlorinase), Lin B (halidohydrolase), Lin C (dehydrogenase), Lin D (reductive dechlorinase), Lin E (ring cleavage dioxygenase) and Lin F (reductase) (Endo et al., 2005). The *lin A* to *lin F* genes of *Sphingobium japonicum* UT26 are dispersed on the three large circular replicons: the *lin A*, *lin B*, and *lin C* genes on the 3.6-Mb chromosome I; the *lin F* gene on the 670-kb chromosome II; and the *lin DE* operon with its regulatory gene (*linR*) on a 185-kb plasmid, pCHQ1 (Nagata et al., 2007). These *lin* genes were also detected in other HCH-degrading bacteria such as *Sphingobium indicum* B90 (Kumari et al., 2002) and B90A (Dogra et al., 2004) from India and *Sphingobium francense* sp. from France (Cérémonie et al., 2006). Most *lin* genes present in these strains are found in close association with an insertion sequence, IS*6100* (Cérémonie et al., 2006).

Nagata et al. (2006) purified a haloalkane dehalogenase from *S. paucimobilis*, that degraded lindane. Dechlorinase synthesis in *Streptomyces* sp. M7 was induced when the microorganism was grown in lindane, as its sole source of carbon and energy (Cuozzo et al., 2009). The degradation pathway of lindane was extensively analyzed in *S. paucimobilis* UT26 (Nagata et al., 2006). The γ-HCH is transformed to 2, 5-dichlorohydroquinone via sequential reactions catalyzed by enzymes Lin A, Lin B, and Lin C. 2,5-dichlorohydroquinone, in turn, is metabolized by enzymes Lin D, Lin E, Lin F, Lin GH, Lin J to succinyl-CoA and acetyl-CoA that are further channeled into and metabolized in the tricarboxylic acid cycle. Mertens et al. (2006) showed a slow-release inoculation approach as a promising bioaugmentation tool for contaminated sites because it can help in removal of pollutants and can extend the degrading activity in contrast with outdated inoculation strategies. Isolating indigenous bacteria that are capable of metabolizing the pesticides received due attention nowadays as they can provide an eco-friendly method of in situ detoxification (Mulchandani et al., 1999). Autochthonous microbial populations have evolved over time in some contaminated environments, so as to adapt to these contaminants (Pahm and Alexander, 1993).

10.3.2.1 The lin *genes*

Aerobic degradation through *lin* genes was initially identified and characterized for UT26 and subsequently recovered from B90A as well. Similar *lin* genes have been identified for all other HCH-degrading sphingomonads (Dogra et al., 2004). In UT26, where the system is best characterized, the pathway is as follows: *lin A* encoding a dehydrochlorinase; *lin B* encoding a haloalkane dehalogenase; *lin C* encoding a dehydrogenase; *lin D* encoding a reductive dechlorinase; *lin E/lin Eb* encoding a ring cleavage oxygenase; *lin F* encoding a maleylacetate reductase; *lin GH* encoding an acyl-CoA transferase; and *lin J* encoding a thiolase, plus *linR/linI*, which are regulatory genes. Thus, *lin A* to *lin C* encode the enzymes responsible for the upper pathway, and *lin D* to *lin J* encode those enzymes for the lower pathway. Evidence across a variety of strains indicate that the *lin A*-encoded HCH dehydrochlorinase (Lin A) and the *lin B*-encoded haloalkane dehalogenase (Lin B) catalyze the dehydrochlorinase and hydrolytic dechlorinase reactions, respectively, in the upper pathway (Saez et al., 2017).

10.4 Future prospects

HCH degradation was originally motivated by environmental and health concerns due to HCH residues. Its aerobic degradation is now proving to be an excellent model for investigating fundamental issues in microbial and molecular evolution. The isomer complexity of HCH adds a further degree of difficulty, with major differences between isomers in the reactions by which their breakdown can be catalyzed under aerobic conditions. A better understanding of the biochemistry of Lin A and Lin B enzymes is detailed with a range of resolved isomers of putative substrates within the pathway. This will be a helpful for understanding its mechanism and the structural and functional relationships supporting that mechanism. One of these involves the mysterious Lin X enzymes, which are distantly related to Lin C and show some Lin C function in vitro but are apparently not essential for gamma-HCH degradation, at least in vivo. The major gap in our current knowledge of the system concerns the extent and biological significance of genetic variation both in the organization of the various *lin* genes and in the coding sequences of the key upstream *lin* A and *lin* B genes. The anaerobic pathway for HCH degradation remains poorly understood. The biostimulation and bioaugmentation approaches implemented thus far for the remediation of HCH-contaminated soils have met with some success for certain isomers. Both technologies are still in the initial developmental stage, and considerable further improvements are still needed, especially for large-scale land cleanup situations where costs must be lower for treatment to be a realistic option. Further work on application protocols, bioaugmentation, and strain development is clearly of the hour.

References

Abhilash, P.C., Singh, N., 2008a. Distribution of hexachlorocyclohexane isomers in soil samples from a small scale industrial area of Lucknow, North India, associated with lindane production. Chemosphere 73, 1011–1015.

Abhilash, P.C., Singh, N., 2008b. Influence of the application of sugarcane bagasse on Lindane mobility through soil columns: implication for bio treatment. Bioresour. Technol. 99, 8961–8966.

Abhilash, P.C., Singh, N., 2009. Seasonal variation in HCH isomers in open soil and lant-rhizospheric soil system of a contaminated environment. Environ. Sci. Pollut. Res. 16, 727−740.

Abhilash, P.C., Singh, N., 2010. *Withania somniferadunal* mediated dissipation of lindane from simulated soil: implications for rhizoremediation of contaminated soil. J. Soils Sediments 10, 272−282.

Abhilash, P.C., Srivastava, S., Singh, N., 2011. Comparative bioremediation potential of four rhizospheric microbial species against lindane. Chemosphere 82, 56−63.

Anupama, K.S., Paul, S., 2009. Ex situ and in situ biodegradation of lindane by *Azotobacter chroococcum*. J. Environ. Sci. Health B 45 (1), 58−66.

Ashizawa, A., Dorsey, A., Wilson, J.D., 2005. Toxicological Profile for Alpha-, Beta-, Gamma-, and Delta-Hexachlorocyclohexane. Agency for Toxic Substances and Disease Registry, Atlanta: GA.

Bala, K., Sharma, P., Lal, R., 2009. *Sphingobium quisquiliarum* sp. nov.,a hexachlorocyclo-hexane (HCH) degrading bacterium isolated from HCH contaminated soil. Int. J. Syst. Evol. Microbiol. 60 (2), 429−433.

Balakrishnan, V., Buncel, E., Vanloon, G., 2005. Micellar catalyzed degradation of fenitro-thion, an organophosphorous pesticide, in solution and soils. Environ. Sci. Technol. 39, 5824−5830.

Battu, R.S., Singh, B., Kang, B.K., Joia, B.S., 2005. Risk assessment through dietary intake of total diet contaminated with pesticide residues in Punjab, India, 1999−2002. Ecotoxicol. Environ. Saf. 62 (1), 132−139.

Benimeli, C.S., Castro, G.R., Chaile, A.P., Amoroso, M.J., 2006. Lindane removal induction by *Streptomyces* sp. M7. J. Basic Microbiol. 46, 348−357.

Boschin, G., D'Agostina, A., Antonioni, C., Locati, D., Arnoldi, A., 2007. Hydrolytic degra-dation of azimsulfuron, a sulfonylurea herbicide. Chemosphere 68, 1312−1317.

Boyle, A.W., Haggblom, M.M., Young, L.Y., 1999. Dehalogenation of lindane (β-hexachlor-ocyclohexane) by anaerobic bacteria from marine sediments and by sulfate-reducing bacteria. FEMS Microbiol. Ecol. 29, 379−387.

Bumpus, J.A., Aust, S.D., 1987. Biodegradation of DDT [1,1,1-trichloro-2,2-bis(4-chlorophenyl)ethane] by the white rot fungus *Phanerochaete chrysosporium*. Appl. Envi-ron. Microbiol. 53 (9), 2001−2008.

Cérémonie, H., Boubakri, H., Mavingui, P., Simonet, P., Vogel, T.M., 2006. Plasmid-encoded γ-hexachlorocyclohexane degradation genes and insertion sequences in *Sphingobium francense* (ex-*Sphingomonas paucimobilis* Sp+). FEMS Microbiol. Lett. 257, 243−252.

Cuozzo, S.A., Rollán, G.G., Abate, C.M., Amoroso, M.J., 2009. Specific dechlorinase activity in lindane degradation by *Streptomyces* sp. M7. World J. Microbiol. Biotechnol. 25, 1539−1546.

Dadhwal, M., Jit, S., Kumari, H., Lal, R., 2009. *Sphingobium chinhatense* sp. nov., a hexa-chlorocyclohexane (HCH)-degrading bacterium isolated from an HCH dumpsite. Int. J. Syst. Evol. Microbiol. 59, 3140−3144.

Datta, J., Maiti, A.K., Modak, D.P., Chakrabartty, P.K., Bhattacharyya, P., Ray, P.K., 2000. Metabolism of γ-hexachlorocyclohexane by *Arthrobacter citreus* strain BI-100: identifi-cation of metabolites. J. Gen. Appl. Microbiol. 46, 59−67.

Dogra, C., Raina, V., Pal, R., Suar, M., Lal, S., Gartemann, K.H., Holliger, C., van der Meer, J.R., Lal, R., 2004. Organization of *lin* genes and IS*6100* among different strains of hexachlorocyclohexane-degrading *Sphingomonas paucimobilis*: evidence for horizon-tal gene transfer. J. Bacteriol. 186, 2225−2235.

Donald, D.B., Block, H., Wood, J., 1997. Role of ground water on hexachlorocyclohexane (lindane) detections in surface water in western Canada. Environ. Toxicol. Chem. 16 (9), 1867–1872.

El-Bestawy, E.A., Abd El-Salam, A.Z., Mansy, A.E., 2007. Potential use of environmental cyanobacterial species in bioremediation of lindane-contaminated effluents. Int. Biodeterior. Biodegrad. 59, 180–192.

Endo, R., Kamakura, M., Miyauchi, K., Fukuda, M., Ohtsubo, Y., Tsuda, M., Nagata, Y., 2005. Identification and characterization of genes involved in the downstream degradation pathway of γ-hexachlorocyclohexane in Sphingomonas paucimobilis UT26. J. Bacteriol. 187, 847–853.

Fianko, J.R., Donkor, A., Lowor, S.T., Yeboah, P.O., 2013. Pesticide residues in fish from the Densu river basin in Ghana. Int. J. Brain Cognit. Sci. 7 (3), 1416–1426.

Fuentes, M.S., Benimeli, C.S., Cuozzo, S.A., Saez, J.M., Amoroso, M.J., 2010. Microorganisms capable to degrade organochlorine pesticides. Curr. Res. Technol. Educ. Topics Appl. Microbiol. Microbial Biotechnol. 2, 1255–1264.

González, R., García-Balboa, C., Rouco, M., Lopez-Rodas, V., Costas, E., 2012. Adaptation of microalgae to lindane: a new approach for bioremediation. Aquat. Toxicol. 109, 25–32.

Gözmen, B., Oturan, M.A., Oturan, N., Erbatur, O., 2003. Indirect electrochemical treatment of bisphenol: a in water via electrochemically generated Fenton's reagent. Environ. Sci. Technol. 37, 3716–3723.

Gupta, A., Kaushik, C.P., Kaushik, A., 2000. Degradation of hexachlorocyclohexane (HCH; α, β, γ and δ) by Bacillus circulans and Bacillus brevis isolated from soil contaminated with HCH. Soil Biol. Biochem. 32, 1803–1805.

Heritage, A.D., MacRae, I.C., 1977. Degradation of lindane by cell-free preparations of Clostridium sphenoides. Appl. Environ. Microbiol. 34, 222–224.

Ikehata, K., Gamal El-Din, M., 2005. Aqueous pesticide degradation by ozonation and ozone-based advanced oxidation processes: a review (Part I). Ozone Sci. Eng. 27 (2), 83–114.

Indira Devi, P., 2010. Pesticides in agriculture-a boon or a curse? A case study of Kerala. Econ. Polit. Wkly. 26, 199–207.

Jagnow, G., Haider, K., Ellwardt, P.C., 1977. Anaerobic dichlorination and degradation of hexachlorocyclohexane isomers by anaerobic and facultative anaerobic bacteria. Arch. Microbiol. 115, 285–292.

Kumari, R., Subudhi, S., Suar, M., Dhingra, G., Raina, V., Dogra, C., Lal, S., van der Meer, J.R., Holliger, C., Lal, R., 2002. Cloning and characterization of lin genes responsible for the degradation of hexachlorocyclohexane isomers by Sphingomonas paucimobilis strain B90. Appl. Environ. Microbiol. 68, 6021–6028.

Kuritz, T., Wolk, C.P., 1995. Use of filamentous cyanobacteria for biodegradation of organic pollutants. Appl. Environ. Microbiol. 61, 234–238.

Li, Y.F., Macdonald, R.W., 2005. Sources and pathways of selected organochlorine pesticides to the Arctic and the effect of pathway divergence on HCH trends in biota: a review. Sci. Total Environ. 342 (1–3), 87–106.

Lodha, B., Bhat, P., Kumar, M.S., Vaidya, A.N., Mudliar, S., Killedar, D.J., Chakrabarti, T., 2007. Bioisomerization kinetics of γ-HCH and biokinetics of Pseudomonas aeruginosa degrading technical HCH. Biochem. Eng. J. 35, 12–19.

Manickam, N., Mau, M., Schlömann, M., 2006. Characterization of the novel HCH-degrading strain, Microbacterium sp. ITRC1. Appl. Microbiol. Biotechnol. 69, 580–588.

Mertens, B., Boon, N., Verstraete, W., 2006. Slow-release inoculation allows sustained biodegradation of gamma-hexachlorocyclohexane. Appl. Environ. Microbiol. 72, 622–627.

Miyazaki, R., Sato, Y., Ito, M., Ohtsubo, Y., Nagata, Y., Tsuda, M., 2006. Complete nucleotide sequence of an exogenously isolated plasmid, pLB1, involved in γ-hexachlorocyclohexane degradation. Appl. Environ. Microbiol. 72 (11), 6923–6933.

Mougin, C., Pericaud, C., Dubroca, J., Asther, M., 1997. Enhanced mineralization of lindane in soils supplemented with the white rot basidiomycetes *Phanerochaete chrysosporium*. Soil Biol. Biochem. 29, 1321–1324.

Mulchandani, A., Kaneva, I., Chen, W., 1999. Detoxification of organophosphate pesticides by immobilized *Escherichia coli* expressing organophosphorus hydrolase on cell surface. Biotechnol. Bioeng. 63, 216–223.

Murthy, H.R., Manonmani, H.K., 2007. Aerobic degradation of technical hexachlorocyclohexane by a defined microbial consortium. J. Hazard Mater. 149 (1), 18–25.

Nagata, Y., Kamakura, M., Endo, R., Miyazaki, R., Ohtsubo, Y., Tsuda, M., 2006. Distribution of γ-hexachlorocyclohexane-degrading genes on three replicons in *Sphingobium japonicum* UT26. FEMS Microbiol. Lett. 256, 112–118.

Nagata, Y., Endo, R., Ito, M., Ohtsubo, Y., Tsuda, M., 2007. Aerobic degradation of lindane (γ-hexachlorocyclohexane) in bacteria and its biochemical and molecular basis. Appl. Microbiol. Biotechnol. 76, 741–752.

Nagpal, V., Paknikar, K.M., 2006. Integrated biological approach for the enhanced degradation of lindane. Int. J. Biotechnol. 15 (Suppl.), 400–406.

Nagpal, V, Srinivasan, M.C., Paknikar, K.M., 2008. Biodegradation of chexachlorocyclohexane (Lindane) by a non- white rot fungus *Conidiobolus* 03-1-56 isolated from litter. Indian J Microbiol 48 (1), 134–141.

Ohisa, N., Yamaguchi, M., Kurihara, N., 1980. Lindane degradation by cell-free extracts of *Clostridium rectum*. Arch. Microbiol. 125, 221–225.

Okeke, B.C., Siddique, T., Arbestain, M.C., Frankenberger, W.T., 2002. Biodegradation of γ-hexachlorocyclohexane (lindane) and α-hexachlorocyclohexane in water and a soil slurry by a *Pandoraea* species. Agric Food Chem. 50 (9), 2548–2555.

Orme, S., Kegley, S., 2006. PAN Pesticide Database. Pesticide Action Network. North America, San Francisco.

Pahm, M.A., Alexander, M., 1993. Selecting inocula for the biodegradation of organic compounds at low concentrations. Microb. Ecol. 25 (3), 275–286.

Pesce, S.F., Wunderlin, D.A., 2004. Biodegradation of lindane by a native consortium isolated from contaminated river sediment. Int. Biodeterior. Biodegrad. 54 (4), 255–260.

Phillips, T.M., Seech, A.G., Lee, H., Trevors, J.T., 2005. Biodegradation of hexachlorocyclohexane (HCH) by microorganisms. Biodegradation 16, 363–392.

Quintero, J.C., Lu-Chau, T.A., Moreira, M.T., Feijoo, G., Lema, J.M., 2007. Bioremediation of HCH presents in soil by the white-rot fungus *Bjerkandera adusta* in a slurry batch bioreactor. Int. Biodeterior. Biodegrad. 60, 319–326.

Renaud, S.M., Thinh, L.V., Parry, D.L., 1999. The gross chemical composition and fatty acid composition of 18 species of tropical Australian microalgae for possible use in mariculture. Aquaculture 170 (2), 147–159.

Rigas, F., Papadopoulou, K., Philippoussis, A., Papadopoulou, M., Chatzipavlidis, J., 2009. Bioremediation of lindane contaminated soil by *Pleurotus ostreatus* in non sterile conditions using multilevel factorial design. Water Air Soil Pollut. 197 (1–4), 121–129.

Saez, J.M., Alvarez, A., Fuentes, M.S., Amoroso, M.J., Benimeli, C.S., 2017. An overview on microbial degradation of lindane. In: Microbe-Induced Degradation of Pesticides. Springer International Publishing, Switzerland, pp. 191–212.

Sagar, V., Singh, D.P., 2011. Biodegradation of lindane pesticide by non white-rots soil fungus *Fusarium* sp. World J. Microbiol. Biotechnol. 27, 1747–1754.

Senoo, K., Wada, H., 1989. Isolation and identification of an aerobic γ-HCH-decomposing bacterium from soil. J. Soil Sci. Plant Nutr. 35 (1), 79–87.

Shukla, G., Kumar, A., Bhanti, M., Joseph, P.E., Taneja, A., 2006. Organochlorine pesticide contamination of ground water in the city of Hyderabad. Environ. Int. 32 (2), 244–247.

Singh, B.K., Kuhad, R.C., 1999. Biodegradation of lindane (γ-hexachlorocyclohexane) by the white-rot fungus *Trametes hirsutus*. Lett. Appl. Microbiol. 28 (3), 238–241.

Singh, B.K., Kuhad, R.C., 2000. Degradation of insecticides lindane (g-HCH) by white-rot fungi *Cyathus bulleri* and *Phanerochaete sordida*. Pest Manag. Sci. 56 (2), 142–146.

Smith, A.G., 1991. In: Hayes Jr., W.J. (Ed.), Handbook of Pesticides Toxicology. Academic Press Inc., New York, pp. 3–6.

Tekere, M., Ncube, I., Read, J.S., Zvauya, R., 2002. Biodegradation of the organochlorine pesticide, lindane by a sub-tropical white rot fungus in batch and packed bed bioreactor systems. Environ. Technol. 23 (2), 199–206.

UNEP, 2005a. Consideration of Chemicals Proposed for Inclusion in Annexes A, B and C of the Convention: Lindane. Lindane Proposal.

UNEP/POPS/POPRC.1/8. 24 August 2005 UNEP, November 2005b. Decision POPRC-1/6: Lindane. UNEP/POPS/POPRC.1/10.

van Doesburg, W., van Eekert, M.H., Middeldorp, P.J., Balk, M., Schraa, G., Stams, A.J., 2005. Reductive dechlorination of β-hexachlorocyclohexane (β-HCH) by a *Dehalobacter* species in coculture with a *Sedimentibacter* sp. Fed. Eur. Microbiol. Soc. Microbiol. Ecol. 54 (1), 87–95.

Zheng, G., Selvam, A., Wong, J.W., 2011. Rapid degradation of lindane (c-hexachlorocyclohexane) at low temperature by *Sphingobium* strains. Int. Biodeterior. Biodegrad. 65, 612–618.

Advances in bioremediation of nonaqueous phase liquid pollution in soil and water

M. Muthukumaran

PG and Research Department of Botany, Ramakrishna Mission Vivekananda College (Autonomous), Affiliated to the University of Madras, Chennai, Tamil Nadu, India

Chapter outline

Biological Approaches to Controlling Pollutants. https://doi.org/10.1016/B978-0-12-824316-9.00006-9

11.1 Introduction

11.1.1 Effects of pollution

Contamination is an unfortunate change in the physical, synthetic, or natural attributes of air, land, and water that wastes or destroys our raw materials assets. Present-day mechanical advancements have expanded the contamination levels beyond the self-cleaning limits of the climate. As of late, one of the significant issues is the danger to human life from the increasing disintegration of the climate (Warrier, 2012). The rapid development of industrial manufacturing and business activities has discharged tons of pollutants into the climate (soil, surface water, or air) and brought about genuine human illnesses due to their poisonousness. Because contamination is anticipated to decline, and the wellsprings of valuable water for agribusiness and other human exercises are restricted, numerous nations have been actively looking for novel water sources. The fast growth of industrialization, urbanization as well as the unforeseen developments in agribusiness has led to an imperative age caused by maltreatment of typical resources to fulfill human requirements, and these needs have contributed much in upsetting the organic adjustment on which the idea of our condition depends (Anton and Mathe-Gaspar, 2005). Organic risk evacuation offers an option in contrast to conventional physicochemical compound techniques. This is considered a manageable innovation with reduced effects on the climate. Bioremediation is a natural instrument for reducing wastes into other structures that can be utilized and reused by different creatures. Bioremediation is profoundly engaged with the corruption, annihilation, immobilization, or detoxification of various chemical wastes and risky materials from the encompassing through the comprehensive and activity of microorganisms. The primary standard is debasing and changing toxins, for example, hydrocarbons, oil, weighty metal, pesticides, colors, etc. (Abatenh et al., 2017). Industrial wastewater (IWW) is an elective water source, and exploration exercises should be centered around creating inventive and contemporary ways to eliminate contaminations from IWW. Bioremediation/phytoremediation/phycoremediation strategies, demonstrated to be effective techniques for eliminating and debasing pollutants of different sorts from contaminated waters and soils, require information on the local plants and related microorganisms (Al-Thani and Yasseen, 2020).

The methodology of bioremediation is a less complicated, more moderate, and normally stable cleanup advancement. The cycles relate to bioremediation as a cleanup or control strategy for the remediation of risky waste substances. Bacterially assisted phytoremediation is offered here for both characteristic and metallic foreign substances to give some comprehension to how these minute living beings help remediation so that future field studies might be initiated (Glick, 2010). Biofiltration is a moderately late air pollution control innovation in which off-gases containing biodegradable volatile organic compounds (VOCs) or inorganic air toxics are vented through an organically dynamic material (Leson and Winer, 1991). A variety of decisions are available while considering a phytoremediation approach, including the utilization of wild plant-microorganism affiliations or the use of explicit planting and refined strategies (DalCorso et al., 2019). Safe microorganisms (for example, microscopic organisms, parasites, and green growth) have been most widely concentrated from this trademark. A few adaptations have been created by microorganisms to manage chromium poisonousness. These apparatuses incorporate biotransformation (decrease or oxidation), bioaccumulation, and biosorption and are considered a choice to eliminate heavy metals (HMs). The point of this audit sums up Cr(VI)-bioremediation innovations arranged on functional applications at larger-scope advancements. Similarly, the most important consequences of a few examinations zeroed in on measure plausibility and the strength of various frameworks (reactors and pilot scale) intended for chromium-expulsion limit are featured (Pablo et al., 2018).

11.1.2 Nonaqueous phase liquid pollution

Contamination of normal assets because of the arrival of a few toxins including hydrocarbons is a key human and environmental well-being concern. Hydrocarbon tainting is a genuine danger to the climate because hydrocarbons are cancer-causing and mutagenic. The principal wellsprings of NAPL toxins are over the aboveground storage tanks or underground storage tanks holding oil or potential petrochemicals are for the most part found close to oil and petroleum gas creation locales, gas stations (USEPA, 2005). Such kinds of contamination from (non) point sources changed the nature of (sub) surface water to a poisonous stream; consequently, the remediation of soil—water assets is expected to decrease the perils to people as well as climate. Different physical, synthetic, and natural practices were utilized to remediate NAPL contaminations in soil—water assets (Farhadian et al., 2008). Soil—water assets defilement coming about because of arrivals of NAPLs in (sub) surface has for some time been known to present dangers to human well-being and the climate through various vehicle pathways. Given their thickness to water, NAPLs are classified as light and dense nonaqueous phase liquids (LNAPLs and DNAPLs), separately, and move descending through the unsaturated zone. LNAPLs are commonly held by the water table because of their lighter thickness to soil—water, whereas DNAPLs infiltrate the water table and move to descend until they are held by an impermeable layer. The ground water table fluctuations alongside high porewater speeds are normal in shallow springs, causing upgraded assembly of NAPLs, and (up) descending development of the crest causes the capture of NAPLs in pore space, which expanded the wide inclusion of the NAPL masses. Subsequently, the remediation of soil—water assets is expected to kill perils to people and additionally the climate (Gupta and Yadav, 2017). Bioremediation is a developing practical strategy that makes no decimation of the environment as connected to the next physicochemical techniques (Yang et al., 2009). Likewise, these practices fundamentally work to remediate the NAPL toxins from soil—water frameworks under a wide scope of subsurface conditions (Yadav and Hassanizadeh, 2011). Attributable to their low fluid dissolvability, they exist as nonwatery stage fluid

Table 11.1 Components of crude oil (Balba et al., 1998; Soccol et al., 2003).

Petroleum hydrocarbons	Classification	Examples
Saturate Compounds	Straight Chain Alkanes (normal alkanes)	Ethane, Ethene, Propane, Paraffinic, Wax
	Branched Alkanes (isoalkanes)	Isobutane
	Cycloalkanes (naphthenes)	Cyclohexane
Aromatic Compounds	Monoaromatic Hydrocarbons (Volatile fraction)	Benzenes, Xylenes, Toluene, Naphthalene, Anthracene
	Polyaromatic Hydrocarbons (naphthenoaromatics)	Phenanthrene, Thiophenes, Dibenzothiophenes
	Aromatic Sulfur Compounds	Thiols, Sulfides, and Disulfides
Resins	Amorphous solids (Truly dissolved in the oil)	Polar macromolecules containing Nitrogen, Sulfur, and Oxygen
Asphaltenes	Large molecules (Colloidally dispersed in the oil)	Bitumens, asphaltene A (AspA), and asphaltene B (AspB)

(NAPL) and the bioavailability of these NAPLs is a significant worry for better biodegradation (Prozorova et al., 2014). In Table 11.1 shows the components of crude oil and hydrocarbon pollutants.

Biodegradation of different LNAPLs sullied soil—water frameworks has been concentrated under different states of low and raised temperatures, fluctuating water table, and changing soil dampness content. A greatly watched biodegradation of LNAPLs under previously mentioned conditions underlines the metabolic capacities of microorganisms to purify LNAPL-contaminated terrains in (semi)- parched beach front climate (Yadav and Hassanizadeh, 2011). Despite the fact that cost and security are essential contemplations, *in situ* bioremediation has additional favorable circumstances. For example, designed in situ bioremediation may get cleanup-together objectives more rapidly than other ordinary techniques—for example, siphon and treat. It can quicken impurity desorption and disintegration, restricting variables for some ordinary strategies, by treating toxins near their source, for example, pollutants inside soil totals and micropores, nonaqueous phase liquids (NAPLs). In this way, bioremediation of characteristic assets dirtied by NAPL toxins is getting expanding consideration, and where important, can help as a profitable sterilization elective (Gupta and Yadav, 2017).

11.1.3 Bioremediation

Bioremediation is a recognizable cycle including activities of micro/macroorganisms on the poisons present in ecological frameworks. The contamination pollution being the middle wherever on over the word needs speedy thought with the objective that our spoiling environmental factors will be remediated. There are several types of waste including homegrown, agrarian, and modern wastes that should be treated to secure the climate and keep it clean. Late advances and accessible bioremediation systems and philosophies as an approach to control and deal with the various wastes with the utilization of biotechnology (Ramírez-García et al., 2019). Bioremediation is an eco-accommodating method that has shown promising results for foreign substances. The basic thing parts in bioremediation/phytoremediation are microorganisms/plants whether terrestrial or oceanic that accept key occupation for remediation of the effects of HMs. Phytoremediation has also been a response to various creating issues (Wani et al., 2017). The equilibrium of unsafe metals has demonstrated solid and growing revenue for the long term. The utilization of biosensor microorganisms is eco-friendly and savvy. Consequently, microorganisms have an assortment of instruments of metal sequestration that hold more noteworthy metal biosorption limits. At last, we give a proposal from microbial apparatuses to eliminate, recoup metals, and metalloids from arrangements utilizing living or dead biomass and their segments (Medfu Tarekegn et al., 2020). Bacterial strains disconnected from the Kakuri channel were described and exposed to different convergences of weighty metal salts, and their capacity to endure the HMs was resolved. This shows their capacity to endure and get by in conditions with elevated levels of substantial metal salts, the *Pseudomonas aeruginosa* possibilities to bioremediation of different HMs (Haroun et al., 2017).

The capability of a couple of plant—animal categories to remediate air contaminations is evaluated regarding both indoor and open-air conditions, explicitly accentuating the design of the plant—soil—microbe framework (Agarwal et al., 2019). Examinations of bioremediation of domesticated animal wastewater through various hydrophyte frameworks, for example, microalgae, duckweed, water hyacinth, developed wetlands, and different hydrophytes are surveyed, and the usage of hydrophytes after administration is likewise discussed. In addition, favorable circumstances and constraints on domesticated animal wastewater on the board using phytotechnologies are underlined (Hao Hu et al., 2020). The study on a very basic level revolves around the specific limits of amphibian plants and as a critical device in phytotechnologies in the organization of foreign substances under seagoing conditions (Ansari et al., 2020). Advances of remediation and bioremediation are ceaselessly being improved utilizing hereditarily changed microorganisms or those normally happening, to clean deposits and sullied territories from poisonous organics. Bioremediation of soils, water, and marine conditions has numerous favorable circumstances and yet it is a test for specialists and architects. Thusly, it is critical to study attainability dependent on pilot testing before beginning a remediation venture to decide the best conditions for the cycle (Soccol et al., 2003).

The preeminent advance in fruitful bioremediation is site portrayal, which builds up the most reasonable and achievable bioremediation method (*ex-situ* or *in-situ*). *Ex-situ* bioremediation strategies will in general be more costly because of extra expenses credited to exhuming and transportation. In any case, they can be utilized to treat a wide scope of poisons in a controlled way. Interestingly, *in situ* procedures have no extra expense credited to uncovering; regardless, cost of on-location establishment of gear, combined with a failure to viably picture and control the subsurface of the dirtied site may deliver some *in situ* bioremediation methods wasteful. Thus, the cost of remediation is not the central point that ought to decide the bioremediation procedure to be applied to any contaminated site. Land qualities of the dirtied site(s) including soil type, contamination profundity and type, site area comparative with human residence, and execution attributes of every bioremediation strategy ought to be joined in choosing the most appropriate and productive technique to adequately treat contaminated locales (Santanu Maitra, 2018). The model portrays the bioremediation of sullied soils, which depends upon the utilization of indigenous microorganisms to corrupt the foreign substance (Di Gregorio et al., 1999). Plants require less management; however, finding the best phytoremediator could be a hard and tedious undertaking. Subordinate bioremediation is a promising methodology. To get all the advantages from this procedure is important to painstakingly choose the most satisfactory catalyst, and to have it all around portrayed. In any case, further examination on biodegradation or biotransformation instruments in plants, microorganisms, parasites or green growth is basic if bioremediation methodologies are to be executed or improved (Bernardino et al., 2012). The cycles of bioremediation normally happen in soil/water climate, whereby mixes are separated into less harmful mixes or potentially natural amicable mixes by microorganisms (Pandey and Fulekar, 2012).

11.2 Materials and methods

The concentration of total petroleum hydrocarbons (TPHs) in the solids and leachate was determined gravimetrically as HEM by the USEPA method 1664 (Environmental Protection Agency, 2010). The microbial activity of soil compost samples was characterized through the oxygen uptake over 7 days determined by manometric OxiTop system (WTW, Germany) at temperature $20 \pm 0.2°C$ (Strotmann et al., 2004; Reuschenbach et al., 2003; Binner et al., 2012). Colorimetric methods are quite widely used for the determination of surfactant concentrations (Salanitro and Diaz, 1995; Zhang et al., 1999; Koga et al., 1999; Hayachi, 1975). Anionic surfactants form ion pairs with methylene blue (MB) and are extracted with chloroform. The concentration of anionic surfactants in the soil and leachate was determined by the spectrophotometric method using MB (Koga et al., 1999). The pH of the leachate and water extracts was measured by a pH meter (SensION1, Hach, USA and Model 3320, Jenway, UK). Soil pH was determined by extracting the soil samples with five volumes of distilled water and measured with a glass electrode. The conductivity (EC) of water extracts was measured by a conductivity meter (Model 4320, Jenway, UK) and the values have been corrected to a constant temperature of $25°C$ and The concentration of nonionic surfactants was determined as cobalt thiocyanate active substances by method 512°C (APHA, 1985). The community of soil used for the biopile experiment was a bioaugmentation of the ex situ soil dense suspension (approx. 1×108 cfu/mL) of the resultant bacteria that were used as an inoculum (Kaszycki et al., 2011). Bioremediation processes in the biopile were monitored by the determined viable bacteria count, pH, and TPH. The number of microorganisms was determined by the method of the serial dilution on the agar plate (Nutrient agar) incubated at 30°C for the total count of bacteria. CFU on Petri dishes were counted as cfu per 1 g of soil (Miller, 1972). The count of TPH in the soil samples was determined gravimetrically after solvent extraction (Mishra et al., 2001a,b) then the percentage of biodegradation was measured (Saxena, 1990). Spill simulations with diesel (D100) in soil were carried out as with Taylor and Jones (2001).

An aliquot of 0.1 mL (at 10^{-4} dilution) of the crude oil soil suspension was seeded onto modified mineral salts medium of Mills et al. (1978). The vapor phase transfer technique of Okpokwasili and Amanchukwu (1988) was adopted, which employs the use of sterile filter paper soaked in crude oil, which served as the carbon and energy sources. The extent of crude oil bioaccumulation by the isolates was determined using the gravimetric analysis method of Odu (1972). The percentage of crude oil degraded after 6 weeks was determined using the equation method of Udeme and Antai (1988). The enrichment procedure was used in the estimation of hydrocarbon utilizers as described by Nwachukwu (2000). Oil agar medium was prepared according to the mineral salts medium composition of Mills et al. (1978) as modified by Okpokwasili and Okorie (1988). The composition and preparation of the crude oil utilization test medium were the same as that of the oil agar medium except that oil was made available via vapor phase transfer (Thijsse and van der Linden, 1961). The biosurfactant extraction from culture broth was carried out as described by Nitschke and Pastore (2006). Biosurfactants were measured in the cell-free culture medium using the phenol—sulfuric acid (DuBois et al., 1956). The model NAPL was prepared as described by Mukherji et al. (1997). The residual hydrocarbons were extracted by addition of an equal volume of ethyl acetate for hexadecane, n-hexane for NAPL, and estimated by gas chromatography for every 24 h till 120 h as per the protocol described by Pasumarthi et al. (2013). The bacterial adhesion to hydrocarbon assay was carried out as described by Rosenberg et al. (1980).

11.3 **Results and discussion**

The crumbling of natural well-being brought about by fast industrialization, urbanization, and expanding population pressure is a significant worry for agricultural nations, including India. The defilement of the climate (soil/water/air) by different harmful toxins delivered from a few common just as anthropogenic exercises and its antagonistic impacts on living elements requires expanded examination pointed toward alleviating and tackling these issues adequately. In contrast with regular remediation draws near, present-day techniques are more predominant and compelling in eliminating many natural and inorganic poisons from tainted sources (Ren and Umble, 2016). The enormous positions and conceivable outcomes of amphibian plants in remediation of tainting in wastewater. It has similarly researched continuous assessment work on the capability of *Salvinia molesta* and *Pistia stratiotes* plants in wastewater remediation and recognized zones for extra examinations as we find stoichiometric homeostatic rundown and resource beat impacts examinations of these plants is principal in wastewater phytoremediation measures (Mustafa and Hayder, 2020). The oceanic plants, this bioaccumulative effect can be abused, pointing a biotechnological and bioengineering application to dispense with metals, called phytoremediation, using floating sea-going macrophytes, which have high potential in light of their properties holding pollutants. Results obtained were complete for a variety of *Eichhornia crassipes* and *Salvinia auriculata* as better phytoremediation masters, exclusively, whereas *Lemna minor* and *Pistia stratiotes* fit better in biomonitoring and have assurance from explicit groupings of metal when related to Album, Hg, Zn, Ni, and Pb (Cleide Barbieri de Souza and Gabriel Rodrigues Silva, 2019). The chance of utilizing sunflower for the phytomanagement of chromium (Cr) substantial metal pollutant soil (Farid et al., 2017). A significant obstruction to the use of transgenic plants for bioremediation in the field is related to the valid or saw danger of even quality exchange to related wild or developed plants (Aken and Doty, 2009). To survey collection capacity of the liverwort of *Pellia endiviifolia* in connection with large scale and minor components take-up from stream water inside the midwoodland headwater environment (Parzych et al., 2018). The expected utilization of floating treatment wetlands planted with *Canna flaccida* for eliminating two drugs (acetaminophen and carbamazepine) and one herbicide (atrazine) from tainted water (Hwang et al., 2020). Plant and soil−water detecting criticism utilizing different warm pressure edges and watering levels can create an ideal harvest reaction for grain Sorghum bicolor (O'Shaughnessy et al., 2020). *Anabaena* sp. (a cyanobacterium), *Pseudomonas spinosa*, *Pseudomonas aeruginosa*, and *Burkholderia* sp. were demonstrated to be acceptable biodegraders of endosulfan (Lee et al., 2003; Hussain et al., 2007). The ligninolytic fungus *Ganoderma australe*, segregated from the stone pine (*Pinus pinea*), is a decent biodegrader of lindane (Rigas et al., 2007). The alga *Chlamydomonas reinhardtii* can bioaccumulate and biodegrade herbicide prometryn (Jin et al., 2012). The populace and kinds of saprophytic and raw petroleum-debasing parasitic genera from dairy animal compost and poultry droppings were explored, there was a critical contrast between bovine manure and poultry droppings in the checks of oil using growths communicated as a rate (%) of absolute saprophytic organisms at $P \leq .05$ (Obire et al., 2008).

In-situ bioremediation is the utilization of bioaugmentation and biostimulation to make anaerobic conditions in groundwater and advance foreign substance biodegradation for the reasons for limiting pollutant relocation or potentially quickening toxin mass evacuation. Bioaugmentation is the expansion of useful microorganisms into groundwater to expand the rate and degree of anaerobic reductive dechlorination to ethane (ITRC, 2005). A substantial portion of ecological poisons, for example, phenolics, nonphenolics, endocrine-disrupting chemicals, and HMs are exceptionally harmful. Governments around the world are carefully supporting the moderation of ecological contamination. Therefore, the alleviation/expulsion of poisons from the climate is of most extreme significance as it advances the feasible improvement of our general public with a negligible ecological effect. The addition exhaustive information on a few techniques for bioremediation (*in situ* and *ex situ*) of perilous poisons (Ren and Umble, 2016). The capacity of three bacterial separates (*Bacillus* spp., *Micrococcus* spp. and *Proteus* spp.) and some parasitic species (*Penicillin* spp., *Aspergillus* spp. what's more, *Rhizopus* spp.) confined from two waterways and treatment facility emanating to debase two Nigerian Crude oils was considered. It was seen that these living beings had the option to use and debase the raw petroleum constituents, with bacterial confines demonstrating increment in cell number and optical thickness as pH diminishes. Single societies were seen to be preferred raw petroleum degraders over the mixed cultures (bacteria or fungi). It was likewise seen that oil-degraders could be confined from a non-oil-dirtied climate, even though those from oil-contaminated conditions have higher debasement possibilities (Okerentugba and Ezeronye, 2003). The bioremediation capability of the treatment choices was appeared by the rate of decrease in unrefined petroleum in the examples. The utilization of 90 g of Cow manure compost demonstrated the best treatment choice with the expulsion of 52.59% of raw petroleum from the example, trailed by 60 g of Cow waste and 30 g of Cow excrement with a rate decrease of 44.35% and 37.06% individually. There was likewise 29.41% expulsion of raw petroleum in the control analysis (nonrevised example). This likewise concurred with the work done and who detailed that the amount of Cow excrement added to unrefined petroleum defiled soil significantly affects the remediation cycle (Osazee et al., 2015). This cycle includes the utilization of microorganisms, green plants (*Typha latifolia*, *Phragmites australis*, *Sagittaria latifolia*, *Phragmites communis*, and *Juncus radiates*, etc.) or their chemicals, to debase/detoxify the ecological toxins, for example, harmful metals, pesticides, azo colors, oil hydrocarbons, plastics, phenols, chlorophenols, biomedical waste, and so on from contaminated soil and water assets (Chandra and Chowdhary, 2015).

11.3.1 Bioremediation techniques: an overview

Bioremediation alludes to any procedure used to dispose of unwanted impacts of contaminations from climate. It is attractive to dispense with contaminations; however, this is not generally conceivable; however, a few living beings could bind or immobilize them. For example, creatures can gather impurities, and lessen their quality and their natural impact, yet do not dispose of them from the climate. Such methodology, if utilized, ought to be incorporated into the "bioremediation" idea. Those creatures ready to bioremediate would be called bioremediators (Bernardino et al., 2012; Kuiper et al., 2004). Bioremediation cycles can be extensively classified into two categories, *ex situ* and *in situ*. *Ex situ* bioremediation innovations incorporate bioreactors, bio-channels, land cultivating and some fertilizing the soil strategies. *In situ* bioremediation innovations incorporate bioventing, biosparging, biostimulation fluid conveyance framework, and some fertilizing the soil techniques. *In situ* medicines will in general be more alluring to merchants

Table 11.2 Classification of bioremediation strategies according to the organism involved (Bernardino et al., 2012).

Bioremediator organism	Strategy
Algae	Phycoremediation
Bacteria	Bacterial remediation
Biomolecules derived from organisms	Derivative bioremediation
Fungi	Mycoremediation
Microorganism	Microbioremediation or Bioremediation
Plants	Phytoremediation
Rhizosphere	Rhizoremediation

and people in question because they require less gear, for the most part, have a lower cost and create fewer aggravations to the climate (Pandey and Fulekar, 2012). Generally, bioremediation has been accomplished by utilizing microorganisms. All things considered, in the past many years, a few reports on bioremediation utilizing plants, parasites, green growth, or catalysts (obtained from living beings) have expanded the extent of bioremediation. Words like phytoremediation or rhizoremediation have been utilized (Bernardino et al., 2012; Kuiper et al., 2004), and maybe it is important to name appropriately every bioremediation strategy regarding the organism used in Table 11.2.

Bioremediation is a gathering of organic cycles utilizing (native) microorganisms under ideal conditions that follow up on poisons to lessen the mass, harmfulness, versatility, volume, or centralization of foreign substances in soil or groundwater assets. Different investigations assessed the bioremediation procedures and a blend of these methods and demonstrated the significant expulsion pace of NAPLs from the dirt water framework. The essential digestion/auxiliary cometabolism of the potential organisms causes the debasement or change of toxins in nontoxic finished results, which is ecologically benevolent and self-maintainable. Bioremediation practices can be ordered as *in situ* and *ex situ*. *In-situ* techniques remediate soils and groundwater set up, whereas *ex situ* rehearses incorporate the disposal of the polluted soil—water assets from the difficult site. *In-situ* bioremediation of spring sullied with NAPLs has been being used for over 40 years and is generally dependent on local microorganisms to diminish poisons. It does not need any mining; hence, it is enhanced by slight or no interruption to soil morphology. Ideally, these strategies ought to be more affordable contrasted with *ex situ* bioremediation, because of no extra cost essential for removal rehearses; by and by, cost of the plan and in-site setting up of some advanced apparatuses to increment microbial development during bioremediation is of significant concern. Some *in situ* bioremediation may be upgraded as biostimulation and phytoremediation, whereas others influence keeps with no type of enhancement (i.e., characteristic bioremediation or common constriction). The local microbial populace is acquainted to debase the contamination being uncovered for quite a while; however, it takes a significantly long effort for cleanup.

The bioremediation rises to quicken the bioremediation utilizing modification of the natural condition and local microflora. This serious bioremediation influences microbial exercises and their neighboring ecological conditions for quickening the act of biodegradation and is classified as biostimulation and bioaugmentation. The biostimulation is upgraded by the adding of supplements, electron acceptors, oxygen, and other significant mixes to the contaminated destinations, which improved the (co) metabolic activities of the microflora. Bioaugmentation is a microorganism cultivating practice for developing the volume of an NAPL degrader by adding likely microbial societies, which are filled autonomously in well-defined conditions. Moreover, the plants additionally quicken NAPL expulsion by advancing the microbial re-foundation in contaminated soils and water because of the steady conveying of oxygen by root zone air circulation and supplements for microbial advancement by fixation and exudation. Additionally, the developed wetlands procedures are simultaneous medicines to dirtied soil−water assets. The nature, profundity, and level of contamination, sort of circumstance, and area are among the choice estimates that are reflected while choosing any bioremediation systems (Azubuike et al., 2016). The spotlight on the arrangement and application status of microbial remediation innovation, and examines the key factors that confine bioremediation impacts, for example, debasing microbial screening and colonization, and improving the bioavailability of oil hydrocarbons (Xia, 2019). Fig. 11.1 shows the bioremediation strategies according to the organisms involved. In addition, execution estimates like a measure of supplement, and further, that control the accomplishment of bioremediation are additionally given significant consideration preceding starting a bioremediation venture. In this part, the artisanship relating to the bioremediation of NAPL-contaminated soil−water assets is given uncommon emphasis in the various bioremediation procedures.

11.3.2 Bioremediation of nonaqueous phase liquid polluted soil−water resources

The biotechnological capability of actinobacteria in the climate was exhibited by their capacity to eliminate natural and inorganic contaminations. This capacity is the motivation behind why actinobacteria have received considerable consideration for use in bioremediation, which has gained significance from the far-ranging arrival of pollutants into the climate. Among natural toxins, pesticides are broadly utilized for bug control, despite the progressively clearer negative effects of these synthetics on the ecological equilibrium. Additionally, the broad utilization of heavy metals (HMs) in modern cycles has led to profoundly debased regions around the world. A few examinations centered on the utilization of actinobacteria for tidying up the climate were acted over the most recent 15 years. Techniques, for example, bioaugmentation, biostimulation, cell immobilization, creation of biosurfactants, plan of characterized blended societies, and the utilization of plant-organism frameworks were created to improve the capacities of actinobacteria in bioremediation (Alvarez et al., 2017). Smells from wastewater treatment plants (WWTPs) have pulled in

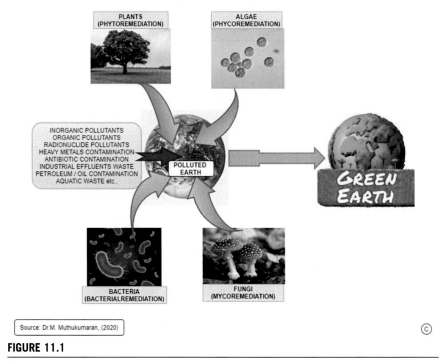

FIGURE 11.1

Bioremediation strategies according to the organism involved.

broad consideration and tough natural guidelines are all the more generally received to diminish scent discharges. Natural smell treatment strategies have more extensive applications than the physical and chemical partners as they are climate cordial, practical, and create low auxiliary wastes. The aqueous activated sludge (AS) measures are among the most encouraging methodologies for the counteraction or end-of-pipe evacuation of smell outflows and can at the same time treat scent and wastewater. Regardless, AS aeration biotechnologies in WWTPs should be further deliberately summed up and sorted, while inside and out conversations on the attributes and basic instruments of AS freshening up measure are as yet inadequate. As of late, significant investigations have been accounted for to explain the microbial digestion systems in scent control and wastewater treatment (Fan et al., 2020). Bioremediation is a viable system for tidying up natural toxins, for example, polycyclic aromatic hydrocarbons (PAHs) and VOCs. Progressed bioremediation suggests that biotic specialists are more effective in completely debasing the pollutants. Dirt/changed earth minerals are compelling adsorbents of PAHs/VOCs, and promptly accessible substrate and territory for microorganisms in the regular soil and residue (Biswas et al., 2015). The utilization of biopolymers-based nanomaterials (NMs) in bioremediation of dirtied waters and soils. We feature late advances on the improvement of novel modified biopolymer-based NMs and cycles for treatment of mechanical and city wastewaters and agrarian soils sullied by NMs, including harmful metal particles, radionuclides, natural and inorganic solutes, microbes, and infections. Furthermore, examine a few difficulties related to the improvement of practical and ecologically adequate utilitarian biopolymers-based NMs for bioremediation (Dhillon et al., 2012). Nanoparticles (NPs) can be either applied straightforwardly for the evacuation of natural pollutants through adsorption or chemical adjustment. It can likewise fill in as a facilitator in microbial remediation of pollutants either by upgrading the microbial development or by immobilizing the remediating operators or through actuated creation of remediating microbial compounds. In addition, nanoparticles initiated upgraded creation of biosurfactants in microorganisms, likewise add to improved dissolvability of hydrophobic hydrocarbons and subsequently, establish a helpful climate for microbial corruption of these mixes in climate (Kumari and Singh, 2016). The point of the examination was to describe the sorption limit of zinc and titanium oxide nanoparticles (TiO_2 NPs 25 and 100 nm; ZnO NPs 50 and 100 nm) and to analyze the bioaccumulation of Zn(II) and Sr(II) particles by sea-going plants *Salvinia natans* and *Elodea canadensis* within the sight of the previously mentioned NPs. The outcomes have demonstrated that NPs can adsorb metal particles present in the water (Asztemborska et al., 2018).

The part of *Dehalococcoides* spp. in chlorinated ethene biodegradation is constrained by energy in complex manners. Accordingly, concentrated *in situ* portrayal and understanding the microbial development on dechlorination assessment are fundamental to build up a reliable and sane designed framework. This is to accomplish fruitful bioremediation systems for destinations polluted with chlorinated ethenes (Saiyari et al., 2018). Vigorous microorganisms like *Pseudomonas*, *Alcaligenes*, *Sphingomonas*, *Rhodococcus*, and *Mycobacterium* to remediate the hydrocarbon and pesticides, the two alkanes, and mixes (Mohan and Ram, 2018). Research center and field pilot considers were done on the bioremediation of soil debased with oil hydrocarbons in the Borhola oil fields, Assam, India. The impacts of air circulation, supplements (for example nitrogen and phosphorus), and immunization of unessential microbial consortia on the bioremediation cycle were examined. The advantageous impacts of these boundaries on the bioremediation rate were acknowledged similarly in a research facility and field pilot tests. The field tests uncovered that up to 75% of the hydrocarbon pollutants were corrupted within 1 year, demonstrating the plausibility of building up a bioremediation convention (Gogoi et al., 2003). Bioremediation is of interest worldwide as a potential oil slick cleanup method under certain geographic and climatic conditions and this section centers around this part of contamination remediation (Speight and El-Gendy, 2018). The microorganisms can burn through oil hydrocarbons. These oil corrupting indigenous microorganisms assumed a huge part in decreasing the general ecological effect of both the *Exxon Valdez* and BP *Deepwater Horizon* oil slicks (Atlas and Hazen, 2011). Arsenic defiled zones are in danger of presentation through drinking water as well. Phytoremediation offers trust in satisfying possibilities for arsenic-influenced conditions (Shukla and Srivastava, 2019).

Nonwarm dielectric obstruction release (DBD) plasma is inspected as a technique for the *ex situ* remediation of NAPL debased soils. A combination of equivalent mass focuses (w/w) of n-decane, n-dodecane and n-hexadecane was utilized as model NAPL. Two soil types varying concerning the level of microheterogeneity were falsely contaminated by NAPL: homogeneous silicate sand and a reasonably heterogeneous loamy sand. The impact of soil heterogeneity, NAPL focus, and energy thickness on soil remediation proficiency was researched by treating NAPL-dirtied tests for different treatment times and three NAPL fixations. The fixation and synthesis of the lingering NAPL in soil were resolved with NAPL extraction in dichloromethane and GC-FID investigation, whereas new oxidized items were related to constricted absolute reflection Fourier change infrared spectroscopy (ATR-FTIR). The exploratory outcomes showed that the general NAPL expulsion productivity increments quickly in early occasions arriving at a level at late occasions, where NAPL is eliminated (Aggelopoulos et al., 2015). Microbial reductive dechlorination is a significant corruption pathway of chlorinated hydrocarbon in anaerobic subsurface conditions. In the bioremediation of chloroethenes and polychlorinated biphenyls (PCBs) dependent on upgraded *in situ* reductive dechlorination, *Dehalococcoides* sp. assume a significant part in the detoxification cycle *in situ* (Xiao et al., 2020). The capability of the indigenously secluded biosurfactant delivering living being, *Paenibacillus* sp. D9 to use wasted searing oils (canola, sunflower, castor, and coconut) were researched as substitute modest substrates for amalgamation of biosurfactant (Jimoh and Lin, 2020). Knowing the digestion of those biodegrader species or strain improves the determination of the bioremediation procedure for each site either by biostimulating the indigenous biodegraders (biostimulation) or adding exogenous to the site (bioaugmentation). Additionally, because of atomic science, the metabolic biodegradation capacity could be moved from a biodegrader to another life form, along these lines improving its corrupting abilities. For example, utilizing hereditary designing, an entire mineralization pathway for paraoxon the oxon metabolite of the organophosphate pesticide parathion was implicit a solitary strain of *Pseudomonas putida* (Mattozzi et al., 2006). Table 11.3 lists the potential microbial organisms for use in the bioremediation of NAPL contaminants and Fig. 11.2 showed the several bioremediation techniques and zone/sources of NAPL polluted places.

11.3.3 Bacterial remediation

Bioremediation studies to eliminate oil pollution regions in an influenced climate by the utilization of the strains of microscopic organisms. Bioremediation through adding microorganisms to influenced regions will assume a significant function later as an earth-safe and practical reaction to oil contamination. Checking the destiny of oil

Table 11.3 Potential of microbial organisms used in bioremediation of NAPL-pollution.

Group	Genera	References
Bacteria	*Acinetobacter* sp.	Margesin and Schinner (2001)
	Achromobacter sp.	Popp et al., (2006)
	Alcaligenes sp.	Samenta et al. (2002)
	Bacillus sp.	Roy et al. (2014)
	Corynebacterium sp.	Balba et al. (1998)
	Methanogenic sp.	Kim et al. (2008)
	Mycobacterium sp.	Bento et al. (2005)
	Paenibacillus sp.	Jimoh and Lin (2020)
	Pseudomonas sp.	Mishra et al., 2001a, b
	Pseudoxanthomonas sp.	Farhadian et al. (2008)
	Rhodococcus sp.	Witzig et al. (2006)
	Stenotrophomonas sp.	Kartik Patel and Mites Patel (2020)
	Xanthomonas sp.	Vinas et al. (2005)
Algae	*Chlamydomonas* sp.	Jin et al. (2012)
	Chlorella spp.	Das et al. (2015); Baldiris-Navarro et al. (2018)
	Chlorococcum sp.	Muthukumaran (2009); Muthukumaran and Sivasubramanian (2017)
	Desmococcus sp.	Muthukumaran (2009); Muthukumaran et al. (2012)
Fungi	*Aspergillus* sp.	Pal and Vimala (2012)
	Candida sp.	Joo et al. (2008)
	Nigrospora sp,	Obire (1988)
	Phanerochaete sp.	Osazee et al. (2015)
	Rhizopus sp.	Thenmozhi et al. (2013)
	Trichoderma sp.	Osazee et al. (2015)
	Trichothecium sp.	Osazee et al. (2015)

hydrocarbons present as pollutants in natural examples is a review way to deal with survey the advancement of biodegradation measure (Imam et al., 2019). Chemoheterotrophic microbes are the main gathering for the corruption of pollutants (ICCS, 2006). To talk about all the elements applicable to the improvement of methodologies for PCB bioremediation (Robinson and Lenn, 1994). In soil, the predominant microorganism populace involved gram-positive microscopic organisms from actinomycete gathering and autochthonous microorganisms that disintegrate hydrocarbons arrived at the most significant level 1.6×10^7 cfu/g at 45 days. In light of this information, we infer *ex-situ* (Biopile) to explore the best procedure, cheap, productive, and earth cordial and may subsequently offer a suitable decision for oil hydrocarbons debased soil remediation (Burghal et al., 2015).

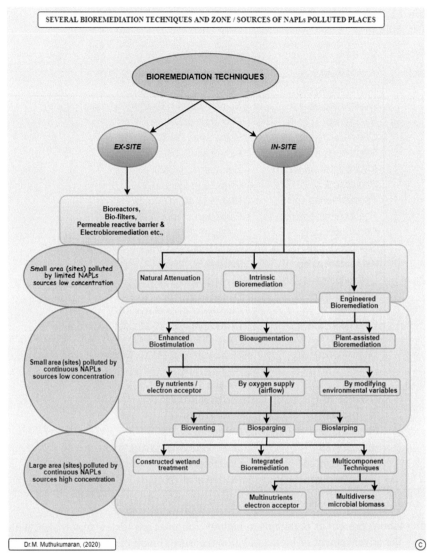

FIGURE 11.2

Several bioremediation techniques and zone/sources of nonaqueous phase liquid polluted places.

Aliphatic and fragrant hydrocarbons are natural toxins of genuine concern. Their bioavailability is the significant restricting variable that makes the bioremediation cycle moderate. Hence, the current investigation centers around biodegradation of NAPL by a halophilic consortium (*Pseudomonas aeruginosa* and *Escherichia fergusonii*) in presence of rhamnolipid just as a rhamnolipid-delivering *Pseudomonas aeruginosa* AMB AS7. The investigation was acted in microcosms, and the remaining hydrocarbons after debasement were assessed by gas chromatography. It was discovered that the corruption of hydrocarbons in NAPL was more in presence of rhamnolipid in correlation with their biotic controls. Be that as it may, among NAPL, the corruption of phenanthrene (37.5%) and octadecane (47.8%) was discovered to be more by coculture of halophilic consortium and rhamnolipid-creating *P. aeruginosa* AMB AS7 (Mariaamalraj et al., 2016). The metataxonomic approach was utilized to depict the bacterial network from a creosote-tainted spring and to get to the potential for *in situ* bioremediation of the PAHs by biostimulation. All these improved taxa have been described as PAH debasing specialists, for example, the variety *Comamonas, Geobacter, Hydrocarboniphaga, Anaerolinea*, and *Desulfomonile*. The pollutants advanced the enhancement of a few gatherings of corrupting microbes in the territory, which fortifies the possibility of applying biostimulation as a spring remediation procedure (Júlio et al., 2019). The outcomes show a great sign of value enhancement for dirtied soil after treated with *Pseudomonas putida*. It is obvious from the obtained outcomes that the utilization of *P. putida* is appropriate as a compelling microorganism and is a potential diesel-soil biodegradable specialist in contaminated soil (Mohamed Sunar et al., 2014). Despite the fact that microbes have ended up being acceptable biodegraders and bioremediators, a few organisms, plants, and green growths could biodegrade pesticides, colors, hydrocarbon as well (Bernardino et al., 2012). Pseudomonas putida have been found to have high potential for biodegradation of unstable oil hydrocarbons (Singh and Fulekar, 2010).

11.3.4 **Biosurfactants**

Surfactants are amphiphilic atoms that will in general segment especially at the interface between periods of various extremity and water holding. Surfactants increment the watery solvency of NAPLs by decreasing their surface/interfacial strain at air-water and water-oil interfaces (Urum and Pekdemir, 2004; Tadros, 2005). Numerous organisms are skilled to combine various kinds of biosurfactants and insoluble substrates (counting hydrocarbons) to actuate the biosurfactant creation (Veenanadig et al., 2000). Surfactant enhanced bioremediation (SEB) of oil is a methodology embraced to conquer the bioavailability imperatives experienced in the biotransformation of NAPL poisons. Fuel oils contain n-alkanes and other aliphatic hydrocarbons, monoaromatics, and PAHs. Despite hydrocarbon-corrupting societies being plentiful, complete biodegradation of oil is seldom accomplished under good ecological conditions because of the basic unpredictability of oil and culture specificities. In addition, the collaboration among societies in a consortium, substrate association impacts during the corruption and capacity of explicit societies to adjust the bioavailability of oil constantly influence the cycle. Although SEB can expand the debasement pace of oil and its constituents, there are various difficulties in the fruitful utilization of this innovation (Mohanty et al., 2013). Bioremediation can be a cheap and successful cycle for treating some dense nonaqueous phase liquid (DNAPL) source zones. Field exhibits have indicated that bioremediation can lessen source life span, decrease DNAPL mass, and diminish convergences of foreign substances (Moretti, 2006). The investigation of fertilizer portions demonstrated that both water and arrangement of Tween 80 decreased the level of oil-based substances in the manure blend by about 20%, and the decrease in hydrocarbon content was produced through biodegradation, as under 0.01% of hydrocarbons were drained out from the dirt segments (Selberg et al., 2013). Ideally, give a premise to the consecutive anaerobic and high-impact biotreatment of PCBs tainted soils and dregs. Plainly, for such a cycle the test is as incredible regarding designing for what it's worth for advancement of skilled microbial consortia (Robinson and Lenn, 1994).

Biosurfactants are surfactants created by microorganisms that advance the breaking of hydrocarbons atoms by micelle development, expanding their versatility, bioavailability, and introduction to microbes, and accordingly prefer hydrocarbon biodegradation. There is an extraordinary variety of microorganisms fit for biodegrading poisons, for example, oil and creating biosurfactants, yet they are not notable. This investigation plans to deliver the issues identified with a progression of boundaries associated with the creation and in the preparation and activity component of biosurfactant monomers in destinations containing hydrocarbons (Souza et al., 2014). The attributes of normal and engineered surfactants and the impacts of biosurfactants on dissolvability, sorption, and biodegradation of hydrophobic natural toxins; just as the impacts of biosurfactants on degrader microorganisms as white-decay growths (Bustamante et al., 2012). To assess the biosurfactant creation capacity of *Stenotrophomonas* sp. S1VKR-26, profiling of its bioremediation capacity to remediate oil processing plant wastewater in a lab-scale bioreactor and evaluation of phytotoxicity of bioremediated oil wastewater (Kartik Patel and Mites Patel, 2020).

11.3.5 Bioaugmentation

The ceaseless arrival of NAPLs causes unfriendly impacts on microbial exercises and a decrease in the microbial populace in (sub)surfaces. To keep up the ideal populace of such likely microorganisms, it is important to seed the local expected organisms. This microbial cultivating method to the dirtied destinations to accomplish the most extreme evacuation of NAPL mass is known as a bioaugmentation strategy (Atlas 1991; Sarkar et al., 2005). This method is ideally utilized in the dirtied zone having persistent delivery with high substrate focuses, fewer microbial tallies, and dynamic environmental factors. Da Silva and Alvarez (2004) examined improved biodegradation of NAPLs in microbial cultivated spring sections and showed expulsion of NAPL mass expanded up to 88%. Joo et al. (2008) utilized the microbial stain of *Candida catenulate* (CM1) to polluted soil and demonstrated that 84% NAPL mass expulsion occurred in 13 days. Likewise, toluene corruption by cultivating *Corynebacterium variabilis* (SVB74) and *Acinetobacter radioresistens* (SVB65) shows the most noteworthy rate in the vadose zone.

11.3.6 **Upgraded biostimulation strategies**

Upgraded biostimulation is an extraordinary instance of designed bioremediation, where the current states of dirtied destinations might be modified by giving a great climate and additionally supplements for the development of possible microbial populaces. The presentation of toxins in the subsurface causes a decrease in oxygen level, supplements, and so forth, which legitimately influences the microbial exercises. In this way, local microorganisms need more (micro) supplements, electron acceptors, and good natural conditions to accomplish total corruption of contaminations. Accordingly, modification in dirtied destinations by giving such basic parts animates microbial development and eventual expulsion of poisons. For the most part, the NAPL-contaminated destinations were modified by giving (1) oxygen sources, (2) supplements, (3) electron acceptors, (4) business items, and (5) good ecological conditions. To upgrade toxin expulsion by expanding exercises of indigenous microorganisms, conveying oxygen to an unsaturated (vadose) zone is the best strategy for NAPL-contaminated destinations. The controlled infusion of airflow in an unsaturated zone to mimic the oxygen level in soil (water) air and to keep up vigorous conditions causes increments in microbial exercises. These strategies are by and large alluded to as bioventing and picked up prevalence among other *in situ* biostimulation methods, particularly in reestablishing destinations contaminated with low fixation LNAPLs. Sui and Li (2011) researched the impacts of air infusion rate on the destiny of an LNAPL-contaminated site by bioventing and saw that a higher airflow rate brought about improved toluene evacuation contrasted with a lower rate. Frutos et al. (2010) examined the adequacy of bioventing treatment in incitement of NAPL-dirtied soil and demonstrated more prominent than 93% NAPL mass expulsion. Essentially, *in situ* air sparging is infused beneath the water table so the NAPL contaminations move in an unsaturated zone, where an overwhelming microbial populace advances the evacuation rate. The air sparging strategy is utilized by and large for profound infiltrating toxins such as DNAPLs in tainted spring frameworks.

Biostimulation by supplement change is a down-to-earth way to deal with improve the pace of biodegradation of NAPLs from (un)saturated zones. Roling et al. (2002) examined the effect of supplement alteration on elements of bacterial networks alongside the NAPL biodegradation and found that the supplement correction significantly expanded the bacterial populace and improved biodegradation up to 92% of NAPL mass evacuation. Yadav et al. (2013) directed a progression of microcosm tests, in which at first the common biodegradation was examined, and they consequently added homegrown wastewater as supplements. The outcomes demonstrated the expansion of wastewater significantly improved the biodegradation rate at room temperature. In addition, Macnaughton et al. (1999) explored the effect of supplements on evacuation rate in groups having NAPL-dirtied soil and demonstrated the higher corruption rate in clusters having extra supplements. The (micro) supplements are additionally accessible as business items, which are utilized generally in basic cases such as substantial contamination load cases.

Franzetti et al. (2008) indicated the expanded evacuation of NAPL mass improved by Brij 56 and Tween 80. Essentially, Asquith et al. (2012) demonstrated 60%−69% and 69%−80% extra expulsion of NAPL mass in RemActiv and Daramend changed bunch frameworks separately. The expansion of electron acceptors is biochemically integral to supplement correction for the improvement of the NAPL corruption. For the most part, the oxygen-delivering intensifies like H_2O_2. MgO_2, O_2, NO_3, SO_4, Mn(IV), and Fe(III) are utilized to invigorate the NAPL-contaminated destinations. These electron acceptors significantly expanded the oxygen level, which helped in keeping up the vigorous condition. Alvarez and Vogel (1995) utilized nitrate as an electron acceptor just as a supplement to hatch NAPL degrader in a bunch framework. The expansion of nitrate to soil is an improvement, as the denitrification causes expanded oxygen levels and significantly corrupts the NAPL mass. It appears in the writing that the use of electron acceptors and supplement revisions quickens the high-impact condition at dirtied locales, causing high NAPL evacuation rates.

Other than the supplement changes and expansion of electron acceptors, keeping up the ecological factors such as temperature, dampness conditions, pH, saltiness, and so on, is the main technique to upgrade the NAPLs' expulsion from contaminated locales. Yadav et al. (2012) inspected the part of fluctuating temperature on biodegradation of LNAPLs under high-impact conditions. To see the occasional effect of temperature, a progression of microcosm tests was directed at three distinctive consistent temperatures: 10, 21, and 30°C. The aftereffects of consistent and fluctuating temperatures in microcosm show that toluene corruption is firmly reliant on temperature level. A very nearly double cross-development in LNAPL debasement time appeared for each 10°C abatement in temperature. Likewise, Coulon et al. (2007) researched the effect of temperature on the biodegradation of NAPLs under expanding temperatures from 4 to 20°C. The outcomes indicated the expanding temperature quicken the corruption rate in the clump framework. Additionally, Dibble and Bartha (1979) confirmed that ideal soil dampness somewhere in the range of 30% to 90% of the soil field limit is appropriate for complete NAPL expulsion from contaminated soil−water frameworks. With regards to pH, microorganisms make do with of a specific scope of pH, and for the biodegradation of NAPLs, that ideal scope is a pH of 6−8. The substrate focus is likewise a significant variable for the incitement of bio-corruption rate, as the high fixation prompts harmfulness hazard. Gupta et al. (2013) examined biodegradation having distinctive substrate groupings of NAPLs and demonstrated an expanded expulsion rate with expanding beginning substrate fixation up to 100 ppm that starts diminishing with higher focus. Exploration has demonstrated that biostimulation is a financially savvy method; however, the locales' orientated supplements, electron acceptors, under ideal natural factors are expected to examine for compelling usage of bioremediation in field conditions.

11.3.7 Phycoremediation

Phycoremediation is described as the usage of green development to kill defilements from the earth or to convey them harmless (Dresback et al., 2001). Olguin (2003) portrays phycoremediation in a much broader sense as the usage of macroalgae or microalgae for the departure or biotransformation of poisons, including supplements and xenobiotics from wastewater and CO_2 from waste air. The effects of high nitrate center around the energy of cell improvement during nitrate and phosphate clearing by a macroalga *Cladophora glomerata*. The algal turn of events and nitrate ejection from media containing beginning nitrate unions of 5 to 400 mg/L were seen in bundle improvement, whereas control media has no additional nitrate (Farahdiba et al., 2020). It was in like manner found that during a support season of 10 days under encompassing temperature conditions, *Cladophora* sp. could chop down arsenic obsession from 6 to\0.1 mg/L, *Chlorodesmis* sp. had the choice to diminish arsenic by 40%−50%, whereas the water hyacinth could reduce arsenic by only 20%. *Cladophora* sp. is as needs be suitable for cotreatment of sewage and arsenic-improved salt water in an algal lake gaining some upkeep experiences of 10 days. The recognized plant species give a clear and down-to-earth method for application in nation regions impacted by arsenic issues (Jasrotia et al., 2017). The progressive appearance of red tides made by the dinoflagellates *Alexandrium minutum* and *Prorocentrum triestinum* reflected high assessments of phytoplankton arranged assortments similarly to changes in the predominant species starting with one then onto the next. This suggests the water idea of the EH may be recovering (Zaghloul et al., 2020).

The effects of specific sole carbon source prompted micropollutants (MPs) cometabolism of *Chlorella* sp. by (1) extracellular polymeric substances, superoxide dismutase, and peroxidase protein creation; (2) MPs expulsion effectiveness and cometabolism rate; (3) MPs' potential corruption items distinguishing proof; and (4) debasement pathways and approval utilizing the EAWAG information database to separate the codigestion of *Chlorella* sp. with different organisms (Vo et al., 2020). The unicellular green alga *Chlorella fusca* var *vacuolata* can biotransform the herbicide metfluorazon by a Cytochrome P450 (Thies et al., 1996). As of late, it has been portrayed that the alga *Chlamydomonas reinhardtii* can bioaccumulate and biodegrade herbicide prometryn (Jin et al., 2012). The capacity of freshwater microalgae *Chlorella vulgaris* to eliminate and biodegrade phenol from wastewater (Baldiris-Navarro et al., 2018). The mixotrophic development of *Chlorella pyrenoidosa* on modern waste phenol could end up being an earth reasonable cycle as it will cause remediation of the harmful material phenol alongside the age of biodiesel feedstock with diminished creation costs unraveling a significant bottleneck in the commercialization of algal biodiesel (Das et al., 2015). The capability of microalgae *Desmococcus olivaceus* and *Chlorococcum humicola* successfully remediated electroplating and oil-penetrating mechanical oozes and effluents of both pilot and scaled-up levels separately (Muthukumaran, 2009; Muthukumaran et al., 2012; Muthukumaran and Sivasubramanian, 2017).

11.3.8 Mycoremediation

The microbial consortium for the bioremediation of refinery effluents. The consortium was made by the white-decay parasite *Phanerochaete chrysosporium*, which delivered the greatest degrees of ligninolytic chemicals alongside disconnected living being for example *Pseudomonas aeruginosa* and *Aspergillus niger* from refinery effluents. *Phanerochaete chrysosporium* in regard to *Pseudomonas aeruginosa* and *Aspergillus niger* indicated a most extreme decolorization zone (31.0 mm) vaccinated with splendid blue color plates within 15 days of hatching (Pal and Vimala, 2012). *Aspergillus* and *Rhizopus* sp were discovered to be prevalent parasites detached from the preowned motor oil tainted soil tests can be misused in the biodegradation of unrefined oil slick and bioremediation of the climate (Thenmozhi et al., 2013). In Nigeria, the organisms detailed as oil-degraders in oceanic conditions of oil delivering regions by Obire (1988) were *Candida, Rhodotorula, Saccharomyces*, and *Sporobolomyces* species, and the molds were *Aspergillus niger, Aspergillus terreus, Blastomyces* sp., *Botryodiplodia theobromae, Fusarium* sp., *Nigrospora* sp., *Penicillium chrysogenum, Penicillium glabrum, Pleurophragmium* sp., and *Trichoderma harzianum*. The importance and effect of the investigation is the usage of local yeast strains detached from the waste water itself having potential for natural bioremediation in oil treatment facilities and petrochemical businesses (Alirezaei, 2020). Soil tests were gathered from the plastic dishes for microbiological examinations. The disconnects were refined to test their capacity to develop on unrefined petroleum. The Fungal Growth as observed by the estimation of absolute contagious populaces, hydrocarbon using parasitic populaces, and the quantitative hydrocarbon misfortunes were additionally decided at various loads of Cow compost. Results demonstrated that the remediation impact was supplement subordinate. The accompanying parasitic gatherings of raw petroleum utilizers were detached, in particular, as *Trichoderma harzianum, Aspergillus flavus, Rhizopus* sp, *Penicillium* sp, *Aspergillus niger, Rhizopus* sp, and *Trichothecium roseum* (Osazee et al., 2015).

11.3.9 Plant-assisted bioremediation strategies

Plant-helped bioremediation alludes to the utilization of particular plant species with the focus on toxins to moderate the harmful impacts and expulsion of contamination mass from the (sub)surface. This method utilized the plant–geochemical connection to alter the contaminated site and flexibly move (micro)nutrients, oxygen, and so on, into the subsurface for better execution of NAPL degrader on focused poisons (Susarla et al., 2002). NAPLs are for the most part eliminated by corruption, rhizoremediation, adjustment, and volatilization, with mineralization being conceivable when a few plants, for example, *Canna generalis* are utilized (Yadav et al., 2013). The plant geochemical cooperation improves the (1) physical and compound properties of locales, (2) supplement gracefully by delivering root exudates (Shimp et al., 1993), (3) air circulation by the move of oxygen (Burken and Schnoor 1998), (4) interference and hindrance of the developments of synthetic compounds, (5) the plant enzymatic change, and (6) protection from the vertical and sidelong relocation of toxins (Narayanan et al. 1998a, b). In addition, the plant microorganism's association expanded mineralization in the rhizosphere and the quantities of degraders and abbreviated the slack stage until vanishing of the compound. Some critical components to consider while picking a plant incorporating a root framework, which might be fibrous or tap subject to the profundity of toxin, harmfulness of poison to plant, plant endurance, and its versatility to winning natural conditions, plant development rate, and protection from illnesses and nuisances. The profound root frameworks of the plant improve air circulation in the subsurface, which keeps up the oxygen level in the profound vadose zone. The root exudates, dead root hair, and fine root fill in as significant wellsprings of the carbon for microbial development (Shimp et al., 1993). The root exudates additionally quicken the compound amalgamation of microbial digestion systems (Dzantor et al., 2007). The part of plants on LNAPL evacuation utilizing a pot-scale arrangement planted with *Canna generalis* under controlled conditions was explored by Basu et al. (2015). The all-out expulsion season of the LNAPLs was discovered to be highest in the unplanted followed by planted mesocosm, and debasement rates were discovered to be higher in planted pot arrangements. A three-dimensional model has been created for the plant-helped remediation of NAPL-contaminated locales (Narayanan et al. 1998a, b). A list of potential plant species mostly used in plant-assisted bioremediation of NAPL polluted places is presented in Table 11.4. Generally speaking, plants assume a vital part in the expulsion of NAPL mass, yet numerous issues identified with plant application to NAPL-dirtied destinations require examination before usage of such strategies. Further, the effects of static and dynamic ecological factors on poison expulsion, the blend of other bioremediation methods to plant-helped bioremediation, and multiscale examination are required.

Table 11.4 Potential of plant species used in plant-assisted bioremediation of nonaqueous phase liquid polluted places.

Plant species	References
Bassia indica	Tsao et al. (1998)
Baumea juncea	Zhang et al. (2010)
Baumea articulata	Zhang et al. (2010)
Canna flaccida	Hwang et al. (2020)
Canna generalis	Boonsaner et al. (2011); Yadav et al. (2013)
Cyperus alternifolius	Langwaldt & Puhakka (2000)
Juncus roemerianus	Lin & Mendelssohn (2009)
Juncus subsecundus	Zhang et al. (2010)
Lolium perenne	Tsao et al. (1998)
Mirabilis jalapa	Boonsaner et al. (2011); Zhang et al. (2010)
Phragmites australis	Ranieri et al. (2013); Vymazal et al. (2009)
Polygonum aviculare	Mohsenzadeh et al. (2010)
Schoenoplectus validus	Zhang et al. (2010)
Scirpus grossus	Bedessem et al. (2007)
Typha domingensis	Shehzadi et al. (2014)
Typha latifolia	Imfeld et al. (2009)

11.3.10 Combined bioremediation strategies of nonaqueous phase liquids

Singular utilization of the previously mentioned bioremediation procedures shows powerful techniques to NAPL-contaminated locales, however, is restricted for destinations having consistent arrival of toxins with high focus levels under quicker subsurface flow and dynamic natural conditions. In this way, the coordinated or joined utilization of these bioremediation procedures is expected to accomplish a more serious level of expulsion of NAPL mass from reasonable conditions. Zu and Lu (2010) coordinated the bioaugmentation, biostimulation, and phytoremediation methods and discovered 100% expulsion of NAPLs from a dirtied soil—water framework. The diverse combined bioremediation strategies and associated dominant bioagents for bioremediation of NAPL polluted sites are listed in Table 11.5 and Fig. 11.3 showed the NAPLs in soil and water as well as its bioremediation.

Table 11.5 Diverse combined bioremediation strategies and associated dominant bioagents for remediation of nonaqueous phase liquid polluted sites.

Integrated strategy	Dominant bioagents	References
Bioaugmentation	*Azoarcus* sp, *Candida* sp., *Cladophialophora* sp. *Corynebacterium variabilis*, *Dehalococcoides* sp., *Mycobacterium* sp., *Pseudomonas* sp., *Pseudomonas aeruginosa*, *Pseudomonas putida*, *Pseudoxanthomonas spadix*, *Rhodococcus* sp.,	Fan and Scow (1993); Joo et al. (2008); Margesin and Schinner (2001); Kim et al. (2008); Karamalidis et al. (2010); Gupta and Yadav (2017)
Biofortification	*Rhodococcus* sp., *Pseudomonas* sp.,	Delille et al. (2007); Lu Si-Jin (2007);Liao You-Gui (2007); Xia, 2019
Biostimulation along with microbial seeding	*Achromobacter* sp., *Acinetobacter* sp., *Alcaligenes* sp., *Bacillus* sp., *Baumannii* sp. *Corynebacterium variabilis*, *Candida catenulate*, *Dehalococcoides* sp., *Pseudomonas* sp., *Trichoderma* sp., *Xanthomonas* sp.,	Mishra et al., 2001a,b; Lendvay and Lo (2003); Van Gestel et al. (2003); Huang et al. (2004); Vinas et al. (2005); Bento et al. (2005); Sarkar et al. (2005); Joo et al. (2008)
Biosurfactant	*Paenibacillus* sp., *Stenotrophomonas* sp.	Souza et al. (2014); Jimoh and Lin (2020); Kartik Patel and Mites Patel (2020)
Enzyme reaction (Kinetic models)	*Dehalococcoides* spp., *Pseudomonas* sp.,	Lu Si-Jin (2007); Haest et al. (2010); Liao You-Gui (2007); Yadav and Hassanizadeh (2011); L€;offler et al. 2013; Xia, 2019; Saiyari et al. (2018)

Table 11.5 Diverse combined bioremediation strategies and associated dominant bioagents for remediation of nonaqueous phase liquid polluted sites.—*cont'd*

Integrated strategy	Dominant bioagents	References
Mycoremediation	*Aspergillus* sp., *Penicillium* sp., *Phanerochaete* sp., *Rhizopus* sp., *Trichothecium* sp., *Trichoderma* sp.,	Pal and Vimala (2012); Thenmozhi et al. (2013); Osazee et al. (2015)
Phycoremediation	*Chlamydomonas* sp., *Chlorella* sp., *Chlorococcum* sp., *Desmococcus* sp.	Muthukumaran (2009); Jin et al. (2012); Muthukumaran et al. (2012); Das et al. (2015); Muthukumaran and Sivasubramanian (2017); Baldiris-Navarro et al. (2018); Vo et al. (2020);
Phytoremediation	*Canna generalis, Juncus subsecundus, Mirabilis jalapa, Scirpus grossus, Polygonum aviculare*	Mohsenzadeh et al. (2010); Boonsaner et al. (2011); Zhang et al. (2010); Yadav et al. (2013)
Plant-enhanced biostimulation	*Achromobacter xylosoxidans, Atriplex centralasiatica, Galega orientalis, Limonium bicolor, Lolium perenne, Rhizobium* sp., *Scorzonera mongolica, Typha domingensis, Vetiver grass*	Yang et al. (2009); Shehzadi et al. (2014); Gupta and Yadav (2017)

11.3.11 Constructed wetland treatment of nonaqueous phase liquids

Constructed wetlands (CWs) are treatment frameworks that utilization normal cycles including wetland vegetation, soils, and their related microbial gatherings to remediate the contaminated destinations. CWs are an easy treatment choice that has been once evaluated mostly at the mesocosm or pilot level for their ability to dispose of NAPLs. In plant-based frameworks, complex physical, chemical, and natural cycles may happen at the same time, including volatilization, sorption, phytodegradation, plant take-up, and gathering, just as microbial corruption. A lot of work has been done in the most recent decade as for the utilization of developed wetlands for remediation of NAPLs in pilot and field-scale frameworks. Treatment wetlands can uphold a blended redox climate that may offer different biochemical pathways for contamination evacuation. This shows that the rhizospheric soil has 10−100 times more organisms in wetland arrangements (Gerhardt et al., 2009; Imfeld et al., 2009). Different vigorous microbial gatherings (heterotrophs, methanotrophs, and smelling salts oxidizers) are dynamic in the wetland rhizosphere and can corrupt different NAPLs by (co)metabolism (Powell et al., 2011).

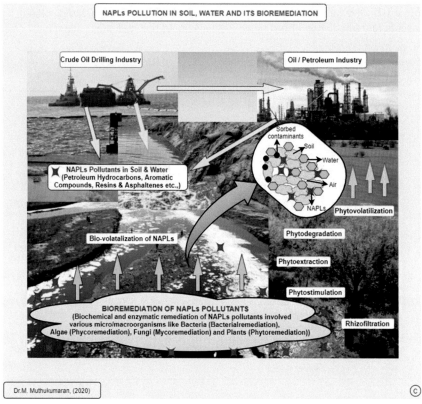

FIGURE 11.3

Nonaqueous phase liquid pollution in soil, water, and its bioremediation.

11.3.12 Nonaqueous phase liquid metabolism and associated kinetics models

Bioremediation energy is one of the significant dynamic elements for the remediation and better administration of NAPL-contaminated sites. The active model aids in the choice of various best procedures for contaminated locales under threatening ecological conditions. In addition, the comprehension of microbial corruption energy will advance the multimicrobial/electron acceptor methodology (Powell et al., 2014). Along these lines, the focal point of this segment is to introduce a review of bioremediation energy models related to NAPL-contaminated locales. The connection between the specific development rate (μ) of the microbial populace and the takeaway fixation (S) is called microbial development energy (Kovarova-kovar and Egli 1998). Microbial development energy meets mass exchange energy and enzymatic energy that brings about a definitive biodegradability of substrates in the NAPLs. An overall articulation of hydrocarbons consumption in soil, in which just microbial densities and the pollutant fixation decide the corrupt energy, can

Table 11.6 Summary of essential microbial growth kinetic models generally mentioned for site-specific bioremediation of nonaqueous phase liquid-contaminated locations (Yadav and Hassanizadeh 2011; Gupta and Yadav, 2017).

Kinetics model	Condition	Equation	Rate constant
Constant or zero-order	$X_0 \gg C_0$; $C_0 \gg K_s$	$-\frac{\partial C}{\partial t} = k_0$	$k_0 = \mu_{max} X_0$
Linear of first order	$X_0 \gg C_0$; $K_s \gg C_0$	$-\frac{\partial C}{\partial t} = k_1 C$	$k_1 = \mu_{max} X_0 / K_s$
Monod	$X_0 \gg C_0$	$-\frac{\partial C}{\partial t} = k_m C \big/ (k_s + C)$	$k_m = \mu_{max} X_0$
Logistic	$K_s \gg C_0$	$-\frac{\partial C}{\partial t} = k_1 C \big/ (C_0 + X_0 - C)$	$k_1 = X_0 / K_s$
Logarithmic	$K_s \gg C_0$	$-\frac{\partial C}{\partial t} = k \big/ (C_0 + X_0 - C)$	$k = \mu_{max}$

be composed as follows (Lyman et al., 1992; Yadav and Hassanizadeh 2011; Gupta and Yadav, 2017):

$$-\frac{\partial C}{\partial t} = \mu_{max} C \left(\frac{C_0 + X_0 - C}{K_s + C} \right)$$ (11.1)

where μ_{max} is the extreme specific growth rate, C is the contaminant concentration at time t, C_0 is the initial concentration and X_0 consistent with the pollutant essential to produce preliminary microbial density, and K_s is the substrate affinities persistent (Yadav and Hassanizadeh 2011). The growth and deprivation spectacles can be defined acceptably with four parameters; the two kinetics parameters are μ_{max} and K_s and correspondingly, Alvarez et al. (1994) examined the first direction kinetics model throughout the biodegradation of NAPLs and premeditated the worth kinetic parameter K_s is equal to 0.01 for BTEX complexes in a sandy aquifer. In the logistics model, the half-saturation is continuous, and the other two stoichiometric constraints $Y_{s/x}$ and S_{min}. Essential biodegradation kinetics models have been commonly used to designate microbial growth in a soil—water system. Table 11.6 provides a summary of essential microbial growth kinetic models generally mentioned for site-specific bioremediation of NAPL-contaminated locations.

Monod's kinetics, in which deprivation rates ensuing from zero-order to first order kinetics to goal substrate concentration, is largely used to define biodegradation duties of NAPLs during remediation of soil—water schemes (Alvarez et al., 1994; Kovarova-kovar and Egli 1998; Littlejohns and Daugulis 2008; Trigureros et al., 2010; Yadav and Hassanizadeh 2011). The equation is written as $k_m C/(K_s + C)$, where $k_m = \mu_{max} X_0$. Alvarez et al. (1994) calculated Monod's coefficient and K_s for benzene and toluene through aerobic biodegradation in a sandy aquifer method. The Monod coefficients were calculated as $k = 8.3$ g benzene/g-cells/day and

$K_s = 12.2$ mg/L for benzene and $k = 9.9$ g benzene g-cells/day and $K_s = 17.4$ mg/L for toluene. The zero-order kinetics embodies the oversimplification in NAPL degradation and frequently used low attentiveness degradation (Yadav and Hassanizadeh 2011; Datta et al., 2014). In zero-order kinetics, the rate of exhaustion of pollutants is occupied as perpetual under conditions $X_0 \gg C_0$ and $C_0 \gg K_s$. In the first-order model, the compulsory conditions are $X_0 \gg C_0$ and $K_s \gg C_0$, and the model is embodied as $k_1 C$, where $k_1 = \mu_{max} X_0 / K_s$. Gupta et al. (2013) examined the first-order model for the NAPL degradation up to 100 ppm substrate concentration is larger than the initial pollute concentration. The discrepancy form of the equation is written as $k_1 C / (C_0 + X_0 - C)$, where $k_1 = X_0 / K_s$. Similarly, the logarithmic model has required a horizontal asymptote as time goes on. In the logarithmic model, the frequency constant K is equal to the maximum growth rate μ_{max}. These models are related to soils taking nonlimiting nutrients, absence of mass transfer restrictions, optimum soil moisture content, and constant physicochemical factors (i.e., temperature, salinity, and pH). Of these influences, some may touch the rate of substrate commitment by microbial accumulation, and others may alter the rate of pollutant conveyance/supply to the microorganisms. The mainstream published papers on dechlorination have used Monod kinetics because it has been confirmed to be able to designate the growth (Eq. 11.2) of *Dehalococcoides* spp. which are known to be complicated in the sequential degradation of chlorinated ethene mixtures (Haest et al., 2010; Saiyari et al., 2018).

$$Rate_n = \frac{k_{max \cdot n} \, C_n X_n}{K_{s,n} + C_n} \tag{11.2}$$

Where,

$k_{max \cdot n}$ is the maximum growth rate ($\mu M \, Cell^{-1} \, d^{-1}$);

X_n is the microbial concentration ($Cell \, L^{-1}$);

$K_{s,n}$ is the half velocity constant μM of the compound, n.

11.4 Conclusion

The quick development of industrial activities and business measures has resulted in the discharging of a great many toxins into the climate through the soil, surface water, and the environment and the bringing about of genuine human infections due to their poisonousness. The hydrocarbons, as NAPLs, are of specific worry under fluctuating ecological conditions due to their high affectability to subsurface changeability, which empowers them to spread broadly. Soils are progressively undermined by spillage of oil-based goods, for example, petroleum, diesel, fuel at petroleum treatment facilities, underground stockpiling tanks, and siphon station pipelines. Remediation utilizing ordinary physical and compound strategies is uneconomical and creates enormous volumes of chemical waste. Biodegradation is a superior option by which hydrocarbon contaminations can be eliminated from the climate. Bioremediation is of interest worldwide as a potential oil slick cleanup method. The bioremediation method for tainted soil with raw petroleum as well as different hydrocarbons is appropriate in the field on account of its minimal effort and because it is climate cordial. Biostimulants and bioaugments in *ex situ* bioremediation for hydrocarbon dirtied soil performed under high-impact conditions end up being a likely strategy for remediation for most hydrocarbon-contaminated soils. *In situ* bioremediation innovations incorporate bioventing, biosparging, biostimulation fluid conveyance framework, and some soil-fertilizing strategies. *In situ* medicines will in general be more alluring to merchants and people in question because they require less gear, by and large have a lower cost, and produce fewer unsettling influences on the climate. Bioremediation innovation has become an examination hotspot because of its colossal financial and natural preferences. Subatomic biology innovation was utilized to screen the development and propagation of extra microorganisms during the time spent arable land rebuilding and the effect on the corruption pace of oil hydrocarbon toxins. Designed bioremediation develops as optimizing techniques to the bioremediation utilizing modifications to the natural condition and potential (local) microflora. This development bioremediation influences microbial exercises and their neighboring natural conditions for quickening the act of NAPL evacuation and is arranged as biostimulation and bioaugmentation. The biostimulation is improved by adding supplements, electron acceptors, oxygen, and other pertinent mixes to the dirtied destinations, which upgraded the (co) metabolic activities of the microflora. Bioaugmentation is a microorganism cultivating practice for developing the volume of a NAPL degrader by adding potential microbial societies that are filled autonomously in well-defined conditions. Moreover, the plants likewise quicken the NAPL evacuation by advancing microbial re-foundation in sullied soils and water because of steady conveyance of oxygen by root zone air circulation and supplements for microbial improvement by fixation and exudation. Hence, the incorporation and progression of multimicrobial and multielectron acceptor methodologies of bioremediation are discussed. The focal point of this part is examined in the various methodologies of bioremediation of NAPL-contaminated soil−water assets. The successful bioremediation of NAPL toxins by utilizing different expected microorganisms, similar to microbes (bacterial remediation), green growth (phycoremediation), parasites (mycoremediation), and some macrophytes of plants (Phytoremediation) have been summed up obviously. The results of this part are of direct use in applying bioremediation methods in the field and for the dynamic identified with arranging of NAPLs contaminating mechanical areas under fluctuating ecological conditions and recognized a compelling (or) possible micro/macroorganisms are reasonable organic specialists of various NAPL toxins for bioremediation of the climate for significant cleanup.

Acknowledgment

Author acknowledged thanks to the Secretary, Principal and HOD of Botany, Ramakrishna Mission Vivekananda College (Autonomous), Chennai − 600004, Tamil Nadu, India for providing the facilities to complete the research work.

References

Abatenh, E., Gizaw, B., Tsegaye, Z., Wassie, M., 2017. The role of microorganisms in bioremediation- a review. Open J. Environ. Biol. 2 (1), 038−046. https://doi.org/10.17352/ojeb.000007.

Agarwal, P., Sarkar, M., Chakraborty, B., Banerjee, T., 2019. Phytoremediation of air pollutants: prospects and challenges. Phytomanag. Polluted Sites 221−241. https://doi.org/10.1016/B978-0-12-813912-7.00007-7.

Aggelopoulos, C.A., Tsakiroglou, C.D., Ognier, S., Cavadias, S., 2015. Non-aqueous phase liquid-contaminated soil remediation by ex situ dielectric barrier discharge plasma. Int. J. Environ. Sci. Technol. 12 (3), 1011−1020. https://doi.org/10.1007/s13762-013-0489-4.

Aken, B. Van, Doty, S.L., 2009. Transgenic plants and associated bacteria for phytoremediation of chlorinated compounds. Biotechnol. Genet. Eng. Rev. 26 (1), 43−64. https://doi.org/10.5661/bger-26-43.

Al-Thani, R.F., Yasseen, B.T., 2020. Phytoremediation of polluted soils and waters by native Qatari plants: future perspectives. Environ. Pollut. 259, 113694. https://doi.org/10.1016/j.envpol.2019.113694.

Alirezaei, A., 2020. Medical Mycology Open Access 2020 Survey Bioremediation of Chlorophenol by Yeast Fungi Isolated from Industrial Wastewaters and Oil, vol. 6, p. 2020.

Alvarez, P.J.J., Vogel, T.M., 1995. Biodegradation of BTEX and their aerobic metabolize by indigenous microorganisms under nitrate reducing conditions. Water Sci. Technol. 3 (1), 15−28.

Alvarez, P.J.J., Anid, P.J., Vogel, T.M., 1994. Kinetics of toluene degradation by denitrifying aquifer microorganisms. J. Environ. Eng. 120, 1327−1336.

Alvarez, A., Saez, J.M., Costa, J.S.D., Colin, V.L., Fuentes, M.S., Cuozzo, S.A., Benimeli, C.S., Polti, M.A., Amoroso, M.J., 2017. Actinobacteria: current research and perspectives for bioremediation of pesticides and heavy metals. Chemosphere 166, 41−62. https://doi.org/10.1016/j.chemosphere.2016.09.070.

Ansari, A.A., Naeem, M., Gill, S.S., AlZuaibr, F.M., 2020. Phytoremediation of contaminated waters: an eco-friendly technology based on aquatic macrophytes application. Egypt. J. Aquat. Res. https://doi.org/10.1016/j.ejar.2020.03.002.

Anton, A., Mathe-Gaspar, G., 2005. Factors affecting heavy metal uptake in plant selection for phytoremediation. Zeitschrift Naturforschung. C J. Biosci. 60 (3−4), 244−246. PMID: 15948590.

APHA, 1985. Standard Methods for the Examination of Water and Wastewater, sixteenth ed.

Asquith, E.A., Geary, P.M., Nolan, A.L., Evans, C.A., 2012. Comparative bioremediation of petroleum hydrocarbon-contaminated soil by biostimulation, bioaugmentation and surfactant addition. J. Environ. Sci. Eng. 1, 637−650.

Asztemborska, M., Bembenek, M., Jakubiak, M., Steborowski, R., Bystrzejewska-Piotrowska, G., 2018. The effect of nanoparticles with sorption capacity on the bioaccumulation of divalent ions by aquatic plants. Int. J. Environ. Res. 12 (2), 245−253. https://doi.org/10.1007/s41742-018-0087-x.

Atlas, R.M., 1991. Microbial hydrocarbon degradation and bioremediation of oil spills. J. Chem. Technol. Biotechnol. 52, 149–156.

Atlas, R.M., Hazen, T.C., 2011. Oil biodegradation and bioremediation: a tale of the two worst spills in U.S. history. Environ. Sci. Technol. 45 (16), 6709–6715. https://doi.org/10.1021/es2013227.

Azubuike, C.C., Chikere, C.B., Okpokwasili, G.C., 2016. Bioremediation techniques- classification based on site of application: principles, advantages, limitations and pro- spects. World J. Microbiol. Biotechnol. 32, 180. https://doi.org/10.1007/s11274-016-2137-x.

Balba, M.T., Al-Awadhi, N., Al-Daher, R., 1998. Bioremediation of oil-contaminated soil: microbiological methods for feasibility assessment and field evaluation. J. Microbiol. Methods 32 (2), 155–164.

Baldiris-Navarro, I., Sanchez-Aponte, J., Gonzalez-Delgado, A., Jimenez, A.R., Acevedo-Morantes, M., 2018. Removal and biodegradation of phenol by the freshwater microalgae *Chlorella vulgaris*. Contemp. Eng. Sci. 11 (40), 1961–1970. https://doi.org/10.12988/ces.2018.84201.

Basu, S., Yadav, B.K., Mathur, S., 2015. Enhanced bioremediation of BTEX contaminated groundwater in pot-scale wetlands. Environ. Sci. Pollut. Control Ser. 22 (24), 20041–20049.

Bedessem, M.E., Ferro, A.M., Hiegel, T., 2007. Pilot-scale constructed wetlands for petroleum-contaminated groundwater. Water Environ. Res. 79 (6), 581–586.

Bento, F.M., Camargo, F.A.O., Okeke, B.C., Frankenberg, W.T., 2005. Comparative bioremediation of soils contaminated with diesel oil by natural attenuation, biostimulation and bioaugmentation. Bioresour. Technol. 96 (9), 1049–1055.

Bernardino, J., Bernardette, A., Ramrez-Sandoval, M., Domnguez-Oje, D., 2012. Biodegradation and bioremediation of organic pesticides. Pesticides - Recent Trends Pesticide Residue Assay 4. https://doi.org/10.5772/48631.

Binner, E., Böhm, K., Lehner, P., 2012. Large scale study on measurement of respiration activity (AT4) by Sapromat and Oxi Top. Waste Manag. 32, 1752–1759.

Biswas, B., Sarkar, B., Rusmin, R., Naidu, R., 2015. Bioremediation of PAHs and VOCs: advances in clay mineral–microbial interaction. Environ. Int. 85, 168–181. https://doi.org/10.1016/j.envint.2015.09.017.

Boonsaner, M., Borrirukwisitsak, S., Boonsaner, A., 2011. Phytoremediation of BTEX contaminated soil by *Canna generalis*. Ecotoxicol. Environ. Saf. 74 (6), 1700–1707.

Burghal, A.A., Al-Mudaffar, N.A., Mahdi, K.H., 2015. Ex situ bioremediation of soil contaminated with crude oil by use of actinomycetes consortia for process bioaugmentation. Pelagia Res. Library Eur. J. Exp. Biol. 5 (5), 24–30. www.pelagiaresearchlibrary.com.

Burken, J.G., Schnoor, J.L., 1998. Predictive relationships for uptake of organic contaminants by hybrid poplar trees. Environ. Sci. Technol. 32 (21), 3379–3385.

Bustamante, M., Durán, N., Diez, M.C., 2012. Biosurfactants are useful tools for the bioremediation of contaminated soil: a review. J. Soil Sci. Plant Nutr. 12 (4), 667–687. https://doi.org/10.4067/s0718-95162012005000024.

Chandra, R., Chowdhary, P., 2015. Properties of bacterial laccases and their application in bioremediation of industrial wastes. Environ. Sci. Processes Impacts 17, 326–342. https://doi.org/10.1039/c4em00627e.

Coulon, F., McKew, B.A., Osborn, A.M., McGenity, T.J., Timmis, K.N., 2007. Effects of temperature and biostimulation on oil-degrading microbial communities in temperate estuarine waters. Environ. Microbiol. 9 (1), 177–186.

Da Silva, M.L.B., Alvarez, P.J.J., 2004. Enhanced anaerobic biodegradation of benzene-toluene-ethylbenzene-xy-lene-ethanol mixtures in bioaugmented aquifer columns. Appl. Environ. Microbiol. 70, 4720–4726.

DalCorso, G., Fasani, E., Manara, A., Visioli, G., Furini, A., 2019. Heavy metal pollution: state of the art and innovation in phytoremediation. Int. J. Mol. Sci. 20 (14) https://doi.org/10.3390/ijms20143412.

Das, B., Mandal, T.K., Patra, S., 2015. A comprehensive study on *Chlorella pyrenoidosa* for phenol degradation and its potential applicability as biodiesel feedstock and animal feed. Appl. Biochem. Biotechnol. 176 (5), 1382–1401. https://doi.org/10.1007/s12010-015-1652-9.

Datta, A., Philip, L., Murty, B.S., 2014. Modeling the biodegradation kinetics of aromatic and aliphatic volatile pollutant mixture in liquid phase. Chem. Eng. J. 241, 288–300.

de Souza, C.B., Silva, G.R., 2019. Phytoremediation of effluents contaminated with heavy metals by floating aquatic macrophytes species. Biotechnol. Bioeng. https://doi.org/10.5772/intechopen.83645.

Delille, D., Coulon, F., Pelletier, E., 2007. Long-term changes of bacterial abundance, hydrocarbon concentration and toxicity during a biostimulation treatment of oil-amended organic and mineral sub-Antarctic soils. J. Polar Biol. 30 (7), 925–933.

Dhillon, G.S., Kaur, S., Verma, M., Brar, S.K., 2012. Biopolymer-based nanomaterials: potential applications in bioremediation of contaminated wastewaters and soils. Compr. Anal. Chem. 59, 91–129. https://doi.org/10.1016/B978-0-444-56328-6.00003-7.

Di Gregorio, S., Serra, R., Villani, M., 1999. Applying cellular automata to complex environmental problems: the simulation of the bioremediation of contaminated soils. Theor. Comput. Sci. 217 (1), 131–156. https://doi.org/10.1016/S0304-3975(98)00154-6.

Dibble, J.T., Bartha, R., 1979. Effect of environmental parameters on parameters on the biodegradation of oil sludge. Appl. Environ. Microbiol. 37, 729–739.

Dresback, K., Ghoshal, D., Goyal, A., 2001. Phycoremediation of trichloroethylene (TCE). Physiol. Mol. Biol. Pla. 7 (2), 117–123.

DuBois, M., Gilles, K.A., Hamilton, J.K., Rebers, P.A., Smith, F., 1956. Colorimetric method for determination of sugars and related substances. Anal. Chem. 28, 350–356. https://doi.org/10.1021/ac60111a017.

Dzantor, E.K., 2007. Phytoremediation: the state of rhizosphere "engineering" for accelerated rhizodegradation of xenobiotic contaminants. J. Chem. Technol. Biotechnol. 232, 228–232.

Environmental Protection Agency, 2010. Method 1664B: N-Hexane Extractable Material and Silica Gel Treated N-Hexane Extractable Material by Extraction and Gravimetry. United States Environmental Protection Agency. https://www.epa.gov/sites/production/files/2015-08/documents/method_1664b_2010.pdf.

Fan, S., Scow, K.M., 1993. Toluene by indigenous microbial populations in biodegradation of trichloroethylene and toluene by indigenous microbial populations in soil. Appl. Environ. Microbiol. 59 (6), 1911–1918.

Fan, F., Xu, R., Wang, D., Meng, F., 2020. Application of activated sludge for odor control in wastewater treatment plants: approaches, advances and outlooks. Water Res. 181, 115915. https://doi.org/10.1016/j.watres.2020.115915.

Farahdiba, A.U., Cahyonugroho, O., Nindhita, S.N., Hidayah, E.N., 2020. Photoinhibition of algal photobioreactor by intense light. J. Phys.: Conf. Ser. 1569, 042095.

Farhadian, M., Vachelard, C., Duchez, D., Larroche, C., 2008. In situ bioremediation of mono-aromatic pollutants in groundwater: a review. Bioresour. Technol. 99 (13), 5296–5308.

Farid, M., Ali, S., Akram, N.A., Rizwan, M., Abbas, F., Bukhari, S.A.H., Saeed, R., 2017. Phyto-management of Cr-contaminated soils by sunflower hybrids: physiological and biochemical response and metal extractability under Cr stress. Environ. Sci. Pollut. Control Ser. 24 (20), 16845−16859. https://doi.org/10.1007/s11356-017-9247-3.

Franzetti, A., Di Gennaro, P., Bestetti, G., Lasagni, M., Pitea, D., Collina, E., 2008. Selection of surfactants for enhancing diesel hydrocarbons-contaminated media bioremediation. J. Hazard Mater. 152 (3), 1309−1316.

Frutos, F.J.G., Escolano, O., García, S., Babín, M., Fernández, M.D., 2010. Bioventing remediation and ecotoxicity evaluation of phenanthrene-contaminated soil. J. Hazard Mater. 183, 806−813.

Gerhardt, K.E., Huang, X.-D., Glick, B.R., Greenberg, B.M., 2009. Phytoremediation and rhizoremediation of organic soil contaminants: potential and challenges. Plant Sci. 176 (1), 20−30.

Glick, B.R., 2010. Using soil bacteria to facilitate phytoremediation. Biotechnol. Adv. 28 (3), 367−374. https://doi.org/10.1016/j.biotechadv.2010.02.001.

Gogoi, B.K., Dutta, N.N., Goswami, P., Krishna Mohan, T.R., 2003. A case study of bioremediation of petroleum-hydrocarbon contaminated soil at a crude oil spill site. Adv. Environ. Res. 7 (4), 767−782. https://doi.org/10.1016/S1093-0191(02)00029-1.

Gupta, P., Yadav, B., July 2017. Bioremediation of nonaqueous phase liquids (NAPLs)-polluted soil-water resources. In: Environmental Pollutants and Their Bioremediation Approaches, pp. 241−256. https://doi.org/10.1201/9781315173351-9.

Gupta, P.K., Ranjan, S., Yadav, B.K., 2013. BTEX biodegradation in soil-water system having different substrate concentrations. Int. J. Eng. Res. Technol. 2 (12), 1765−1772. ISSN: 2278-0181.

Haest, P.J., Springael, D., Smolders, E., 2010. Dechlorination kinetics of TCE at toxic TCE concentrations: assessment of different models. Water Res. 44, 331e9.

Haroun, A.A., Kamaluddeen, K.K., Alhaji, I., Magaji, Y., Oaikhena, E.E., 2017. Evaluation of heavy metal tolerance level (MIC) and bioremediation potentials of *Pseudomonas aeruginosa* isolated from a-Kakuri industrial drain in Kaduna, Nigeria. Eur. J. Exp. Biol. 7 (5), 3−6. https://doi.org/10.21767/2248-9215.100028.

Hayachi, K., 1975. A rapid determination of sodium dodecyl sulfate with methylene blue. Anal. Biochem. 67, 503−506.

Hu, H., Xiang, L., Wu, S., Yang, C., 2020. Sustainable livestock wastewater treatment via phytoremediation: current status and future perspectives. Bioresour. Technol. 315, 123809. https://doi.org/10.1016/j.biortech.2020.123809.

Huang, X.D., El-Alawi, Y., Penrose, D.M., Glick, B.R., Greenberg, B.M., 2004. A multi-process phytoremediation system for removal of polycyclic aromatic hydrocarbons from contaminated soils. Environ. Pollut. 130 (3), 465−476.

Hussain, S., Arshad, M., Saleem, M., Khalid, A., 2007. Biodegradation of alpha- and betaendosulfan by soil bacteria. Biodegradation 18 (6), 731−740.

Hwang, J.I., Li, Z., Andreacchio, N., Ordonez Hinz, F., Wilson, P.C., 2020. Potential use of floating treatment wetlands established with *Canna flaccida* for removing organic contaminants from surface water. Int. J. Phytoremediation 22 (12), 1304−1312. https://doi.org/10.1080/15226514.2020.1768511.

ICCS, 2006. In: Manual for Biological Remediation Techniques, vol. 81. International Centre for Soil and Contaminated Sites.

Imam, A., Suman, S.K., Ghosh, D., Kanaujia, P.K., 2019. Analytical approaches used in monitoring the bioremediation of hydrocarbons in petroleum- contaminated soil and sludge. Trac. Trends Anal. Chem. 118, 50−64. https://doi.org/10.1016/j.trac.2019.05.023.

Imfeld, G., Braeckevelt, M., Kuschk, P., Richnow, H.H., 2009. Monitoring and assessing processes of organic chemicals removal in constructed wetlands. Chemosphere 74 (3), 349−362.

ITRC, October 2005. Overview of in Situ Bioremediation of Chlorinated Ethene DNAPL Source Zones. www.itrcweb.org.

Jasrotia, S., Kansal, A., Mehra, A., 2017. Performance of aquatic plant species for phytoremediation of arsenic-contaminated water. Appl. Water Sci. 7, 889−896. https://doi.org/10.1007/s13201-015-0300-4.

Jimoh, A.A., Lin, J., 2020. Bioremediation of contaminated diesel and motor oil through the optimization of biosurfactant produced by *Paenibacillus* sp. D9 on waste canola oil. Bioremed. J. 24 (1), 21−40. https://doi.org/10.1080/10889868.2020.1721425.

Jin, Z.P., Luo, K., Zhang, S., Zheng, Q., Yang, H., 2012. Bioaccumulation and catabolism of prometryne in green algae. Chemosphere 87 (3), 278−284. Epub 2012/01/26.

Joo, H.S., Ndegwa, P.M., Shoda, M., Phae, C.G., 2008. Bioremediation of oil-contaminated soil using *Candida catenulata* and food waste. Environ. Pollut. 156 (3), 891−896.

Júlio, A.D.L., de Cássia Mourão Silva, U., Medeiros, J.D., Morais, D.K., dos Santos, V.L., 2019. Metataxonomic analyses reveal differences in aquifer bacterial community as a function of creosote contamination and its potential for contaminant remediation. Sci. Rep. 9 (1), 1−14. https://doi.org/10.1038/s41598-019-47921-y.

Karamalidis, A.K., Evangelou, A.C., Karabika, E., Koukkou, A.I., Drainas, C., Voudrias, E.A., 2010. Laboratory scale bioremediation of petroleum-contaminated soil by indigenous microorganisms and added *Pseudomonas aeruginosa* strain Spet. Bioresour. Technol. 101 (16), 6545−6552.

Kartik, P., Mitesh, P., 2020. Improving bioremediation process of petroleum wastewater using biosurfactants producing *Stenotrophomonas* sp. S1VKR-26 and assessment of phytotoxicity. Bioresour. Technol. 315, 123861. https://doi.org/10.1016/j.biortech.2020.123861.

Kaszycki, P., Petryszak1, P., Pawlik1, M., Kołoczek1, H., 2011. *Ex-situ* bioremediation of soil polluted with oily waste: the use of specialized microbial consortia for process bioaugmentation. Ecol. Chem. Eng. 18, 83−92.

Kim, J.M., Le, N.T., Chung, B.S., Park, J.H., Bae, J.W., Madsen, E.L., Jeon, C.O., 2008. Influence of soil components on the biodegradation of benzene, toluene, ethylbenzene, and o-, m-, and p-xylenes by the newly isolated bacterium *Pseudoxanthomonas spadix* BD-a59. Appl. Environ. Microbiol. 74 (23), 7313−7320.

Koga, M., Yamamichi, Y., Nomoto, Y., 1999. Rapid determination of anionic surfactants by improved spectrophotometric method using Methylene Blue. Anal. Sci. 15, 563−568.

Kovarova-kovar, K., Egli, T., 1998. Growth kinetics of suspended microbial cells: from single-substrate controlled growth to mixed-substrate kinetics. Microbiol. Mol. Biol. Rev. 62 (3), 646.

Kuiper, I., Lagendijk, E.L., Bloemberg, V.G., Lugtenberg, B.J.J., 2004. Rhizoremediation: a beneficial plant-microbe interaction. Mol. Plant. Microbe Interact. 17 (1), 6−15. https://doi.org/10.1094/MPMI.2004.17.1.6.

Kumari, B., Singh, D.P., 2016. A review on multifaceted application of nanoparticles in the field of bioremediation of petroleum hydrocarbons. Ecol. Eng. 97, 98−105. https://doi.org/10.1016/j.ecoleng.2016.08.006.

Langwaldt, J.H., Puhakka, J.A., 2000. On-site biological remediation of contaminated groundwater: a review. Environ. Pollut. 107 (2), 187–197. https://doi.org/10.1016/S0269-7491(99)00137-2.

Lee, S.E., Kim, J.S., Kennedy, I.R., Park, J.W., Kwon, G.S., Koh, S.C., 2003. Biotransformation of an organochlorine insecticide, endosulfan, by *Anabaena* species. J. Agric. Food Chem. 51 (5), 1336–1340.

Lendvay, J.M., Lo, F.E., 2003. Bioreactive barriers: a comparison of bioaugmentation and biostimulation for chlorinated solvent remediation. Environ. Sci. Technol. 37 (7), 1422–1431.

Leson, G., Winer, A.M., 1991. Biofiltration: an innovative air pollution control technology for voc emissions. J. Air Waste Manag. Assoc. 41 (8), 1045–1054. https://doi.org/10.1080/10473289.1991.10466898.

Liao, Y.-G., 2007. High-efficiency petroleum-degrading bacteria remove oil from oil-contaminated soil. Human 25–43. Master of Science in Chemical Engineering, Xiangtan University.

Lin, Q., Mendelssohn, I.A., 2009. Potential of restoration and phytoremediation with *Juncus roemerianus* for diesel-contaminated coastal wetlands. Ecol. Eng. 35 (1), 85–91.

Littlejohns, J.V., Daugulis, A.J., 2008. Kinetics and interactions of BTEX compounds during degradation by a bacterial consortium. Process Biochem. 43 (10), 1068–1076.

Lu, S.-J., 2007. Study on Enhanced Technology of Microbial Degradation of Soil Diesel Pollutants (Doctoral dissertation). Institute of Water Sciences, Beijing Normal University, Beijing, pp. 30–53.

Lyman, W.J., Reidy, P.J., Levy, B., 1992. Mobility and Degradation of Organic Contaminants in Subsurface Environments. C.K. Smoley, Chelsea, p. 395.

L€offler, F.E., Yan, J., Ritalahti, K.M., Adrian, L., Edwards, E.A., Konstantinidis, K.T., 2013. *Dehalococcoides mccartyi* gen. nov., sp. nov., obligately organohalide-respiring anaerobic bacteria relevant to halogen cycling and bioremediation, belong to a novel bacterial class, *Dehalococcoidia classis* nov., order *Dehalococcoidales* ord. nov. and family *Dehalococcoidaceae fam.* nov., within the phylum *Chloroflexi.* Int. J. Syst. Evol. Microbiol. 63, 625e35.

Macnaughton, S.J., Stephen, J.R., Venosa, A.D., Davis, G.A., Chang, Y., White, D.C., 1999. Microbial population changes during bioremediation of an experimental oil spill. Appl. Environ. Microbiol. 65 (8), 3566.

Maitra, S., 2018. In-situ bioremediation - an overview. RJLBPCS 4 (6), 576–598. https://doi.org/10.26479/2018.0406.45.

Margesin, R., Schinner, F., 2001. Biodegradation and bioremediation of hydrocarbons in extreme environments. Appl. Microbiol. Biotechnol. 56 (5–6), 650–663.

Mariaamalraj, S.K., Pasumarthi, R., Achary, A., Mutnuri, S., 2016. Effect of rhamnolipid on biodegradation of hydrocarbons in non-aqueous-phase liquid (NAPL). Ann. Finance 20 (3), 183–193. https://doi.org/10.1080/10889868.2016.1212807.

Mattozzi, M.D.L.P., Tehara, S.K., Hong, T., Keasling, J.D., 2006. Mineralization of paraoxon and its use as a sole C and P source by a rationally designed catabolic pathway in *Pseudomonas putida*. Appl. Environ. Microbiol. 72 (10), 6699–6706. Epub 2006/10/06.

Medfu Tarekegn, M., Zewdu Salilih, F., Ishetu, A.I., 2020. Microbes used as a tool for bioremediation of heavy metal from the environment. Cogent Food gric. 6 (1) https://doi.org/10.1080/23311932.2020.1783174.

Miller, J.H., 1972. Experiments in Molecular Genetics. Cold Spring Harbor Laboratory, Cold Spring Harbor, New York.

Mills, A.L., Breuil, C., Colwell, R.R., 1978. Enumeration of petroleum-degrading marine and estuarine microorganisms by the most probable number method. Can. J. Microbiol. 24, 552–557.

Mishra, S., Jyot, J., Kuhad, R.C., Lal, B., 2001a. Evaluation of inoculum addition to stimulate in-situ bioremediation of oily-sludge-contaminated soil evaluation of inoculum addition to stimulate in-situ bioremediation of oily-sludge-contaminated soil. Appl. Environ. Microbiol. 67 (4), 1675–1681. https://doi.org/10.1128/AEM.67.4.1675.

Mishra, S., Jyoti, J., Kuhad, R.C., Lal, B., 2001b. In-situ bioremediation potential of an oily sludge-degrading bacterial consortium. Curr. Microbiol. 43 (5), 328–335. https://doi.org/10.1007/s002840010311.

Mohamed Sunar, N., Emparan, Q.A., Abdul Karim, A.T., Noor, S.F.M., Maslan, M., Mustafa, F., Khaled, N., 2014. The effectiveness of bioremediation treatment for diesel-soil contamination. Adv. Mater. Res. 845, 146–152. https://doi.org/10.4028/www.scientific.net/AMR.845.146.

Mohan, M.T.M., Ram, S., 2018. Bioremediation of groundwater: an overview. Int. J. Appl. Eng. Res. 13 (24), 16825–16832. http://www.ripublication.com.

Mohanty, S., Jasmine, J., Mukherji, S., 2013. Practical considerations and challenges involved in surfactant enhanced bioremediation of oil. BioMed Res. Int. https://doi.org/10.1155/2013/328608.

Mohsenzadeh, F., Nasseri, S., Mesdaghinia, A., Nabizadeh, R., Zafari, D., Khodakaramian, G., Chehregani, A., 2010. Phytoremediation of petroleum-polluted soils: application of Polygonum aviculare and its root-associated (penetrated) fungal strains for bioremediation of petroleum-polluted soils. Ecotoxicol. Environ. Saf. 73 (4), 613–619.

Moretti, L., 2006. In situ bioremediation of DNAPL source zones. Hazard. Waste Consult. 24 (1).

Mukherji, S., Peters, C.A., Weber, W.J., 1997. Mass transfer of polynuclear aromatic hydrocarbons from complex DNAPL mixtures. Environ. Sci. Technol. 31, 416–423. https://doi.org/10.1021/es960227n.

Mustafa, H.M., Hayder, G., 2020. Recent studies on applications of aquatic weed plants in phytoremediation of wastewater: a review article. Ain Shams Eng. J. https://doi.org/10.1016/j.asej.2020.05.009.

Muthukumaran, M., 2009. Studies on the Phycoremediation of Industrial Effluents and Utilization of Algal Biomass (Ph.D. thesis). University of Madras.

Muthukumaran, M., Thirupathi, P., Chinnu, K., Sivasubramanian, V., 2012. Phycoremediation efficiency and biomass production by micro alga *Desmococcus olivaceus* (Persoon et Acharius) J.R. Laundon treated on chrome-sludge from an electroplating industry-A open raceway pond study. Int. J. Curr. Sci. Special Issue 52–62.

Muthukumaran, M., Sivasubramanian, V., 2017. Microalgae cultivation for biofuels: cost, energy balance, environmental impacts and future perspectives (Chapter 15). In: Singh, R.S., Pandey, A., Gnansounou, E. (Eds.), Biofuels: Production and Future Perspectives. To be published by Taylor & Francis Group, CRC Press, USA, pp. 363–411. ISBN 9781498723596.

Narayanan, M., Tracy, J.C., Davis, L.C., Erickson, L.E., 1998a. Modeling the fate of toluene in a chamber with alfalfa plants 1. Theory and modeling concepts. J. Hazard. Substance Res. 1, 5a-1–5a-30.

Narayanan, M., Tracy, J.C., Davis, L.C., Erickson, L.E., 1998b. Modeling the fate of toluene in a chamber with alfalfa plants 2. Numerical results and comparison study. J. Hazard. Substance Res. 1, 5b-1–5b-28.

Nitschke, M., Pastore, G.M., 2006. Production and properties of a surfactant obtained from *Bacillus subtilis* grown on cassava wastewater. Bioresour. Technol. 97, 336–341. https://doi.org/10.1016/j.biortech.2005.02.044.

Nwachukwu, S.C.U., 2000. Enhanced rehabilitation of tropical aquatic environments polluted with crude petroleum using *Candida utilis*. J. Environ. Biol. 21, 241–250.

Obire, O., 1988. Studies on the biodegradation potentials of some microorganisms isolated from water systems of two petroleum producing areas in Nigeria. Niger. J. Bot. 1, 81–90.

Obire, O., Anyanwu, E.C., Okigbo, R.N., 2008. Saprophytic and crude oil degrading fungi from cow dung and poultry droppings as bioremediating agents. J. Agric. Technol. 4 (2), 81–89.

Odu, C.T.I., 1972. Microbiology of soils contaminated with petroleum hydrocarbons. I. Extent of contamination and some soil and microbial properties after contamination. J. Inst. Petrol. 58, 201–208.

Okerentugba, P.O., Ezeronye, O.U., 2003. Petroleum degrading potentials of single and mixed microbial cultures isolated from rivers and refinery effluent in Nigeria. Afr. J. Biotechnol. 2 (9), 312–319.

Okpokwasili, G.C., Amanchukwu, S.C., 1988. Petroleum hydrocarbon degradation by *Candida* species. Environ. Int. 14, 243–347.

Okpokwasili, G.C., Okorie, B.B., 1988. Biodeterioration potentials of microorganisms isolated from car engine lubricating oil. Tribol. Int. 21 (4), 215–220.

Olguin, E., 2003. Phycoremediation: key issues for cost-effective nutrient removal processes. Biotechnol. Adv. 22 (1–2), 81–91. https://doi.org/10.1016/j.biotechadv.2003.08.009.

Osazee, E., Yerima, M.B., Shehu, K., 2015. Bioremediation of crude oil contaminated soils using cow dung as bioenhancement agent. Ann. Biol. Sci. 3 (2), 8–12.

O'Shaughnessy, S.A., Kim, M., Andrade, M.A., Colaizzi, P.D., Evett, S.R., June 2020. Site-specific irrigation of grain sorghum using plant and soil water sensing feedback - Texas High Plains. Agric. Water Manag. 240, 106273. https://doi.org/10.1016/j.agwat.2020.106273.

Pablo, M.F., Viñarta, S.C., Bernal, A.R., Cruz, E.L., Figuero, L.I.C., 2018. Bioremediation strategies for chromium removal: current research, scale-up approach and future perspectives. Chemosphere 208, 139–148. https://doi.org/10.1016/j.chemosphere.2018.05.166.

Pal, S., Vimala, Y., 2012. Bioremediation and decolorization of distillery effluent by novel microbial consortium. Eur. J. Exp. Biol. 2 (3), 496–504.

Pandey, B., Fulekar, M.H., 2012. Bioremediation technology: a new horizon for environmental clean-up. Biol. Med. 4 (1), 51–59.

Parzych, A., Jonczak, J., Sobisz, Z., 2018. *Pellia endiviifolia* (dicks.) dumort. Liverwort with a potential for water purification. Int. J. Environ. Res. 12 (4), 471–478. https://doi.org/10.1007/s41742-018-0105-z.

Pasumarthi, R., Chandrasekaran, S., Mutnuri, S., 2013. Biodegradation of crude oil by *Pseudomonas aeruginosa* and *Escherichia fergusonii* isolated from the Goan coast. Mar. Pollut. Bull. 76, 276–282. https://doi.org/10.1016/j.marpolbul.2013.08.026.

Popp, N., Schlömann, M., Mau, M., 2006. Bacterial diversity in the active stage of a bioremediation system for mineral oil hydrocarbon-contaminated soils. Microbiology 152 (Pt 11), 3291–3304.

Powell, C.L., Nogaro, G., Agrawal, A., 2011. Aerobic co-metabolic degradation of trichloroethene by methane and ammonia oxidizing microorganisms naturally associated with *Carex comosa* roots. Biodegradation 22, 527–538.

Powell, C.L., Goltz, M.N., Agrawal, A., 2014. Degradation kinetics of chlorinated aliphatic hydrocarbons by methane oxidizers naturally-associated with wetland plant roots. J. Contam. Hydrol. 170, 68−75.

Prozorova, G.F., Pozdnyakov, A.S., Kuznetsova, N.P., Korz- hova, S.A., Emel'yanov, A.I., Ermakova, T.G., Fadeeva, T.V., Sosedova, L.M., 2014. Green synthesis of water- soluble nontoxic polymeric nanocomposites containing silver nanoparticles. Int. J. Nanomed. 9, 1883−1889. https://doi.org/10.2147/IJN.S57865.

Ramírez-García, R., Gohil, N., Singh, V., 2019. Recent advances, challenges, and opportunities in bioremediation of hazardous materials. Phytomanag. Pollut. Sites 517−568. https://doi.org/10.1016/B978-0-12-813912-7.00021-1.

Ranieri, E., Gikas, P., Tchobanoglous, G., 2013. BTEX removal in pilot-scale horizontal subsurface flow constructed wetlands. Desalin. Water Treat. 51 (13−15), 3032−3039.

Ren, Z.J., Umble, A.K., 2016. Water treatment: recover wastewater resources locally. Nature 529, 25. https://doi.org/10.1038/529025b.

Reuschenbach, P., Pagga, U., Strotmann, U., 2003. A critical comparison of respirometric biodegradation tests based on OECD 301 and related test methods. Water Res. 37, 1517−1582.

Rigas, F., Papadopoulou, K., Dritsa, V., Doulia, D., 2007. Bioremediation of a soil contaminated by lindane utilizing the fungus *Ganoderma australe* via response surface methodology. J. Hazard Mater. 140 (1−2), 325−332.

Robinson, G.K., Lenn, M.J., 1994. The bioremediation of polychlorinated biphenyls (PCBS): problems and perspectives. Biotechnol. Genet. Eng. Rev. 12 (1), 139−188. https://doi.org/10.1080/02648725.1994.10647911.

Röling, W.F., Milner, M.G., Jones, D.M., Lee, K., Daniel, F., Swannell, R.J., Head, I.M., 2002. Robust hydrocarbon degradation and dynamics of bacterial communities during nutrient-enhanced oil spill bioremediation. Appl. Environ. Microbiol. 68 (11), 5537−5548.

Rosenberg, M., Gutnick, D., Rosenberg, E., 1980. Adherence of bacteria to hydrocarbons: a simple method for measuring cell-surface hydrophobicity. FEMS Microbiol. Lett. 9, 29−33. https://doi.org/10.1111/j.1574-6968.1980.tb05599.x.

Roy, A.S., Baruah, R., Borah, M., Singh, A.K., Deka Boruah, H.P., Saikia, N., Chandra Bora, T., 2014. Bioremediation potential of native hydrocarbon degrading bacterial strains in crude oil contaminated soil under microcosm study. Int. Biodeterior. Biodegrad. 94, 79−89.

Saiyari, D.M., Chuang, H.P., Senoro, D.B., Lin, T.F., Whang, L.M., Chiu, Y.T., Chen, Y.H., 2018. A review in the current developments of genus Dehalococcoides, its consortia and kinetics for bioremediation options of contaminated groundwater. Sustain. Environ. Res. 28 (4), 149−157. https://doi.org/10.1016/j.serj.2018.01.006.

Salanitro, J.P., Diaz, L.A., 1995. Anaerobic biodegradability testing of surfactants. Chemosphere 30, 813−830.

Samanta, S.K., Singh, O.V., Jain, R.K., 2002. Polycyclic aromatic hydrocarbons: environmental pollution and bioremediation. Trends Biotechnol. 20 (6), 243−248.

Sarkar, D., Ferguson, M., Datta, R., Birnbaum, S., 2005. Bioremediation of petroleum hydrocarbons in contaminated soils: comparison of biosolids addition, carbon supple- mentation, and monitored natural attenuation. Environ. Pollut. 136 (1), 187−195.

Saxena, M.M., 1990. Environmental Analysis: Water, Soil and Air. Agro Botanical Pub., pp. 172−174.

Selberg, A., Juuram, K., Budashova, J., Tenno, T., 2013. Biodegradation and leaching of surfactants during surfactant-amended bioremediation of oil-polluted soil. In: Applied Bioremediation - Active and Passive Approaches. https://doi.org/10.5772/56908.

Shehzadi, M., Afzal, M., Khan, M.U., Islam, E., Mobin, A., Anwar, S., Khan, Q.M., 2014. Enhanced degradation of textile effluent in constructed wetland system using *Typha domingensis* and textile effluent-degrading endophytic bacteria. Water Res. 58, 152−159.

Shimp, J.F., Tracy, J.C., Davis, L.C., Lee, E., Huang, W., Erickson, L.E., Schnoor, J.L., 1993. Beneficial effects of plants in the remediation of soil and groundwater contaminated with organic materials. Crit. Rev. Environ. Sci. Technol. 23 (1), 41−77.

Shukla, A., Srivastava, S., 2019. A review of phytoremediation prospects for arsenic contaminated water and soil. Phytomanag. Pollut. Sites 243−254. https://doi.org/10.1016/B978-0-12-813912-7.00008-9.

Singh, D., Fulekar, M.H., 2010. Biodegradation of petroleum hydrocarbons by *Pseudomonas putida* strain MHF 7109 isolated from cow dung microbial consortium. Clean Soil, Air, Water 38 (8), 781−786.

Soccol, C.R., Vandenberghe, L.P.S., Woiciechowski, A.L., Thomaz-Soccol, V., Correia, C.T., Pandey, A., 2003. Bioremediation: an important alternative for soil and industrial wastes clean-up. Indian J. Exp. Biol. 41 (9), 1030−1045.

Souza, E.C., Vessoni-Penna, T.C., de Souza Oliveira, R.P., 2014. Biosurfactant-enhanced hydrocarbon bioremediation: an overview. Int. Biodeterior. Biodegrad. 89, 88−94. https://doi.org/10.1016/j.ibiod.2014.01.007.

Speight, J.G., El-Gendy, N.S., 2018. Bioremediation of marine oil spills. Introduct. Petroleum Biotechnol. 419−470. https://doi.org/10.1016/B978-0-12-805151-1.00011-4.

Strotmann, U., Reuschenbach, P., Schwarz, H., Pagga, U., 2004. Development and evaluation of an online CO_2 evolution test and a multicomponent biodegradation test system. Appl. Environ. Microbiol. 70, 4621−4628.

Sui, H., Li, X., 2011. Modeling for volatilization and bioremediation of toluene contaminated soil by bioventing. Chin. J. Chem. Eng. 19, 340−348.

Susarla, S., Medina, V.F., McCutcheon, S.C., 2002. Phytoremediation: an ecological solution to organic chemical contamination. Ecol. Eng. 18 (5), 647−658.

Tadros, T.F., 2005. Applied Surfactants: Principles and Applications. Wiley-VCH.

Taylor, L.T., Jones, D.M., 2001. Bioremediation of coal tar PAH in soil using biodiesel. Chemosphere 44, 1 131−1 136.

Thenmozhi, R., Arumugam, K., Nagasathya, A., Thajuddin, N., Paneerselvam, A., 2013. Studies on Mycoremediation of used engine oil contaminated soil samples. Pelagia Res. Library Adv. Appl. Sci. Res. 4 (2), 110−118. www.pelagiaresearchlibrary.com.

Thies, F., Backhaus, T., Bossmann, B., Grimme, L.H., 1996. Xenobiotic biotransformation in unicellular green algae. Involvement of cytochrome P450 in the activation and selectivity of the pyridazinone pro-herbicide metflurazon. Plant Physiol. 112 (1), 361−370. Epub 1996/09/01.

Thijsee, G.J.E., Van der Linden, A.C., 1961. Iso-alkane oxidation by a *Pseudomonas*. Antonie van Leuwenhoek 27, 171−179.

Trigueros, D.E.G., Modenes, A.N., Kroumov, A.D., Espinoza-Quinones, F.R., 2010. Modeling of biodegradation process of BTEX compounds: kinetic parameters estimation by using particle swarm global optimizer. Process Biochem. 45 (8), 1355−1361.

Tsao, C., Song, H., Bartha, R., 1998. Metabolism of benzene, toluene, and xylene hydrocarbons in soil. Appl. Environ. Microbiol. 64 (12), 4924.

Udeme, J.J., Antai, S.P., 1988. Biodegradation and mineralization of crude oil by bacteria. Niger. J. Biotechnol. 5, 77−85.

Urum, K., Pekdemir, T., 2004. Evaluation of biosurfactants for crude oil contaminated soil washing. Chemosphere 57, 1139−1150.

USEPA, 2005. *A Decision-Making Framework for Cleanup of Sites Impacted with Light Non Aqueous Phase Liquids* (LNAPL). EPA 542-R-04-011. U.S. Environmental Protection Agency, Washington, D.C.

Van Gestel, K., Mergaert, J., Swings, J., Coosemans, J., Ryckeboer, J., 2003. Bioremediation of diesel oil-contaminated soil by composting with biowaste. Environ. Pollut. 125 (3), 361−368.

Veenanadig, N.K., Gowthaman, M.K., Karanth, N.G.K., 2000. Scale up studies for the production of biosurfactant in packed column bioreactor. Bioprocess Eng. 22, 95−99.

Vinas, M., Sabate, J., Espuny, M.J., Anna, M., Vin, M., 2005. Bacterial community dynamics and polycyclic aromatic hydrocarbon degradation during bioremediation of heavily creosote- contaminated soil. Appl. Environ. Microbiol. 71 (11), 7008.

Vo, H.N.P., Ngo, H.H., Guo, W., Nguyen, K.H., Chang, S.W., Nguyen, D.D., Liu, Y., Liu, Y., Ding, A., Buig, X.T., 2020. Micropollutants cometabolism of microalgae for wastewater remediation: effect of carbon sources to cometabolism and degradation products. Water Res. 183, 115974. DOI: 0.1016/j.watres.2020.115974.

Vymazal, J., 2009. The use constructed wetlands with horizontal sub-surface flow for various types of wastewater. Ecol. Eng. 35 (1), 1−17.

Wani, R.A., Ganai, B.A., Shah, M.A., Uqab, B., 2017. Heavy metal uptake potential of aquatic plants through phytoremediation technique - a review. J. Biorem. Biodegrad. 08 (04) https://doi.org/10.4172/2155-6199.1000404.

Warrier, R.R., 2012. Phytoremediation for environmental clean up. Forestry Bull. 12 (2), 1−7.

Witzig, R., Junca, H., Hecht, H., Pieper, D.H., 2006. Assessment of toluene/biphenyldioxygen- ase gene diversity in benzene-polluted soils: links between benzene biodegradation and genes similar to those encoding isopropyl benzene dioxygenases assessment of toluene/biphenyl dioxygenase. Appl. Environ. Microbiol. 72 (5), 3504−3514.

Xia, L., 2019. Status of microbial remediation technology in petroleum contaminated land. IOP Conf. Ser. Earth Environ. Sci. 300 (5) https://doi.org/10.1088/1755-1315/300/5/052050.

Xiao, Z., Jiang, W., Chen, D., Xu, Y., 2020. Bioremediation of typical chlorinated hydrocarbons by microbial reductive dechlorination and its key players: a review. Ecotoxicol. Environ. Saf. 202, 110925. https://doi.org/10.1016/j.ecoenv.2020.110925.

Yadav, B.K., Hassanizadeh, S.M., 2011. An overview of biodegradation of LNAPLs in coastal (Semi)-arid environment. Water Air Soil Pollut. 220 (1−4), 225−239. https://doi.org/10.1007/s11270-011-0749-1.

Yadav, B.K., Hassanizadeh, S.M., Rajbhandari, S., 2012. Biodegradation of toluene under seasonal and diurnal fuctuations of soil-water temperature conditions. Water, Air, Soil Pollut. 223 (7), 3579−3588.

Yadav, B.K., Ansari, F.A., Basu, S., Mathur, A., 2013. Remediation of LNAPL contaminated groundwater using plant-assisted biostimulation and bioaugmentation methods. Water, Air, Soil Pollut. 225 (1), 1793−1799.

Yang, S.Z., Jin, H.J., Wei, Z., He, R.X., Ji, Y.J., Li, X.M., Yu, X.P., 2009. Bioremediation of oil spills in cold environments: a review. Pedosphere 19, 371–381.

Zaghloul, F.A.R., Khairy, H.M., Hussein, N.R., 2020. Assessment of phytoplankton community structure and water quality in the Eastern Harbor of Alexandria, Egypt. Egypt. J. Aquat. Res. 46 (2) https://doi.org/10.1016/j.ejar.2019.11.008.

Zhang, C., Valsaraj, K.T., Constant, W.D., Roy, D., 1999. Aerobic biodegradation kinetics of four anionic and non ionic surfactants at sub- and supra-critical micelle concentrations (CMCs). Water Res. 33, 115–224.

Zhang, B.Y., Zheng, J.S., Sharp, R.G., 2010. Phytoremediation in engineered wetlands: mechanisms and applications. Proc. Environ. Sci. 2, 1315–1325.

Zu, Y., Lu, M., 2010. Bioremediation of crude oil-contaminated soil: comparison of different biostimulation and bioaugmentation treatments. J. Hazard Mater. 183 (1–3), 395–401.

Advances in bioremediation of organometallic pollutants: strategies and future road map

12

K.S. Vinayaka[1], Supreet Kadkol[2]

[1]*Plant Biology Lab., Department of Botany, Sri Venkataramana Swamy College, Dakshina Kannada, Karnataka, India;* [2]*Department of Zoology, Sri Venkataramana Swamy College, Dakshina Kannada, Karnataka, India*

Chapter outline

12.1 Introduction

Pollution is a threat to our health, and it damages the environment; it also has effects on the biota and sustainability of our planet. Organometallic compounds are a major group of environmental contaminants that are inorganic chemical compounds (Manisalidis et al., 2020). Organometallic compounds are one of the major categories of pollutants that enter our environment and finally accumulate in sediments. Their occurrence raises major concerns for human health, especially during various activities, and continue to largely have combined adverse effects.

Biological Approaches to Controlling Pollutants. https://doi.org/10.1016/B978-0-12-824316-9.00012-4

Pollution of the environment with organometallic compounds, mostly due to industrial activities, has become a major environmental issue. The current situation is getting worse because of the continued release of large amounts of chemical pollutants from industries and spread and contamination in nature. Organometallic contaminants are recognized as potential threats to human and ecosystem health. Because of various human activities, organometallic pollutants are found in water and soil. In the present scenario there is need of urgent attention is being given to the potential health hazards presented by organometallic pollutants in our environment (Kuppusamy et al., 2017). Because of modern technology and applications, various industries used organometallic substances as raw materials for the manufacture of various goods—disposing of these chemicals into soil and water bodies seriously affects the biota and human health.

Bioremediation is not a new technique, but as our knowledge of the underlying microbial reactions grows, our ability to use them to our advantage increases. The field of bioremediation has experienced a dynamic and intense period of development during the last decade. Bioremediation involves characterization of new microbes and phytons with novel activities, which involves interaction of biotic and abiotic components within an efficient remediation process. Bioremediation is a process by which organic wastes are biologically degraded under controlled conditions to an innocuous state or to levels below concentration limits established by regulatory authorities (Mueller et al., 1996). Bioremediation of toxic compounds is often a result of the actions of multiple organisms. Bioaugmentation is when microorganisms are imported to a contaminated site to enhance degradation. The bioremediation process uses microorganisms to reduce pollution through the biological degradation of pollutants into nontoxic substances. This can involve aerobic or anaerobic microorganisms, which often use this breakdown as an energy source. For the bioremediation to be effective, microorganisms must enzymatically attack the pollutants and convert them to harmless products. As bioremediation can be effective only where environmental conditions permit microbial growth and activity, its application often involves the manipulation of environmental parameters to allow microbial growth and degradation to proceed at a faster rate.

Bioremediation can be tailored to the needs of the polluted site in question, and the specific microbes needed to be break down the organometallic pollutants are encouraged by selecting the limiting factor needed to promote their growth. This tailoring may be further improved by using synthetic biological tools to preadapt microbes to organometallic pollution in the environment to which they are to be added.

In this chapter, we focus on the current scientific knowledge, practical experiences, and future perspective with a review and compilation of the bioremediation of organometallic pollutants.

12.2 **Properties of organometallic compounds**

These organometallic compounds resemble organic rather than inorganic compounds in their physical properties. Many of them possess discrete molecular structures. Commonly, organometallic compounds are soluble in weakly polar organic solvents, and their chemical properties vary widely. Organometallic compounds exist at ordinary temperatures as low melting crystals, liquids, or gases. These compounds have wide differences in their kinetic stability under oxidation. These chemicals are classified by the type of metal−carbon bonding they contain.

12.3 **Sources of organometallic pollutants**

Sources of organometallic pollutants in our environment include volcanoes and fossil fuels.

12.4 **Toxicity and effects of organometallic pollutants**

Organometallic pollutants reveal their toxicity following biotransformation to toxic metabolites that can be bonded covalently to cellular macromolecules such as DNA, RNA, and proteins. Exposure to organometallic pollutants causes suppression of the immune system.

12.5 **Bioremediation factors**

Control and optimization of the bioremediation process is a complex system with many factors. Important factors are the existence of a microbial population capable of degrading the pollutants, the availability of the contaminant to the microbial population, and environmental factors.

12.6 **Bioremediation process**

Recalcitrant compounds present one of the most pressing problems for the biotreatment of contaminated soils or sediments. These compounds constitute a broad class of chemicals that appear as persistent contaminants in soils and sediments.

Such compounds have low solubility, low volatility, low intrinsic reactivity, and typically exhibit very slow release rates from soil or sediment (Steinberg et al., 1987; Pignatello and Xing, 1996). These characteristics make biotreatment difficult in any geologic setting.

12.7 Current strategies in the field of organometallic pollutants

Current methods for assessing the sorption and sequestration of recalcitrant compounds in soils and sediments do not provide a basic understanding of what is attainable by biostabilization, the bioavailability of recalcitrant compounds, or information to aid interpretation of results of ecotoxicological testing of residuals after biotreatment. Whether residual recalcitrant compounds remaining after biotreatment represent an acceptable endpoint requires understanding of the mechanisms that bind contaminant recalcitrant compounds within soil or sediment. Research is needed that will assess the fundamental character of the binding of recalcitrant compounds at the microscale level in parallel with the development of bioslurry treatment and ecotoxicological testing, to show how the association of polycyclic aromatic hydrocarbon with soils and sediments relates to biostabilization and achievable treatment endpoints.

Research is needed to identify those factors affecting the bioavailability of recalcitrant compounds in soils and sediments and the development of a technical basis for enhancing natural recovery processes involved in the in situ biotreatment of soils and sediments contaminated with recalcitrant compounds. A special focus is needed on improving the mechanistic understanding of the sequestration and bioavailability of recalcitrant compounds in soils and sediments. Such research could result in guidelines for assessment and prediction of the bioavailability of recalcitrant compounds for in situ biotreatment (Kuppusamy, 2017).

12.8 Future road map for reducing organometallic pollutants

Bioremediation is an effort to control pollution by utilizing various biological processes and parts of environmental biotechnological sciences (Hardiani et al., 2011). Some of the bioremediation method's advantages compared with conventional methods are relatively low cost, high efficiency, relatively less use of chemicals, less sludge, less biosorbent regeneration, and possible metals recovery (Kratochvil and Volesky, 1998). Naturally, the provision of contaminants from the environment is completed by indigenous bacteria that are often called natural bioremediation bacteria (natural attenuation). Bioremediation by spatial planning for environmental resource management is called technical bioremediation or often "simply" bioremediation. Bioremediation at a site depends on the metabolic capacity of indigenous microorganisms and environmental conditions (Rodríguez-Rodríguez et al., 2010). Indigenous bacteria in many cases cannot effectively exclude pollutant compounds, so the alternative is to add bacteria that can set aside contaminants on contaminated land. This approach is called bioaugmentation (Pimmata et al., 2013; Gupta and Ali, 2004). Bioremediation studies globally have been conducted over the last 4 decades, including those based on bioremediation purpose research; type of pollutant; aspects of microorganisms as bioremediation agents; bioremediation strategies; and various techniques for bioremediation. Those studies consist of laboratory and field research. Several studies have been conducted by combining laboratory and field research.

Bioremediation research is conducted to exclude a pollutant from groundwater, surface water, or soil/land. Groundwater bioremediation studies have made up a very small portion of research until now LISTNUM —one example is work by Farhadian et al. (2008) that examined the performance of bioremediation (in situ) in removing polluted hydrocarbon from groundwater. Groundwater bioremediation is a strategic study because groundwater is potentially polluted with pollutants and is also mobile, so the area of spreading pollutants becomes quite large. Bioremediation of surface water is a research area of wastewater bioremediation, either existing in natural reactors such as water bodies or artificial reactors. Research conducted by Dobler (2003) and Zahoor and Rechman (2009) has examined bacterial capabilities in the bioremediation of contaminated water. Mercury and chrome are examples of wastewater bioremediation using artificial reactors (bioreactors). The term "soil" used by researchers in laboratory and land research is also used in field research. Bioremediation with land recovery is highly similar because of the large number of bioremediation studies of land compared with other types of bioremediation research. Especially for water pollution, research from the existing literature tends not to use the word bioremediation as a keyword or research title, although such research is included in bioremediation research. Initially, bioremediation research aimed to clear land of heavy metals produced by industrial and mining activities. Until the 2000s, the amount of research on the bioremediation of heavy metals was always higher than for other pollutants. Several studies of the bioremediation of heavy metals in this period were conducted by Jeyasingh and Philip (2005), Umrania and Valentina (2006), Jaishankar et al. (2014), and Choudhary and Sar (2009) in India.

12.8.1 **Future challenges**

Future bioremediation research worldwide should lead to more in situ field research so that bioremediation methods developed in the laboratory can directly add value to pollution control. In situ bioremediation research can be done with community participation so that many cases of pollution in Indonesia can be resolved.

12.9 **Conclusion**

There is a large gap between laboratory research and field research in land bioremediation research in the world. Bioremediation studies of land polluted by various pollutants, especially heavy metals and hydrocarbon, are mostly carried out in the laboratory to assess their microbial potential and examine bioremediation strategies for the best performance. But in the last 10 years, field research has been initiated, especially with the in situ method.

The bioremediation strategy for hydrocarbon-contaminated land in the last decade has been a trending topic in bioremediation research worldwide. Although microbial aspects of research are still widely practiced, there was a decrease in line with increasing bioremediation laboratory research, field research, or a combination of both. Bioremediation research on heavy metal contaminated soil in various periods of the year had become a hot topic because of the nature of the pollutants, while the potential bioremediation research topic that has been developing for future research is polluted groundwater bioremediation research by the in situ method. For each topic and subtopic of bioremediation research, there has been an upward trend over the last year, suggesting that research interest in developing bioremediation methods, either in the laboratory or on-site, such as in situ, has been increasing. Bioremediation methods are increasingly believed to solve various current contamination cases.

References

Choudhary, S., Sar, P., 2009. Characterization of a metal resistant Pseudomonas sp. isolated from uranium mine for its potential in heavy metal (Ni2+, Co2+, Cu2+, and Cd2+) sequestration. Bioresour Technol 100 (9), 2482–2492.

Dobler, W., 2003. Pilot plant for bioremediation of mercury-containing industrial wastewater. Appl. Microbiol. Biotechnol. 124–133.

Farhadian, M., Duchez, D., Larroche, C., 2008. In situ bioremediation of monoaromatic pollutants in groundwater: a review. Bioresour. Technol. 99, 5296–5308.

Gupta, N., Ali, A., 2004. Mercury volatilization by R factor systems in Escherichia coli isolated from aquatic environments of India. Curr. Microbiol. 48, 88–96.

Hardiani, H., Kardiansyah, T., Sugesty, S., 2011. Bioremediasi logam Timbal (Pb) dalam tanah terkontaminasi limbah sludge industri kertas proses deinking. J. Selulosa 1 (1), 31–41.

Jaishankar, M., Tseten, T., Anbalagan, N., Mathew, B.B., Beeregowda, K.N., 2014. Toxicity, mechanism and health effects of some heavy metals. Interdiscip. Toxicol. 7, 60–72.

Jeyasingh, J., Philip, L., 2005. Bioremediation of chromium contaminated soil: optimization of operating parameters under laboratory conditions. J Hazard Mater 118 (1-3), 113–120.

Kratochvil, D., Volesky, B., 1998. Advances in the biosorption oh Heavy Metals. Trends in Biotechnology 16 (7), 77–83.

Kuppusamy, S., Palanisami, T., Kadiyala, V., Lee, Y.B., Ravi, N., Mallavarapu, M., 2017. Remediation approaches for polycyclic aromatic hydrocarbons (PAHs) contaminated soils: technological constraints, emerging trends and future directions. Chemosphere 168, 944–968.

Manisalidis, I., Stavropoulou, E., Stavropoulos, A., Bezirtzoglou, E., 2020. Environmental and health impacts of air pollution: a review. Front Public Health 8, 14.

Mueller, J.G., Cerniglia, C.E., Pritchard, P.H., 1996. Bioremediation of environments contaminated by polycyclic aromatic hydrocarbons. In: Crawfold, R.L., Crawfold, D.L. (Eds.), Bioremediation: principles and applications. Cambridge University Press, UK, pp. 125–194.

Pignatello, J.J., Xing, B., 1996. Mechanisms of slow sorption of organic chemicals to natural partcles. Environmental Science and Technology 30, 1–11.

Pimmata, P., Reungsang, Alissara, Plangklang, Pensri, 2013. Comparative bioremediation of carbofuran contaminated soil by natural attenuation, bioaugmentation and biostimulation. International Biodeterioration & Biodegradation 85, 196−204.

Rodríguez-Rodríguez, C.E., Marco-Urrea, E., Caminal, G., 2010. Degradation of naproxen and carbamazepine in spiked sludge by slurry and solid-phase Trametes versicolor systems. Bioresour Technol 101, 2259−2266.

Steinberg, S.M., Pignatello, J.J., Sawhney, B.L., 1987. Persistence of 1,2-dibromoethane in soils: entrapment in interparticle micropores. Environmental Science and Technology 21, 1201−1208.

Valentina, Umrania V., 2006. Bioremediation of toxic heavy metals using acidothermophilic autotrophes. Bioresource technology 97 (12), 78−84.

Zahoor, A., Rehman, A., 2009. Isolation of Cr (VI) reducing bacteria from industrial effluents and their potential use in bioremediation of chromium containing wastewater. J Environ Sci 21, 814−820.

Further reading

Adeyemi, A.O., 2009. Biological immobilization of lead from lead sulphide by *Aspergillus niger* and *Serpula himantioides*. Int. J. Environ. Res. 3 (201373), 477−482.

Alexandrino, M., Macías, F., Costa, R., Gomes, N.C.M., Canário, A.V.M., Costa, M.C., 2011. A bacterial consortium isolated from an Icelandic fumarole displays exceptionally high levels of sulfate reduction and metals resistance. J. Hazard Mater. 187 (1−3), 362−370.

Appanna, V.D., Hamel, R., 1996. Aluminum detoxification mechanism in *Pseudomonas fluorescens* is dependent on iron. Fed. Eur. Microbiol. Soc. Microbiol. Lett. 143, 223−228.

Beškoski, V.P., Milic, J., Ilic, M., Miletic, S., Šolevic, T., Gojgic, G., 2011. "Ex- situ bioremediation of a soil contaminated by Mazut (heavy residual fuel oil) − a field experiment". Chemosphere 83, 34−40.

Chai, L., Huang, S., Yang, Z., Peng, B., Huang, Y., Chen, Y., 2009. Cr (VI) remediation by indigenous bacteria in soils contaminated by chromium-containing slag. J. Hazard Mater. 167, 516−522.

Chemlal, R., Tassist, A., Drouiche, M., Lounici, H., Drouiche, N., Mameri, N., 2012. Microbiological aspects study of bioremediation of diesel-contaminated soils by biopile technique. Int. Biodeterior. Biodegrad. 75, 201−206.

Fang, H., Zhou, W., Cao, Z., Tang, F., Wang, D., Liu, K., Yu, Y., 2012. Combined remediation of DDT congeners and cadmium in soil by sphingobacterium sp . D-6 and sedum *Alfredii hance*. J. Environ. Sci. 24 (6), 1036−1046.

Gaonkar, T., Bhosle, S., 2013. "Effect of metals on A siderophore producing bacterial isolate and its implications on microbial assisted bioremediation of metal contaminated soils. Chemosphere 93 (9), 1835−1843.

Ge, S., Ge, S., Zhou, M., Dong, X., 2015. Bioremediation of Hexavalent chromate using permeabilized *Brevibacterium* sp. and *Stenotrophomonas* sp. cells. J. Environ. Manag. 157, 54−59.

Gomez, F., Sartaj, M., 2013. Field scale ex-situ bioremediation of petroleum contaminated soil under cold climate conditions. Int. Biodeterior. Biodegrad. 85, 375−382.

Hamdi, H., Benzarti, S., Manusadzianas, L., Aoyama, I., Jedidi, N., 2007. Solid-phase bioassays and soil microbial activities to evaluate PAH-spiked soil ecotoxicity after a Long-Term bioremediation process simulating landfarming. Chemosphere 70, 135−143.

Bioremediation of polycyclic aromatic hydrocarbons from contaminated dumpsite soil in Chennai city, India

Sancho Rajan[1], V. Geethu[2], Paromita Chakraborty[1,3]

[1]*Department of Civil Engineering, SRM Institute of Science and Technology, Kancheepuram, Tamil Nadu, India;* [2]*Department of Civil Engineering, New Horizon College of Engineering, Bangalore, Karnataka, India;* [3]*SRM Research Institute, SRM Institute of Science and Technology, Kancheepuram, Tamil Nadu, India*

Chapter outline

Biological Approaches to Controlling Pollutants. https://doi.org/10.1016/B978-0-12-824316-9.00014-8

13.1 Introduction

Polycyclic aromatic hydrocarbons (PAHs) are a class of pollutants, 16 of which are listed as priority pollutants by the United States Environmental Protection Agency (USEPA). Their elevated levels in the environment are a result of tremendous modernization, population growth, and industrialization. Apart from natural activities, human activities have led to a large increase in their levels. Their persistent nature and resistance to degradation classifies them as a very toxic class of pollutants.

The USEPA classified these PAHs as probable human carcinogens in 2008. Lung, skin, and bladder cancers are the common forms of cancers reported related to PAH exposure. PAHs are considered major toxicants because the reactive metabolites of some major PAHs bind with DNA and cellular proteins and induce toxic effects. This can lead to mutations, tumors, malformations of cells, and cancer due to biochemical destruction and cell damage (Armstrong et al., 2004). Exposure to PAHs can lead to breathing difficulty, chest pain, chest congestion, cough, and throat pain. These symptoms can often be observed in workers of industrial units. PAHs can cause cataracts, kidney and liver damage, and jaundice in humans, whereas reproductive effects are common in animals exposed to PAHs (ATSDR, 1995).

PAHs are highly mobile in the environment and are distributed across different matrixes, namely air, water, soil, and sediment. The levels of the PAHs in different matrixes are given in Table 13.1. Everyday activities such as residential wood and coal combustion, traffic conditions, tobacco smoke, etc. can lead to emission of PAHs into the atmosphere. They can also affect indoor and outdoor air quality. They enter the water through runoff, wastewater discharge from industries, and wastewater treatment plants. PAHs are detected in all surface soils. Plant synthesis, volcanic activity, and forest fires contribute greatly to increasing PAH levels. The presence of PAHs in sediments is detrimental to aquatic plants and animals and can indirectly affect humans, too.

Bioremediation is considered as a cost-effective process in which the transformation or degradation of organic contaminants is carried out by indigenous microbes available in the contaminated site. These organic pollutants become completely mineralized, leading to CO_2, H_2O, and biomass formation (Das and Chandran, 2011). Microbes can degrade PAHs because they act as a sole carbon source for their growth, which leads to the breaking down of the compounds (Winquist et al., 2014). Naphthalene (Nap) and phenanthrene (Phe) catabolism by many microbial species have been studied extensively. Indigenous microbes such as *Pseudomonas aeruginosa, Pseudomonas fluorescens, Mycobacterium* spp., *Hemophilus* spp., *Rhodococcus* spp., *Paenibacillus sp* are a few microbial species capable of degrading PAHs (Bisht et al., 2015). Alquati et al. (2005) reported that six groups of microorganisms were extracted from the soil samples of a polluted petroleum site because of their ability to degrade Nap, the key pollutant present at the site. Due to the ability of selected bacteria to metabolize hydrocarbons by producing gentisic acid and

Table 13.1 Comparison table of polycyclic aromatic hydrocarbons (PAHs) levels in different environmental matrixes. Here "n" depicts the number of priority PAHs reported in each study.

Location	Matrix	PAHs (n)	Range	References
Coimbatore, India	Air	13	20–172 ng/m^3	Mohanraj and Azeez (2004)
Chennai, India	Air	11	326–791 ng/m^3	Mohanraj et al. (2011)
Taiyuan, China	Air	8	696.6–2765 ng/m^3	Peng et al. (2003)
Chennai, India	Surface sediment	16	136–2063 ng/g	Rajan et al. (2019)
Chennai, India	Surface sediment	16	105–1710 ng/g	Rajan et al. (2019)
Sundarban Mangrove Wetland, India	Core sediment	16	132–2938 ng/g	Domínguez et al. (2010)
New Delhi, Kolkata, Mumbai, Chennai	Surface soil	16	1029 ng/g	Chakraborty et al. (2019)
Kaohsiung Harbor, Taiwan	Sediment	17	472–16,201 ng/g	Chen and Chen (2011)
Daliao River watershed, China	Surface water, suspended particulate matter (SPM) and sediment	18	946.1–13,448.5 ng/L (surface water), 317.5–238518.7 ng/g (SPM) 61.9–840.5 ng/g (sediments)	Guo et al. (2007)
Brahmaputra and Hooghly Rivers, India	Riverine sediment	16	Hooghly riverine sediment avg value, 445 ng/g and Brahmaputra riverine sediment avg, 169 ng/g	Khuman et al. (2018)
Isfahan metropolis, Iran	Soil	16	57.70–11,730.08 µg/kg	Moore et al. (2015)
Black and Ashtabula Rivers, Ohio	River sediment	17		Christensen and Bzdusek (2005)

catechol as intermediates, the study of Nap-degrading species highlighted the metabolic differences. Nine of the 13 selected Nap-degrading bacteria of the genera *Rhodococcus*, *Arthrobacter*, *Nocardia* and *Pseudomonas* showed PCR fragment amplification with specific primers of naphthalene dioxygenase (Alquati et al., 2005). A study by Pathak et al., 2009 revealed that Nap-reducing bacteria were isolated from the polluted sediments of Amlakadi Channel, Gujarat. *Pseudomonas* sp. were among the isolates. HOB1 demonstrated that Nap could degrade 2000 ppm within 24 h and can tolerate Nap up to 60,000 ppm. Maximum Nap degradation was observed at 35−37°C (Pathak et al., 2009). Lin et al., 2010 revealed that the strains isolated from sewage sludge oil was used in the cultivated medium to biodegrade Nap. The *Bacillus fusiformis* (BHN) strain was identified using sequence analysis of the 16S rDNA gene. The optimal conditions for Nap biodegradation were 300°C, pH 7.0%, 0.2% inoculum volume, and C/N ratio of 1. Under these conditions and the initial concentration of 50 mg/L Nap, more than 99.1% was removed within 96 h. Among the factors that affect Nap biodegradation, salinity and inoculum concentration are of utmost importance (Lin et al., 2010). The experiments carried out by Mohandass et al. (2012) revealed that PAHs pose a threat to human and animal life. Strategies are constantly being found to reduce the amount of PAH in the atmosphere. The consortia isolated from petrochemical soil can use BaP as their sole carbon source (Mohandass et al., 2012). Based on morphological characterization as well as 16S rDNA gene sequence analysis, the isolates were identified as *Bacillus cereus* and *Bacillus vireti* and the consortium removed approximately 59% of BaP at a concentration of 500 mg/L in a carbon-free minimal medium (Mohandass et al., 2012). The study reveals that the bacterial consortium is a new bacterial tool for the biodegradation of BaP (Mohandass et al., 2012). The study by Maiti et al. (2012) explained that the hydrophobic essence of PAHs is less soluble in water and cannot be easily accessed or broken down by bacteria. *Bacillus thuringiensis* strain NA2 isolated from a contaminated oil field in India degraded some of the most toxic PAH components, such as Fluranthene (Flu) and Pyrene (Pyr). Various bioenhancement parameters such as temperature, pH, and glucose supply as co-substrates, Tween 80, H_2O_2, or organic solvents have been designed to maximize this degradation. The findings are suitable for bioremediation and are very important. The contaminated soil or sludge can be used directly to verify the effectiveness of bacteria (Maiti et al., 2012). Similarly, research conducted by (Ping et al., 2014) revealed that PAHs are the key pollutants present in the environment that are difficult to remediate. Degrading PAHs in the ecosystem is becoming more important and necessary (Ping et al., 2014). This study was based on morphological and physiological characteristics as well as 16S rDNA sequencing, and the strain PL1 was isolated from soil and identified as *Klebsiella pneumoniae*. Same study also revealed that the PL1 strain decreased Pyr by 63.4% and BaP to 55.8% within 10 days. The degradation ability of the PL1 strain varied was based on soil type—e.g., Pyr degradation in paddy soil, red earth, and fluvoaquic soil took 16.9, 24.9, and 88.9 days respectively, whereas BaP's half-life and degradation with PL1 was 9.5, 9.5, and 34 days in paddy soil, red soil, and fluvoaquic soil respectively. The results show that PL1 was fairly well suited to degrade Pyr and BaP in paddy soil. In recent studies, the use of consortia for the removal of PAHs from contaminated sites has been considered a newer approach in remediation experiments because it is regarded as an efficient and cost-effective technique (Ghazali et al., 2004; Janbandhu and Fulekar, 2011).

The objective of this study was to use isolated indigenous microbial consortia from dumpsite soils of Chennai city, reported elsewhere (Rajan et al., 2021) for the bioremediation of PAHs including the effect of co-substrate and report a semi-microcosm study.

13.2 Materials and methods

13.2.1 Enrichment of indigenous microbes

Enrichment of indigenous microbes was based on a previous study (Pathak et al., 2009). The indigenous microbes isolated from dumpsite soil taken from Kodungaiyur dumpsite (KDS) and Perungudi dumpsite (PDS) of Chennai city have been reported elsewhere (Rajan et al., 2021). Bushnell Haas broth (BHB) was supplemented with Nap and Phe because these compounds were abundant throughout the sampling sites (Rajan et al., 2021). Fig. 13.1 explains the outline of the enrichment process (Rajan et al., 2021).

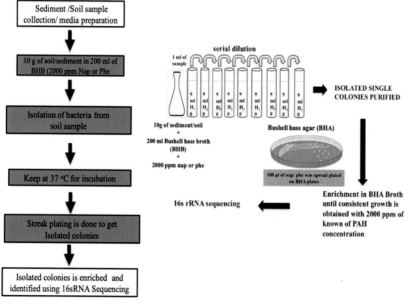

FIGURE 13.1

Schematic representation of enrichment technique of indigenous microbes has been reported elsewhere (Rajan et al., 2021).

13.2.1.1 Degradation and growth study of indigenous microbes from soil samples

13.2.1.1.1 Effect of temperature

Three experimental setups were designed. We have used 1, 2 for Control setups and 3 for Nap, Phe setup, respectively. The setup 3 was 15 conical flasks (250 mL) containing 200 mL of BHB, as shown in Fig. 13.2. The broth was supplemented with 1 mL of Nap and Phe (500 ppm) as the sole carbon source, and 1 mL of the enriched consortia was also added to the setup 3. Setup 1 was maintained as a negative control with only Nap and Phe and no inoculum. Setup 2 was maintained as another negative

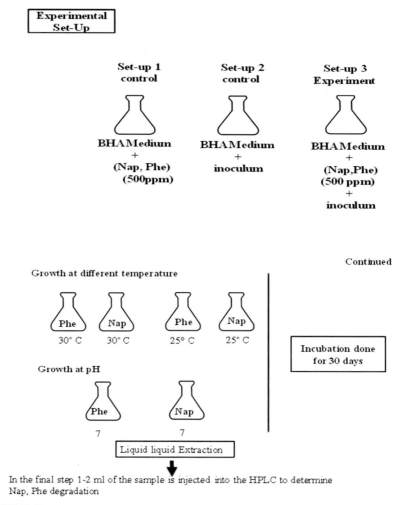

FIGURE 13.2

Schematic representation of the bioremediation experiment of indigenous microbes.

control with only inoculum. This experimental setup was incubated at two different temperatures, viz., 25 and 30°C, for 30 days. For quantification, samples were collected on alternate days for bacterial growth at an optical density (OD_{600}) using ELISA Tecan Infinite 200 PRO, and Nap and Phe uptake were quantified using high performance liquid chromatography (HPLC).

13.2.2 Effect of co-substrates on isolates from soil samples

The consortia for Nap (N2, N3 and N4) and Phe (S1 and S2) pregrown on salicylic acid (2 g/L), succinic acid (2 g/L), and starch (2 g/L) were inoculated in setups (A, B, and C) and (A′, B′, and C′), respectively, and supplemented with Nap and Phe (500 ppm) in BHB medium, as shown in Fig. 13.3. In addition, one control setup (D, D′) with no cosubstrates was used with consortia, which is supplemented with Nap and Phe (500 ppm) as the only source of carbon in the BHB medium. All these flasks were incubated at 25°C, under shaking conditions of 150 rpm. Samples were collected and analyzed on the 10th, 15th, and 30th day with HPLC.

13.2.3 Experimental setup for semimicrocosm study

Simulated microcosms were prepared by method proposed elsewhere (Pathak et al., 2009; Filonov et al., 1999) with some modifications. Simulated microcosms were set up in Petri plates (90 mm diameter). The soil moisture content was maintained at 18%, while soil pH was 7. The soil was dry and sieved with 2.0 mm. For 1 h at 121°C under 15 lbs, the soil samples were autoclaved to produce sterile microcosms. A Nap and Phe (500 ppm) solution was formulated in the oil ether and sprayed into the soil; 10 g soil from the prepared soil mixture was taken from a

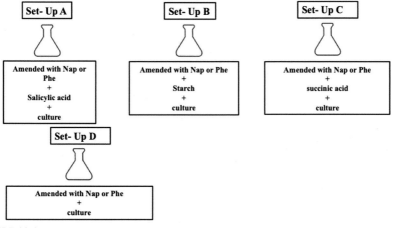

FIGURE 13.3

Schematic representation of co-substrate study.

Petri dish. BHB consortia of 500 ppm of Nap and Phe elevated the inoculum to late log level (Pathak et al., 2009). The cells were harvested for 10 min at 6000 rpm by centrifugation, washed with sterile deionized water thrice, and resuspended to O.D. of 1.0 (Pathak et al., 2009). Each of the five sets of soil models with three replicates was inoculated around 5 mL of this suspension (Fig. 13.4). Experimental set-ups were: (Set A)—sterile soil modified to use Nap to assess the degradation of Nap and Phe; (Set B)—sterile soil modified with Nap and inoculated with consortia to determine the mineralization ability of Nap and Phe consortia without indigenous microbes; (Set C)—nonsterile soil modified with Nap and Phe, both in the presence of indigenous microbe and of isolated biodegrade consortia of Nap and Phe; (Set D)— nonsterile soil modified and inoculated with consortia for testing whether the strain has increased deterioration of Nap and Phe while in competition with indigenous microorganisms to determine the capacity of strain consortia for soil nutrient development and colonization; and (Set E)—sterile soil (without Nap and Phe) inoculated with consortia. To obtain a moisture level of 18%, water was weighed and added. Fig. 13.4 displays a representation of the scheme of the semimicrocosm analysis. The soil samples have been homogenized with an inoculum spatulation. Incubated into ambient conditions and collected at 24,144 and 192 h to monitor degradation of Nap and Phe, all experimental setup were prepared for the semimicrocosm test.

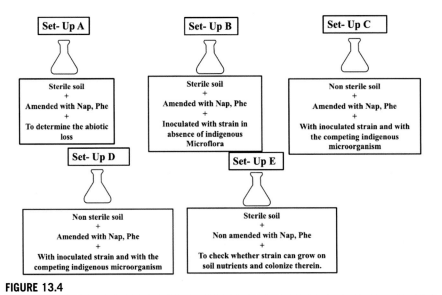

FIGURE 13.4

Schematic representation for semimicrocosm study using microbial species isolated from dumpsite soil (Rajan et al., 2021).

13.2.4 Instrumental analysis

13.2.4.1 High performance liquid chromatography

For Phe and Nap extraction, 60 mL of ethyl acetate was used, and the samples were extracted by liquid−liquid extraction. After extraction, the samples were evaporated to dryness on a rotary evaporator and made up to 1 mL with methanol. Then these samples were passed through a 0.45 μm filter syringe prior to injection in HPLC. These extracts with residual Nap and Phe were quantified using HPLC (Shimadzu Japan, Binary Pump: LC 20 AD). 20 μL of this solution were injected in HPLC, C-18RP column (250 × 4.6 mm packed with 100 RP-18 5 micron). The solvent system was made up of 95% methanol, 5% H_2O (vol/vol), and 0.1% formic acid. The run time was 30 min for a flow rate of 0.8 mL/min in isocratic mode, during which the column oven was at ambient temperature. The instrument used a photodiode array detector.

13.3 Results and discussion

13.3.1 Overview of the bioremediation process

The study focuses on utilization of indigenous microbes from the soils of contaminated sites and study bioremediation of Nap and Phe using those microbes. Finally we saw the effect of the co-substrates and conducted a semimicrocosm study on how to use the microbes present in the contamination sites to degrade the particular known concentrations of Nap and Phe, respectively. These compounds have been used as a model compound in PAH bioremediation studies (Bhawana et al., 2016; Samanta et al., 2002). It is the simplest and most abundant compound with the ability to induce cancer in humans and other animals owing to its lipophilic nature (Nagabhushanam et al., 1991; Santos et al., 2008; Ghosal et al., 2016). Several studies reported that Nap and Phe are dominant in most sites and in the previous studies, as stated elsewhere (Rajan et al., 2019; Chakraborty et al., 2019). A hypothesis of the remediation experiment performed in this study is given below in Fig. 13.5.

13.3.2 Screening and isolation of microbes from dumpsite soil

After repetitive streaking and purification in a carbon-free medium, cultures were supplemented with Nap and Phe as the only carbon source. Later those cultures that could grow in Nap- and Phe-supplemented media were taken for identification process by 16s rRNA process, and these species were N2 (*Pseudomonas plecoglossicida* sp.), N3 (*Pseudomonas alcaligenes* sp.), and N4 (*Pseudomonas stutzeri* sp.) for Nap uptake, and S1 (*Agromyces indicus* sp.) and S2 (*Pseudomonas resinovorans* sp.) for Phe uptake have been reported elsewhere (Rajan et al., 2021).

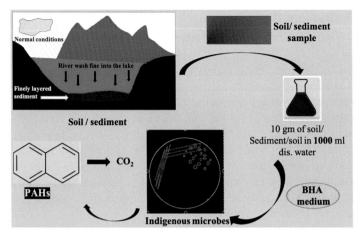

FIGURE 13.5

Schematic representation of bioremediation of polyaromatic hydrocarbons.

13.3.3 Degradation of napthalene by microbial species isolated from soil samples

In setup (A), Nap-degrading consortia N2 (*Pseudomonas plecoglossicida* sp.), N3 (*Pseudomonas alcaligenes* sp.), and N4 (*Pseudomonas stutzeri* sp.) were incubated at 30°C and 25°C. In this study, as shown in Fig. 13.6A and B, we observed that on the second day itself, 98% removal of Nap was visible at both temperatures (Table 13.2). The consortia were able to degrade 100% of Nap on the fourth day. Previous study reports of (Rajan et al., 2019) state that Nap degradation with single

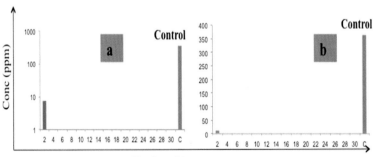

FIGURE 13.6

Column graph showing the trend of napthalene degradation at 30°C (A) and 25°C (B) for 30 days.

Table 13.2 Removal percentage of polyaromatic hydrocarbons using indigenous microbes.

Compounds	Temperature	Time	Removal %
Nap	30 and 25°C	24 h	98%
Phe	30°C	30 d	92%
Phe	25°C	8 d	100%

species took 10 d, whereas in the present study, by using consortia, we were able to degrade Nap (500 ppm) within 4 d. It was observed that in control setup (B), on the 30th day without the consortia at 30°C, the abiotic loss was 28%, and for 25°C, 27% loss of Nap was observed as shown in Fig. 13.6A and B, respectively, thus implying that much less abiotic loss of Nap was observed in the initial days and it also proves that the selected consortia were found superior to degrade Nap(500 ppm) completely within 4 days.

The bacterial growth rate (Fig. 13.7A and B) at both temperatures 30°C and 25°C, correlated with the degradation rate. Consortia showed negligible lag phase till complete removal of Nap was observed. Similar results for Nap degradation were observed by Rajan et al. (2019). An increase in growth was observed even in the last phase, which might be due to biomass production on Nap intermediates. This observation was found similar to previous studies on Nap degradation (Pathak et al., 2009; Filonov et al., 1999).

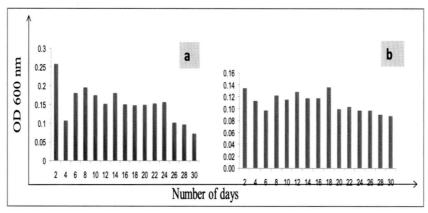

FIGURE 13.7

Column graph showing the growth rate of naphthalene-degrading consortia at 30°C (a) and 25°C (b) for 30 days.

13.3.4 Degradation of phenanthrene by microbial species isolated from the soil sample

In setup (A′), the consortia containing S1 (*Agromyces indicus* sp.) and S2 (*Pseudomonas resinovorans* sp.), which was incubated at 30°C, showed Phe degradation of 92% in 30 days. From Fig. 13.8A, we observed that the degradation was at a lower rate initially, but after the sixth day, we observed a drastic decrease in Phe concentration.

As shown in Fig. 13.9A, the growth rate of the consortia also correlated with the degradation pattern in the initial days indicated that the growth was slow. Hence, the degradation also was at a lower rate, after the sixth day, an exponential growth phase was observed. Hence, a drastic decrease in Phe concentration was observed in this study. The control sample in setup (B′) showed a loss of 8% of Phe, which may be due to abiotic losses.

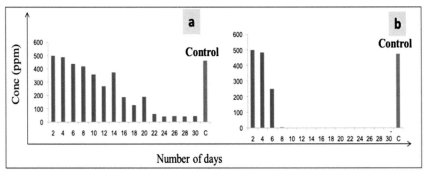

FIGURE 13.8

Column graph showing phenanthrene degradation at 30°C (A) and 25°C (B) for 30 days.

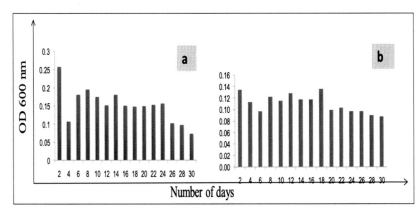

FIGURE 13.9

Column graph showing the growth rate of phenanthrene degrading consortia at 30°C (A) and 25°C (B) for 30 days.

The same consortia in setup (A′) was incubated at 25°C showed 100% removal of Phe within 8 days (Table 13.2). The control sample showed an abiotic loss of 5.2% in setup (B′). The growth rate at 25°C was also correlating with the Phe degradation pattern, as shown in Fig. 13.8B. The rate of growth was less in the initial days but gradually increased with a negligible lag phase until complete degradation of Phe on the eighth day, which is shown in Fig. 13.9B. During the last phase shown in the figure, it can be seen that growth decreases after the eighth day, but the growth is visible; this might be due to the formation of intermediates of Phe biomass. This mixed consortia, S1 (*Agromyces indicus* sp.) and S2 (*Pseudomonas resinovorans* sp.) in the present study were found superior to the mixed microbial consortia containing *Sphingobacterium* sp., *Bacillus cereus* sp., and *Achromobacter insolitus* sp. that could degrade 250 ppm of Phe with 56.9% efficiency after 14 days (Janbandhu and Fulekar, 2011). However, the single species such as *B. fungorum* (FM-2) and *Bacillus thuringiensis* (FQ₁) isolated from oil-contaminated sites was observed to degrade Phe in less than 5 days were found to be superior to the consortia in this study (Jiang et al., 2015; Liu et al., 2019).

13.3.5 Effect of co-substrates on napthalene degradation

Nap-degrading consortia (N2 (*Pseudomonas plecoglossicida* sp.), N3 (*Pseudomonas alcaligenes* sp.) and N4 (*Pseudomonas stutzeri* sp.) were subjected with Nap (500 ppm), and different co-substrates in setups A, B, and C, respectively, showed 100% removal of Nap on the 10th day. Even the control setup (D) without any co-substrates showed complete removal of Nap on the 10th day. Hence, this indicates that Nap was degraded by the consortia even in the presence of other co-substrates, implying that other cosubstrates did not affect Nap degradation.

13.3.6 Effect of co-substrates on phenanthrene degradation

Phe-degrading consortia S1 (*Agromyces indicus* sp.) and S2 (*Pseudomonas resinovorans* sp.) were inoculated in similar setups (A′, B′, and C′) and subjected to Phe (500 ppm) and co-substrates. The results showed that for salicylic acid in setup A′, the removal percentages of Phe were 36%, 28%, and 58% on the 10th, 20th, and 30th days, respectively; for succinic acid (setup B′), the removal percentages of Phe were 43.2%, 64.2%, and 73% on the 10th, 20th, and 30th days, respectively; for starch (setup C′), the removal percentages of Phe were 28%, 67%, and 63% on the 10th, 20th, and 30th days, respectively (Fig. 13.10). Finally, for the control setup (D′) without any co-substrates, the removal percentage was 82%. The second setup shows that the degradation period was extended in the presence of salicylic acid; however, in the setup containing succinic acid and starch, the degradation percentages were 73% and 63%, respectively, as shown in Fig. 13.10. Grim and Harwood suggested that succinic acid acts as an energy source for the microbes that increases strain mobility (Pathak et al., 2009; Grimm and Harwood, 1997).

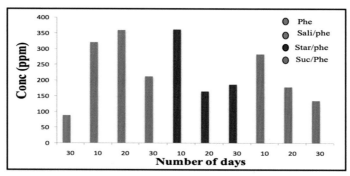

FIGURE 13.10

Column graph showing the effect of co-substrate on phenanthrene degradation.

13.3.7 Semimicrocosm study

A semimicrocosm study was conducted to examine the potential of indigenous consortia to carry out the bioremediation of selected PAHs. This study was done to evaluate the selected consortia's potential to grow with available nutrients in natural soil and to remediate xenobiotic compounds such as PAHs with the presence of various co-substrates. In this study, in the absence of both indigenous microbes and isolated consortia in setup A (sterile soil + Phe), the Phe loss due to evaporation was 12.4%, 7.2%, and 4% at 24, 144, and 192 h, respectively, as observed in Fig. 13.11. For setup B (sterile soil + consortia + Phe) in the presence

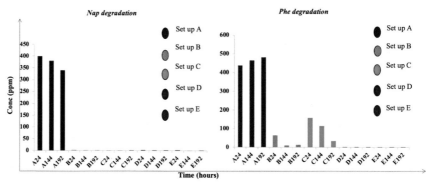

FIGURE 13.11

Column graph showing degradation by consortia simulated microcosm studies for (1) phenanthrene (Phe) and (2) napthalene (Nap). Each setup indicates (A) sterile soil with Nap and Phe, (B) sterile soil with Nap and Phe inoculated with consortia, (C) nonsterile soil with Nap and Phe, (D) nonsterile soil with Nap and Phe inoculated with consortia, (E) sterile soil (without Nap and Phe) inoculated with consortia.

of consortia in sterile soil, the removal percentages of Phe were 87.4%, 98%, and 97.4% at 24, 144, and 192 h, respectively (Fig. 13.11 (1)). For setup C (nonsterile soil + consortia + indigenous microflora + Phe), nonsterile soil in the presence of both indigenous microbes and isolated consortia, the removal percentages of Phe were 68.4%, 77.2%, and 93.2% at 24, 144, and 192 h, respectively, which implies that with the presence of other indigenous microflora, the degradation was taking place at a slightly slower rate than with setup (B) (Fig. 13.11 (1)). In setup D (nonsterile soil + consortia + indigenous microflora), nonsterile soil without the presence of Phe spiked the consortia, and the indigenous microflora were able to grow and colonize the soil (Fig. 13.11 (1)). Similarly, Nap loss in setup A (sterile soil + Nap) could be attributed to abiotic loss or the absorption of Nap onto the soil. In setup B (sterile soil + consortia + Nap), in the presence of consortia in sterile soil, the removal percentage of Nap was 99%. Setup C (nonsterile soil + consortia + indigenous microflora + Nap) nonsterile soil in the presence of both indigenous microbes and isolated consortia, the removal percentage of Nap was 99%. Here the Nap degradation was not affected by the presence of other indigenous microbes as affected with Phe uptake. In setup D (nonsterile soil + consortia + indigenous microflora) and setup E (sterile soil + consortia), no external Nap was spiked, and the analysis revealed that a very low concentration of Nap was present in the soil naturally that was not taken up the consortia, perhaps indicating that it was not bioavailable to the consortia. A similar observation was reported earlier (Silva et al., 2009), wherein authors reported that the residual Nap concentration of 25 and 29 ppm were not bioavailable to the consortia. Further, it was observed that in setup D (nonsterile soil + consortia + indigenous microflora) and setup E (sterile soil + consortia), the results were similar to both Nap and Phe consortia that they were able to colonize the soil better without the presence of indigenous microbes. These observations were compared with the studies done reported earlier (Pathak et al., 2009; Filonov et al., 1999), where complete removal of Nap was observed in 3−8 d and 24 h, respectively. The study results are shown in Fig. 13.11(2).

13.4 **Conclusion**

In the present study, two indigenous species, S1 (*Agromyces indicus* sp.) and S2 (*Pseudomonas resinovorans* sp.), were used for Phe degradation, and three indigenous species, N2 (*Pseudomonas plecoglossicida* sp.), N3 (*Pseudomonas alcaligenes* sp.), and N4 (*Pseudomonas stutzeri* sp.), for Nap degradation were isolated and identified from dumpsite soils (Rajan et al., 2021). All isolated indigenous consortia were found to degrade Phe and Nap more efficiently at 25°C than at 30°C. The consortia degraded Phe and Nap even in the presence of co-substrates. Finally, the consortia colonized the soil successfully in the semimicrocosm study. This research work yields a cost-effective remediation technique that can remove pollutants, such as PAHs, from the environment.

Acknowledgments

This work was supported by the Ministry of Environment and Forest and Climate Change, Grant No. Q-14011/43/2013- CPW (EHC). The authors would also like to thank the graduate students of SRM University who extended their help in sample collection.

References

Alquati, C., Papacchini, M., Riccardi, C., Spicaglia, S., Bestetti, G., 2005. Diversity of naphthalene-degrading bacteria from a petroleum contaminated soil. Ann. Microbiol. 55 (4), 237.

Armstrong, B., Hutchinson, E., Unwin, J., Fletcher, T., 2004. Lung cancer risk after exposure to polycyclic aromatic hydrocarbons: a review and meta-analysis. Environ. Health Perspect. 112 (9), 970.

ATSDR, 1995. Toxicological profile for polycyclic aromatic hydrocarbons. Agency for Toxic Substances and Disease Registry Atlanta, GA.

Bhawana, P., Shiv, G., Gaherwal, S., Kiran, B., Soni, R., 2016. Isolation and identification of naphthalene degrading microbe from petroleum contaminated soil. Int. J. Res. Chem. Environ. 6 (4), 26−29.

Bisht, S., Pandey, P., Bhargava, B., Sharma, S., Kumar, V., Sharma, D., 2015. Bioremediation of polyaromatic hydrocarbons (PAHs) using rhizosphere technology. Braz. J. Microbiol. 46 (1), 7−21.

Chakraborty, P., Sampath, S., Mukhopadhyay, M., Selvaraj, S., Bharat, G.K., Nizzetto, L., 2019. Baseline investigation on plasticizers, bisphenol A, polycyclic aromatic hydrocarbons and heavy metals in the surface soil of the informal electronic waste recycling workshops and nearby open dumpsites in Indian metropolitan cities. Environ Pollut. 248, 1036−1045.

Chen, C.W., Chen, C.F., 2011. Distribution, origin, and potential toxicological significance of polycyclic aromatic hydrocarbons (PAHs) in sediments of Kaohsiung Harbor, Taiwan. Mar. Pollut. Bull. 63, 5−12.

Christensen, E.R., Bzdusek, P.A., 2005. PAHs in sediments of the Black River and the Ashtabula River, Ohio: source apportionment by factor analysis. Water Res. 39, 511−524.

Das, N., Chandran, P., 2011. Microbial degradation of petroleum hydrocarbon contaminants: an overview. Biotechnol. Res. Int. 2011.

Domínguez, C., Sarkar, S.K., Bhattacharya, A., Chatterjee, M., Bhattacharya, B.D., Jover, E., Albaigés, J., Bayona, J.M., Alam, M.A., Satpathy, K.K., 2010. Quantification and source identification of polycyclic aromatic hydrocarbons in core sediments from Sundarban Mangrove Wetland, India. Arch. Environ. Contam. Toxicol. 59, 49−61.

Filonov, A.E., Puntus, I.F., Karpov, A.V., Gaiazov, R.R., Kosheleva, I.A., Boronin, A.M., 1999. Growth and survival of *Pseudomonas putida* strains degrading naphthalene in soil model systems with different moisture levels. Process Biochem. 34 (3), 303−308.

Ghazali, F.M., Rahman, R.N.Z.A., Salleh, A.B., Basri, M., 2004. Biodegradation of hydrocarbons in soil by microbial consortium. Int. Biodeterior. Biodegrad. 54 (1), 61−67.

Ghosal, D., Ghosh, S., Dutta, T.K., Ahn, Y., 2016. Current state of knowledge in microbial degradation of polycyclic aromatic hydrocarbons (PAHs): a review. Front. Microbiol. 7, 1369.

Grimm, A.C., Harwood, C.S., 1997. Chemotaxis of Pseudomonas spp. to the polyaromatic hydrocarbon naphthalene. Appl. Environ. Microbiol. 63 (10), 4111−4115.

Guo, W., He, M., Yang, Z., Lin, C., Quan, X., Wang, H., 2007. Distribution of polycyclic aromatic hydrocarbons in water, suspended particulate matter and sediment from Daliao River watershed, China. Chemosphere 68, 93−104.

Janbandhu, A., Fulekar, M.H., 2011. Biodegradation of phenanthrene using adapted microbial consortium isolated from petrochemical contaminated environment. J. Hazard Mater. 187 (1−3), 333−340.

Jiang, J., Liu, H., Li, Q., Gao, N., Yao, Y., Xu, H., 2015. Combined remediation of Cd−phenanthrene co-contaminated soil by *Pleurotus cornucopiae* and *Bacillus thuringiensis* FQ1 and the antioxidant responses in *Pleurotus cornucopiae*. Ecotoxicol. Environ. Saf. 120, 386−393.

Khuman, S.N., Chakraborty, P., Cincinelli, A., Snow, D., Kumar, B., 2018. Polycyclic aromatic hydrocarbons in surface waters and riverine sediments of the Hooghly and Brahmaputra Rivers in the Eastern and Northeastern India. Sci. Total Environ. 636, 751−760.

Lin, C., Gan, L., Chen, Z.L., 2010. Biodegradation of naphthalene by strain *Bacillus fusiformis* (BFN). J. Hazard Mater. 182 (1−3), 771−777.

Liu, X.X., Hu, X., Cao, Y., Pang, W.J., Huang, J.Y., Guo, P., Huang, L., 2019. Biodegradation of phenanthrene and heavy metal removal by acid-tolerant *Burkholderia fungorum* FM-2. Front. Microbiol. 10, 408.

Maiti, A., Das, S., Bhattacharyya, N., 2012. Bioremediation of high molecular weight polycyclic aromatic hydrocarbons by *Bacillus thuringiensis* strain NA2. J. Sci. 1 (4), 72−75.

Mohandass, R., Rout, P., Jiwal, S., Sasikala, C., 2012. Biodegradation of benzo [a] pyrene by the mixed culture of Bacillus cereus and *Bacillus vireti* isolated from the petrochemical industry. J. Environ. Biol. 33 (6), 985.

Mohanraj, R., Azeez, P.A., 2004. Polycyclic aromatic hydrocarbons in PM 10 of urban and suburban Coimbatore, India. Fresenius Environ. Bull. 13, 332−335.

Mohanraj, R., Solaraj, G., Dhanakumar, S., 2011. Fine particulate phase PAHs in ambient atmosphere of Chennai metropolitan city, India. Environ. Sci. Pollut. Res. 18 (5), 764−771.

Moore, F., Akhbarizadeh, R., Keshavarzi, B., Khabazi, S., Lahijanzadeh, A., Kermani, M., 2015. Ecotoxicological risk of polycyclic aromatic hydrocarbons (PAHs) in urban soil of Isfahan metropolis, Iran. Environ. Monit. Assess. 187, 1−14.

Nagabhushanam, R., Machale, P.R., Katyayani, R.V., Reddy, P.S., Sarojini, R., 1991. Erythrophoretic responses induced by naphthalene in freshwater prawn, *Caridina rajadhari*. J. Ecotoxicol. Environ. Monit. 1 (3), 185−191.

Pathak, H., Kantharia, D., Malpani, A., Madamwar., D., 2009. Naphthalene degradation by *Pseudomonas* sp. HOB1: in vitro studies and assessment of naphthalene degradation efficiency in simulated microcosms. J. Hazard Mater. 166 (2−3), 1466−1473.

Peng, R.H., Xiong, A.S., Xue, Y., Fu, X.Y., Gao, F., Zhao, W., Tian, Y.S., Yao, Q.H., 2008. Microbial biodegradation of polyaromatic hydrocarbons. FEMS Microbiol. Rev. 32 (6), 927−955. https://doi.org/10.1111/j.1574-6976.2008.00127.x.

Ping, L., Zhang, C., Zhang, C., Zhu, Y., He, H., Wu, M., Tang, T., Li, Z., Zhao, H., 2014. Isolation and characterization of pyrene and benzo [a] pyrene-degrading *Klebsiella pneumonia* PL1 and its potential use in bioremediation. Appl. Microbiol. Biotechnol. 98 (8), 3819−3828.

Rajan, S, Geethu, V., Sampath, S., Chakraborty, P., 2019. Occurrences of polycyclic aromatic hydrocarbon from Adayar and Cooum riverine sediment in Chennai city, India. Int. J. Environ. Sci. Technol. 16 (12), 7695−7704.

Rajan, S., Rex, K.R., Pasupuleti, M., Muñoz, J.A., Jiménez, B., Chakraborty, P., 2021. Soil concentrations, compositional profiles, sources and bioavailability of polychlorinated dibenzo dioxins/furans, polychlorinated biphenyls and polycyclic aromatic hydrocarbons in open municipal dumpsites of Chennai city, India. Waste Manage. 131, 331–340.

Samanta, S.K., Singh, O.V., Jain, R.K., 2002. Polycyclic aromatic hydrocarbons: environmental pollution and bioremediation. Trends Biotechnol. 20 (6), 243–248.

Santos, E.C., Jacques, R.J., Bento, F.M., Maria do Carmo, R.P., Selbach, P.A., Sa, E.L., Camargo, F.A., 2008. Anthracene biodegradation and surface activity by an iron-stimulated *Pseudomonas* sp. Bioresour. Technol. 99 (7), 2644–2649.

Silva, I.S., dos Santos, E.D.C, de Menezes, C.R., de Faria, A.F., Franciscon, E., Grossman, M., Durrant, L.R., 2009. Bioremediation of a polyaromatic hydrocarbon contaminated soil by native soil microbiota and bioaugmentation with isolated microbial consortia. Bioresour. Technol. 100 (20), 4669–4675.

Winquist, E., Björklöf, K., Schultz, E., Räsänen, M., Salonen, K., Anasonye, F., Cajthaml, T., Steffen, K.T., Jørgensen, K.S., Tuomela, M., 2014. Bioremediation of PAH-contaminated soil with fungi—From laboratory to field scale. Int. Biodeterior. Biodegrad. 86, 238–247.

Advances in bioremediation of biosurfactants and biomedical wastes

14

Shreya Sharma, Akhilesh Dubey

*Department of Biological Sciences and Engineering, Netaji Subhas University of Technology,
Delhi, India*

Chapter outline

14.1 Introduction

Biomedical waste is characterized by wastes comprising whole or parts of human or animal tissues, bodily fluids, hazardous chemicals including antibiotics, anesthetics, and cytotoxic agents, cultures, disinfectants, swabs or dressings, pharmaceutical compounds, needles, or other sharp instruments. These types of emerging contaminants may prove perilous to the environment or to any organism coming in contact with them unless rendered safe (Allen, 2014; Stalder et al., 2013). Apprehension regarding their appropriate disposal is aggravating rapidly due to the scare of the spread of lethal pathogens as well as about likely exposure to toxic materials and products. The dimension of the problem is massive worldwide. Approximately 5.2 million people in the world die from medical waste pollution each year, among which 4 million of them are children (Babanyara et al., 2013). The United States is among the top medical waste-producing nations with 10.7 kg of biomedical waste generated per bed per day. The amount goes from 1.18 to 4.4 kg per bed per day in Europe, 0.36−5.34 kg per bed per day in Asia, and 0.44−1.1 kg per bed per day in Africa (Mannocci et al., 2020; Minoglou et al., 2017). The indecorous treatment of biomedical waste engenders egregious environmental issues in terms of air, water, and land pollution. Pathogens present in the waste can be a significant risk factor for disease transmission as they may invade and persist in the atmosphere for a long period. Burning of biomedical waste through incinerators or open burning releases chemical pollutants (emissions of dioxins, furans, and particulate matter) into the atmosphere that may be dangerous to human health. Open burning of biomedical waste is the most detrimental exercise and should uncompromisingly abstain from. Dumping of wastes in low-lying areas or into lakes and water bodies can engender acute water contamination. Water pollution can be caused by biological, chemical, or radioactive substances. The pathogens present in wastes can leach out from waste dumping sites into groundwater. Harmful chemicals present in biomedical waste like heavy metals can also adulterate water bodies. Land Pollution is inevitable as the final disposal of all biomedical waste including liquid effluent after treatment is on land. Open dumping of medical waste is the greatest cause of land pollution (Manzoor and Sharma, 2019).

Biological processes involving microorganisms play a major role in the removal of pollutants. Microbes can succor environmental rejuvenation by binding, oxidizing, volatilizing, immobilizing, or otherwise transmuting contaminants. There is conspicuous inquisitiveness in such microbially conciliated bioremediation as it commits to being simpler, cheaper, and are environment friendly in comparison with the more commonly used nonbiological options (Lovley, 2003). The metabolic processes that enable the degradation and expedite the use of disparate toxic compounds are classified as aerobic and anaerobic. Absolute mineralization in an ideal bioremediation process may yield carbon dioxide and water as end products without amassing the intermediates (Agrawal et al., 2020). Interest has lately transpired in engineering microbial consortia—multiple interacting microbial populations—to effectuate intricate functions and to be sturdier to environmental variations. The degradation ability of a single microorganism is limited to certain compounds whereas microbial consortia enable a higher level of degradation due to their coalescing activities (Sumathi, 2020). Nevertheless, low adaptability, lack of competitiveness of microbes, low bioavailability to target pollutants, and time-consuming techniques are major drawbacks to the process. Successful implications of system biology and metabolic engineering approaches fields of life sciences make it tempting for environmental scientists to utilize these approaches for bioremediation (Dangi et al., 2019). Fig. 14.1 categorizes different types of biomedical wastes and gives an illustration of emerging biological processes exploiting microbes for waste management.

Surface-active compounds of biological origin have evoked interest and their prevalence appears to expand in a regular and even manner. The certitude that biosurfactants are characterized by a vast structural diversity and divulge into a wide array of properties may expound why the class of molecules allures scientific curiosity. While biosurfactants are customarily equally efficacious in terms of solubilization and emulsification, they are also contemplated to exhibit low toxicity, high

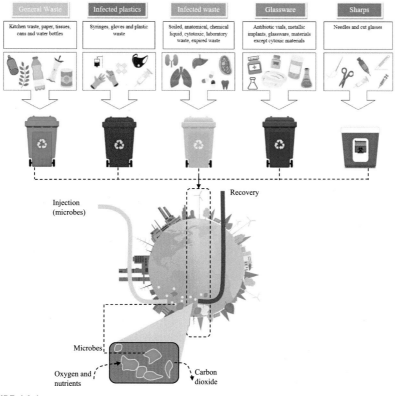

FIGURE 14.1

Segregation of biomedical waste and microbial treatment for effective management.

biodegradability, and environmental compatibility, and thus by far, making them potential substitutes for synthetic surfactants (Marcelino et al., 2020; Ławniczak et al., 2013). The positive influence of biosurfactants on bioremediation may be dedicated to the exceedingly eulogized environmental soundness amalgamated with the capability to solubilize hydrophobic compounds. Biosurfactants tend to associate with abysmally soluble pollutants and ameliorate their transference into the aqueous phase. On the other hand, biosurfactants may likewise prompt alterations in the properties of cellular membranes, bringing about expanded microbial adherence. Another remarkable utilization includes their capacity to complex heavy metal ions, improving their expulsion or extraction.

The overarching objective of the chapter is to highlight the recent innovations and unconventional ideas to determine the current status and potential of bioremediation and biosurfactants to get rid of biomedical wastes. We briefly discuss the role of microbial community structure and composition to give a practical approach to the subject. The challenges, advances, and direction of the research and future perspectives in the field based on recent literature have also been addressed.

14.2 Life cycle assessment of biomedical waste

Life cycle assessment, a systematized and plausible stratagem for quantifying the environmental impact of targeted processes, activity, or product throughout its life cycle is comprehensively applied in ecological design, product enhancement, decision making, and policy formulation. So far, biomedical disposal is principally enacted globally using steam sterilization, compacting, shredding, incineration, chemical disinfection, microwave, and landfill. Hong et al. investigated the environmental and economic impacts of pyrolysis, steam sterilization, and chemical disinfectants. Overall environmental impact is predominantly provoked from fossil depletion and global warming categories due to emission of greenhouse gases and coal utilization during the energy consumption stage. The influence of respiratory organic and inorganic, water depletion, carcinogens, and terrestrial acidification categories proffer auxiliary contributions in steam sterilization and pyrolysis because of emission of nitrogen oxide, sulfur dioxides, particulates, dibenz(a,h)anthracene, and mercury effused from the energy and sodium hydroxide production stage. The environmental and economic lucrative setting was perceived in the electricity and sodium hydroxide production processes in the case of pyrolysis; chlorine dioxide, landfill, and electricity processes in the case of chemical disinfection; and electricity and infrastructure processes for the steam sterilization of medical waste. Furthermore, the development of an incineration system was suggested for the reduction in employment of sodium oxide and chlorine oxide and initiating energy recovery (Hong et al., 2018). Kythavone et al. analyzed the life cycle of municipal solid waste incinerator for infectious medical waste. Results revealed steel as the principal contributor to climate change followed by all plastics. Machine transport, Steel, glass wool insulation manufactured from bromine or chlorine groups and aerosols caused by particulate matter formation in the atmosphere from ammonia, sulfur oxide, nitrogen oxides were reported as the main cause degradation of human health. Deposition of nutrients including nitrates, phosphates, and sulfates in acidifying form resulted in terrestrial acidification whereas materials such as brass, plastics, steel, and copper were reported responsible for freshwater and marine eutrophication (Kythavone and Chaiyat, 2020). On the other hand, a study on biodegradation of medical waste utilizing biotransformation and bioremediation methods evoked interest in microbial community structure and diversity as the main contributing factor for the performance of biological medical waste treatment mechanisms. The kismet and effacing procedure of the pollutants relied on pH, stereochemical structure, redox potential, and chemical properties of both sorbent and sorbed molecules.

The biodegradability is reported to be regulated by the complexity and stability of different pharmaceutical compounds and in some cases even rely on their chemical structure (Tiwari et al., 2017).

14.3 **Bioremediation**

Microbes are copious and ubiquitous and can even survive in environments of extreme pressure and temperature. The potentiality of microbes to exploit pollutants as substrates is dependent on their competence to spawn on specific metabolic and enzymatic responses in a polluted environment (Agrawal et al., 2020). Under specified conditions, microorganisms degrade pollutants and acquire energy to grow and reproduce. Breakage of chemical bonds and the transfer of released free electrons to an electron acceptor, for example, oxygen results in the production of energy. The fundamental process involves an oxidation-reduction reaction encompassing the oxidation of organic matter through the loss of electrons. The compound that accepts the electron (anaerobes: iron, sulfate, and nitrate; aerobes: oxygen) is said to be reduced. The complete process for redox reaction is governed by designated gene clusters and for different types of pollutants, the respective degrading gene cluster or pathway lies within either the genome or plasmid of organisms (Vishwakarma et al., 2020).

14.3.1 **Bioremediation techniques**

Bioremediation is categorized based on the site of the contaminant: *in situ* bioremediation in the place treatment of a contaminated site, and *ex situ* remediation carried elsewhere when contaminants are bored from the polluted area. *In situ* bioremediation techniques including bioventing, biosparging, and bioslurping are based on proffering oxygen, nutrients, and stimulating conditions that are indispensable for the growth of microbial communities. Bioventing entails administered stimulation of airflow by the delivery of oxygen to the vadose (unsaturated) zone to enhance bioremediation by surging activities of indigenous microorganisms. Biosparging follows the same technique by injecting air into the saturated surface. The efficacy of the technique relies on pollutant biodegradability and soil permeability. Bioslurping involves the amalgamation of vacuum-enhanced pumping, soil vapor extraction, and bioventing. The technique utilizes a "slurp" that stretches into the free product layer and levers up the fluid in a manner comparable to how a straw draws liquid from any vessel. *Ex situ* bioremediation techniques involving windrows, biopiles, bioreactors, and land farming are routinely appraised based on treatment cost, type, degree, and depth of pollution, the geology of the contaminated site. Windrows turn the contaminants to exalt bioremediation by escalating the degradation activities of transient or indigenous hydrocarbonoclasticus bacteria present. Above ground piling of excavated polluted soil, accompanied by nutrient amendment, and intermittently aeration to escalate bioremediation by predominantly augmenting microbial activities. Aeration, irrigation, nutrient, leachate collection systems, and a treatment bed form the main integrants of the system. Bioreactors aid in the conversion of raw materials to specific products. Different operating modes include batch, fed-batch, sequencing batch, continuous and multistage. Land farming is categorized into both *ex situ* and *in situ* remediation technique based on the site of treatment. Pollution depth plays a significant part as to whether land farming can be bolstered out *in situ* or *ex situ*. When excavated contaminants are treated on-site, it is regarded as *in situ*; else, it is *ex situ* as it has more in common with other *ex situ* bioremediation methods (Azubuike et al., 2016).

14.3.2 Bioremediation of medical waste: state of the art

Proper biomedical waste management is the linchpin of hospital cleanliness, hygiene, and maintenance activities. Pertinent systems for medical waste treatment are a crucial integrant of quality assurance in hospitals. Antibiotics are one of the most remarkable innovations and have brought a paradigm shift in the field of medicine for human therapy. Antibiotics additionally have also been employed in agriculture and animal husbandry as growth promoters, or to ameliorate the feeding efficiency. Increased exploitation of all known antibiotics infers consistent and recurrent release into the environment and natural ecosystem. Antibiotics found in soil, hospital wastewater, wastewater treatment plant effluents, surface water, groundwater, drinking water, biota, and sediments have posed a major global threat because of the observed high degree of toxicity, and growing antibiotic resistance (Kumar et al., 2019a,b). Afolabi et al. exploited white rot fungi to bioremediate streptomycin in an aqueous solution. Even low concentrations of antibiotics revealed efficient bioremediation (Afolabi). Bacterial species *Serratia* sp. R1 was reported to degrade antibiotic penicillin G to the final concentration of 84% by hydrolysis carried out by esterases (Kumar et al., 2019a,b). Gahlawat et al. observed a 90% and 70% degradation rate for aspirin and tetracycline degradation exploiting the microalgae *Brassica juncea* (Gahlawat and Gauba, 2016).

Biomedical waste ash produced from the incineration of biomedical waste contains large amounts of heavy metals and polycyclic aromatic hydrocarbons. Excessive landfilling may infer detrimental effects on the environment. Heera et al. demonstrated the bacterial efficiency in reducing the toxicity level of incinerator ash based on hardness, alkalinity, metal, and chloride content (Heera and Rajor, 2014). A reduced concentration of endocrine-disrupting chemicals was observed in wastewater effluents as a result of bioremediation by bacterial, fungal, and microalgal species (Roccuzzo et al., 2020). A very cheap, easily available, and effective stratagem exploiting fungus *Periconiella* sp. isolated from cow dung was reported to efficiently degrade the biomedical waste materials without leaving any evidence of detrimental repercussions on the population (Pandey, 2008). Bacterial strains *A. faecalis*, *B. paramycoides* D6, *and B. paramycoides* D7 were reported to degrade the pharmaceutical contaminants including diazepam, octadecene, caffeine, salicylic acid, naproxen, and phenol into lidocaine and butalbital, ticlopidine, tetradecane, and griseofulvin, respectively presenting a low-cost and low tech alternative (Rashid et al., 2020). The increase in metoprolol detection in wastewater due to the prevalence of hypertension and cardiovascular diseases exacts associated environmental risks and tremendous health burdens. Fungal species *Trametes versicolor* in combination with UV/H_2O_2 were recently used to achieve total compound removal (Jaén-Gil et al., 2020). Biomedical waste comprises a higher concentration of organic matter and consequently hydrolytic bacteria producing protease, lipase, and amylase should be copious in such waste making it plausible for isolation and utilization as candidates of bioremediation. Ethica et al. demonstrated a development method to procure pure isolates or consortium of indigenous hydrolytic bacteria from the liquid biomedical waste reservoir of hospitals potentiality for bioremediation of the liquid biomedical waste directing feasible facilities to test extracellular enzyme production of bacteria, encompassing techniques to examine their pathogenicity and capability to reduce parameter values of liquid organic waste (Ethica et al., 2018).

Other bioremediation techniques that can be soundly utilized include the following:

(1) *Phytoremediation* utilizes plants to eliminate pollutants from water or soil. Phytoremediation may help eliminate metals from water. It is exceptionally viable in remediating soil and organic compounds. A few investigations manifest that plantation of trees like neem, which has air-cleansing activity and phytoremediation activity is a wonderful practice to control contamination (Erakhrumen and Agbontalor, 2007).

(2) *Composting* can be embraced to dispose of herbal medicine wastes. The proof is available in databases concerning the exploitation of techniques for compelling bioremediation of natural pollutants and toxins in soil (Marin et al., 2005).

(3) *Vermicomposting* can likewise be embraced as an alternative for the removal of natural squanders from Ayurveda emergency clinics. The exploitation of techniques for managing wastewaters, remediating dirtied soils, ameliorating agrarian profitability are logically demonstrated (Kästner and Miltner, 2016).

(4) *Biofiltration* treats effluents from the pharmaceutical industry containing exorbitant concentrations of phenols. Biofiltration techniques can be used for the evacuation of phenolic deposits (Neves et al., 2006).

(5) *Rhizofiltration* utilizes plants to eliminate metals in water (Dushenkov et al., 1995)

(6) *Administration of oil-eating microorganisms* may aid in the treatment of used oils, one of the principal sources of waste in ayurvedic hospitals (Atlas Ronald, 1991).

14.4 **Biosurfactants**

Biosurfactants are characterized as amphiphilic moieties possessing the capability to lessen surface tensions over the interface of archetypal polar materials, for instance, oil and water, consequently demonstrating emulsification properties. Biosurfactants stand out from manufactured surfactants principally on account of their biological and accordingly sustainable origins, being overwhelmingly produced by microbial species. Contrasted with their engineered partners' biosurfactants have more prominent emulsification exercises, work over a more extensive scope of temperature conditions and, above all, they have been demonstrated to show a fundamentally low level of cytotoxicity (Smith et al., 2020). The vast range of microbial processes and strains utilized for the production of biosurfactants gives rise to numerous chemical structures and resulting interfacial properties. This specialized functionality implies that biosurfactants can be delivered to meet exceptionally explicit needs. Microbial biosurfactants are grouped into two classifications: low molecular weight surfactants including glycolipids and lipopeptides; and high molecular weight polymeric compounds including polysaccharides, proteins, or joint forms of lipoproteins or lipopolysaccharides. High molecular weight biosurfactants can unequivocally cling to different surfaces and act as bioemulsifiers. Low molecular weight biosurfactants encompass rhamnolipids, which are disaccharides acetylated with long-chain unsaturated fat or hydroxy greasy acids. These biosurfactants are valuable for lessening surface and interfacial tension. Both fungi and bacteria, colonizing water, soil, and extreme environments engender surface-active particles, for instance, surfactin (derived from *B. subtilis* and related species), sophorolipids (derived from *Candida* and related species), rhamnolipids (derived from mainly *Pseudomonas*), and so forth. The concentration of ions in the medium, salinity, temperature, pH, bacterial strains, and culture conditions are the deciding factors for produced biosurfactants and their physicochemical properties. Having structural diversity and utilitarian properties, biosurfactants play a critical role in emulsification, bioremediation, wetting, foaming, detergency, lubrication, dispersion, wetting, solubilization of hydrophobic compounds. They are nontoxic to the environment and speak to potential options for manufactured surfactants in the creation of clothing cleansers (Jahan et al., 2020; Naughton et al., 2019).

14.4.1 Biosurfactants as useful tools in bioremediation

The biosynthesis of surfactants mainly as secondary metabolites, especially when grown on a water-insoluble substrate can be either extracellular or as cell-bound molecules. Extracellular production of biosurfactants exhibits emulsifying activity toward the substrate. Contradictorily, biosurfactants are produced as part of the cell membrane of the microbial function in expediting the passage of substrates through the membrane (Olasanmi and Thring, 2018). A prospective solution for the remediation of soil sullied with metals and oil consists of the utilization of surfactants. The ionic nature, biodegradability, low toxicity, and phenomenal surface properties, forge these surfactant molecules as budding candidates for the remediation of heavy metals from soil and residue (da Rocha Junior et al., 2019). The endurance of oil hydrocarbons in the earth relies upon a few components, for example, concentration, chemical structure, and dispersion. Biological processes promise clean decontamination technologies, as they combine simplicity and cost-effectiveness, bioremediation emerges as the least aggressive and the most suitable method for keeping the ecological balance (Lopes et al., 2018). Biosurfactant-mediated bioremediation also aids in remediating/destroying certain microorganisms released into waste streams from hospitals/laboratories (Fig. 14.2). For instance, the amphiphilic nature of biosurfactants implies that their hydrophobic area can communicate with the lipid film of the viral particles, while at the same time communicating with other hydrophilic substances, for example, water. This property permits them to distort the virus structure and in this way deactivation of

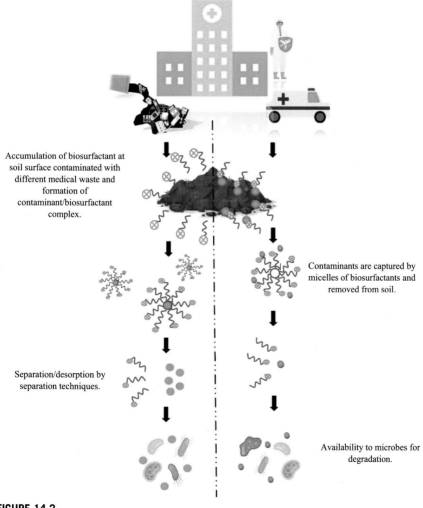

FIGURE 14.2

Mechanism of action of biosurfactant in bioremediation of biomedical waste.

Table 14.1 Application of biosurfactants in bioremediation of pollutants.

S. No	Application	Result	References
1.	Degradation of aromatic amine compound 2,4-diaminotoluene (2,4-DAT).	Biosurfactants raised the degrading capability of bacterial strains for 2,4-DAT by 14.54, 27.15, and 35.79–135.79 mg/L.	Femina Carolin et al. (2020)
2.	Cleaning of oil	Biosurfactant enhanced the degradation of the oil in the sand to 90%. With regard to the removal of oil on seawater, it was observed that removal percentages were around 85%.	Santosa et al. (2020)
3.	Degradation of polycyclic aromatic hydrocarbons	Promoted the biodegradation of slow-desorption polycyclic aromatic hydrocarbons.	Posada-Baquero et al. (2019)
4.	Removal of heavy metals	Promoted removal of heavy metals depending on the factors like time and concentration of inoculums used.	Gomaa and El-Meihy (2019)
5.	Cleaning of gasoline contaminated soil	Enhanced bioremediation of gasoline	Rahman et al. (2002)
6.	Degradation of polychlorinated biphenyls	Enhanced bioremediation of polychlorinated biphenyls (saponin—55% and rhamnolipids—60%)	Singer et al. (2000)
7.	Inactivation of viruses	Disruption of the viral lipid membrane and partially of the capsid	Vollenbroich et al. (1997)
8.	Removal of Cu and Pb	Crude biosurfactant also removed 30%–40% of Cu and Pb from standard sand, while the isolated biosurfactant removed ~30% of the heavy metal.	Santos et al. (2017)

the viral particles. Biosurfactants also have a fascinating attribute with regards to which they can shape micellar structures around their basic micelle fixation, a worth that varies enormously between the diverse biosurfactant types. This structure will be critical in legitimately focusing on the released viral particle, affecting its overall emulsification action (Smith et al., 2020). Table 14.1 summarizes the role and effect of biosurfactants on bioremediation.

14.5 Conclusion

The quest for cheap and environment friendly techniques to treat biomedical waste is paramount and obligatory. Worldwide investigations on the management of biomedical waste are in progress, as conventional strategies are viewed as costly and not environment friendly. Bioremediation is a substitute strategy for the management and degradation of waste. Bioremediation techniques have many benefits including low energy consumption, large scale-up potential, ecosystem conservation, and suitability for locations without sophisticated waste management facilities. It is eco-friendly and more affordable than other customary techniques. Many bacterial, fungal, and algal species exhibit propitious bioremediation potential for the removal of biomedical wastes. The information regarding life patterns and evolution of microbes utilizing omics study (proteome, metabolome, epigenome, and transcriptome) and through cutting-edge sequencing could aid in the anticipation and control of infectious diseases. The implantation of biological methodologies, system, and system biology approaches could proffer superior comprehension of the environmental factors of disease-emergence and drug-resistance mechanisms.

Academia has shifted its attention to the choice of microorganisms delivering biosurfactants, primer identification of the chemical composition of the bioproduct, and the proposed utilization. Aspects of extraordinary biosurfactants identified with experimental and enormous scope approaches should be given a more prominent effort to progress further with biotechnological applications. The utilization of sustainable and cheap substrates, advanced biosurfactant profitability, and improved and less costly bioprocesses for bioproduction, recuperation, and purification, notwithstanding hereditary building strategies and bioinformatics instruments, are a few techniques that ought to be further created to propel this rising innovation.

Acknowledgment

The authors would like to pay gratitude toward the Department of Biotechnology, Netaji Subhas University of Technology, Delhi, India.

References

Afolabi, K. Bioremediation of Aminoglycoside Antibiotic (Streptomycin) in Water by White Rot Fungi (*Ceriporia lacerata* and Trametes versicolor) (Doctoral dissertation), Evergreen State College.

Agrawal, K., Bhatt, A., Chaturvedi, V., Verma, P., January 1, 2020. Bioremediation: an effective technology toward a sustainable environment via the remediation of emerging environmental pollutants. In: Emerging Technologies in Environmental Bioremediation. Elsevier, pp. 165–196.

Allen, R., April 1, 2014. Disposing of clinical and dental waste. BDJ Team 1, 14038.

Atlas Ronald, M., 1991. Microbial hydrocarbon degradationbioremediation of oil spills. J. Chem. Technol. Biotechnol. 52 (2), 149e56.

Azubuike, C.C., Chikere, C.B., Okpokwasili, G.C., November 1, 2016. Bioremediation techniques—classification based on site of application: principles, advantages, limitations and prospects. World J. Microbiol. Biotechnol. 32 (11), 180.

Babanyara, Y.Y., Ibrahim, D.B., Garba, T., Bogoro, A.G., Abubakar, M.Y., October 5, 2013. Poor Medical Waste Management (MWM) practices and its risks to human health and the environment: a literature review. Int. J. Environ. Health Sci. Eng. 11 (7), 1−8.

da Rocha Junior, R.B., Meira, H.M., Almeida, D.G., Rufino, R.D., Luna, J.M., Santos, V.A., Sarubbo, L.A., August 15, 2019. Application of a low-cost biosurfactant in heavy metal remediation processes. Biodegradation 30 (4), 215−233.

Dangi, A.K., Sharma, B., Hill, R.T., Shukla, P., January 2, 2019. Bioremediation through microbes: systems biology and metabolic engineering approach. Crit. Rev. Biotechnol. 39 (1), 79−98.

Dushenkov, V., Kumar, P.B., Motto, H., May 1995. Raskin I Rhizofiltration: the use of plants to remove heavy metals from aqueous streams. Environ. Sci. Technol. 29 (5), 1239e45.

Erakhrumen, A.A., Agbontalor, A., July 2007. Phytoremediation: an environmentally sound technology for pollution prevention, control and remediation in developing countries. Educ. Res. Rev. 2 (7), 151−156.

Ethica, S.N., Saptaningtyas, R., Muchlissin, S.I., Sabdono, A., September 1, 2018. The development method of bioremediation of hospital biomedical waste using hydrolytic bacteria. Health Technol. 8 (4), 239−254.

Femina Carolin, C., Senthil Kumar, P., Janet Joshiba, G., Ramamurthy, R., Varjani, S.J., July 1, 2020. Bioremediation of 2, 4-diaminotoluene in aqueous solution enhanced by lipopeptide biosurfactant production from bacterial strains. J. Environ. Eng. 146 (7), 04020069.

Gahlawat, S., Gauba, P., September 1, 2016. Phytoremediation of aspirin and tetracycline by Brassica juncea. Int. J. Phytoremediation 18 (9), 929−935.

Gomaa, E.Z., El-Meihy, R.M., December 1, 2019. Bacterial biosurfactant from Citrobacter freundii MG812314. 1 as a bioremoval tool of heavy metals from wastewater. Bull. Natl. Res. Cent. 43 (1), 69.

Heera, S., Rajor, A., 2014. Bacterial treatment and metal characterization of biomedical waste ash. J. Waste Manag. 2014.

Hong, J., Zhan, S., Yu, Z., Hong, J., Qi, C., February 10, 2018. Life-cycle environmental and economic assessment of medical waste treatment. J. Clean. Prod. 174, 65−73.

Jaén-Gil, A., Buttiglieri, G., Benito, A., Mir-Tutusaus, J.A., Gonzalez-Olmos, R., Caminal, G., Barceló, D., Sarrà, M., Rodriguez-Mozaz, S., July 31, 2020. Combining biological processes with UV/H$_2$O$_2$ for metoprolol and metoprolol acid removal in hospital wastewater. Chem. Eng. J. 126482.

Jahan, R., Bodratti, A.M., Tsianou, M., Alexandridis, P., January 1, 2020. Biosurfactants, natural alternatives to synthetic surfactants: physicochemical properties and applications. Adv. Colloid Interface Sci. 275, 102061.

Kästner, M., Miltner, A., April 1, 2016. Application of compost for effective bioremediation of organic contaminants and pollutants in soil. Appl. Microbiol. Biotechnol. 100 (8), 3433−3449.

Kumar, M., Jaiswal, S., Sodhi, K.K., Shree, P., Singh, D.K., Agrawal, P.K., Shukla, P., March 1, 2019a. Antibiotics bioremediation: perspectives on its ecotoxicity and resistance. Environ. Int. 124, 448−461.

Kumar, M., Sodhi, K.K., Singh, D.K., December 1, 2019b. Bioremediation of *Penicillin* G by *Serratia* sp. R1, and enzymatic study through molecular docking. Environ. Nanotechnol. Monit. Manag. 12, 100246.

Kythavone, L., Chaiyat, N., March 16, 2020. Life cycle assessment of a very small organic rankine cycle and municipal solid waste incinerator for infectious medical waste. Thermal Sci. Eng. Prog. 100526.

Ławniczak, Ł., Marecik, R., Ł, C., March 1, 2013. Contributions of biosurfactants to natural or induced bioremediation. Appl. Microbiol. Biotechnol. 97 (6), 2327−2339.

Lopes, P.R., Montagnolli, R.N., Cruz, J.M., Claro, E.M., Bidoia, E.D., 2018. Biosurfactants in improving bioremediation effectiveness in environmental contamination by hydrocarbons. In: Microbial Action on Hydrocarbons. Springer, Singapore, pp. 21−34.

Lovley, D.R., October 2003. Cleaning up with genomics: applying molecular biology to bioremediation. Nat. Rev. Microbiol. 1 (1), 35−44.

Mannocci, A., di Bella, O., Barbato, D., Castellani, F., La Torre, G., De Giusti, M., Cimmuto, A.D., 2020. Assessing knowledge, attitude, and practice of healthcare personnel regarding biomedical waste management: a systematic review of available tools. Waste Manag. Res. 38 (7), 717−725, 0734242X20922590.

Manzoor, J., Sharma, M., October 2, 2019. Impact of biomedical waste on environment and human health. Environ. Claims J. 31 (4), 311−334.

Marcelino, P.R., Gonçalves, F., Jimenez, I.M., Carneiro, B.C., Santos, B.B., da Silva, S.S., January 23, 2020. Sustainable production of biosurfactants and their applications. Lignocellulosic Biorefining Technol. 159−183.

Marin, J.A., Hernandez, T., Garcia, C., June 1, 2005. Bioremediation of oil refinery sludge by landfarming in semiarid conditions: influence on soil microbial activity. Environ. Res. 98 (2), 185−195.

Minoglou, M., Gerassimidou, S., Komilis, D., February 2017. Healthcare waste generation worldwide and its dependence on socio-economic and environmental factors. Sustainability 9 (2), 220.

Naughton, P.J., Marchant, R., Naughton, V., Banat, I.M., July 2019. Microbial biosurfactants: current trends and applications in agricultural and biomedical industries. J. Appl. Microbiol. 127 (1), 12−28.

Neves, L.C., Miyamura, T.T., Moraes, D.A., Penna, T.C., Converti, A., 2006. Biofiltration methods for the removal of phenolic residues. In: Twenty-Seventh Symposium on Biotechnology for Fuels and Chemicals. Humana Press, pp. 130−152.

Olasanmi, I.O., Thring, R.W., December 2018. The role of biosurfactants in the continued drive for environmental sustainability. Sustainability 10 (12), 4817.

Pandey, G.H.S., July 2008. Role of the fungus e Periconiella sp. in destruction of biomedical waste. J. Environ. Sci. Eng. 50 (3), 239e40.

Posada-Baquero, R., Grifoll, M., Ortega-Calvo, J.J., June 10, 2019. Rhamnolipid-enhanced solubilization and biodegradation of PAHs in soils after conventional bioremediation. Sci. Total Environ. 668, 790−796.

Rahman, K.S., Banat, I.M., Thahira, J., Thayumanavan, T., Lakshmanaperumalsamy, P., January 1, 2002. Bioremediation of gasoline contaminated soil by a bacterial consortium amended with poultry litter, coir pith and rhamnolipid biosurfactant. Bioresour. Technol. 81 (1), 25−32.

Rashid, A., Mirza, S.A., Keating, C., Ali, S., Campos, L.C., 2020. Indigenous Bacillus paramycoides and Alcaligenes faecalis: potential solution for the bioremediation of wastewaters. Environ. Technol. 1−14.

Roccuzzo, S., Beckerman, A.P., Trögl, J., February 28, 2020. New perspectives on the bioremediation of endocrine disrupting compounds from wastewater using algae-, bacteria-and fungi-based technologies. Int. J. Environ. Sci. Technol. 1−8.

Santos, D.K., Resende, A.H., de Almeida, D.G., Soares da Silva, R.D., Rufino, R.D., Luna, J.M., Banat, I.M., Sarubbo, L.A., May 1, 2017. Candida lipolytica UCP0988 biosurfactant: potential as a bioremediation agent and in formulating a commercial related product. Front. Microbiol. 8, 767.

Santosa, E.M., Liraa, I.R., Filhoa, A.A., Meiraa, H.M., Farias, C.B., Rufinoa, R.D., Sarubboa, L.A., de Luna, J.M., 2020. Application of the biosurfactant produced by Candida sphaerica as a bioremediation agent. Chem. Eng. 79.

Singer, A.C., Gilbert, E.S., Luepromchai, E., Crowley, D.E., December 1, 2000. Bioremediation of polychlorinated biphenyl-contaminated soil using carvone and surfactant-grown bacteria. Appl. Microbiol. Biotechnol. 54 (6), 838−843.

Smith, M.L., Gandolfi, S., Coshall, P.M., Rahman, P.K., 2020. Biosurfactants: a covid-19 perspective. Front. Microbiol. 11.

Stalder, T., Alrhmoun, M., Louvet, J.N., Casellas, M., Maftah, C., Carrion, C., Pons, M.N., Pahl, O., Ploy, M.C., Dagot, C., July 16, 2013. Dynamic assessment of the floc morphology, bacterial diversity, and integron content of an activated sludge reactor processing hospital effluent. Environ. Sci. Technol. 47 (14), 7909−7917.

Sumathi, S., March 18, 2020. Chapter six the microbial rescue of the environment from biomedical waste. Microb. Biodiv. 70.

Tiwari, B., Sellamuthu, B., Ouarda, Y., Drogui, P., Tyagi, R.D., Buelna, G., January 1, 2017. Review on fate and mechanism of removal of pharmaceutical pollutants from wastewater using biological approach. Bioresour. Technol. 224, 1−2.

Vishwakarma, G.S., Bhattacharjee, G., Gohil, N., Singh, V., January 1, 2020. Current status, challenges and future of bioremediation. In: Bioremediation of Pollutants. Elsevier, pp. 403−415.

Vollenbroich, D., Özel, M., Vater, J., Kamp, R.M., Pauli, G., September 1, 1997. Mechanism of inactivation of enveloped viruses by the biosurfactant surfactin from Bacillus subtilis. Biologicals 25 (3), 289−297.

Can algae reclaim polychlorinated biphenyl—contaminated soils and sediments?

15

Muhammad Kaleem[1], Muhammad Zaffar Hashmi[2], Abdul Samad Mumtaz[1]

[1]*Department of Plant Sciences, Quaid-i-Azam University, Islamabad, Pakistan;* [2]*Department of Chemistry, COMSATS University Islamabad, Islamabad, Pakistan*

Chapter outline

15.1 Introduction

Polychlorinated biphenyls (PCBs) are ubiquitous and incessant organic pollutants (Williams et al., 2020). PCBs are synthetic in nature and produced by the chlorination of biphenyls. The biphenyl moiety carries up to 10 chlorine atoms, and thus PCBs attain an extraordinarily stable structure and persist in the environment for a substantially long time. Despite their adverse traits, PCBs are used in large quantity in industry for diverse commercial applications such as hydraulic lubricants, plasticizers, high thermal capacity fluids, dielectric fluids for capacitors and transformers, adhesives, cutting oils, carbonless reproducing paper, dedusting agents, flame retardants, organic diluents, and wax extenders (Chatel et al., 2017). Mixtures of PCBs are commercially utilized under the trade names Aroclor, Pheclor, Delor, and Clophen (Chun et al., 2019). The consequent adverse impacts of PCBs on the environment include toxicity, bioaccumulation, and global persistence. Their production was formally put to a halt in the early 1980s (Reddy et al., 2019).

Despite bans, PCBs are still used in sealants and paints and being sporadically discharged into the environment (Stuart-Smith and Jepson, 2017). PCBs enter the atmosphere through evaporation of paints, plastics, and coatings, downright leakage into streams and sewage, and tipping in open garbage sites followed by typical disposal methods—for instance, ocean dumping (Huang et al., 2011). Hence, PCBs persist worldwide throughout marine and terrestrial environments and are often found in the food chain as a complex concoction. One incidence of elevated quantities of PCBs has been reported in the long-lived odontocetes that are nourished at the top trophic level, causing suppression of reproductive and immune systems and population decline (Desforges et al., 2018; Murphy et al., 2015). PCBs also enter the marine environment through various phenomena—for instance, atmospheric transport, dredging, and terrestrial runoff (Jartun, 2011). In the soil, PCBs emanate from deposition of industrial wastes, biomass combustion, and incinerators. PCBs become adsorbed on soil particles and so persist even longer (Kumar et al., 2014).

Numerous sources and reservoirs of PCB contaminants exist including soils and sediments. According to estimates, the average concentration of PCBs in soil may be from 100 to 1000 pg/g, while the airborne average concentration ranges from 120 to 170 pg/m^3 (Batterman et al., 2009). Soils and sediments are important reservoirs for PCBs because these xenobiotics are persistent (Meijer et al., 2003). A survey of PCB concentrations estimated as much as 21,000 tons of PCB congeners present in soils and sediments globally (Dalla Valle et al., 2005). Furthermore, PCBs are easily accumulated in paddy soils (Moeckel et al., 2008). Thus, many PCB ridden-residues have been confirmed in the soil used to grow food, such as rice as reported in South China (Shen et al., 2008), where its concentration is as high as 1636.8 ng/g (An et al., 2006). A recent study has also revealed that PCB concentrations in rice paddies may range from 41.1 to 132.4 ng/g in Zhejiang province, where the daily intake of PCBs by people is about 12,372.9 ng/day through rice consumption (Zhao et al., 2009).

15.1.1 Phycoremediation of polychlorinated biphenyls

Bioremediation is a promising technology with several manifestations such as bioaugmentation, biostimulation, rhizoremediation and phytoremediation. Coupled with genetically modified microorganisms and transgenic plants, it has proven revolutionary in the bioremediation of PCBs (Sharma et al., 2018). Some microorganisms can degrade PCBs aerobically and anaerobically under various conditions (Pieper and Seeger, 2008). This technology has provided a paradigm shift solution in reducing toxic wastes. Algae, being photosynthetic, harvest light energy and carbon dioxide from the atmosphere for biomass production (Demirbas, 2010). These organisms range from unicellular to multicellular in form, microscopic to gigantic kelp in size, and spherical to filamentous in shape (Amaro et al., 2011; Demirbas and Demirbas, 2011). Mostly phytoplanktonic, they play an important role in remediating persistent organic pollutants (POPs) such as PCBs at the base of benthic and pelagic food chains (Lynn et al., 2007). Organic pollutant degradation through algae is a biological process that is likely to have a softer environmental effect than the chemical, physical and mechanical approaches that have been tested on POPs. Among the ancillary benefits of algae-based bioremediation are higher biomass production to accumulate, degrade, or detoxify pollutants and xenobiotics (Baghour, 2019). Fortuitously, as ubiquitous organisms efficient in degrading chlorinated organic pollutants ex situ, microalgae and cyanobacteria have shown great capacity and application in bioremediation (Papazi et al., 2012). Studies suggest that algae and other microorganisms adopt diverse strategies for their existence in contaminated environments. These mechanisms may include like approaches termed bioremediation strategies per se (Gadd, 2000; Lim et al., 2003; Malik, 2004). However, it is not clear what strategies are adopted specifically and most frequently. This warrants a detailed study with a clear focus on species. Studies on organic xenobiotics (either through biodegradation or bioaccumulation) in green algae are significant for environmental health as well as being relevant in the agricultural sector with a huge economic impact.

The biomass produced during the bioremediation process is significant as feedstock in biofuel production (Huang et al., 2010). Photoautotrophic microorganisms are among the most important bioresources and have gained tremendous fame for their capability to grow at a faster rate, requiring nonarable lands and wastewater only (Baghour, 2017). Phycoremediation is eco-friendly, cost-effective, and comparatively safe and thereby efficient in lowering the nutrient load and total dissolved solids in wastewater. Intriguingly, the potential to degrade organic pollutants has been found in several algal types (Baghour, 2019). Green types include *Scenedesmus*, *Chlorella*, *Chlamydomonas*, *Desmodesmus*, and *Botryococcus*; blue-green types include *Spirulina*, *Oscillatoria*, *Phormidium*, *Arthrospira*, *Cyanothece*, and *Nodularia* (Dubey et al., 2011; Rawat et al., 2011); and marine macroalgae include *Kappaphycus alvarezii* and *Ulva lactuca*.

Phycoremediation is potentially applicable to several pollutant types. Some of the most important are heavy metals, petroleum hydrocarbons, polycyclic aromatic hydrocarbons (PAHs), pentachlorophenol (PCP), other chlorinated solvents, nutrients, and radiation-emitting atomic moieties, with some of these posing serious threats to aquatic environments (Jin et al., 2012). Freshwater submergent angiosperms such as parrot's feather, water milfoil, hyacinth, and duckweed have exhibited great potential for accumulation of copper, lead, selenium, mercury, and cadmium (Kamal et al., 2004). Likewise, the potential of micro- and macroalgae in the uptake of radionuclides and heavy metals is promising. However, little information is available about their ability to assimilate PCBs and PAHs. It is established that under normal growth conditions, macroalgae can assimilate nutrients in their tissues. Thus, culturing macroalgae has long been described as an auspicious approach in reducing harmful impacts of heavy metals (Han et al., 2017). Likewise, both marine macro- and microalgae (e.g., *Chlorella*, *Ascophyllum*, *Ulva*, and *Fucus*) and macroalgae in fresh water (e.g., *Spirogyra*, *Nitella*, and *Ulva*) have shown closely related capacities to uptake and concentrate copper, lead, strontium, cadmium, nickel, mercury, chromium, and uranium (U (VI)) (Davis et al., 2003; Rybak et al., 2012).

15.1.2 Modes of phycoremediation

Generally, two distinct modes of phycoremediation have been reported in algae. The contextual description of these is provided in the discussion that follows.

15.1.2.1 Bioaccumulation

In marine types, Sieburth and Jensen (1969) originally reported the potential of the brown seaweed *Fucus vesiculosus* to exude humic substances (also called gelbstoff). These phenolics precipitate to form organic aggregates. Developing on this further, Lara et al. (1989) studied the potential of the brown algae *Caepidium antarcticum* and *Desmarestia* sp. to link these exudates with hydrophobic pollutants such as PCBs. Likewise, organic macromolecules exuded from *Ascophyllum nodosum* and *Fucus* sp. in seawater samples were found to incorporate organic compounds such as fatty acids, sugars, and amino acids. Lately, Lauze and Hable (2017) estimated that *Fucus vesiculosus* remediates 20%−30% of PCBs through bioaccumulation. PCBs are normally stocked up in the lipid-based storage molecules of algae. Berglund et al. (2001) found that lipid content describes the variation in PCB concentration in phytoplankton.

Likewise, in free living freshwater types, PCB subcellular accumulation has been studied in the thylakoids of *Chlamydomonas reinhardtii*. All PCB congeners amass in the membrane of thylakoids irrespective of their water partition constant (K_{ow} value), and steric hindrance plays only a trivial role in controlling bioaccumulation of PCBs. Kinetic limitation is associated with the accumulation process when observed on a scale of days to weeks (Jabusch and Swackhamer, 2004). The highest increase in the concentration of lipid-based PCB congeners in thylakoids takes place within the initial phase of the experiment. Altogether, there is a decreasing trend for its uptake over time. The mode of direct uptake, on average, is manifold higher than that of consecutive intervals, a pattern analogous to a reversible first-order process.

The subcellular organelles involved in the uptake of PCBs, especially the thylakoids, act as internal lipid pools. However, this requires huge kinetic energy and therefore remains a slower process compared with the much more direct process of total cellular uptake. This signifies an operational distinction of exterior and interior subcellular lipid pools within a cell. Thus, on a time scale followed for accumulation, it was concluded that uptake is largely controlled kinetically with no physical blocking observed in the cell wall (Jabusch and Swackhamer, 2004).

15.1.2.2 Degradation

POPs have been reported as potential degradable targets for algae. By virtue of the Stockholm Convention, countries should reduce their production and use of key POPs. These pollutants also include PCBs in addition to PAHs, phenolics, and pesticides. Most chemicals are produced in industry and used extensively in agriculture and thus are spilled into the environment where they endure for exceptionally long periods as they continue to accumulate and pass through the trophic levels of the food chain. Research has revealed that besides bacteria and fungi, algal cultures too can degrade phenolics and other organic pollutants besides their intermediary metabolites under mixotrophic conditions. However, scientific information about organic pollutant degradation by microalgae is far less extensive than for bacteria. Nonetheless, a recent study on the blue-green alga *Anabaena*, known as strain *PD*-1, produced encouraging evidence of degradation of 85% of Aroclor 1254 (PCB) and 40%—68% of dioxin-like PCBs (Zhang et al., 2015).

Generally, algae only consume pesticides if they are exposed within admissible limits. Butler et al. (1975) detected that several algal species degrade as little as 1 ppm of diazinon and carbaryl and only 0.01 ppm of methoxychlor and of 2,4-D. A similar study reported degradation of organophosphorus insecticides by very few green and blue-green types when exposed to 5—50 ppm concentrations of pesticides for 30 days (Megharaj et al., 1987). By contrast, the cyanobacterial types *Aulosira fertilissima* and *Anabaena* sp. accumulated chlorpyrifos, DDT, and fenitrothion considerably (Lal et al., 1987). This evidence is encouraging enough to characterize more algal types with promising phycoremediation potential in soils and sediments.

15.1.3 **Factors that affect phycoremediation**

Several factors should be considered to improve phycoremediation efficiency. Among the most important are the choice of algal strain (free living or filamentous types); response rate of microalgae; and growth rate, biomass, and cost-effective culturing conditions. How these factors affect algal efficiency has been described in some recent studies. For instance, Lynn et al. (2007) investigated the effect of different nutrient treatments on the uptake of PCB congeners by *Stephanodiscus minutulus* (a diatom) and transferred to a zooplankton (*Daphnia pulicaria*). It was observed that the availability of nutrients significantly changed the uptake of PCB by phytoplankton, which later affected the transfer to zooplankton, potentially through lipid assimilation and grazing rate.

Schoeny et al. (1988) reported that species of *Selenastrum, Scenedesmus*, and *Ankistrodesmus* converted up to 98% of benzo[*a*]pyrene (BaP) to 3,6-quinones under white light and consequently inhibited the growth rate of these species. A similar experiment with warm light enhanced the growth and consequently helped catabolize the pollutant-producing benzo[*a*]pyrene-diols. Intriguingly, white light had no such influence on *Euglena gracilis* (a euglenoid alga), *Anabaena flos-aquae* (cyanobacterium), *Chlamydomonas reinhardtii* (green alga), and *Ochromonas malhamensis* (yellow-green alga). Hence, this provided conclusive evidence to suggest that degradation of BaP by algae is a function of light energy being absorbed and emitted, with optimum metabolic rates achieved by *Selenastrum capricornutum* under warm light. Similarly, in contrasting trophic exposures *Chlorella protothecoides* responded quite differently. For instance, under normal autotrophic conditions, removal of 20% anthracene was observed, whereas in heterotrophic conditions, the capacity of *C. protothecoides* grew to ca. 33.5% (Yan et al., 2002), while this alga under mixed trophic conditions accumulated 80% of anthracene. Furthermore, *S. capricornutum* at a preliminary cell surface density of 1×10^7 cells/mL removed

100% of fluoranthene, 100% of pyrene, and 96% of phenanthrene over a 4-day interval (Chan et al., 2006). Hence, several factors may affect phycoremediation efficiency.

15.1.4 The phycoremediation promise of genetically modified algae

Having both eukaryotic and prokaryotic photoautotrophs, green and blue-green algae are instrumental in understanding the mechanism(s) to counter pollution (Siripornadulsil et al., 2002) merely by deploying algae-based genetic resources in their natural manner. Hitherto, genetic engineering tests on algal systems to develop transgenic resources for phycoremediation are still awaited. We know that bacteria, fungi, and algae possess several enzyme-coding genes that catabolically degrade soil- and sediment-born pollutants (Pinyakong et al., 2000; Potin et al., 2004; Semple et al., 1999). These novel genes known for pollutant degradation can be used to transform the promising mixotrophic algae to assess expression behavior, taking the precedence of earlier reports of successful transfer of bacterial genes into mixotrophic algae. For instance, the *lin*A gene of *P. paucimobilis* UT26 transferred to *Anabaena* sp. demonstrated enhanced efficiency in lindane degradation in the absence of nitrates (Kuritz et al., 1997). Increased lindane degradation capacity has also been observed in transgenic *Nostoc ellipsosporum* with *lin*A gene (Kuritz et al., 1997). Furthermore, Rajamani et al. (2007) reported buoyancy in tolerance to pollutants in transgenic microalgae. Such types and strains, if developed and characterized, will be readily available for phycoremediation.

Takahashi et al. (2008) proposed that *pmg*A has a role in carbon flow between the oxidative pentose phosphate pathway and the Calvin cycle. The *pmg*A gene has been recounted in several strains of *Geobacter violaceus*, *S. elongatus*, *Cyanothece* sp., and *Nostoc* sp. Ostensibly, this gene is conserved in algae and enhances the capacity of such strains to assume a mixotrophic mode of nutrition. Hihara and Ikeuchi (1997) reported that the *pmg*A functional gene is necessary for the photomixotrophic growth of algae. One important consideration for any such work is the poor social recognition and acceptance of the release of genetically altered microbes into the environment because of the unprecedented outcomes. However, investigative hazard assessments of such living modified organisms in terms of ethical, ecological, and social aspects will reduce the supposed risks and ready their use in the remediation of contaminants (Megharaj et al., 2011; Tiedje et al., 1989).

Algal DNA engineering manifests only infancy (León-Bañares et al., 2004; Stevens and Purton, 1997), mainly because of the rare codon arrangement phenomena that pose bias in the expression of foreign genes within algal cells (Heitzer et al., 2007). Despite this, the availability of TN transposons, cloning vectors, and other methods of mutagenesis and genome-sequencing platforms provide alternative opportunities to explore cyanobacteria and microalgae with objectivity (Koksharova and Wolk, 2002). Despite these difficulties, active research in developing transgenic microalgae is gaining attention (Walker et al., 2005). Recent nuclear-based transformations for metabolic control and chloroplast-based transformations, especially for high-level protein expression (León-Bañares et al., 2004; Rosenberg et al., 2008), have received ample attention. Applications of more robust methods such as metagenomics and whole-genome sequencing focused on algal species should provide further comprehensive knowledge about metabolic processes—for instance, the protein expression and associated evolutionary developments we see in algae holding key ecological niches.

15.2 Conclusion

With a huge capacity to grow under stress and eutrophication, algae still offer a tremendous tendency to accumulate or biodegrade organic pollutants. Thylakoid membranes in algae are important sites for accumulating PCBs and play a leading role in bioremediation if exposed. Algal cultures degrade phenolics and other organic pollutants under mixotrophic conditions, which is a capacity not seen in fungi. Hence, molecular-guided identification of species and desirable genetic resources ought to be applied to design and deliver the promise of mixotrophic algae to reclaim polluted soils and sediments.

Funding

Higher Education Commission of Pakistan NRPU projects 7958 and 7964. Thanks to Pakistan Science Foundation project PSF/Res/CP/C-CUI/Envr (151) provides the funding. Further, thanks to Pakistan Academy of Sciences for funding.

References

Amaro, H.M., Guedes, A.C., Malcata, F.X., 2011. Advances and perspectives in using microalgae to produce biodiesel. Appl. Energy 88, 3402—3410.

An, Q., Dong, Y.-h., Wang, H., Ceng, F., Zhang, J.-q., 2006. Residues of PCBs in agricultural fields in the Yangtze Delta, China. Environ. Sci. 27, 528—532.

Baghour, M., 2017. Effect of Seaweeds in Phyto-Remediation, pp. 47—83.

Baghour, M., 2019. Algal degradation of organic pollutants. In: Torres Martínez, L.M., Kharissova, O.V., Kharisov, B.I. (Eds.), Handbook of Eco-Materials. Springer Nature Switzerland, pp. 565—586.

Batterman, S., Chernyak, S., Gouden, Y., Hayes, J., Robins, T., Chetty, S., 2009. PCBs in air, soil and milk in industrialized and urban areas of KwaZulu-Natal, South Africa. Environ. Pollut. 157, 654—663.

Berglund, O., Larsson, P., Ewald, G., Okla, L., 2001. The effect of lake trophy on lipid content and PCB concentrations in planktonic food webs. Ecology 82, 1078—1088.

Butler, G.L., Deason, T., O'Kelley, J., 1975. Loss of five pesticides from cultures of twenty-one planktonic algae. Bull. Environ. Contam. Toxicol. 13, 149—152.

Chan, S.M., Luan, T., Wong, M.H., Tam, N.F., 2006. Removal and biodegradation of polycyclic aromatic hydrocarbons by *Selenastrum capricornutum*. Environ. Toxicol. Chem. 25, 1772−1779.

Chatel, G., Naffrechoux, E., Draye, M., 2017. Avoid the PCB mistakes: a more sustainable future for ionic liquids. J. Hazard Mater. 324, 773−780.

Chun, S.C., Muthu, M., Hasan, N., Tasneem, S., Gopal, J., 2019. Mycoremediation of PCBs by *Pleurotus ostreatus*: possibilities and prospects. Appl. Sci. 9, 4185.

Dalla Valle, M., Jurado, E., Dachs, J., Sweetman, A.J., Jones, K.C., 2005. The maximum reservoir capacity of soils for persistent organic pollutants: implications for global cycling. Environ. Pollut. 134, 153−164.

Davis, T.A., Volesky, B., Mucci, A., 2003. A review of the biochemistry of heavy metal biosorption by brown algae. Water Res 37, 4311−4330.

Demirbas, A., 2010. Use of algae as biofuel sources. Energy Convers. Manag. 51, 2738−2749.

Demirbas, A., Demirbas, M.F., 2011. Importance of algae oil as a source of biodiesel. Energy Convers. Manag. 52, 163−170.

Desforges, J.-P., Hall, A., McConnell, B., Rosing-Asvid, A., Barber, J.L., Brownlow, A., De Guise, S., Eulaers, I., Jepson, P.D., Letcher, R.J., 2018. Predicting global killer whale population collapse from PCB pollution. Science 361, 1373−1376.

Dubey, S.K., Dubey, J., Mehra, S., Tiwari, P., Bishwas, A., 2011. Potential use of cyanobacterial species in bioremediation of industrial effluents. Afr. J. Biotechnol. 10, 1125−1132.

Gadd, G.M., 2000. Bioremedial potential of microbial mechanisms of metal mobilization and immobilization. Curr. Opin. Biotechnol. 11, 271−279.

Han, T., Qi, Z., Huang, H., Fu, G., 2017. Biochemical and uptake responses of the macroalga *Gracilaria lemaneiformis* under urea enrichment conditions. Aquat. Bot. 136, 197−204.

Heitzer, M., Eckert, A., Fuhrmann, M., Griesbeck, C., 2007. Influence of codon bias on the expression of foreign genes in microalgae. In: Transgenic Microalgae as Green Cell Factories. Springer, pp. 46−53.

Hihara, Y., Ikeuchi, M., 1997. Mutation in a novel gene required for photomixotrophic growth leads to enhanced photoautotrophic growth of *Synechocystis* sp. PCC 6803. Photosynth. Res. 53, 243−252.

Huang, G., Chen, F., Wei, D., Zhang, X., Chen, G., 2010. Biodiesel production by microalgal biotechnology. Appl. Energy 87, 38−46.

Huang, J., Matsumura, T., Yu, G., Deng, S., Yamauchi, M., Yamazaki, N., Weber, R., 2011. Determination of PCBs, PCDDs and PCDFs in insulating oil samples from stored Chinese electrical capacitors by HRGC/HRMS. Chemosphere 85, 239−246.

Jabusch, T., Swackhamer, D., 2004. Subcellular accumulation of polychlorinated biphenyls in the green alga *Chlamydomonas reinhardii*. Environ. Toxicol. Chem. 23, 2823−2830.

Jartun, M., 2011. Building materials: an important source for polychlorinated biphenyls (PCBs) in urban soils. In: Mapping the Chemical Environment of Urban Areas, pp. 128−133.

Jin, Z.P., Luo, K., Zhang, S., Zheng, Q., Yang, H., 2012. Bioaccumulation and catabolism of prometryne in green algae. Chemosphere 87, 278−284.

Kamal, M., Ghaly, A., Mahmoud, N., Cote, R., 2004. Phytoaccumulation of heavy metals by aquatic plants. Environ. Int. 29, 1029−1039.

Koksharova, O., Wolk, C., 2002. Genetic tools for cyanobacteria. Appl. Microbiol. Biotechnol. 58, 123−137.

Kumar, B., Verma, V.K., Singh, S.K., Kumar, S., Sharma, C.S., Akolkar, A.B., 2014. Poly-chlorinated biphenyls in residential soils and their health risk and hazard in an industrial city in India. J. Public Health Res. 3.

Kuritz, T., Bocanera, L.V., Rivera, N.S., 1997. Dechlorination of lindane by the cyanobacterium *Anabaena* sp. strain PCC7120 depends on the function of the nir operon. J. Bacteriol. 179, 3368–3370.

Lal, S., Lal, R., Saxena, D.M., 1987. Bioconcentration and metabolism of DDT, fenitrothion and chlorpyrifos by the blue-green algae *Anabaena* sp. and *Aulosira fertilissima*. Environ. Pollut. 46, 187–196.

Lara, R., Wiencke, C., Ernst, W., 1989. Association between exudates of brown algae and polychlorinated biphenyls. J. Appl. Phycol. 1, 267.

Lauze, J.F., Hable, W.E., 2017. Impaired growth and reproductive capacity in marine rockweeds following prolonged environmental contaminant exposure. Bot. Mar. 60, 137–148.

León-Bañares, R., González-Ballester, D., Galván, A., Fernández, E., 2004. Transgenic microalgae as green cell-factories. Trends Biotechnol. 22, 45–52.

Lim, P.E., Mak, K., Mohamed, N., Noor, A.M., 2003. Removal and speciation of heavy metals along the treatment path of wastewater in subsurface-flow constructed wetlands. Water Sci. Technol. 48, 307–313.

Lynn, S.G., Price, D.J., Birge, W.J., Kilham, S.S., 2007. Effect of nutrient availability on the uptake of PCB congener 2, 2′, 6, 6′-tetrachlorobiphenyl by a diatom (*Stephanodiscus minutulus*) and transfer to a zooplankton (*Daphnia pulicaria*). Aquat. Toxicol. 83, 24–32.

Malik, A., 2004. Metal bioremediation through growing cells. Environ. Int. 30, 261–278.

Megharaj, M., Ramakrishnan, B., Venkateswarlu, K., Sethunathan, N., Naidu, R., 2011. Bioremediation approaches for organic pollutants: a critical perspective. Environ. Int. 37, 1362–1375.

Megharaj, M., Venkateswarlu, K., Rao, A., 1987. Metabolism of monocrotophos and quinalphos by algae isolated from soil. Bull. Environ. Contam. Toxicol. 39, 251–256.

Meijer, S.N., Ockenden, W., Sweetman, A., Breivik, K., Grimalt, J.O., Jones, K.C., 2003. Global distribution and budget of PCBs and HCB in background surface soils: implications for sources and environmental processes. Environ. Sci. Technol. 37, 667–672.

Moeckel, C., Nizzetto, L., Guardo, A.D., Steinnes, E., Freppaz, M., Filippa, G., Camporini, P., Benner, J., Jones, K.C., 2008. Persistent organic pollutants in boreal and montane soil profiles: distribution, evidence of processes and implications for global cycling. Environ. Sci. Technol. 42, 8374–8380.

Murphy, S., Barber, J.L., Learmonth, J.A., Read, F.L., Deaville, R., Perkins, M.W., Brownlow, A., Davison, N., Penrose, R., Pierce, G.J., 2015. Reproductive failure in UK harbour porpoises *Phocoena phocoena*: legacy of pollutant exposure? PLoS One 10, e0131085.

Papazi, A., Andronis, E., Ioannidis, N.E., Chaniotakis, N., Kotzabasis, K., 2012. High yields of hydrogen production induced by meta-substituted dichlorophenols biodegradation from the green alga *Scenedesmus obliquus*. PLoS One 7, e49037.

Pieper, D.H., Seeger, M., 2008. Bacterial metabolism of polychlorinated biphenyls. J. Mol. Microbiol. Biotechnol. 15, 121–138.

Pinyakong, O., Habe, H., Supaka, N., Pinpanichkarn, P., Juntongjin, K., Yoshida, T., Furihata, K., Nojiri, H., Yamane, H., Omori, T., 2000. Identification of novel metabolites in the degradation of phenanthrene by *Sphingomonas* sp. strain P2. FEMS Microbiol. Lett. 191, 115–121.

Potin, O., Rafin, C., Veignie, E., 2004. Bioremediation of an aged polycyclic aromatic hydrocarbons (PAHs)-contaminated soil by filamentous fungi isolated from the soil. Int. Biodeterior. Biodegrad. 54, 45–52.

Rajamani, S., Siripornadulsil, S., Falcao, V., Torres, M., Colepicolo, P., Sayre, R., 2007. Phycoremediation of heavy metals using transgenic microalgae. In: Transgenic Microalgae as Green Cell Factories. Springer, pp. 99–109.

Rawat, I., Kumar, R.R., Mutanda, T., Bux, F., 2011. Dual role of microalgae: phycoremediation of domestic wastewater and biomass production for sustainable biofuels production. Appl. Energy 88, 3411–3424.

Reddy, A.V.B., Moniruzzaman, M., Aminabhavi, T.M., 2019. Polychlorinated biphenyls (PCBs) in the environment: recent updates on sampling, pretreatment, cleanup technologies and their analysis. Chem. Eng. J. 358, 1186–1207.

Rosenberg, J.N., Oyler, G.A., Wilkinson, L., Betenbaugh, M.J., 2008. A green light for engineered algae: redirecting metabolism to fuel a biotechnology revolution. Curr. Opin. Biotechnol. 19, 430–436.

Rybak, A., Messyasz, B., Łeska, B., 2012. Freshwater Ulva (Chlorophyta) as a bioaccumulator of selected heavy metals (Cd, Ni and Pb) and alkaline earth metals (Ca and Mg). Chemosphere 89, 1066–1076.

Schoeny, R., Cody, T., Warshawsky, D., Radike, M., 1988. Metabolism of mutagenic polycyclic aromatic hydrocarbons by photosynthetic algal species. Mutat. Res. Fund Mol. Mech. Mutagen 197, 289–302.

Semple, K.T., Cain, R.B., Schmidt, S., 1999. Biodegradation of aromatic compounds by microalgae. FEMS Microbiol. Lett. 170, 291–300.

Sharma, J.K., Gautam, R.K., Nanekar, S.V., Weber, R., Singh, B.K., Singh, S.K., Juwarkar, A.A., 2018. Advances and perspective in bioremediation of polychlorinated biphenyl-contaminated soils. Environ. Sci. Pollut. Control Ser. 25, 16355–16375.

Shen, C., Huang, S., Wang, Z., Qiao, M., Tang, X., Yu, C., Shi, D., Zhu, Y., Shi, J., Chen, X., 2008. Identification of Ah receptor agonists in soil of e-waste recycling sites from Taizhou area in China. Environ. Sci. Technol. 42, 49–55.

Sieburth, J.M., Jensen, A., 1969. Studies on algal substances in the sea. II. The formation of Gelbstoff (humic material) by exudates of Phaeophyta. J. Exp. Mar. Biol. Ecol. 3, 275–289.

Siripornadulsil, S., Traina, S., Verma, D.P.S., Sayre, R.T., 2002. Molecular mechanisms of proline-mediated tolerance to toxic heavy metals in transgenic microalgae. Plant Cell 14, 2837–2847.

Stevens, D.R., Purton, S., 1997. Genetic engineering of eukarygtic algae: progress and prospects. J. Phycol. 33, 713–722.

Stuart-Smith, S.J., Jepson, P.D., 2017. Persistent threats need persistent counteraction: responding to PCB pollution in marine mammals. Mar. Pol. 84, 69–75.

Takahashi, H., Uchimiya, H., Hihara, Y., 2008. Difference in metabolite levels between photoautotrophic and photomixotrophic cultures of *Synechocystis* sp. PCC 6803 examined by capillary electrophoresis electrospray ionization mass spectrometry. J. Exp. Bot. 59, 3009–3018.

Tiedje, J.M., Colwell, R.K., Grossman, Y.L., Hodson, R.E., Lenski, R.E., Mack, R.N., Regal, P.J., 1989. The planned introduction of genetically engineered organisms: ecological considerations and recommendations. Ecology 70, 298–315.

Walker, T.L., Purton, S., Becker, D.K., Collet, C., 2005. Microalgae as bioreactors. Plant Cell Rep. 24, 629–641.

Williams, R.S., Curnick, D.J., Barber, J.L., Brownlow, A., Davison, N.J., Deaville, R., Perkins, M., Jobling, S., Jepson, P.D., 2020. Juvenile harbor porpoises in the UK are exposed to a more neurotoxic mixture of polychlorinated biphenyls than adults. Sci. Total Environ. 708, 134835.

Yan, X., Yang, Y., Li, Y., Sheng, G., Yan, G., 2002. Accumulation and biodegradation of anthracene by *Chlorella protothecoides* under different trophic conditions. Ying yong sheng tai xue bao/J. Appl. Ecol. 13, 145−150.

Zhang, H., Jiang, X., Lu, L., Xiao, W., 2015. Biodegradation of polychlorinated biphenyls (PCBs) by the novel identified cyanobacterium Anabaena PD-1. PLoS One 10, e0131450.

Zhao, G., Zhou, H., Wang, D., Zha, J., Xu, Y., Rao, K., Ma, M., Huang, S., Wang, Z., 2009. PBBs, PBDEs, and PCBs in foods collected from e-waste disassembly sites and daily intake by local residents. Sci. Total Environ. 407, 2565−2575.

Bacterial remediation to control pollution

16

Swati Srivastava, Sunil Kumar

Faculty of Biosciences, Institute of Biosciences and Technology, Shri Ramswaroop Memorial
University, Barabanki, Uttar Pradesh, India

Chapter outline

16.1 Introduction

Many major incidents have occurred in the past that reveal the necessity to prevent the escape of effluents into the environment, such as the Exxon Valdez oil spill, the Union Carbide (Dow) Bhopal disaster, large scale contamination of the Rhine River, the progressive deterioration of the aquatic habitats and conifer forests in the Northeastern US, Canada and parts of Europe, or the release of radioactive material in the Chernobyl accident, etc. The conventional techniques used for remediation have been to excavate a contaminated site and remove it to a landfill or to cap and close

Biological Approaches to Controlling Pollutants. https://doi.org/10.1016/B978-0-12-824316-9.00017-3

the contaminated areas of a site. The methods have some drawbacks. A better approach than these traditional methods is to completely destroy the pollutants if possible, or at least to transform them into harmless substances. Some technologies that have been used are high-temperature incineration and various types of chemical decomposition. Bioremediation is an option that offers the possibility to destroy or render harmless various contaminants using natural biological activity (Gupta and Mahapatra, 2003). As such, it uses relatively low cost, low technology techniques, which generally have a high public acceptance and can often be carried out on-site. To successfully protect the environment from different types of pollution such as soil, water, air, and heavy metals, it is required that the sources of pollution are known along with the quantity and uniqueness of the pollutants and their harmful effects. As the number and type of pollutants are unlimited, they frequently change because of the use of applied technologies and degree of urbanization, etc. The term pollutant encompasses any physical, chemical, biological, or radioactive gaseous, liquid, or solid matter that decrease the normal quality of water, soil, or air. The wide distribution of microorganisms on the biosphere is due to their impressive metabolic ability and fast and easy growth in a wide range of environmental conditions. The nutritional flexibility of microorganisms can also be exploited for the biodegradation of pollutants. This kind of process is termed bioremediation. The process involves naturally occurring bacteria and fungi or plants to degrade or detoxify substances dangerous to human health and/or the environment. The microorganisms may be native to a contaminated area or they may be isolated from elsewhere and transported to the contaminated site. From an ecological point of view, the term bioremediation involves the interactions between three factors that is substrate (pollutant), organisms, and environment. The two processes that are primarily affected by the interaction of these three factors are biodegradability and bioavailability of pollutants, along with the physiological requirements of microbes, which plays a vital role in the assessment of the possibility of bioremediation (Fig. 16.1). Biodegradability defines whether any chemical can be degraded by microbes or not, whereas bioavailability refers to the availability of a pollutant to

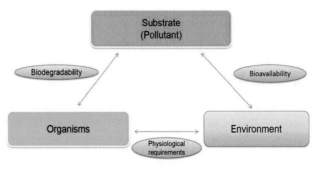

FIGURE 16.1

Bioremediation from an environmental perspective.

organisms that are capable of degrading it. As the effectiveness of the process mainly depends on the growth of microbes and their activity, its application often involves the manipulation of environmental parameters so that the growth of the microbes and their degradation occurs at a faster rate (Kumar et al., 2011).

Microorganisms need a number of nutrients such as carbon, nitrogen, and phosphorous to survive and continue their microbial activities. This can be justified with the help of several examples where we have observed that the nonavailability of nutrients has limited the degradation of pollutants. For example, the degradation of hydrocarbon is also limited if these nutrients are present in small concentrations. The biodegradation process in cold environments and the metabolic activity of microorganisms can be enhanced by the addition of an appropriate quantity of nutrients (Couto et al., 2014; Phulia et al., 2013). Biodegradation in an aquatic environment is limited by the availability of nutrients (Thavasi et al., 2011). The oil-eating microbes also require nutrients for optimal growth and development, similar to the nutritional needs of other organisms. Today, there is a need to control different types of pollution like air pollution, water pollution, soil pollution, and heavy metal pollution, which are posing a great threat to our environment. These are of special interest, and drastic changes in the implementations of various techniques to control this pollution have been taken place for a long time. The process of bioremediation has been classified into two main groups on the basis of contamination site. If the process of bioremediation is performed at the original site of contamination, it is called in situ bioremediation (ISB) as there is no excavation or removal of polluted soil/groundwater to any secondary location for conduction of the remediation process. In situ remediation includes techniques such as bioventing, biosparging, bioslurping, and phytoremediation, along with physical, chemical, and thermal processes (Fig. 16.2). *In situ* bioremediation has the potential to provide advantages such as complete obliteration of the contaminants, lower risk to site workers, and the operating cost or equipment cost is also lowered (Koning et al., 2000; Vidali, 2001). Chemotaxis is important to the study of *in situ* bioremediation because microbial organisms with chemotactic abilities can move into an area containing contaminants. So if the chemotactic abilities of the cells are enhanced, in situ bioremediation will become a safer method in degrading harmful compounds. Alternatively, *ex situ* bioremediation involves excavation of polluted soil or pumping of groundwater in a lined above-ground treatment area to make possible microbial removal of contaminants. *Ex situ* remediation includes techniques such as Land-farming, biopiling, and processing by bioreactors along with thermal, chemical, and physical processes (Koning et al., 2000) (Fig. 16.2).

In addition to the above techniques, biofiltration is an emerging technology applied to waste gas purification in order to control the volatile organic, inorganic compounds, aromatic compounds, toxic compounds, and any odorous compounds. This novel technique has been widely used in the treatment of volatile organic compounds (VOCs), the toxic air emissions from commercial processes, and moreover, it employed the use of microorganisms for the removal and oxidation of compounds from contaminated air. It has proved to be a safe, eco-friendly "green"

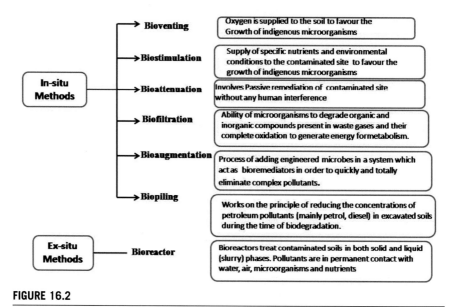

FIGURE 16.2

Types of in situ and ex situ bioremediation involving microorganisms.

technology as it is based on a process that occurs naturally in soils and water. No hazardous by-products are emitted or produced in this process, and the technique is more cost-effective and relevant for air pollution control. The two types of pollutants in the atmosphere that are mainly degraded by the microorganisms are the solid ones, entrenched in airborne particulate matter, and gaseous pollutants, such as ozone (O_3) nitrogen oxides (NOx) and VOCs (Guerreiro et al., 2013). When we talk of water pollution, it is caused mainly by the contaminants (including major nutrients such as nitrogen and phosphorus) that are transported from the soil to surface waters and groundwater, causing great damage to the environment through eutrophication and direct human health issues that are caused by contaminated drinking water. The biodiversity of soil and the services provided by its microorganisms are also affected by pollutants, as they directly harm soil microorganisms and larger soil-dwelling organisms. Both anthropogenic and natural sources contribute significantly to heavy metal pollution. Their occurrence in soil and discharge due to soil weathering is an important source of heavy metal pollution. In the past few decades, a vast range of xenobiotic compounds has been found to be prone to microbial mineralization. Although bacterial degradation capabilities in many cases have been proved to be efficient for remediation of contaminated sites, bacterial chemotaxis toward pollutants has received less attention, and it has helped us to find the role of different species of bacteria in bioremediation applications. The capacity of a compound to bring out a positive or negative chemotactic response is at times related to its nutritional properties. This can be best illustrated by an example of *E. coli* and *Pseudomonas putida*, which differ distinctly in their nutritional properties, i.e., react differently toward the presence of benzoate and salicylate in their microenvironment. These two pollutants are chemorepellents for *E. coli* (Tso and Adler, 1974). So we can say that the same compound can act as an attractant to a microorganism, and it utilizes the compound as a growth substrate, and when it acts as a repellent for another microorganism, it becomes toxic.

16.2 Bacterial remediation

Bacteria have never been thought of as man's best friend, but they are making a path in that direction by providing safe, natural, cost-effective cleanups of polluted sites with everything from fuels to heavy metals. Bacteria have survived on the planet for two-and-a-half billion years to become one of the most beneficial and deadly organisms on Earth solely because of their adaptability. They have evolved an ability to degrade most naturally occurring organic compounds and have proven equally responsive to manufactured compounds. Bacteria play a critical role in the process of bioremediation since it breaks down the dead materials into organic matter and nutrients. Several types of contaminants, such as chlorinated pesticides, etc., can be easily digested by bacteria. Bioremediation encompasses different types of treatment technologies that use bacteria as a preferable microorganism for the treatment of pollutants. Some of the basic bioremediation methods are Bio-stimulation, bioleaching, biofiltration, bioventing, biosorption, biopiling and bacterial chemotaxis (Table 16.1). While some soil cleaning procedures require the addition of new microbes, a technique called 'bio-stimulation' enhance the growth of microbes already present and increase the process of natural degradation. Natural biodegradation processes can be restricted by many factors, together with nutrient availability, temperature, or moisture content in the soil. Bio-stimulation techniques trounce these limitations, providing microbes with the resources they need, which increases their abundance and leads to an increased rate of degradation. When microorganisms are used in the process of metal extraction from ores, it is called bioleaching. It is one of several applications within biohydrometallurgy, and several methods are used to recover Cu, Zn, Pb, As, Ni, Mo, Au, Ag, and Co. Heterotrophic bacteria are widely used in the study of bacterial leaching of manganese from manganese dioxide ores, and glucose or other organic compounds are used as a source of energy, rendering their commercial utilization uneconomic (Cornu et al., 2017). Until the 1980s, intensive advancement was limited to Western Europe and the United States. Since then, biofiltration has been explored and employed for the degradation of toxic volatile chemicals and in different industrial applications using various types of filters and microorganisms. In the process of biofiltration, a biofilter or active filter is used through which the polluted air is passed, and the microorganisms used to convert the air pollutants into harmless by-products that are primarily carbon dioxide and water. In compost biofilters, activated seeded cultures can be used because microbes acclimatize easily as they are already exposed to pollutants. People think atmosphere as a nonsuitable habitat for bacteria, but recent studies

Table 16.1 Bioremediation techniques using bacteria to treat pollutants.

Techniques	Elucidation
Bio-stimulation	Nutrient elements like nitrogen and phosphorus incorporated to stimulate indigenous bacteria
Bioleaching	It involves the use of living organisms like microbes (bacteria, fungi, etc.) in the metal extraction from their ores
Biofiltration	Biofilm producing bacteria used to filter contaminants from polluted water
Bioventing	Insertion of some oxidants (oxygen, sulfate, nitrate, etc.) into the wells to stimulate resident bacteria usually sulfate oxidizing/reducing bacteria
Bio-sorption	Biosorption can remove contaminants even in dilute concentrations and has special relevance with respect to heavy metal removal. Microorganisms (live and dead) and other industrial and agriculture by-products can be used as biosorbents for the process of biosorption.
Biopiling	The technique involves forming piles with the contaminated soil and stimulating the microbial communities through aeration and/or by adding nutrients and water
Bacterial Chemotaxis	In conditions of limited carbon and energy sources, it is possible that chemotaxis might have been selected as an advantageous behavior in bacteria along with xenobiotic degradation capabilities after exposure to such compounds.

point toward the presence of diverse bacterial phyla, such as *Firmicutes*, *Actinobacteria*, *Proteobacteria*, and *Bacteroidetes*, which may form active bacterial communities in the atmosphere. Bioventing works on the principal that by emitting oxygen through the soil, growth of natural or introduced bacteria and fungus in the soil is stimulated and it has been found functional in aerobically degradable compounds. Bioventing uses low airflow rates to supply only adequate oxygen to carry on microbial activity. Bioventing has been used by many researchers as a successful bioremediation technique for the treatment of petroleum contaminated soil (Lee et al., 2006; Samuel and Latinwo, 2015). The discovery that both Gram-negative bacteria, such as *Burkholderia*, *Comamonas*, *Pseudomonas*, *Achromobacter*, *Alcaligenes*, and Gram-positive bacteria, such as *Bacillus*, *Corynebacterium*, and *Rhodococcus*, are able to degrade some polychlorinated biphenyl (PCB) congeners unlocked the door to put into service different biological technologies. The use of bacteria in biosorption has emerged as an inexpensive and competent technique to remove pollutants, together with nonbiodegradable elements, like heavy metals, from wastewater. The living and nonliving bacterial biomass have modified and developed mechanisms for metals ion resistance and remediation for their continued existence. The use of bacterial agents in the bioremediation of heavy metal ions has been widely researched. Bacterial biomass, including species like *Pseudomonas*, *Desulfovibrio*, *Bacillus*, and *Geobacter* have been used for bioremediation and bring about quick removal of metals such as Cu, Zn, Pb, Cd, and Cr. Bacterial isolates and their combined effects in the treatment of different parameters that increase the pollution level of wastewater by brewery effluents have been studied and found successful. Biopiles are mounds of contaminated soils that are kept aerated by pumping air into piles of soil through an injection system (US EPA, 2006; "Biopiles"). Biopiling (also known as biocells, bioheaps, biomounds, and compost piles) is an in situ remediation method that works on the principle of reducing the concentrations of petroleum pollutants (mainly petrol, diesel) in excavated soils during the time of biodegradation. In this process, air is supplied to the biopile system through piping and pumps that either forces air into the pile under positive pressure or draws air through the pile under negative pressure (Dellile et al., 2008). The microbial activity is enhanced through microbial respiration that result in the increased degradation of adsorbed petroleum pollutant (Emami et al., 2012). Texture of the soil influences soil permeability, water content, and soil density. Highly permeable soils are the most easily aerated and, therefore, are the most adequate to be used for biopiles. The movement of bacteria under the influence of a chemical gradient, either toward (positive chemotaxis) or away (negative chemotaxis) from the gradient helps bacteria to find optimum conditions for their growth and survival. However, the role of microorganisms with the chemotactic ability toward different xenobiotic compounds have been isolated and characterized (Bhushan et al., 2000; Grimm and Harwood, 1997; Harwood et al., 1990). Studies have shown that the chemoattractant is a compound that serves as a carbon and energy source, whereas a chemorepellent is lethal for the bacteria. It is necessary for successful biodegradation that bacteria can access the target compounds. They can do so either by suspension of the target compounds in the aqueous phase or by adhesion of the bacteria directly to the NAPL water interface. Chemotaxis has been shown to play an important role in biofilm formation in several microorganisms (O'Toole and Kolter, 1998; Pratt and Kolter, 1998; Prigent-Combaret et al., 1999; Watnick and Kolter, 1999) that may guide a bacterium to swim toward nutrients (hydrophobic pollutants) adsorbed to a surface, followed by attachment using its flagella.

In addition to abovementioned techniques of bacterial remediation, the potential of genetically engineered microorganisms (GEMs) is used nowadays for bioremediation applications in soil, groundwater, and activated sludge environments, as they exhibit improved degradative capacity encompassing a wide array of chemical contaminants. A single chromosome may carry all the essential genes of bacteria, but genes specifying enzymes that are necessary for the catabolism of some of these remarkable substrates may be carried on plasmids. Therefore GEMs can be used successfully for biodegradation and leads to represent/point toward a research frontier with broad implication in the upcoming time (Kulshrestha, 2013). The different shapes of bacteria include rods (*Bacillus*), cocci (*Streptococcus*), filamentous (*Actinomyces*) and spiral (*Vibrio cholera*) which are known to be present in the environment. Bioremediation technologies using the degradation capacity of bacteria have been proved to be an environmentally inexpensive alternative approach to physicochemical processes to eradicate diffusive contamination of persistent organic pollutants (POPs) in various environmental milieu, e.g., soil, sediments, and sludges (Fig. 16.3).

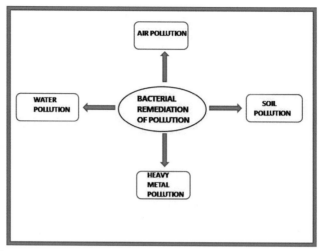

FIGURE 16.3

Types of pollution undergoing bacterial remediation.

16.3 Types of pollutants subjected for bacterial remediation

The various pollution causing agents known around the globe are fertilizers, garbage, oil spills, toxic chemicals, and sewage disposals that are likely to cause soil, water, air, and heavy metal pollution. Soil and water resources are being polluted badly with different organic and inorganic pollutants, which has caused a decline of both soil and food quality. The pollutants, including polycyclic aromatic hydrocarbons (PAHs), pesticides, PCB, explosives, metals, metalloids, and radionuclides, are well reported in soils (Zhu and Shaw, 2000; Kumar et al., 2012; Testiati et al., 2013; Vane et al., 2014). These pollutants from soil are added into underground and surface water by leaching and runoff, respectively, thus polluting water bodies. Their presence in the soil and water resources is undesirable due to their toxic nature and the anomalies they create. For example, when hydrocarbons and metals are present in the soil it adversely affects seed germination, plant growth (Smith et al., 2006; Ahmad et al., 2014), soil microbial population and activities (Guo et al., 2012; Alrumman et al., 2015) In order to look at the suitability of microbial treatment at a particular contaminated site, cities should conduct a thorough site assessment to establish if conditions necessary for success are present. The outcome of the assessment should include a depiction of the facility; identification of the contaminants; determination of the level of contamination; determination of physical/chemical properties of the contaminants; a description of the chemical and biochemical processes in the groundwater in the nearby vicinity of the site; and a systematic description of the site hydrogeology. The various pollutants that cause soil and water pollution are classified into organic and inorganic pollutants. In addition, various pollutants causing air pollution, marine pollution, and bacterial remediation has been discussed here in detail.

16.3.1 **Bacterial remediation of organic pollutants**

During recent years several bacterial strains belonging to different genera have been reported to degrade a variety of toxic organic compounds, including pesticides, synthetic dyes, PCBs, PAHs etc. under laboratory conditions (Glick, 2010; Hussain et al., 2013; Imran et al., 2015). Despite that, several bacterial strains have shown their potential to degrade a variety of organic compounds under laboratory conditions, such transformations have repeatedly been found difficult to be achieved in the environment (Glick, 2010). Another group of pollutants that are highly soluble in lipids and least soluble in the water are persistent in the environment are called POPs. POPs can diffuse and move to very long distances in the atmospheric layers and settle down. These produce an adverse effect on both human health and the related environment (Farshid and Mahsa, 2017; Rajmohan et al., 2018). The POPs present in the environment are categorized into two main groups: (a) intentionally released POPs and (b) unintentionally released POPs. The POPs that are products of chemical reactions, specifically the compounds that include a chlorine atom in the reaction, are known as intentionally released POPs. One of the known examples is an organochlorine compound which is also lipid soluble. Some groups of POP are also found to be neurotoxic in nature. The second category of POPs includes those which are the by-product of chlorine compounds produced during the combustion process. The pollutants coming under this category are furans, dioxin, and PAHs. POPs can be further classified into two types based on their application: organochlorine pesticides and industrial chemicals. Organochlorine pesticides include hexachlorobenzene, dichlorodiphenyltrichloroethane (DDT), heptachlor, chlordane, endrin, dieldrin, and aldrin, etc. Industrial chemicals mainly include polychlorinated biphenyls (PCBs) (El-Shahawi, 2010; Gaur and Narasimhulu, 2018). These chemicals are not used recurrently because of their harmful impact on the environment. Microorganisms, plants, and some lower eukaryotes have shown potential biodegradation activity for organic pollutants. The strategy used by different types of bacteria alone or in combination with plants and other microorganisms in the degradation of organic pollutants is described here in detail and listed in Table 16.2.

16.3.1.1 Pesticides

Monocrotophos is an organophosphate insecticide, which is a type of pesticide. These kinds of pesticides are known to be neurotoxins, which affect the work of neurons in the body. Bacterial degradation of monocrotophos pesticides through different species include *Arthrobacter atrocyaneus, Bacillus megaterium* and *Pseudomonas mendocina*, and they showed 80%−90% degradation of monocrotophos at a maximum initial concentration of 500 mg/L in a synthetic medium within 48 h. Moreover, one of the most highly toxic organochlorine pesticides is hexachlorocyclohexane (HCH), which is known to have been released into the environment in the last few decades. Many Gram-negative bacteria have been reported to have metabolic abilities to attack HCH. For example, several *Sphingomonas* strains have been reported to degrade the pesticide. The most widespread mechanism for

Table 16.2 Bacteria having biodegradation potential for different organic pollutants (Chatterjee et al., 2008).

Pollutants	Bacterial species	References
Benzene, anthracene, hydrocarbons, PCBs	*Pseudomonas* spp.	Kapley et al. (1999) and Cybulski et al. (2003)
Aromatics, long-chain alkanes, phenol, cresol	*Bacillus* spp.	Cybulski et al. (2003)
Halogenated hydrocarbons, phenoxyacetate	*Corynebacterium* spp.	Jogdand (1995)
Phenoxyacetate, halogenated hydrocarbon diazinon	*Streptomyces* spp.	Jogdand (1995)
Naphthalene, biphenyl	*Rhodococcus* spp.	Dean-Ross et al. (2002)
Aromatics, branched hydrocarbons benzene, cycloparaffins	*Mycobacterium* spp.	Sunggyu (1995)

pesticide biodegradation is performed to some extent through the metabolization of pesticides by microbial cultures that is done either by utilization of the compounds as a source of energy or nutrients or by cometabolism with other substrates supporting microbial growth (De Schrijver and de Mot, 1999). The different strains of HCH-bioremediating aerobic bacteria that are isolated or discovered in various countries are *Micromonospora* sp. and several *Streptomyces* spp. strains (Argentina), *Bacillus* sp. and *Pseudomonas* sp. (Canada); *Sphingobium* sp. MI1205 (China); *Rhodanobacter lindaniclasticus* RP555 and *Sphingobium francense* sp. (France).

16.3.1.2 Hydrophobic toxic environmental pollutants
These pollutants include PAHs and PCBs:

➤ *Polycyclic Aromatic Hydrocarbons:* PAHs are a group of hydrophobic compounds that contain two or more aromatic benzene rings in their molecule and are considered among the most recalcitrant petroleum compounds. To overcome the hazardous effects of organic compounds belonging to this group, several bacteria have been isolated for biodegradation of PAHs (Chaudhry et al., 2005). However, it has been observed that PAHs having only low and intermediate molecular weight are metabolized by several bacterial strains but require some readily available carbon source for biodegradation of high molecular weight PAHs (Bouchez et al., 1999; Chaudhry et al., 2005). Various bacteria have been found to degrade PAHs, in which degradation of naphthalene and phenanthrene has been most widely studied. Principally, bacteria favor aerobic conditions for degradation of PAHs via oxygenase-mediated metabolism (involving either monooxygenase or dioxygenase enzymes). Usually, the first step in the aerobic bacterial degradation of PAHs is the hydroxylation of an aromatic ring via a dioxygenase, with the formation of a *cis*-dihydrodiol, which gets rearomatized to a diol intermediate by the action of a dehydrogenase. The intradiol or extradiol ring-cleaving dioxygenases then cleave the diol intermediates through either an ortho-cleavage or metacleavage pathway, leading to intermediates such as catechols that are ultimately converted to TCA cycle intermediates (Mueller et al., 1996; Mallick et al., 2011). Dioxygenase is a multicomponent enzyme generally consisting of reductase, ferredoxin, and terminal oxygenase subunits (Mallick et al., 2011). The alternative methods through which bacteria can degrade PAH *include* the cytochrome P450-mediated pathway, with the production of *trans*-dihydrodiols (Moody et al., 2004) or under anaerobic conditions, mainly under nitrate-reducing conditions (Carmona et al., 2009).

➢ *Polychlorinated biphenyls (PCBs):* Being at the crossing point between the cell and the environment, the cytoplasmic membrane is the first site of contact between the cell and the contaminant. The primary effect of hydrophobic organic pollutants is observed as a change in fluidity of bacterial membrane that can lead to a significant disturbance of physiological function and apoptosis. Although many bacteria have the ability to metabolize, e.g., PCBs, high concentrations of these chemicals act as an environmental stress factor and inhibit cell survival and then ability to metabolize these pollutants. One of the observed membrane adaptation mechanisms is the increase of saturation of bacterial membrane lipids. The two supportive bioremediation strategies that are usually applied, when natural attenuation is not fast enough are biostimulation and bioaugmentation. PCBs are usually subjected to both aerobic and anaerobic metabolism of bacteria. It is generally known that under aerobic conditions, biphenyl dioxygenase attacks biphenyl core and transforms PCB congeners into the respective chlorobenzoate and a pentanoic acid derivative. Under anaerobic conditions, PCB congeners are subjected to reductive dechlorination resulting in the intact biphenyl and some lower chlorinated PCB congeners. Both metabolic pathways are effective only when the environmental conditions are optimal for the indigenous or introduced bacteria. For example, a naturally occurring soil bacterium *Agrobacterium tumefaciens* has been reported not only to induce root proliferation but also to improve the ability of the plant to absorb greater amounts of PCBs and other pollutants from soil and groundwater along with an increase in uptake of nutrients (Stomp et al., 1994; Chaudhry et al., 2005).

16.3.1.3 Explosives

One of the known bacterial species capable of degrading different types of explosives is a pink-pigmented symbiotic bacterium that was isolated from hybrid poplar tissues (*Populus deltoides* × nigra DN34). The bacterium was identified by 16S and 16S−23S intergenic spacer ribosomal DNA analysis as a *Methylobacterium* sp. (strain BJ001). The use of methanol as the sole source of carbon and energy is a specific characteristic of the genus *Methylobacterium*. The bacterium in pure culture was shown to degrade the toxic explosives 2,4,6-trinitrotoluene (TNT), hexahydro-1,3,5-trinitro-1,3,5-triazine (RDX), and octahydro-1,3,5,7-tetranitro-1,3,5-tetrazocine (HMX). The distribution of the genus *Methylobacterium* in a wide diversity of natural environments and its association with plants promote the process of either natural attenuation or in situ biodegradation (including phytoremediation) of explosive-contaminated sites (Ahmad et al., 2016). Several indigenous strains that have the potential to degrade TNT and PETN were identified as *Planomicrobacterium flavidum*, *Pseudomonas aeruginosa*, *Enterobacter asburiae*, *Azospirillum*, *Rhizobium*, *Methylobacterium*, and *Pseudomonas denitrificans* strains.

16.3.1.4 Volatile organic compounds

The most used microorganisms for VOCs are heterotrophs and for the treatment of inorganic influent gases, are chemoautotrophs. The use of activated seeded cultures in compost biofilters is done because microbes are already adapted to pollutants, and moreover they take less time for acclimatization. Sole inorganic media can also be seeded with activated sludge cultures. The treatment of complex waste is performed by seeding with specially cultured microbial strains. In biofiltration, the polluted air is passed through a biologically active filter or biofilter. Microorganisms in the biofilter convert the air pollutants into harmless by-products that are primarily carbon dioxide and water.

16.3.2 Bacterial remediation of inorganic pollutants

A number of synthetic chemicals are released into the environment either intentionally or accidentally through soil and water bodies every day. Some of them are essential to having a specific function while most of them are challenging owing to their long persistence and resistance to degradation. This group mainly encompasses heavy metals, metalloids, and radionuclides. These chemicals are linked with these processes (like mining, fossil fuel burning, municipal solid waste, industrial effluent, and phosphate fertilizers application) through which they enter into the food chain. In addition to this, these chemicals are known to suppress plant growth. Plant-bacteria interaction is useful as in situ remediation technique because bacteria either tolerate high concentrations of metals leading to their bioaccumulation or improved plant growth biomass in contaminated soil or ease the process of metal uptake (Gadd, 2010; Glick et al., 2014; Ahmad et al., 2016). The bacterial remediation of different types of inorganic pollutants is described here in detail.

16.3.2.1 Heavy metals

Heavy metals constitute the group of pollutants that have been found to affect cellular fractions and organelles. With increasing knowledge about the persistence, character, and toxic effects of heavy metals, new technologies are developed to remediate this contamination. The bioaccumulation of heavy metals in crops and associated food chains has increased due to the application of heavy metal-contaminated water in agricultural fields. Since heavy metals cannot be destroyed biologically, they are only transformed from one oxidation state to another. In addition, bacteria are also competent in heavy metals bioremediation. The various mechanisms developed by microorganisms to protect themselves from heavy metal

Table 16.3 Biosorption of heavy metals (inorganic pollutant) by bacteria.

Metals as pollutants	Bacterial species	References
Cr	*Bacillus cereus*	Kanmani et al. (2012)
Cd and Pb	*Burkholderia species*	Jiang et al. (2008)
Cu	*Kocuria flava*	Achal et al. (2011)
Cd, Zn and Cu	*Pseudomonas veronii*	Vullo et al. (2008)
As	*Sporosarcina ginsengisoli*	Achal et al. (2012)
Au	*Stenotrophomonas* spp.	Song et al. (2008)
Cu	*Arthrobacter* sp.	Hasan and Srivastava (2009)
Cd and Cu	*Bacillus laterosporus*	Zouboulis (2004)
Cr, Ni	*Bacillus licheniformis*	Zhou et al. (2007)
As, Cd, Cr, Co, Cu, Hg, Ni, Pb	*Pseudomonas fluorescens*	Lopez et al. (2000)
Cd, Cu, Pb, Zn	*Pseudomonas putida*	Pardo et al. (2003)

toxicity include adsorption, uptake, methylation, oxidation, and reduction. Moreover, microorganisms take up heavy metals actively by the process of bioaccumulation and/or passively by adsorption. Many bacteria have been reported to use the process of microbial methylation as one of the key tools in heavy metals bioremediation, as methylated compounds are often volatile. For example, Mercury, Hg (II) can be biomethylated by a number of different bacterial species *Alcaligenes faecalis, Bacillus pumilus, Bacillus* sp., *P. aeruginosa* and *Brevibacterium iodinum* to gaseous methyl mercury (Jaysankar et al., 2008). The various bacteria that are strategically used in bioremediation treatments for heavy metals include the following: *Arthrobacter* spp. (Roane et al., 2001), *Pseudomonas veronii* (Vullo et al., 2008), *Burkholderia* spp. (Jiang et al., 2008), *Kocuria flava* (Achal et al., 2011), *Bacillus cereus* (Kanmani et al., 2012), and *Sporosarcina ginsengisoli* (Achal et al., 2012). The bioremediation strategy is based on the high metal-binding ability of biological agents, which can remove heavy metals from polluted sites with high efficiency (Table 16.3).

16.3.2.2 Metalloids

Bacteria in association with plants are used in the degradation of different metalloids. Bacteria associated with the roots of *Cirsium arvense* were able to reduce the toxicity of As because these bacteria show ACC deaminase, IAA, and siderophore production that promotes plant growth in As contaminated soil (Cavalca et al., 2010). In another study, the production of biomass of sunflower grown in As-contaminated soil is increased with the help of bacterial strain isolated from root nodules of a leguminous plant (Reichman, 2007). Shagol et al. (2014) isolated or identified six genera *Brevibacterium, Paenibacillus, Pseudomonas, Microbacterium, Rhodococcus,* and *Rahnella*, which have the potential to tolerate high concentrations of arsenite [As(III)] and arsenate [As(V)]. From the abovementioned bacterial strains, most of the strains showed reduction of Ass(V) whereas only one strain, *Pseudomonas koreensis* JS123 showed oxidation and reduction of As. Certain bacteria residing in the rhizosphere show a detoxification mechanism as they convert some metals/metalloids (Hg, Te, Tn, Se, As, Pb, Sb) into volatile form. Rhizobacteria are involved in Se accumulation in plants and its subsequent volatilization into the atmosphere (De Souza et al., 1999). Similarly, degradation of methyl mercury and reduction of Hg^{+2} into Hg0 is the mediated reactions of rhizobacteria assisted plant volatilization (Barkay et al., 2003).

16.3.2.3 Radionuclides

The sources that release radionuclides into the environment include the nuclear power plants, waste disposal sites of nuclear power plants, accidental release (U, Pu) and as fission product (Cs, Sr). Similar to other metals, their presence also threatens life on the planet. Fortunately, nature has blessed this planet with such plants and microbes, which have the ability to transform, solubilize, and accumulate radionuclides (Francis, 2006). Moreover, sulfur- and iron-oxidizing bacteria amplify the dissolution of U because these bacteria acidify the medium by producing sulfuric acid and oxidizing agent ferric sulfate. Biosorption is another mechanism through which some bacteria show the ability to accumulate U on their surfaces. It occurs actually due to the affiliation of U with carboxyl, hydroxyl, and phosphate functional groups present in the cell wall of bacteria. Another bacterial species *Clostridium* mediate the reduction of U(VI) into U(IV) (Francis, 2006).

16.3.3 Bacterial remediation of perchlorate

The chemical, called perchlorate, comes from rocket fuel, munitions, and fireworks. It is dangerous to humans because it can impair thyroid function. Its adverse effects have also been observed on the functioning of the thyroid gland in freshwater animals like fish and amphibians and alteration of gonad development in some animals. But certain bacteria, including several species of Dechloromonas and Azospira, have evolved to use perchlorate to make energy-storing molecules. In the process, they break it down into harmless chloride and oxygen. Contaminated ground or surface water gets pumped into the bioreactors, which are full of these bacteria. Once the bacteria have broken down the perchlorate, the water is filtered and sterilized to remove bacteria. The decontaminated water, which is suitable for drinking, can then be sent to consumers or pumped back into the ground.

16.3.4 Bacterial remediation of xenobiotics and aromatic compounds as pollutants

The movement of bacteria to different niches that are optimum in terms of oxygen concentration is possible with the help of aerotaxis (Shoi et al., 1998; Taylor, 1983). This phenomenon could play a significant role in oxidative biodegradation of diverse xenobiotics since the molecular oxygen for their activity requires different oxygenase in degradation pathways. Whereas anaerobic/microaerophilic microorganisms move deeper into the soils and sediments (anaerobic/microaerophilic conditions) and degrade toxic contaminants. Chemotaxis has been selected as an advantageous behavior in bacteria in conditions of limited carbon and energy sources. The different soil bacteria such as *Rhizobium* sp. (Dharmatilake and Bauer, 1992), *Bradyrhizobium* sp. (Parke et al., 1985), *Pseudomonas* sp. (Harwood et al., 1984), and *Azospirillum* sp. (Lopez-de-Victoria and Lowell, 1993) have been shown to be chemotactically attracted toward different aromatic hydrocarbons. Many of these compounds are present in soils, sediments, and rhizosphere, and they serve as the basis of carbon and energy for microorganisms. Aromatic acids such as benzoate, *p*-hydroxybenzoate (PHB), methylbenzoates, the *m*-, *p*-, and *o*-toluates, salicylate, DL-mandelate, phenylpyruvate, and benzoylformate have been reported to be attractants for *Pseudomonas putida* (Harwood et al., 1984).

16.3.5 Bacterial remediation of brewery effluents and pollutants in wastewater

The property of microbes to adapt more quickly to an environment has been used efficiently to break down organic materials contained in the waste (Komarawidjaja, 2007). The potential of different bacterial isolates was assessed in the treatment of brewery effluents, and the removal of pollutants was confirmed by conducting a germination test on seeds of beetroot (*Beta vulgaris*). *Pseudomonas* is a common bacterium proficient in degrading pollutants (Anggraeni et al., 2014) although for the treatment of wastewater and other isolates, it is not found effective. The maximum pollutant removal occurred in brewery effluents inoculated with pooled bacterial isolates for all parameters indicating their synergistic effect on the degradation of wastes (Oljira et al., 2018). Different studies have used bacterial isolates in the treatment of wastewater. The treatment of lipid-rich wastewater was done using mixed bacterial culture comprising *Pseudomonas aeruginosa*, *Bacillus* sp., and *Acinetobacter calcoaceticus* (Mongkolthanaruk and Dharmsthiti, 2002). Similarly, (Surti,2016) has reported that bacterial strains of *Pseudomonas aeruginosa*, *Bacillus subtilis*, *Enterobacter aerogenes,* and mixed cultures of these three bacterial strains are used for COD reduction of wastewater from the pharmaceutical industry.

16.4 Future prospects of bacterial remediation of pollutants

Pollution is hazardous to our health and spoils the environment, disturbing wildlife and the sustainability of our planet. Bioremediation can assist in reducing and eradicating the pollution we produce so that we can get clean water, clean air, and healthy soils for future generations. Bioremediation is not a universal remedy but rather a "natural process" alternative to traditional physicochemical and chemical treatment methods. The presence of an active microbial degrader population differentiates bioremediation from other remediation techniques. Cleaning up oil-polluted soil is an example where stimulating microbial growth can be used to remediate soil pollution. Research has shown that the natural growth rate of oil-degrading bacteria can be stimulated with the help of poultry droppings that provides nitrogen and phosphorous to the system. These methods may prove cheaper and more environmentally friendly than current chemical treatment options. The most important and largest of the bioremediation project are the sewage treatment plants. In the United Kingdom, 11 billion liters of wastewater are collected and treated every day. Bioremediation can be customized to the needs of the polluted site in problem, and the specific microbes needed to break down the pollutant are encouraged by opting for the limiting factor needed to enhance their growth. The efforts must be concentrated toward faster and economic restoration of pesticide-contaminated soil to protect the soil quality and food and health of mankind. Future research should examine the potential of recombinant and indigenous bacteria, with and without supplements, in the bioremediation of pesticides and different types of pollutants that are posing serious threats to our environment. Though we have seen that different strains of bacteria use different mechanisms to degrade the pollutants, there still is a need to develop and focus on different methods of bioremediation and how we can effectively use them to make our environment pollutant free.

16.5 Conclusion

Microorganisms are restoring the original natural surroundings and preventing further pollution. Today, successful environmental decontamination is done by using biological approaches that mainly involve bacterial strains that can degrade one or more contaminants. Moreover, for efficient functioning, such strains must be able to live and acclimatize to an unpleasant environment. The role of potential bacterial isolates and their mixed cultures as and effective treatment method for pollutant removal has been studied widely. The contamination of environmental sections like soil and river by discharge from industrial wastewater requires urgent involvement of governmental agencies as well as regular monitoring and remediation using suitable methods. Due to certain limitations of conventional treatment methods, they should be replaced by efficient, cost-effective, and eco-friendly techniques such as bioremediation which make use of different microbial population. Despite which aspect of bioremediation is used, this technology offers an efficient and cost-effective way to treat contaminated sites such as groundwater and soil. Its advantages generally prevail over the disadvantages, which is marked by the number of sites that choose to use this technology and its growing popularity. Even though bioremediation of soil has been established as a very useful, cost-efficient technology, the use of airborne microbes for the bioremediation of polluted air has barely been studied. Many studies have shown the resistance of bacterial species to high concentrations of air pollutants. The atmosphere possesses a varied quality of microorganisms, and we therefore expect a wide range of biodegradation mechanisms possessed by different microorganisms for the dissimilar air pollutants.

Coordinated regulation of bacterial chemotaxis toward almost all toxic compounds and their respective mineralization and/or transformation indicate that this phenomenon might be an essential feature of degradation. The degradation of environmental pollutants such as naphthalene, toluene, and nitroaromatic compounds have proved the role of chemotaxis and its role in the metabolism of chemoattractants. Lastly, the bioremediation systems can be improved with the help of a technique called biomolecular engineering, where we can make use of the capabilities of the bacteria or enzymes for degrading different pollutants, which will always be in favor of the environment.

References

Achal, V., Pan, X., Fu, Q., Zhang, D., 2012. Biomineralization based remediation of as (III) contaminated soil by *Sporosarcina ginsengisoli*. J. Hazard Mater. 201–202, 178–184.

Achal, V., Pan, X., Zhang, D., 2011. Remediation of copper-contaminated soil by *Kocuria flava* CR1, based on microbially induced calcite precipitation. Ecol. Eng. 37 (10), 1601–1605.

Ahmad, I., Akhtar, M.J., Zahir, Z.A., Naveed, M., Mitter, B., Sessitsch, A., 2014. Cadmium-tolerant bacteria induce metal stress tolerance in cereals. Environ. Sci. Pollut. Res. 21, 11054–11065.

Ahmad, I., Akhtar, M.J., Asghar, H.N., Ghafoor, U., Shahid, M., 2016. Differential effects of plant growth-promoting rhizobacteria on maize growth and cadmium uptake. J. Plant Growth Regul. 35, 303–315.

Alrumman, S.A., Standing, D.B., Paton, G.I., 2015. Effects of hydrocarbon contamination on soil microbial community and enzyme activity. J. King Saud Univ. Sci. 27, 31–41.

Anggraeni, P., Gunam, I., Kawuri, R., 2014. Potential bacterial consortium to increase the effectiveness of beer wastewater treatment. Curr. World Environ. 9, 312–320.

Barkay, T., Miller, S.M., Summers, A.O., 2003. Bacterial mercury resistance from atoms to ecosystems. Fed. Eur. Microbiol. Soc. Microbiol. Rev. 27, 355–384.

Bhushan, B., Samanta, S.K., Chauhan, A., Chakraborti, A.K., Jain, R.K., 2000. Chemotaxis and degradation of 3-methyl-4-nitrophenol by *Ralstonia* sp. SJ98. Biochem. Biophys. Res. Commun. 275, 129–133.

Bouchez, M., Blanchet, D., Bardin, V., Haeseler, F., Vandecasteele, J.P., 1999. Efficiency of defined strains and of soil consortia in the biodegradation of polycyclic aromatic hydrocarbon (PAH): mixtures. Biodegradation 10, 429–435.

Carmona, M., Zamarro, M.T., Blazquez, B., Durante-Rodriguez, G., Juarez, J.F., Valderrama, J.A., 2009. Anaerobic catabolism of aromatic compounds: a genetic and genomic view. Microbiol. Mol. Biol. Rev. 73, 71–133. https://doi.org/10.1128/MMBR.00021-08.

Cavalca, L., Zanchi, R., Corsini, A., Colombo, M., Romagnoli, C., Canzi, E., Andreoni, V., 2010. Arsenic-resistant bacteria associated with roots of the wild *Cirsium arvense* (L.) plant from an arsenic polluted soil, and screening of potential plant growth-promoting characteristics. Syst. Appl. Microbiol. 33, 154–164.

Chatterjee, S., Chattopadhyay, P., Roy, S., Sen, S.K., 2008. Bioremediation: a tool for cleaning polluted environments. J. Appl. Biosci. 11, 594–601.

Chaudhry, Q., Blom-Zandstra, M., Gupta, S., Joner, E.J., 2005. Utilizing the synergy between plants and rhizosphere microorganisms to enhance breakdown of organic pollutants in the environment. Environ. Sci. Pollut. Res. 12, 34−48.

Cornu, J.Y., Huguenot, D., Jezequel, K., Lollier, M., Lebeau, T., 2017. Bioremediation of copper-contaminated soils by bacteria. World J. Microbiol. Biotechnol. 33, 26.

Couto, N., Fritt-Rasmussen, J., Jensen, P.E., Højrup, M., Rodrigo, A.P., 2014. Suitability of oil bioremediation in an Artic soil using surplus heating from anincineration facility. Environ. Sci. Pollut. Res. 21, 6221−6227. Link: https://goo.gl/eXKYxm.

Cybulski, Z., Dzuirla, E., Kaczorek, E., Olszanowski, A., 2003. The influence of emulsifiers on hydrocarbon biodegradation by *Pseudomonadacea* and *Bacillacea* strains. Spill Sci. Technol. Bull. 8, 503−507.

De Schrijver, A., de Mot, R., 1999. Degradation of pesticides by actinomycetes. Crit. Rev. Microbiol. 25, 85−119.

De Souza, M.P., Huang, C.P.A., Chee, N., Terry, N., 1999. Rhizosphere bacteria enhance the accumulation of selenium and mercury in wetland plants. Planta 209, 259−263.

Dean-Ross, D., Moody, J., Cerniglia, C.E., 2002. Utilization of mixtures of polycyclic aromatic hydrocarbonsby bacteria isolated from contaminated sediment. FEMS Microbiol. Ecol. 41, 1−7.

Delille, D., Duval, A., Pelletier, E., 2008. Highly efficient pilot biopiles for on-site fertilization treatment of diesel oil-contaminated sub-Antarctic soil. Cold Reg. Sci. Technol. 54, 7−18. Link: https://goo.gl/Zrf9FT.

Dharmatilake, A.J., Bauer, W.D., 1992. Chemotaxis of *Rhizobium meliloti* towards nodulation gene-inducible compounds from alfalfa roots. Appl. Environ. Microbiol. 58, 1153−1158.

El-Shahawi, M., Hamza, A., Bashammakh, A., 2010. An overview on the accumulation, distribution, transformations, toxicity and analytical methods for the monitoring of persistent organic pollutants. Talanta 80, 1587−1597.

Emami, S., Pourbabaee, A.A., Alikhani, H.A., 2012. Bioremediation principles and techniques on petroleum hydrocarbon contaminated soil. Technical J. Eng. Appl. Sci. 2, 320−323. Link: https://goo.gl/2iCBqU.

Farshid, G., Mahsa, M., 2017. Application of peroxymonosulfate and its activation methods for degradation of environmental organic pollutants: review. Chem. Eng. J. 310, 41−62.

Francis, A.J., 2006. Microbial transformations of radionuclides released from nuclear fuel reprocessing plants. In: International Symposium on Environmental Modeling and Radioecology, Rakkasho, Aomori, Japan October 18−20.

Gadd, G.M., 2010. Metals: minerals and microbes: geomicrobiology and bioremediation. Microbiology 156, 609−643.

Gaur, N., Narasimhulu, K.Y.P., 2018. Recent advances in the bio-remediation of persistent organic pollutants and its effect on environment. J. Clean. Prod. 198, 1602−1631.

Glick, B.R., 2010. Using soil bacteria to facilitate phytoremediation. Biotechnol. Adv. 28, 367−374.

Glick, B.R., 2014. Bacteria with ACC deaminase can promote plant growth and help to feed the world. Microbiol. Res. 169, 30−39.

Grimm, A.C., Harwood, C.S., 1997. Chemotaxis of *Pseudomonas* sp. to the polycyclic aromatic hydrocarbon, naphthalene. Appl. Environ. Microbiol. 63, 4111−4115.

Guerreiro, C.F., de Leeuw, V., Foltescu, J., Schilling, J., Aardenne, van, 2013. Air Quality in Europe − 2012 Report. European Environment Agency.

Guo, H., Yao, J., Cai, M., Qian, Y., Guo, Y., Richnow, H.H., Blake, R.E., Doni, S., Ceccanti, B., 2012. Effects of petroleum contamination on soil microbial numbers, metabolic activity and urease activity. Chemosphere 87, 1273−1280.

Gupta, R., Mahapatra, H., 2003. Microbial biomass:An economical alternative for removal of heavy metals from waste water. Indian J. Exp. Biol. 41, 945−966.

Harwood, C.S., Rivelli, M., Ornston, L.N., 1984. Aromatic acids are chemoattractants for *Pseudomonas putida*. J. Bacteriol. 160, 622−628.

Harwood, C.S., Parales, R.E., Dispensa, M., 1990. Chemotaxis of *Pseudomonas putida* toward chlorinated benzoates. Appl. Environ. Microbiol. 42, 263−287.

Hasan, S.H., Srivastava, P., August 2009. Batch and continuous biosorption of Cu(2+) by immobilized biomass of *Arthrobacter* sp. J. Environ. Manag. 90 (11), 3313−3321. Available from: https://doi.org/10.1016/j.jenvman.2009.05.005.

Hussain, S., Maqbool, Z., Ali, S., Yasmin, T., Imran, M., Mehmood, F., Abbas, F., 2013. Biodecolorization of Reactive Black-5 by a metal and salt tolerant bacterial strain *Pseudomonas* sp. RA20 isolated from Paharang Drain effluents in Pakistan. Ecotoxicol. Environ. Saf. 98, 331−338.

Imran, M., Arshad, M., Khalid, A., Hussain, S., Mumtaz, M.W., Crowley, D.E., 2015. Decolorization of reactive black-5 by *Shewanella* sp. in the presence of metal ions and salts. Water Environ. Res. 87, 579−586.

Jaysankar, D., Ramaiah, N., Vardanyan, L., 2008. Detoxification of toxic heavy metals by marine bacteria highly resistant to mercury. Mar. Biotechnol. 10, 471−477. Link: https://goo.gl/JBfhJp.

Jiang, C.Y., Sheng, X.F., Qian, M., Wang, Q.Y., 2008. Isolation and characterization of heavy metal resistant *Burkholderia* species from heavy metal contaminated paddy field soil and its potential in promoting plant growth and heavy metal accumulation in metal polluted soil. Chemosphere 72, 157−164.

Jogdand, S.N., 1995. Environ Biotechnol, first ed. Himalaya Publishing House, Bombay, pp. 104−120.

Kanmani, P., Aravind, J., Preston, D., 2012. Remediation of chromium contaminants using bacteria. Int. J. Environ. Sci. Technol. 9, 183−193.

Kapley, A., Purohit, H.J., Chhatre, S., Shanker, R., Chakrabarti, T., 1999. Osmotolerance and hydrocarbon degradation by a genetically engineered microbial consortium. Bioresour. Technol. 67, 241−245.

Komarawidjaja, W., 2007. Aerobic microbial role in textile wastewater treatment. J. Environ. Eng. 8 (3), 223−228 (Indonesian).

Koning, M., Hupe, K., Stegmann, R., 2000. Thermal processes, scrubbing/extraction, bioremediation and disposal, 11b, 306−317.

Kulshreshtha, S., 2013. Genetically engineered microorganisms: a problem solving approach for bioremediation. J. Biorem. Biodegrad. 4, 1−2. Link: https://goo.gl/JkdMBV.

Kumar, A., Bisht, B.S., Joshi, V.D., Dhewa, T., 2011. Review on bioremediation of polluted environment: a management tool. Int. J. Environ. Sci. 1, 1079−1093. Link: https://goo.gl/P6Xeqc.

Kumar, B., Gaur, R., Goel, G., Mishra, M., Singh, S.K., Prakash, D., Sharma, C.S., 2012. Residues of pesticides and herbicides in soils from agriculture areas of Delhi region, India. Elec. J. Env. Agricult. Food Chem. 11, 328−338.

Lee, T.H., Byun, I.G., Kim, Y.O., Hwang, I.S., Park, T.J., 2006. Monitoring biodegradation of diesel fuel in bioventing processes using in situ respiration rate. Water Sci. Technol. 53, 263−272. Link: https://goo.gl/6Sn481.

Lopez-de-Victoria, G., Lowell, C.R., 1993. Chemotaxis of *Azospirillum* species to aromatic compounds. Appl. Environ. Microbiol. 59, 2951−2955.

Lopez, A., Lazaro, N., Priego, J., Marques, A., 2000. Effect of pH on the biosorption of nickel and other heavy metals by *Pseudomonas* fluorescens 4F39. J. Ind. Microbiol. Biotechnol. 24 (2), 146−151. Available from: https://doi.org/10.1038/sj.jim.2900793.

Mallick, S., Chakraborty, J., Dutta, T.K., 2011. Role of oxygenases in guiding diverse metabolic pathways in the bacterial degradation of low-molecular- weight polycyclic aromatic hydrocarbons: a review. Crit. Rev. Microbiol. 37, 64—90. https://doi.org/10.3109/1040841X.2010.512268.

Mongkolthanaruk, W., Dharmsthiti, S., 2002. Biodegradation of lipid-rich wastewater by a mixed bacterial consortium. Int. Biodeterior. Biodegrad. 50, 101—105.

Moody, J.D., Freeman, J.P., Fu, P.P., Cerniglia, C.E., 2004. Degradation of benzo[a]pyrene by *Mycobacterium van baalenii* PYR-1. Appl. Environ. Microbiol. 70, 340—345. https://doi.org/10.1128/AEM.70.1.340-345.2004.

Mueller, J.G., Cerniglia, C.E., Pritchard, P.H., 1996. Bioremediation of environments contaminated by polycyclic aromatic hydrocarbons,. In: Crawford, R.L., Crawford, L.D. (Eds.), Bioremediation: Principles and Applications. Cambridge University Press, Cambridge, pp. 125—194.

O'Toole, G.A., Kolter, R., 1998. Flagellar and twitching motility are necessary for *Pseudomonas aeruginosa* biofilm development. Mol. Microbiol. 30, 295—304.

Oljira, T., Muleta, D., Jida, M., 2018. Potential applications of some indigenous bacteria isolated from polluted areas in the treatment of brewery effluents. Biotechnol. Res. Int. 2018, 9745198 https://doi.org/10.1155/2018/9745198.

Pardo, R., Herguedas, M., Barrado, E., Vega, M., 2003. Biosorption of cadmium, copper, lead and zinc by inactive biomass of *Pseudomonas putida*. Anal. Bioanal. Chem. 376 (1), 26—32. Available from: https://doi.org/10.1007/s00216-003-1843-z.

Parke, D., Rivelli, M.L., Ornston, L.N., 1985. Chemotaxis to aromatic and hydroaromatic acids: comparison of *Bradyrhizobium japonicum* and *Rhizobium trifolii*. J. Bacteriol. 163, 417—422.

Phulia, V., Jamwal, A., Saxena, N., Chadha, N.K., Muralidhar, 2013. Technologies in aquatic bioremediation 65—91.

Pratt, L.A., Kolter, R., 1998. Genetic analysis of *Escherichia coli* biofilm formation: roles of flagella, motility, chemotaxis and type I pili. Mol. Microbiol. 30, 285—293.

Prigent-Combaret, C., Vidal, O., Doral, C., Lejeune, P., 1999. Abiotic surface sensing and biofilm-dependent regulation of gene expression in *Escherichia coli*. J. Bacteriol. 181, 5993—6002.

Rajmohan, K.S., Gopinath, M., Chetty, R., 2018. Bioremediation of Nitrate-Contaminated Wastewater and Soil. Bioremediation: Applications for Environmental Protection and Management. Energy, Environment, and Sustainability, first ed. Springer, Singapore, pp. 387—409.

Reichman, S.M., 2007. The potential of the legume-rhizobium symbiosis for the remediation of arsenic contaminated sites. Soil Biol. Biochem. 39, 2587—2593.

Roane, T.M., Josephson, K.L., Pepper, I.L., 2001. Dual-bioaugmentation strategy to enhance remediation of cocontaminated soil. Appl. Environ. Microbiol. 67 (7), 3208—3215.

Samuel, A., Latinwo, G.K., 2015. Biodegradation of diesel oil in soil and its enhancement by application of bioventing and amendment with brewery waste effluents as Biostimulation-Bioaugmentation agents. J. Ecol. Eng 16, 82—91. Link: https://goo.gl/eGLYrf.

Shagol, C.C., Krishnamoorthy, R., Kim, K., Sundaram, S., Sa, T., 2014. Arsenic-tolerant plant-growth-promoting bacteria isolated from arsenic-polluted soils in South Korea. Environ. Sci. Pollut. Res. 21, 9356—9365.

Shoi, J., Dang, C.V., Taylor, B.L., 1998. Oxygen as attractant and repellent in bacterial chemotaxis. J. Bacteriol. 169, 3118—3123.

Smith, M.J., Flowers, T.H., Duncan, H.J., Alder, J., 2006. Effects of polycyclic aromatic hydrocarbons on germination and subsequent growth of grasses and legumes in freshly contaminated soil and soil with aged PAHs residues. Environ. Pollut. 141, 519−525.

Song, H.P., Li, X.G., Sun, J.S., Xu, S.M., Han, X., 2008. Application of a magnetotatic bacterium, *Stenotrophomonas* sp. to the removal of Au(III) from contaminated wastewater with a magnetic separator. Chemosphere 72, 616−621.

Stomp, A.M., Han, K.H., Wilbert, S., Gordon, M.P., Cunningham, S.D., 1994. Genetic strategies for enhancing phytoremediation. Ann. N. Y. Acad. Sci. 721, 481−492.

Sunggyu, L., 1995. Bioremediation of polycyclic aromatic hydrocarbon-contaminated soil. J. Clean. Prod. 3, 255.

Surti, H., 2016. Physico-chemical and microbial analysis of waste water from different industry and cod reduction treatment of industrial waste water by using selective microorganisms. Int. J. Curr. Microbiol. Appl. Sci. 5, 707−717.

Taylor, B.L., 1983. How do bacteria find the optimum concentration of oxygen? Trends Biochem. Sci. 8, 438−441.

Testiati, E., Parinet, J., Massiani, C., Laffont-Schwob, I., Rabier, J., Pfeifer, H.R., Prudent, P., 2013. Trace metal and metalloid contamination levels in soils and in two native plant species of a former industrial site: evaluation of the phytostabilization potential. J. Hazard Mater. 248, 131−141.

Thavasi, R., Jayalakshmi, S., Banat, I.M., 2011. Application of biosurfactant produced from peanut oil cake by *Lactobacillus delbrueckii* in biodegradation of crude oil. Bioresour. Technol. 102, 3366−3372. Link: https://goo.gl/FWUrD3.

Tso, W.-W., Adler, J., 1974. Negative chemotaxis in *Escherichia coli*. J. Bacteriol. 118, 560−576.

Vane, C.H., Kim, A.W., Beriro, D.J., Cave, M.R., Knights, K., Moss-Hayes, V., Nathanail, P.C., 2014. Polycyclic aromatic hydrocarbons (PAH) and polychlorinated biphenyls (PCB) in urban soils of Greater London, UK. Appl. Geochem. 51, 303−314.

Vidali, M., 2001. Bioremediation: an overview. J. Appl. Chem. 1163−1172.

Vullo, D.L., Ceretti, H.M., Daniel, M.A., Ramírez, S.A., Zalts, A., 2008. Cadmium, zinc and copper biosorption mediated by *Pseudomonas veronii* 2E. Bioresour. Technol. 99 (13), 5574−5581.

Watnick, P.I., Kolter, R., 1999. Steps in the development of a *Vibrio cholerae* EI Tor biofilm. Mol. Microbiol. 34, 586−595.

Zhou, M., Liu, Y., Zeng, G., Li, X., Xu, W., Fan, T., 2007. Kinetic and equilibrium studies of Cr(VI) biosorption by dead *Bacillus licheniformis* biomass. World J. Microbiol. Biotechnol. 23 (1), 43−48. Available from: https://doi.org/10.1007/s11274-006-9191-8.

Zhu, Y.G., Shaw, G., 2000. Soil contamination with radionuclides and potential remediation. Chemosphere 41, 121−128.

Zouboulis, A.I., Loukidou, M.X., Matis, K.A., 2004. Biosorption of toxic metals from aqueous solutions by bacteria strains isolated from metal-polluted soils. Process Biochem. 39 (8), 909−916. Available from: https://doi.org/10.1016/S0032-9592(03)00200-0.

Role of lower plants in the remediation of polluted systems

17

Lini Nirmala[1], Shiburaj Sugathan[2]

[1]*Department of Biotechnology, Mar Ivanios College, Thiruvananthapuram, Kerala, India;*
[2]*Department of Botany, University of Kerala, Thiruvananthapuram, Kerala, India*

Chapter outline

17.1 Introduction

Industrial processes, agricultural practices, and many other anthropogenic activities release hazardous chemicals into the environment that pose detrimental effects to human beings and the ecosystem (Rajkumar et al., 2010). Hydrocarbon pollution in terrestrial and aquatic systems and massive amounts of metal pollution is seen as a primary global concern. It is vital to reduce pollution concentrations to satisfy ever-growing legislative requirements. Recently, there has been a growing emphasis on cleaning and remediating pollutants using green technologies. Pollution control

Biological Approaches to Controlling Pollutants. https://doi.org/10.1016/B978-0-12-824316-9.00008-2

using plants is one of these. Lower plants are a group of plants and plantlike organisms defined primarily by their lack of vascular tissue. These groups represent some of the oldest organisms on earth. They play significant roles in ecosystems as primary producers and nutrient and water recyclers. This group includes bryophytes, lichens, algae, and fungi that are very sensitive to environmental changes. Their peculiar morphological, physiological, and biochemical characteristics make them excellent model organisms to study ecological pollution. Lower plants exhibit distinctive damage symptoms from different environmental pollutants and accumulate environmental pollutants in varying degrees, which may be used to identify the specific contaminant and to control pollution within various habitats.

17.2 Bryophytes

Bryophytes are small, nonvascular, seedless plants that include mosses, liverworts, and hornworts. Most bryophyte species are susceptible to environmental changes. The leaves and thallus of bryophytes lack a cuticle, and the tissues are readily permeable to water solutions. Bryophytes absorb and accumulate heavy metals and environmental pollutants from the atmosphere, the substrate, or both. They are distributed widely throughout the world, and their morphologies do not vary with the seasons. Their habitat diversity, high metal accumulation efficiency, and morphological and physiological characteristics make them an ideal organism for controlling environmental pollution (Govindapyari et al., 2010). They have a high cationic exchange capacity and high surface-to-volume or surface-to-weight ratio that favor pollutant accumulation. Expansion of pollution levels can differ significantly from site to site depending on geological and climatic conditions. Bryophytes can concentrate heavy metals in larger amounts than vascular plants can (Zvereva and Kozlov, 2011). They can accumulate heavy metals such as Al, Ba, Cr, Cu, Fe, Ga, Ni, Pb, Ag, Ti, Vi, Zn, and Zr and are reported to concentrate rare earth elements (Govindapyari et al., 2010).

17.2.1 Use of bryophytes in controlling air pollution

Air pollution is one of the most severe environmental health risks worldwide. Bryophytes show a wide range of visible symptoms to various pollutants. They are easy to handle, making them good candidates for the biomonitoring of air pollution. The use of bryophytes as a biomonitoring tool started in the 1950s to characterize fluorine emissions from industries (Markert et al., 1996). Pollutants interact with bryophytes as gas, solid particles, or aqueous solutions (Chakrabortty and Paratkar, 2006). The sensitivity and accumulation capacity of bryophytes can vary based on their morphology and physiology. Profusely branched mosses are more efficient accumulators of pollutants than unbranched and erect Acrocarpi are (Govindapyari et al., 2010). Among the mosses, *Atrichum undulatum* is a better bioindicator of air pollution than *Ceratodon purpureus*. Bryophytes also vary their specificity of accumulation of heavy metals. These properties should be considered when using bryophytes as pollution control agents. There are different techniques used for biomonitoring with bryophytes (Govindapyari et al., 2010). The "moss bag technique" is one of the most popular approaches for biomonitoring. Some important bryophytes that are promising candidates for air pollution control in different environments are shown in Table 17.1.

17.2.2 Use of bryophytes in controlling water pollution

Aquatic bryophytes are frequently used for biomonitoring water quality. There are two approaches in water biomonitoring: (1) passive biomonitoring, including the observation and analysis of native bryophytes, and (2) active biomonitoring, which focuses on species transplantation for a fixed period of exposure. Aquatic bryophytes accumulate more heavy metals from contaminated water than vascular plants do. Their unique characteristics allow them to be used as biosorbents to clean polluted rivers and industrial wastewaters. Bryophytes like thallus liverwort, *Marchantia polymorpha*, can adsorb Cu, Zn, and Cd (Ares et al., 2018). Chlorinated hydrocarbons and lead were also accumulated by adsorption on to aquatic bryophytes (Mouvet et al., 1993). The aquatic mosses *Fontinalis antipyretica* and *Platyhypnidium riparioides* are the most commonly used biomonitors in the Northern Hemisphere for river quality assessment.

17.2.3 Use of bryophytes in controlling soil pollution

Plants that can accumulate and tolerate extraordinarily high concentrations of heavy metals (hyperaccumulators) can be used for phytoremediation (removal of contaminants from soils) or phytomining (growing a crop of plants to harvest metals). Several bryophyte species were found to accumulate high levels of pollutants and heavy metals and can be utilized as phytoremediators. Ectohydric mosses such as *Pleurozium schreberi*, *Hylocomium splendens*, *Hypnum cupressiforme*, *Scleropodium purum*, and *Dicranella heteromalla* are more widely used for biomonitoring than endohydric mosses are.

17.3 Lichens

Lichens are dual organisms that emerge from algae or cyanobacteria and live in a symbiotic relationship with fungal filaments. Lichens cover more than 6% of the land surface of the earth. They have been used to monitor the degree of pollution in particular environments. Lichens have no cuticle, and the free exchange of both gases and solutions occurs through their entire thallus (Turetsky, 2003). Due to their habitat diversity and morphological features, lichens function as useful indicators of pollution. Lichens are profitable accumulators of heavy metals, including nickel, cobalt, cadmium, copper, lead, mercury, and silver. Some important lichens used in controlling pollution are shown in Table 17.2.

Table 17.1 Use of bryophytes in controlling pollution.

Sl.No.	Bryophyte	Pollutant which is absorbed	Source	Reference
1.	Atrichum undulatum	Hg	Air	Govindapyari et al. (2010)
2.	Ceratodon purpureus	Cu	Soil	Elvira et al. (2020)
3.	Dicranum scoparium	Hg	Soil	Atwood and Buck (2020)
4.	Conocephalum conicum, Marchantia polymorpha Pellia epiphylla	Ba, Cd, Co, Cr, Cu, Fe, Hg, Ni, Pb, Sr, V, and Zn and the macroelements N, P, K, ca. and Mg	Soil	Samecka-Cymerman et al. (1997)
5.	Grimmia laevigata	Sn	Soil	Govindapyari et al. (2010)
6.	Atrichum angustatum, Polytrichum commune	Ag	Soil, water	Govindapyari et al. (2010)
7.	Mielichhoferia, Ptychostomum capillare, Scopelophila cataractae	Cu, Pb, and B	Soil	Antreich et al. (2016)
8.	Hypnum yokohamae	SO$_2$	Soil	Antreich et al. (2016)
9.	Amblystegium riparium	S	Air	Saxena (2004)
10.	Funaria hygrometrica	Fe, Zn, Cu, Pb, Cr, Mn, Ni, Cd.	Soil, air	(Adie et al., 2014)
11.	Marchantia polymorpha	Al, Cd, Cu, Fe, and Zn	Water	(Vásquez et al., 2019)
12.	Pseudoscleropodium purum	Cd, Fl, PAH	Soil, air	(Aboal et al., 2020; Saxena, 2004)
13.	Sphagnum papillosum, S. magellanicum, S. capillifolium, S. tenellum	N$_2$	Soil	Saxena (2004)

Table 17.2 Use of lichens in controlling pollution.

Sl.No	Lichen	Pollutant which is absorbed	Source	Reference
1.	Usnea articulate	SO_2	Air	Henderson (1996)
2.	Parmelia caperata, Lepraria incana	SO_2	Air	Nimis et al. (1990)
3.	Cladonia digitata, Cladonia portentosa	N_2	Air	Maslov and Pavlova (2005)
4.	Dermatocarpon luridum, Hymenelia lacustris	Cu, SO_2, O_3	Water	Maslov and Pavlova (2005)
5.	Xanthoria parietina	Al, As, Co, Cd, Cu, Fe, Hg, Mn, Ni, Pb, Ti, Tl, V, and Zn	Air	Demiray et al. (2012)
6.	Hypogymnia physodes	N_2,S	Air	Bruteig (1993)
7.	Usnea and Protousnea species	Ca, K, Se, Mg, Mn, S	Air near volcanic site	(Bubach et al., 2020)
8.	Pseudevernia furfuracea	Bisphenol	Aqueous solution	Senol et al. (2020)

17.3.1 Use of lichen in controlling air pollution

Lichens have been widely used as a biomonitoring tool for air pollution. Lichens lack a protective cuticle and vascular root system. They absorb mineral nutrients and trace elements, including metals, from the environment throughout their thallus. Air pollution's sensitivity can be measured in various ways: morphological, anatomical, and physiological changes, a decline in diversity, absence of sensitive species, etc. They have been used extensively to document the long-term deposition of metal ions around point sources and urban areas. It is recognized that a wide range of pollutants like sulfur dioxide, hydrogen fluoride, nitrogen oxides, and ozone ammonia, fluorine, eutrophication, alkaline dust, metal and radionuclide, chlorinated hydrocarbons, and acid rain can be detected and monitored using lichens. Various heavy metals such as Pd, Cd, Ni, Hg, Zn, Cu, and Cr, considered toxic for many other living organisms, may be accumulated simultaneously in one lichen specimen, which may appear to be intact in many cases. So lichens have an extended history of use as a biological indicator of air quality.

17.3.2 Use of lichens in controlling water pollution

Lichen biomass has been reported as an inexpensive and effective adsorbent for removing fluoride, Pb(II), and Cr(III) ions from aqueous solutions. This property can be utilized to control pollution in wastewater and other aquatic forms (Mondal et al., 2007; Uluozlu et al., 2009). At sea, various pollutants such as hydrocarbons and anionic surfactants spread by forming a thin film of a few micrometers on the sea surface, and pollutants can reach lichens on the coast. Lichens such as *Dermatocarpon luridum* are used as bioaccumulators of metallic elements in water. Lichen-based biosensors have been developed for assessing the presence of benzene in aquatic environments (Antonelli et al., 2005).

17.3.3 Use of lichen in controlling soil pollution

Some lichen species can grow on soil containing metallic elements, which is indicative of that particular pollutant in the ground. *Diploschistes muscorum*, *Cladonia* sp., and *Stereocaulon* sp. tolerate high levels of metals in the soil. *Verdean leprosa* is a species particularly vulnerable to zinc presence because it is often found near road safety zinc slides. Biosorption of radio cobalt by *Hypogymnia physodes* opens up a new possibility to utilize lichens in these areas. Especially, epiphytic lichens are most commonly used for biosorption.

17.4 Algae

Algae are simple nonflowering photosynthetic eukaryotic organisms ranging in size from single-celled diatoms (microalgae) to giant multicellular forms such as kelp or seaweed (macroalgae). Algae play an important role in aquatic ecosystem balances by forming the energy base of the food web. Algal communities possess many attributes as biological indicators of spatial and temporal environmental changes. Algae are useful in biomonitoring ecosystem conditions because they respond quickly in species composition and densities to a wide range of pollutants. Algae are present in diverse environmental conditions. They are adapted to saline and freshwater, even to wetlands. As they mostly live in an environment connected with water where the concentration of essential nutrients is usually low, they have had to develop mechanisms to accumulate nutrients and several inorganic ions up to high levels. This capability can be used in the processes of pollutant removal. Metal ions are an important group of hazardous contaminants in industrial wastewater that can be removed by biosorption. The use of algae for the degradation of pollutants is vast. They can assimilate NH^{4+} ions, a bioproduct of biodegradation of organic compounds, and decrease the concentration of nitrogen in polluted water. Algae have great potential in pollution control and the processes applied to decrease pollution in the environment. Some important algae used in controlling pollution are shown in Table 17.3.

17.4.1 Use of algae in controlling air pollution

The use of algae in controlling air pollution is limited because they are mostly aquatic. Spirulina is capable of reducing carbon dioxide (CO_2), nitrogen dioxide

Table 17.3 Use of algae in controlling pollution.

S.No.	Algae	Pollutant which is absorbed	Source	Reference
1.	Chlorella vulgaris, Aphanothece microscopica	CO_2	Air, water	Malińska and Zabochnicka-Ś
2.	Cladophora, Microcystis, Scenedesmus	NO_2, PO_4	Water	Gökçe (2016)
3.	Spirogyra sp. Palmaria palmata	Mn, Fe, Cu, and Zn	Water	Rajfur and Klos (2015)

(NO_2), and sulfur dioxide (SO_2) in the air and can generate oxygen (Malińska and Zabochnicka-Ś). Microalgae are used to capture carbon dioxide emitted by industrial sources such as fossil-fueled power plants and fermentation processes to reduce CO_2 emissions (Sayre, 2010). Microalgae can assimilate carbon dioxide into organic material such as carbohydrates, proteins, and lipids for conversion into valuable materials.

17.4.2 Use of algae in controlling water pollution

Algae can be either marine or freshwater and have higher photosynthetic efficiencies than terrestrial plants. Algae are the leading primary producers in all kinds of water bodies. They are involved in water pollution in many significant ways. Firstly, enrichment of nutrients in water through organic effluents may selectivity stimulate the growth of algal species producing massive surface growths or blooms that, in turn, reduce water quality and affect its use. However, certain algae that flourish in water polluted with organic wastes play an essential role in the "self-purification of water bodies." Their temporal and spatial distribution, sensitivity to environmental change, and availability throughout the year make this an ideal organism to monitor aquatic pollution. Many kinds of algae are excellent water-quality indicators, and many lakes are characterized by their dominant phytoplankton groups. Many desmids are known to be present in oligotrophic waters. At the same time, a few species frequently occur in eutrophic bodies of water.

Similarly, many blue-green algae live in nutrient-poor waters, and others grow well in organically polluted waters. The ecosystem approach to water-quality assessment also includes diatom species and associations used as organic pollution indicators. Five algal species were selected as indicators of the degree of pollution in rivers in England. *Stigeoclonium tenue* is present at the downstream margin of the heavily polluted part of a river. *Nitzschia palea* and *Gomphonema parvulum* always appear to be dominant in the mild pollution zone. Simultaneously, *Cocconeis* and *Chamaesiphon* are reported to occur in unpolluted parts of the stream or repurified zone. *Navicula accomoda* is stressed to be a good indicator of sewage or organic pollution as the species comfortably happen in the most heavily polluted zones in which other species cannot occur. The same holds for species and varieties of Gomphonema commonly found in highly organically polluted water. *Amphora ovalis* and *Gyrosigma attenuatum* have also been indicated as good examples of diatoms affected by high organic water content.

17.4.3 Use of algae in controlling soil pollution

Algae are often used as indicators to evaluate the ecotoxicity, genotoxicity, and environmental risk of pollutants, both in soil and sediments, due to their sensitivity to the presence of toxic chemicals. Given the sensitivity of algae to pollutants, variation in the algal species composition can serve as a useful bioindicator of pollution and should be used in conjunction with bioassays and chemical analysis for toxicological estimations of samples of various habitat. An illustrative example is that intracellular dehydrogenase and urease activity is a standard method of estimating the total microbial activity in soils. They are susceptible to the presence of various pollutants. The species composition and diversity of algae and cyanobacteria are altered in polluted soil. This can be utilized to study the pollution rate in that particular area. Thus, algae could be useful as bioindicators of pollution.

17.5 Fungi

Fungi, which include molds, yeasts, and mushrooms, are a eukaryote but are primarily different from plants because they have different cellular compositions and other ancestors. Fungi play a crucial role in many ecosystems' biological processes as decomposers of river organic matter, parasites or symbionts, and food sources for higher trophic organisms. However, fungi are sensitive to environmental change. In the last decade, the role of fungi in bioremediation has been increasingly recognized. Mycoremediation is a form of bioremediation in which fungi-based technology is used to remove toxic environmental pollutants. Mycoremediation has been proven to be a cheap, effective, and environmentally sound way to clean up a wide array of toxins from polluted environments. Fungal-mediated bioremediation generally uses extracellular enzymes to break down the pollutants and reduce or remove toxic waste (Kulshreshtha et al., 2013). Both filamentous fungi (molds) and macrofungi (mushrooms) possess enzymes for degradation. Microfiltration using fungal mycelium can also work as a filter network in buffer zones around streams to filter the runoff from farms in suburban zones (Chiu et al., 2000). The production of versatile extracellular ligninolytic enzymes, fast growth, and the network of filaments, high surface area-to-volume ratio, resistance to heavy metals, adaptability to fluctuating pH and temperature, and the presence of metal-binding proteins make fungi an ideal candidate for the remediation of various pollutants (Akhtar and Mannan, 2020). Fungi can be also be used for in situ remediation of multiple contaminants such as dyes, herbicides, and pharmaceutical drugs released by various industries. Some important fungi used in controlling pollution are shown in Table 17.4.

Table 17.4 Use of fungi in controlling pollution.

S.No	Fungi	Pollutant which is absorbed	Source	Reference
1.	*Aspergillus oryzae*	Ar	Soil	Singh et al. (2015)
2.	*Trametes polyzona*	PAH, phenols	Water (sewage)	(Batista-García et al., 2017)
3.	*Pleurotus ostreatus*	Plastic degradation, DDT, phenols, PCBs, pesticides such as Endosulfan	Soil	Stella et al. (2017)
4.	*Phanerochaete chrysosporium*, *Lentinula edodes*	Dyes such as fuchsin, malachite green, 2,4 dichlorophenol	Water	Deshmukh et al. (2016)
5.	*Pleurotus pulmonarius*	Degradation of radioactive cellulosic-based water, crude oil	Soil	Deshmukh et al. (2016)
6.	*Penicillium simplicissimum*	Pd, Cd ions		Deshmukh et al. (2016)
7.	*Rhizopus arrhizus*	Cr(VI) ions	Soil, water	Preetha and Viruthagiri (2007)
8.	*Coriolus versicolor* MKACC 5249	PAH	Water	Deshmukh et al. (2016)
9.	*Pestalotiopsis* sps.	Polyurethane	Soil	Russell et al. (2011)
10.	*Aspergillus niger*	Nigrosin, basic fuchsin	Water	Rani et al. (2014)
11.	*Pleurotus citrinopileatus*	Industrial wastes such as cardboard	Water	Kulshreshtha et al. (2013)
12.	*P. Crustosum*	Boron	Soil, water	Ertit Taştan et al. (2016)
13.	*Lentinus connotus*	Industrial effluents such as paddy straw	Water	Deshmukh et al. (2016)
14.	Mycorrhizal association between *Rhizophagus intraradices* and *Robinia*	Pd	Soil	Yang et al. (2016)
15.	Mycorrhizal association between *Rhizophagus irregularis* and *Phragmites australis*	Benzene and ammonium salts	Wetland	Fester (2013)
16.	*Fomes fasciatus*	Cu(II) ions	Soil	Deshmukh et al. (2016)
17.	*Aspergillus sclerotiorum* CBMAI 849	Pyrene, benzopyrene	Soil	Passarini et al. (2011)

Continued

Table 17.4 Use of fungi in controlling pollution.—*cont'd*

S.No	Fungi	Pollutant which is absorbed	Source	Reference
18.	*Trametes maxima*	Atrazine	Clayey soil	Chan-Cupul et al. (2016)
19.	*Aspergillus flavus* KRP1	Hg (II) ions	Water	Kurniati et al. (2014)
20.	*Trichoderma harzianum*	Ni^{+2}	Soil, water	Cecchi et al. (2017)

17.5.1 Use of fungi in controlling air pollution

The use of fungi in air pollution monitoring is significantly less studied than for other lower plants. Few metal bioaccumulation studies using macrofungi concerning urban and industrial sources have been reported. Inorganic materials are often colonized, as they absorb dust and serve as suitable growth substrates for fungus such as *Aspergillus fumigatus* and *A. versicolor* (Samet and Spengler, 2003). Species of macrofungi can be used as (they are ideal for) biomonitoring of air pollution in the environment. In some studies using macrofungi, *Ganoderma* species contained more Ni, As, and Pb than other species did. In contrast, *Coriolis* species had more Ti, Cr, and Co. *Canoparmelia texana* is used for passive biomonitoring of environmental pollutants.

17.5.2 Use of fungi in controlling water pollution

Fungi have been reported to accumulate heavy metals and pesticides in aquatic and marine environments (Wu et al., 2016). Aquatic fungi such as *Mucor hiemalis* can accumulate and degrade cyanotoxins (Balsano et al., 2017). Fungi are capable of bioremediation of river water contaminated with *E. coli*. Lead and cadmium uptake by marine fungi have been reported. Mushrooms can be an excellent tool for mycoremediation owing to their high accumulation potential and short life-span. Mushrooms such as *Russula* spp., *Boletus* spp., *Armillaria* spp., *Agaricus* spp., *Termitomyces* spp., and *Polyporus* spp. have been studied for the accumulation of heavy metals (Raj et al., 2011). Microfiltration reduces the concentration of harmful pollutants frequently found in urban stormwater runoff, such as heavy metals and polycyclic aromatic hydrocarbons. Mycoremediation can eliminate pathogenic bacteria from pet wastes and waterfowl. Introducing mushroom spores to pollution sites is an inexpensive and effective method.

17.5.3 Use of fungi in controlling soil pollution

An extended hyphal network, the possibility of using pollutants as nutrient substrate, and the low specificity of extracellular enzymatic complexes make filamentous fungi an excellent model organism for controlling soil pollution compared with other microorganisms. The biomass content of fungi has a high percentage of cell wall materials that act as a cation exchanger and have excellent metal-binding capacity. Fungi use immobilized heavy metals as a source of energy and convert them into soluble substances. Basidiomycetes and the ecological groups of saprotrophic and biotrophic fungi are the most suitable fungi used in soil remediation. *Trichoderma* species are commonly found in all types of soils and are the right candidate for mycoremediation. The fungus can remove and concentrate ions—Pb, Cu, Zn, Cd, and Ni, herbicides, insecticides, PAH, etc.—from soil. Bioremediation using *Pleurotus pulmonarius* has been reported in soils polluted with cement and battery waste (Bosco and Mollea, 2019). Basidiomycetes have been used to accumulate and recycle uranium (Aksu, 2001; Wang and Zhou, 2005). Some of the most representative basidiomycetes that can degrade pollutants include *Pleurotus ostreatus*, *Lentinula edodes*, *Agaricus bisporus*, *Trametes versicolor*, and *Phanerochaete chrysosporium*. Bioremediation by macrofungi is more advantageous than with others. Along with bioremediation, the soil becomes enriched with organic matter and nutrients.

17.6 Summary and conclusion

Anthropogenic activities are increasingly adding contaminants to our environment. Phytoremediation is a cost-effective and environmentally safe approach solving the problems of soil and water pollution. Using lower plants is more cost-effective than using higher plants to achieve the goals of sustainable development. However, more research needs to be conducted for their effective utilization in large-scale applications. The mechanisms of plant uptake and subsequent enzymatic metabolism have been well studied in recent years. The efficiency of phytoremediation by lower plants, including algae and fungi, can be substantially improved using genetic engineering technologies. The recombinant expression of genes encoding proteins involved in metal uptake, transport, and sequestration will enhance the efficiency of phytoremediation. Detailed knowledge of enzyme degradation of hazardous organic compounds and the biochemical pathways involved in processing xenobiotic compounds will open up new possibilities in phytoremediation.

References

Aboal, J.R., Concha-Graña, E., De Nicola, F., Muniategui-Lorenzo, S., López-Mahía, P., Giordano, S., Capozzi, F., Di Palma, A., Reski, R., Zechmeister, H., 2020. Testing a novel biotechnological passive sampler for monitoring atmospheric PAH pollution. J. Hazard Mater. 381, 120949.

Adie, P.A., Torsabo, S.T., Uno, U.A., Ajegi, J., 2014. Funaria hygrometrica moss as bioindicator of atmospheric pollution of heavy metals in Makurdi and environs, North Central Nigeria. Res. J. Chem. Sci. 4, 10—17.

Akhtar, N., Mannan, M.A., 2020. Mycoremediation: expunging environmental pollutants. Biotechnol. Rep. e00452.

Aksu, Z., 2001. Equilibrium and kinetic modelling of cadmium (II) biosorption by *C. vulgaris* in a batch system: effect of temperature. Separ. Purif. Technol. 21, 285−294.

Antonelli, M.L., Campanella, L., Ercole, P., 2005. Lichen-based biosensor for the determination of benzene and 2-chlorophenol: microcalorimetric and amperometric investigations. Anal. Bioanal. Chem. 381, 1041−1048.

Antreich, S., Sassmann, S., Lang, I., 2016. Limited accumulation of copper in heavy metal adapted mosses. Plant Physiol. Biochem. 101, 141−148.

Ares, Á., Itouga, M., Kato, Y., Sakakibara, H., 2018. Differential metal tolerance and accumulation patterns of Cd, Cu, Pb and Zn in the liverwort *Marchantia polymorpha* L. Bull. Environ. Contam. Toxicol. 100, 444−450.

Atwood, J.J., Buck, W.R., 2020. Recent literature on bryophytes—123 (2). Bryol. 123, 333−362.

Balsano, E., Esterhuizen-Londt, M., Hoque, E., Lima, S.P., 2017. Responses of the antioxidative and biotransformation enzymes in the aquatic fungus *Mucor hiemalis* exposed to cyanotoxins. Biotechnol. Lett. 39, 1201−1209.

Batista-García, R.A., Kumar, V.V., Ariste, A., Tovar-Herrera, O.E., Savary, O., Peidro-Guzmán, H., González-Abradelo, D., Jackson, S.A., Dobson, A.D., del Rayo Sánchez-Carbente, M., 2017. Simple screening protocol for identification of potential mycoremediation tools for the elimination of polycyclic aromatic hydrocarbons and phenols from hyperalkalophile industrial effluents. J. Environ. Manag. 198, 1−11.

Bosco, F., Mollea, C., 2019. Mycoremediation in soil. In: Biodegradation Processes. IntechOpen.

Bruteig, I.E., 1993. The epiphytic lichen *Hypogymnia physodes* as a biomonitor of atmospheric nitrogen and sulphur deposition in Norway. Environ. Monit. Assess. 26, 27−47.

Bubach, D.F., Catán, S.P., Messuti, M.I., Arribére, M.A., Guevara, S.R., 2020. Bioaccumulation of trace elements in lichens exposed to geothermal and volcanic activity from copahue-caviahue volcanic complex, patagonia, Argentina. Ann. Environ. Sci. Toxicol. 4, 005−015.

Cecchi, G., Roccotiello, E., Di Piazza, S., Riggi, A., Mariotti, M.G., Zotti, M., 2017. Assessment of Ni accumulation capability by fungi for a possible approach to remove metals from soils and waters. J. Environ. Sci. Health B 52, 166−170.

Chakrabortty, S., Paratkar, G.T., 2006. Biomonitoring of trace element air pollution using mosses. Aerosol Air Qual. Res. 6, 247−258.

Chan-Cupul, W., Heredia-Abarca, G., Rodríguez-Vázquez, R., 2016. Atrazine degradation by fungal co-culture enzyme extracts under different soil conditions. J. Environ. Sci. Health B 51, 298−308.

Chiu, S.-W., Law, S.-C., Ching, M.-L., Cheung, K.-W., Chen, M.-J., 2000. Themes for mushroom exploitation in the 21st century: sustainability, waste management, and conservation. J. Gen. Appl. Microbiol. 46, 269−282.

Demiray, A.D., Yolcubal, I., Akyol, N.H., Çobanoğlu, G., 2012. Biomonitoring of airborne metals using the lichen *Xanthoria parietina* in Kocaeli Province, Turkey. Ecol. Indicat. 18, 632−643.

Deshmukh, R., Khardenavis, A.A., Purohit, H.J., 2016. Diverse metabolic capacities of fungi for bioremediation. Indian J. Microbiol. 56, 247−264.

Elvira, N.J., Medina, N.G., Leo, M., Cala, V., Estébanez, B., 2020. Copper content and resistance mechanisms in the terrestrial moss *Ptychostomum capillare*: a case study in an Abandoned copper mine in Central Spain. Arch. Environ. Contam. Toxicol. 79, 49–59.

Ertit Taştan, B., Çakir, D.N., Dönmez, G., 2016. A new and effective approach to boron removal by using novel boron-specific fungi isolated from boron mining wastewater. Water Sci. Technol. 73, 543–549.

Fester, T., 2013. *Arbuscular mycorrhizal* fungi in a wetland constructed for benzene-, methyl tert-butyl ether-and ammonia-contaminated groundwater bioremediation. Microbial Biotechnol. 6, 80–84.

Gökçe, D., 2016. Algae as an Indicator of Water Quality. Algae—Organisms for Imminent Biotechnology. InTech, pp. 81–101.

Govindapyari, H., Leleeka, M., Nivedita, M., Uniyal, P.L., 2010. Bryophytes: indicators and monitoring agents of pollution. NeBIO 1, 35–41.

Henderson, A., 1996. Literature on air pollution and lichens XLIII. Lichenol. 28, 279–285.

Kulshreshtha, S., Mathur, N., Bhatnagar, P., 2013. Mycoremediation of paper, pulp and cardboard industrial wastes and pollutants. In: Fungi as Bioremediators. Springer, pp. 77–116.

Kurniati, E., Arfarita, N., Imai, T., Higuchi, T., Kanno, A., Yamamoto, K., Sekine, M., 2014. Potential bioremediation of mercury-contaminated substrate using filamentous fungi isolated from forest soil. J. Environ. Sci. 26, 1223–1231.

Malińska, K., Zabochnicka-Świątek, M., 2010. Biosystems for air protection. Air Pollut. 177–194.

Markert, B., Herpin, U., Berlekamp, J., Oehlmann, J., Grodzinska, K., Mankovska, B., Suchara, I., Siewers, U., Weckert, V., Lieth, H., 1996. A comparison of heavy metal deposition in selected Eastern European countries using the moss monitoring method, with special emphasis on the 'Black Triangle'. Sci. Total Environ. 193, 85–100.

Maslov, A.I., Pavlova, E.A., 2005. Simple method for fractionation of *Parmelia sulcata* lichen thalli. Russ. J. Plant Physiol. 52, 271–274.

Mondal, M.K., Das, T.K., Biswas, P., Samanta, C.C., Bairagi, B., 2007. Influence of dietary inorganic and organic copper salt and level of soybean oil on plasma lipids, metabolites and mineral balance of broiler chickens. Anim. Feed Sci. Technol. 139, 212–233.

Mouvet, C., Morhain, E., Sutter, C., Couturieux, N., 1993. Aquatic mosses for the detection and follow-up of accidental discharges in surface waters. Water Air Soil Pollut. 66, 333–348.

Nimis, P.L., Castello, M., Perotti, M., 1990. Lichens as biomonitors of sulphur dioxide pollution in La Spezia (Northern Italy). Lichenol. 22, 333–344.

Passarini, M.R., Rodrigues, M.V., da Silva, M., Sette, L.D., 2011. Marine-derived filamentous fungi and their potential application for polycyclic aromatic hydrocarbon bioremediation. Mar. Pollut. Bull. 62, 364–370.

Preetha, B., Viruthagiri, T., 2007. Batch and continuous biosorption of chromium (VI) by *Rhizopus arrhizus*. Separ. Purif. Technol. 57, 126–133.

Raj, D.D., Mohan, B., Vidya Shetty, B.M., 2011. Mushrooms in the remediation of heavy metals from soil. Int. J. Environ. Pollut. Control Manag. 3, 89–101.

Rajfur, M., Klos, A., 2015. Use of algae in active biomonitoring of surface waters. Ecol. Chemi. Eng. S 21, 561–576.

Rajkumar, M., Ae, N., Prasad, M.N.V., Freitas, H., 2010. Potential of siderophore-producing bacteria for improving heavy metal phytoextraction. Trends Biotechnol. 28, 142–149.

Rani, B., Kumar, V., Singh, J., Bisht, S., Teotia, P., Sharma, S., Kela, R., 2014. Bioremediation of dyes by fungi isolated from contaminated dye effluent sites for bio-usability. Braz. J. Microbiol. 45, 1055–1063.

Russell, J.R., Huang, J., Anand, P., Kucera, K., Sandoval, A.G., Dantzler, K.W., Hickman, D., Jee, J., Kimovec, F.M., Koppstein, D., 2011. Biodegradation of polyester polyurethane by endophytic fungi. Appl. Environ. Microbiol. 77, 6076–6084.

Samecka-Cymerman, A., Marczonek, A., Kempers, A.J., 1997. Bioindication of heavy metals in soil by liverworts. Arch. Environ. Contam. Toxicol. 33, 162–171.

Samet, J.M., Spengler, J.D., 2003. Indoor environments and health: moving into the 21st century. Am. J. Publ. Health 93, 1489–1493.

Saxena, D.K., 2004. Uses of bryophytes. Resonance 9, 56–65.

Sayre, R., 2010. Microalgae: the potential for carbon capture. Bioscience 60, 722–727.

Şenol, Z.M., Gul, Ü.D., Gurkkan, R., 2020. Bio-sorption of bisphenol a by the dried-and inactivated-lichen (*Pseudoevernia furfuracea*) biomass from aqueous solutions. J. Environ. Health Sci. Eng. 1–12.

Singh, M., Srivastava, P.K., Verma, P.C., Kharwar, R.N., Singh, N., Tripathi, R.D., 2015. Soil fungi for mycoremediation of arsenic pollution in agriculture soils. J. Appl. Microbiol. 119, 1278–1290.

Stella, T., Covino, S., Čvančarová, M., Filipová, A., Petruccioli, M., D'Annibale, A., Cajthaml, T., 2017. Bioremediation of long-term PCB-contaminated soil by white-rot fungi. J. Hazard Mater. 324, 701–710.

Turetsky, M.R., 2003. The role of bryophytes in carbon and nitrogen cycling. Bryologist 106, 395–409.

Uluozlu, O.D., Tuzen, M., Mendil, D., Soylak, M., 2009. Assessment of trace element contents of chicken products from Turkey. J. Hazard Mater. 163, 982–987.

Vásquez, C., Calva, J., Morocho, R., Donoso, D.A., Benítez, Á., 2019. Bryophyte communities along a tropical urban river respond to heavy metal and arsenic pollution. Water 11, 813.

Wang, M., Zhou, Q., 2005. Single and joint toxicity of chlorimuron-ethyl, cadmium, and copper acting on wheat *Triticum aestivum*. Ecotoxicol. Environ. Saf. 60, 169–175.

Wu, M., Xu, Y., Ding, W., Li, Y., Xu, H., 2016. Mycoremediation of manganese and phenanthrene by *Pleurotus eryngii* mycelium enhanced by Tween 80 and saponin. Appl. Microbiol. Biotechnol. 100, 7249–7261.

Yang, Y., Liang, Y., Han, X., Chiu, T.-Y., Ghosh, A., Chen, H., Tang, M., 2016. The roles of arbuscular mycorrhizal fungi (AMF) in phytoremediation and tree-herb interactions in Pb contaminated soil. Sci. Rep. 6, 20469.

Zvereva, E.L., Kozlov, M.V., 2011. Impacts of industrial polluters on bryophytes: a meta-analysis of observational studies. Water Air Soil Pollut. 218, 573–586.

Higher plant remediation to control pollutants

18

Pankaj Kumar Jain[1], Prama Esther Soloman[1], R.K. Gaur[2]

[1]*Indira Gandhi Centre for Human Ecology, Environmental and Population Studies, Department of Environmental Science University of Rajasthan, Jaipur, Rajasthan, India;* [2]*Department of Biotechnology, Deen Dayal Upadhyay University, Gorakhpur, Uttar Pradesh, India*

Chapter outline

Biological Approaches to Controlling Pollutants. https://doi.org/10.1016/B978-0-12-824316-9.00005-7

18.1 **Introduction**

Presently, the pollution of air, water, and soil is recognized as one of the most serious problems worldwide. Our Mother Earth is a beautiful landscape, but over the last century, it has been ruthlessly exploited and destroyed by humans to fulfill their greed. The Industrial Revolution exponentially added several million tons of hazardous pollutants to the environment. Such environmental pollutants are toxic and hazardous. They pose potential risks and are responsible for several types of diseases in humans and other living beings.

There are diverse strategies to mitigate environmental pollution. By applying these methods, we can reduce the deleterious impacts of pollutants on the environment. One method is planting trees, herbs, and shrubs in urban areas, near industries, and on highways to reduce pollution and improve environmental quality. In earlier times, we believed only in the aesthetic value of plants, but now it is has been proven that plants are also useful in reducing pollution levels in the soil, air, and water. Plants are an effective and well-recognized way to reduce the harmful impacts of pollutants on human health and the ecosystem resulting from polluted sites. In this treatment, we use the inherent ability of living plants to reduce contaminant levels in soil and water. It is a cost-effective, solar-driven, and eco-friendly technology and an alternative or complementary cleanup technology that can be used solely or along with other physicochemical methods.

Plants can remove many types of contaminants including pesticides, herbicides, heavy metals, explosives, and petroleum oil, etc. These are also helpful in preventing wind, rain, and groundwater from carrying pollutants away from contaminated sites to other areas. Thus, we can utilize these living plants to remove contaminants from the environment. When we use plants for the removal of contaminants from the environment, it is known as phytoremediation. In this technology, we treat contaminants at the point of their origin, so this technology is also known as in situ remediation technology. Phytoremediation is a broad term that has been used since 1991 to describe the use of plants to reduce the volume, mobility, or toxicity of contaminants in soil, groundwater, or other contaminated media (USEPA, 2000). This technology is a nondestructive and cost-effective in situ method used for the cleanup of contaminated sites. The potential for this technology in the tropics is high due to prevailing climatic conditions that favor plant growth and stimulate microbial activity (Zhang et al., 2013).

The objective of this chapter is to discuss the various plant species that can be used as potential resources to reduce/remove pollutants from the environment.

18.2 **Heavy metal pollutants**

Heavy metal pollution is considered a great menace to living beings. Several physical and chemical treatment technologies are available to decontaminate these pollutants from soil and water (Fig. 18.1). These are incineration, thermal desorption (removal of organic or volatile contaminants), immobilization, ion exchange, vitrification, soil washing/flushing (for the removal of organic or inorganic contaminants, by-products of laboratory processes), precipitation, particle-size separation (removal of the soil fraction that holds the greater portion of the contaminants), membrane filtration, adsorption, and excavation, and disposal (complete removal of the contaminated material, which is packaged and sent to a landfill) (Bondada and Ma, 2003). In the excavation, contaminated materials that cause environmental disruption must be disposed of in suitable landfills. Methods that treat contaminated soil usually generate secondary waste (e.g., wash solution) and produce soil that has lost its fertility. Thus, these methods do not solve the problem but rather only relocate it. In addition, these methods usually are expensive, energy-intensive, and vulnerable to equipment breakdown. Physicochemical methods are not considered environmentally safe and cannot be used long term.

Heavy metals are a group of hazardous inorganic chemicals found at polluted sites. These include arsenic (As), zinc (Zn), lead (Pb), chromium (Cr), cadmium

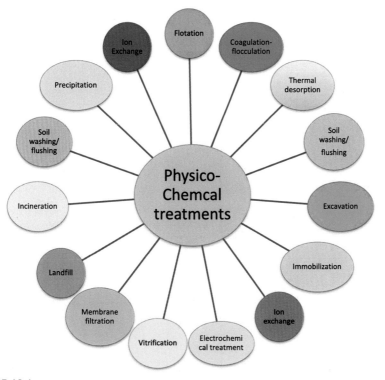

FIGURE 18.1

Various physicochemical methods of treating pollutants.

(Cd), copper (Cu), mercury (Hg), and nickel (Ni) (GWRTAC, 1997). The presence of heavy metals at a particular site depends on the source of the individual pollutants (Iqbal et al., 2008). Weathering of minerals, erosion, volcanoes, forest fires, biogenic sources, and particles released by vegetation are the natural sources of heavy metals contamination. Anthropogenic sources, uses, and impacts of heavy metals on living beings are given in Table 18.1.

Table 18.1 Heavy metals sources in environment and their impacts.

S.No.	Heavy metals	EPA regulatory limit (ppm)	Sources	Impacts
1.	Arsenic (As)	0.01 (USEPA, 2009)	• Agricultural pesticides, herbicides, cosmetics, paints, fungicides, insecticides, wood preservatives, cotton desiccants, semiconductors, light emitting diodes, and components of lasers and microwave circuits. • Contamination of drinking water from natural geological sources. (Nriagu, 1994)	• Arsenic toxicity inactivates up to 200 enzymes, most notably those involved in cellular energy pathways, DNA replication, and repair. • Nausea, vomiting, abdominal pain, diarrhea, skin rash, renal failure, respiratory failure, and pulmonary edema encephalopathy etc. are the other features of arsenic toxicity. (Ratnaike, 2003).
2.	Zinc (Zn)	0.5 (USEPA, 2009)	• Smelting, refining, electroplating industry.	• Occupational exposure to high levels of zinc oxide is associated with metal-fume fever. Excessive absorption of zinc can suppress copper and iron absorption.

Table 18.1 Heavy metals sources in environment and their impacts.—*cont'd*

S.No.	Heavy metals	EPA regulatory limit (ppm)	Sources	Impacts
3.	Lead (Pb)	15 (USEPA, 2009)	• Batteries, cable covers, pigments, mining and smelting, packaging or storage containers (including lead soldered cans), cosmetics, jewelry, toys, candy, glassware, cookware, cigarettes, paints, and leaded gasoline etc. (Wani et al., 2015).	• Highly poisonous affecting almost every organ in the body but nervous system is mostly affected. • Longtime exposure to lead cause anemia, high blood pressure, mainly in old and middle-aged people. • It causes severe damage to the brain and kidneys in both adults and children. • In pregnant women, high exposure to lead may cause miscarriage. It also reduces fertility in males (Sokol and Berman, 1991).
4.	Chromium (Cr)	0.1 (USEPA, 2009)	• It is widely used in industry as steel manufacturing, metal finishing, alloying, electroplating, leather tanning, inhibition of water corrosion, textile dyes, mordents, pigments, ceramic glazes, refractory bricks, etc. (Avudainayagam et al., 2003).	• It causes bronchial asthma, lung and nasal ulcers, cancers, skin allergies, and reproductive and developmental problems. • It is carcinogenic in nature, when taken in excess it may cause death also (Chatterje, 2015).

Continued

Table 18.1 Heavy metals sources in environment and their impacts.—*cont'd*

S.No.	Heavy metals	EPA regulatory limit (ppm)	Sources	Impacts
5.	Cadmium (Cd)	5.0 (USEPA, 2009)	• Cd is used in steel industry, zinc production, television screens, lasers, batteries, paint pigments, cosmetics, as a barrier in nuclear fission, phosphate fertilizer manufacturing, cigarette smoking etc. (Hutton, 1983; Bernhoft, 2013).	• Lung damage, kidney damage, bone damage, and Itai-Itai disease. Reproductive failure and possibly even infertility (Johannes et al., 2006).
6.	Copper (Cu)	1.3 (USEPA, 2009)	• Smelters, phosphate fertilizer production, mining, smelting, refining industries, products from copper such as wire, pipes, and sheet metal.	• Acute copper toxicity is not observed. There is inadequate evidence that copper plays a direct role in the development of cancer in humans. • In sensitive human populations, the major target of chronic copper toxicity is the liver. In Wilson disease, neurological toxicity also occurs (ATSDR, 2004).

Table 18.1 Heavy metals sources in environment and their impacts.—*cont'd*

S.No.	Heavy metals	EPA regulatory limit (ppm)	Sources	Impacts
7.	Mercury (Hg)	2.0 (USEPA, 2009)	• Used in variety of industrial, medical, and consumer products. • Released by burning of coal, oil, wood, wastes that contain mercury. Emissions from power plants, batteries, fluorescent light bulbs, thermometers, thermostats, amalgam in dental fillings and blood pressure instruments (sphygmomanometers). • Used as preservatives in vaccines. • Used in barometers, hydrometers, flame sensors etc. (USEPA, 2017)	• It can harm the brain, heart, kidneys, lungs, and immune system on high level exposure. • Methyl mercury may harm children's developing nervous systems and affect their ability to think and learn (USEPA, 2017).
8.	Nickel (Ni)	0.2 (WHO, 2014)	• Used for stainless steel production. Its alloys are used in the metallurgical, chemical, and food processing industries, especially as catalysts and pigments (Cempel and Nikel, 2005).	• Higher chances of development of cancer, lung embolism, respiratory failure, allergic reactions such as skin rashes, asthma, and chronic bronchitis, birth defects, heart disorders, sickness and dizziness (Cempel and Nikel, 2005).

18.3 **Phytoremediation technology**

One of the new and most promising technologies is phytoremediation, which utilizes plants (terrestrial and aquatic) to remove contaminates from air, water, and soil (Lasat, 2000). The term "phytoremediation" consists of the Greek prefix *phyto* (plant), and Latin *remedium* means (to correct/remove an evil) (Erakhrumen and Agbontalor, 2007; USEPA, 2017). This is an efficient and green emerging technology that can achieve an acceptable level of decontamination at a reasonable cost with minimal environmental disruption. This technology is important in controlling pollution through plants by the utilization of organic toxins and heavy metals. It is a natural process of growing plants to remediate soil, air, and water that is also known as green remediation, botanoremediation, agroremediation, and vegetative remediation (Mirza et al., 2014). In this method, contaminated areas are cultivated with special plants capable of removing contaminants or rendering them harmless because it is similar to usual agricultural practices (Negri and Hinchman, 1996). The success of this technology is dependent on the selection of plant species because it is based on the biological process, anatomy, and physiology of plants. Several plant species have been identified that can remove contaminants from the air, soil, and water efficiently (Luqman et al., 2013). The phytoremediation process is also affected by the types of media, contaminants, and plant species (Mahmood et al., 2007).

We know that trees help maintain a healthy environment because they can filter, absorb, and break down the toxic pollutants of the environment (Aronsson and Perttu, 1994; Glimmerveen, 1996; Beckett et al., 1998; USEPA, 2000). Trees take up heavy metals through their root systems (Bose et al., 2008). Plants that can phytoremediate are classified as accumulators, geobotanical indicators, excluders, and hyperaccumulators (Table 18.2) based on the way they take up and translocate toxic elements to aril parts of the plant (Baker, 1981).

Table 18.2 Classification of plants based on heavy metal accumulation.

S. No.	Types of plants	Growth and metal accumulation characteristics
1.	Accumulators	• These plants accumulate high levels of toxic metals by absorbing them from contaminated soil or water. • These metabolize the pollutants and translocate to aboveground tissues (arial part). • These plants concentrate high levels of toxic elements relatively independently of the soil concentration. Such plant species possess an enormous plant biomass and accumulation of heavy metals. • A huge biomass is noteworthy and permits the immobilization of metal in the aboveground parts of the tree (Peuke and Rennenberg, 2005). • These plants are tolerant to abiotic factors. So these need low investment for pest management. • The total metal extraction can be higher in nonhyperaccumulator than in hyperaccumulator plant species because of the relationship between metal accumulation in aerial parts and the growth potential of the species. • Some species of woody plants for example Willow, poplar (Zacchini et al., 2009) and Brazilian leguminous trees (*Mimosaceae pinifolia, Erythrina speciosa,* and *Schizolobium parahyba*) can remediate Pb-contaminated soil.

Table 18.2 Classification of plants based on heavy metal accumulation.—*cont'd*

S. No.	Types of plants	Growth and metal accumulation characteristics
2.	Geobotanical Indicators	• These plants composition reflects the composition of the parent soil. • Uptake and transport of metal to shoot increases with soil content. • These plants can be used to locate mineral deposits.
3.	Nonaccumulators/ Excluders	• These plants restrict the contaminant uptake into their biomass. So elemental concentration in the shoot is maintained constant. • These plants restrict the amount of a toxic element that is transferred to the aboveground biomass to a certain concentration at which the mechanism breaks down, with subsequent uncontrolled translocation.
4.	Hyperaccumulators	• These plants are rare plants that take up and accumulate higher amounts of heavy metals in aboveground biomass regardless to soil concentration. • Hyperaccumulation depends on the plant's ability to transport/translocate and accumulation of large amounts of metals into the aerial parts of the plant. • Heavy metal concentration in aerial parts varying from 1000 to 10,000 mg kg-1 in plant shoots (Krämer, 2010). For example, in the case of cadmium (Cd), the observed concentration is considerably lower (100 mg per kg) (Krämer, 2010) due to high toxicity of this metal. • Another criteria, developed by Baker et al. (1994), He consider the element concentration ratio between aerial parts and roots, a ratio above 1.0 indicates that the contaminant is mainly accumulated in aerial parts, which means it is a hyperaccumulator plant. • These plant species are unfeasible to phytoremediation because these species produce little biomass and are slow-growing plants. • Example of hyperaccumulator plant species is *Noccaea caerulescens* (Monsant et al., 2011). It is a Brassicaceae hyperaccumulator of zinc (Zn) (Baker et al., 1994).

Plants species used for phytoremediation should have the following characteristics (Cunningham and Ow, 1996; Ali et al., 2013; Arslan et al., 2017; Burges et al., 2018):

1. Fast growth rate;
2. High biomass yield;
3. Capacity to take up a large volume of heavy metals (hyperaccumulator);
4. Ability to transport metals to other parts of the plant;
5. Mechanism to tolerate metal toxicity;

6. Root exudates that degrade the contaminants in the soil;
7. Plants should have the ability to reduce pH and oxygenate sediment, which helps increase the bioavailability of heavy metals. We can also increase the bioavailability of metals by adding chelating agents and micronutrients (Bieby, 2011). Roy et al. (2005) observed positive results by applying a synthetic chelating agent at 5 mmol/kg yield.
8. Plants are screened and selected for those that can hyperaccumulate heavy metals. These can produce large amounts of biomass using established crop production and management practices.

Environmental conditions that support the growth of plants include soil pH/water availability, temperature, sunlight, nutrients, and salinity also influence the phytoremediation potential (Reeves et al., 2018; Tewes et al., 2018).

18.3.1 **Phytoremediation mechanism**

Phytoremediation includes different types of mechanisms during uptake or accumulation of contaminates (Rahman et al., 2011; Sarwar et al., 2017). The various remediation mechanisms are represented in Fig. 18.2. The remedial process is directly proportional to the absorptive and accumulative potential of plants. The following features provide more tolerance to plants:

1. Phytochelators are proteins that can bind to heavy metals and form complexes. These complexes are transported to the vacuoles to reduce the toxicity of metals (Bücker-Neto et al., 2017).

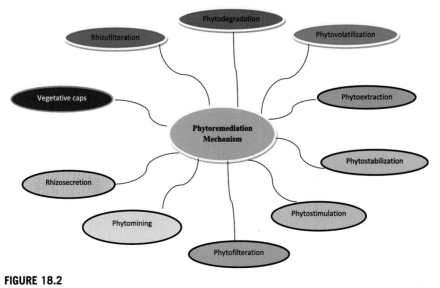

FIGURE 18.2

Various types of phytoremediation mechanisms.

2. Additional thickness of the root wall tissue (exoderm and endoderm) also increases the metal-retaining capacity of plants.
3. Mycorrhizal associations (occurrence of fungi in the root region) secrete root exudates such as organic acids that form complexes with metal cations.
4. Heat shock proteins are associated with the protection of the plasma membrane and chelation by ligands such as metallothionein and organic acids that reduce the activity of the metal in the cytosol, which when ligated with the same, are transported to vacuoles and stored.

By these attributes, a plant can become a good source for phytoremediation. Broadly, five types of remediation mechanisms are used by plants. These mechanisms are recognized as phytoextraction, phytodegradation, phytostabilization, phytovolatilization, and rhizofiltration (Mahmood et al., 2007; Santiago and Bolan, 2010) (Table 18.3).

18.3.2 Plant species used in phytoremediation

Various plant species have been identified that can uptake and store high levels of metallic organic compounds without any toxic impacts. These plant species have heavy metal tolerance that provides them to grow in the presence of high heavy metal polluted ecosystems. These plants uptake and accumulate heavy metals, such as zinc, nickel, chromium, copper, and radionuclides in their roots and shoots (Bollag et al., 1994; Pradhan et al., 1998). Such plant species have specific characteristics that support their growth in contaminated sites, so these plant species have high phytoremediation efficiency and have been excessively utilized for the treatment of heavy metal-contaminated water air, and soil. Plant species can remove heavy metals by the above phytoremediation mechanism is represented in Table 18.4.

Nickel can be remediated through approximately 290 plant species of the Brassicaceae, Cunoniaceae, Euphorbiaceae, Flacourtiaceae, Violaceae family. It is observed that *Adiantum* sp., *Alyssum* sp., *Berkheya coddii*, *Cochlearia aucheri*, *Dicoma niccolifera*, *Geissois pruinosa*, *Pearsonia metallifera*, *Phidiasia lindavii*, *Phyllanthus* sp. (41 taxa), *Pimelea leptospermoides*, *Psychotria douarrei*, *Rinorea bengalensis*, *Sebertia acuminata*, *Stackhousia tryonii*, *Streptanthus polygaloides* plant species are useful in nickel phytoremediation. Cobalt can be remediated through approximately 26 plant species of the Lamiaceae, Scrophulariaceae families. The important plants that can perform phytoremediation are *Cyanotis longifolia*, *Haumaniastrum robertii*, *Hibiscus rhodanthus*, etc. Copper can be remediated through approximately 24 plant species of the Cyperaceae, Lamiaceae, Poaceae, and Scrophulariaceae families. The important plants that can perform phytoremediation are *Aeollanthus subacaulis*, *Eragrostis racemosa*, *Haumaniastrum Katangese*, *Ipomoea alpine*, *Pandiaka metallorum*, *Vigna dolomitica*. Selenium can be remediated by 19 species of the Fabaceae family. *Acacia cana*, *Astragalus bisulcatus*, *Astragalus racemosus*, *Atriplex confertifolia*, *Lecythis ollaria*, *Machaeranthera*

Table 18.3 Types of phytoremediation mechanisms and their applications.

S. No.	Mechanism type	Mechanism and its application	References
1.	Phytoextraction/Phytomining/Phytoaccumulation:	• Plants take up contaminants from soil/water by roots and then translocate and accumulate in the usable tissues/aboveground-level portion of the plant (shoot). • After this, plants are harvested and burned to remove or recycling the contaminant. • **Medium:** Soils • **Contaminant removal:** **Metals:** Ag, Cd, Co, Cr, Cu, Hg, Mn, Mo, Ni, Pb, Zn **Radionuclides:** Sr, Cs, Pu, U **Advantages:** • Phytoextraction is a suitable and inexpensive phytoremediation technique for the removal of heavy metals from wastewater, sediments, and soil permanently. **Disadvantages:** • Hyperaccumulator plants are generally slow growing with a small biomass and shallow root systems. • Proper disposal of the biomass is required after harvesting and metal reclamation • **Plants:** Indian mustard, Pennycress, Alyssum, Sunflowers, Hybrid Poplars.	Kumar et al. (1995); USEPA, 2000; Prasad, 2004; Ali et al. (2013); Erakhrumen (2017); Chandra et al. (2018).
2.	Phytodegradation/Phytotransformation	• Break down of organic pollutants by the metabolic activity within the plant or external to the plant by various microorganisms and released into the rhizosphere. • **Medium:** Surface water and groundwater • **Contaminant removal:** Organic compounds: chlorinated, solvents, phenols, herbicides, **Advantages:** • Bacteria present in the rhizosphere also help in the biotransformation of the contaminant, which ultimately increases the rate of phytovolatilization. • Various contaminates can be removed from the environment, including solvents in groundwater, petroleum and aromatic compounds in soils, and volatile compounds in the air. • **Plants:** Algae, Stonewort, Hybrid Poplar, Black Willow, Bald Cypress.	Burken and Schnoor (1997), 1998; Lee and Reynolds (2004)
3.	Rhizofiltration	• Plant's root system absorbs/adsorbs/precipitates the contaminants onto their roots and restricts the movement of the contaminants in ground water. • In this process roots of plant species uptake, accumulate, and sediment the heavy metals from polluted water and soils. Once the plant uptakes all the contaminants, they can easily be harvested and disposed. • These plants are characterized by longer and fibrous root systems covered with root hairs, which increase their absorptive, surface areas.	Dushenkov et al. (1995); Zhu et al. (1999); Henry (2000) Shammas (2009); Abhilash et al. (2009); Sreelal and Jayanthi (2017);

		• **Contaminant removal:** Organic compounds (TPH, PAHs, pesticides chlorinated solvents, PCBs) • **Medium:** Surface water and water pumped through troughs **Advantages:** • It can be used for both water and soil for either in situ or ex situ applications. • Contaminants are not translocated to the shoots. • Heavy metals that are mostly retained in the soil or water such as cadmium, lead, chromium, nickel, zinc, and copper can be effectively treated through this method. • This method is used for treatment of the agricultural runoff, industrial discharge, radioactive contaminant, and metals. **Disadvantages:** • Plants used for rhizofiltration need to grow in a greenhouse or nursery. • **Plants:** Sunflowers, Indian Mustard, Water Hyacinth	Benavides et al. (2018); Galal et al. (2018)
4.	Phytostabilization/Place deactivation	• Plants reduce the potentially harmful contaminants availability by immobilization of contaminants/accumulation by root/precipitation within the rhizosphere and restrict the movement of contaminants in the water and soil. So this method is best suited for restoration of soil and water quality because it cuts short the movement of the contaminants. • It is not a permanent solution because only the movement of contaminants is restricted, but they continue to stay in the soil and water. • **Medium:** Soils, groundwater, and mine tailings • **Contaminant removal:** As, Cd, Cr, Cu, Pb, Zn **Advantages:** • The disposal of hazardous biomass is not required. **Disadvantages:** • Contaminant left behind in the soil and amendments, mandatory monitoring of soil is required. • **Plants:** Indian mustard, hybrid poplars, grasses	Smith and Bradshaw (1972); Henry (2000); Jadia and Fulekar (2009); Zeng et al. (2018)
5.	Phytovolatilization	• Plants uptake the pollutants from the soil or water and convert pollutants into volatile nature. Then these pollutants released into the atmosphere by in a modified or unmodified form through stomata. • **Medium:** Soils and groundwater • **Contaminant removal:** Chlorinated solvents, some inorganics (Se, Hg, and As) **Advantages:** • Contaminants could be transformed to less toxic forms, such as elemental mercury and dimethyl selenite gas. • It does not require any additional management once the plantation is done. • This method minimizes the soil erosion because here harvesting and disposal of plant biomass is not required, **Disadvantages:** • The metabolite might accumulate in vegetation such as fruit. • **Plants:** Poplars, Alfalfa Black Locust, Indian mustard	Bañuelos et al. (1997); Burken J.G., and Schnoor J.L. 1998; Cristaldi et al. (2017)

Table 18.4 Plant species involved in remediation of heavy metals.

S.No.	Name of plant species	Pollutant removal	Phytoremediation mechanism	References
1.	Eichhornia crassipes	Cd and Pb at a concentration of more than 6000 ppm in plant tissues	Rhizofiltration	Sela et al. (1989); Sela et al. (1990)
2.	Azolla filiculoides Lam	Cd, Cu, Ni, and Zn from the soil solution at levels of 10,000, 9000, 9000, and 6500 ppm	Rhizofiltration	Startford et al. (1984); Demirezen et al. (2004)
3.	Nelumbo nucifera Nymphaea alba L.	Cr up to 3000 mg Cr kg	Rhizofiltration	Vajpayee et al. (2000)
4.	Reed (Phragmites australis)	Zn, Ni, Pb, As, and Cd	Rhizofiltration	Papazogloua (2005); Miao et al. (2012)
5.	Indian mustard (Brassica juncea L.)	Pb	Phytoextraction	Weis and Weis (2004); Wang (2009)
6.	Poplar tree (Populus deltoides)	Unspecified	Vegetative cover	Gardea-Torresdey (2003)
7.	Willow (Salix species)	Ni, Cd, Cu, and Zn	Phytoextraction/ Phytostabilization	
8.	Brassica juncea	Cd and Pb	Phytoextraction	
9.	Grasses	Cd, Pb, Cu, and Zn	Phytostabilization	
10.	Tobacco (Nicotiana spp.)	Cd, Cu, and Zn	Phytostabilization	
11.	Hybrid polar grasses (Populus deltoids × P. balsamifera)	Several heavy metals	Phytostabilization	
12.	Scirpus validus	NaCl	Evapotranspiration	Negri and Hinchman (1996)
13.	Spartina alterniflora			
14.	Agropyron,	Organic compounds, PAHs, PCB	Rhizoremediation	McCutcheon and Schnoor (2004)
15.	Cynodon dactylon,			
16.	Festuca arundinacea,			
17.	Festuca rubra,			
18.	Lolium perenne,			
19.	Lotus corniculatus,			
20.	Trifolium pretense,			
21.	Melilotus officinalis,			
22.	Helianthus annus			

glabriuscula, Neptunia amplexicaulis plants are useful in selenium-contaminated soil bioremediation. Manganese (Mn) is phytoremediated through 11 species *of* Apocynaceae, Cunoniaceae, Proteaceae family. *Alyxia rubricaulis, Eugenia clusioides, Macadamia angustifolia* have a good capacity to remediate Mn-contaminated soil (Warrier, 2012). The above species of plants are useful in the phytoremediation of heavy metal-contaminated soil. These plants can tolerate high levels of metal elements in root and shoot and accumulate several metals. These plants have a fast uptake rate for the metals and translocate the metals from roots to shoots at high rates. In normal plants, metal concentration is usually low but in hyperaccumulator plants shoot metal concentration is higher than root levels (Warrier and Saroja, 2008).

Several terrestrial and aquatic plants are known to be natural hyperaccumulators of metals (Warrier and Saroja, 2008), but because these tend to be slow growers, researchers have turned to other species, which are recently identified or selected, as more promising commercial candidates (Table 18.5). Deep-rooted trees such as poplar, willow, and cottonwood are most commonly used for applications requiring withdrawal of large amounts of water from the subsurface, whereas several different plants, trees, and grasses are used to stimulate microbial degradation of organic contaminants in soil. Some plants species, for example, are barley (*Hordeum vulgare* L.), bermudagrass (*Cynodon dactylon*), tall fescue (*Festuca arundinacea* Schreb), kenaf (*Hibiscus cannabinus* L.), brassica (*Brassica napus*), alfalfa (*Medicago sativa*), channel grass (*Vallisneria spiralis*), water hyacinth (*Eichhornia crassipes*), sunflower (*Helianthus annuus* L.), Indiangrass (*Sorghastrum nutans*), *Acer, Arundo, Astragalus, Betula, Brassica, Cannabis, Castor, Eucalyptus, Helianthus, Jatropha, Linum, Miscanthus, Phalaris, Pisum, Populus, Quercus, Ricinus, Robinia, Salix, Sarcocornia, Sorghum, Zea mays*, and many others have been identified as potential heavy metal remediators.

18.4 Air pollutants and their remediation

Global air quality is degrading quickly and has led to a plethora of health problems worldwide. Because air pollutants can travel long distances, the populations affected by them are not localized. The possible adverse impacts of air pollutants on human health include chronic obstructive pulmonary diseases, acute respiratory infections, lung cancer, enhanced mortality (WHO, 2014; Banerjee et al., 2017), kidney disorders, dry-eye syndrome, peptic ulcers, intestinal disorder, and other acute health problems (Kumar et al., 2015). Apart from outdoor air pollution, indoor air pollution has also become a major concern. Technological advancement has helped to solve the issue, but because most affected populations reside in low- or middle-income countries, most of these technologies are not affordable.

Table 18.5 Plant species that can remove pollutants.

S. No.	Name of plant species	Metal's name	References
1.	*Sebertia acuminata*	than 20% nickel, on a dry-weight basis	Miao et al. (2012)
2.	Penny cress (*Thlaspi calaminare*)	10% zinc accumulation	Kabata-Pendias and Pendias (1991)
3.	Alyssum (*Alyssum bertolonii*)	10% nickel accumulation	
4.	*Pimela suteri*	3% chromium	
5.	Broom tea tree (*Leptospermum scoparium*)	3% chromium	
6.	*Uncinia leptostachya*	3% uranium	
7.	*Coprosma arborea*	3% uranium	
8.	Paper birch (*Betula papyrifera*)	1% mercury	
9.	*Helianthus annuus*	Cr 74%, Cd 42.2%, and Pb 62%,	Eswara et al. (2019)
10.	*Macrotyloma uniflorum*	Cr, Cd, and Pb 72%, 27.81%, and 37.39%,	
11.	*Vigna radiata*	Cr, Cd, and Pb 72.54%, 27.66%, and 47.43%	
12.	Salix spp. *Salix viminalis* *Salix fragilis*	Cd, Cu, Pb, Zn	Pulford and Watson (2003) Volk et al. (2006) Ruttens et al. (2011)
13.	Castor (*Ricinus communis*)	Cd	Huang et al. (2011)
14.	Corn (*Zea mays*)	Cd, Pb, Zn	Meers et al. (2010)
15.	Populus spp. *Populus deltoids* *Populus nigra* *Populus trichocarpa*	Cd, Cu, Pb, Zn	Ruttens et al. (2011)
16.	Jatropha (*Jatropha curcas* L.)	Cd, Cu, Ni, Pb	Abhilash et al. (2009); Jamil et al. (2009)
17.	*Populus deltoids*	Hg	Che et al. (2003)
18.	*Brassica juncea, Astragalus bisulcatus*	Se	Bitther et al. (2012)
19.	*Populus canescens*	Zn	Bittsanszkya et al. (2005)

On the other hand, the use of plants as natural pollutant sinks to reduce pollution is a more efficient and cost-effective technique. Thus, phytoremediation would be more economical, eco-friendly, and sustainable way of combating air pollution (Weyens et al., 2015).

Polluted air is a complex mixture of several pollutants that may undergo series of reactions under specific environmental conditions. Gases such as CO, O_3, sulfur compounds (SO_2, SO_3, H_2S, mercaptans), nitrogen compounds (NO_2, NO, NH_3), organic compounds (e.g., aldehydes, hydrocarbons), and some radioactive gases (e.g., radon) have a short residence time in the atmosphere as they are reactive. Some of these compounds such as volatile organic compounds (VOCs), aldehydes, peroxyacetyl nitrate, and formaldehyde, are highly reactive in the atmosphere and lead to the production of secondary air pollutants. On the other hand, gases such as CO_2, N_2O, and long-lived greenhouse gases (e.g., CFCs, HCFCs, halons, CCl_4, and CF_4), polychlorinated biphenyls (PCBs), polybrominated diphenyl ethers (PBDEs), persist in the environment for quite a long time. Apart from the gases both natural and anthropogenic sources give out airborne particulates (e.g., $PM_{2.5}$ and PM_{10}) which are of major concern in the urban areas. This particulate matter (PM) also acts as carriers of phthalates used as plasticizers in PVC plastics that are gradually released into the air and cause several hormonal problems in humans (Gawronski et al., 2017).

18.4.1 Phytoremediation of air pollutants

The common route for entry of various environmental substances in plants is through either soil or water but the majority of air pollutants enter a plant through the aerial parts i.e., stomata and cuticles present on leaves. Particles are removed from the atmosphere by either dry or wet deposition. After intake, assimilation of pollutants inside the plant cells is governed by physicochemical properties of the pollutant, particular plant species, and environmental factors (Agarwal et al., 2018). The enormous biologically active surface area of plants is a boon for phytoremediation (Gawronski et al., 2017). The surface area of aboveground plant parts on earth is estimated to be no less than 200 million km^2 which even surpasses the earth land area of 149 million km^2 (Varma et al., 2008). Leaves, needles, and young twigs of plants contribute the most to this surface area and are exposed to the air pollutants directly.

Plants have an elaborated gas exchange system that acts as their life support arrangement and can be applied to remove gaseous pollutants from the atmosphere through adsorption or absorption (Gawronski et al., 2017; Agarwal et al., 2018). Thus, phytoremediation that involves the use of plants, or plants and microbial association, can be effectively used to degrade, detoxify, contain, stabilize or sequester some air pollutants (Cunningham, 1995; Harvey et al., 2002). Various plant species involved in the remediation of air pollutants are listed in Table 18.6.

Phylloremediation, a term first used by Sandhu et al. (2007), is the natural process of bioremediating air pollutants through plant leaves and leaf-associated microbes (Wei et al., 2017). Leaf structures such as leaf folding, hairs (pubescence), trichomes (epidermal outgrowths), and wax layers (cuticle) have specific roles in pollutant removal (Gawronski et al., 2017; Wei et al., 2017). The cuticle helps prevent the evaporation of water from the surfaces of the leaf also acts as a primary barrier for the penetration of xenobiotics. The cuticle has a waxy coating, and its volume and composition vary by plant species. Deposition of PM on the waxy cuticle is a much more efficient process for removing pollutants from the air than any other dry deposition process (McDonald et al., 2007). Besides the waxy cuticle, trichomes, hairs, and veins are helpful in the accumulation of PM on the leaf surface (Gawronski et al., 2017).

The aboveground parts of plants that are colonized by a variety of bacteria, yeasts, and fungi are known as phyllosphere (Last, 1955; Bringel and Couee, 2015). The plant—microbe interaction has been widely studied to determine their efficiency in removing atmospheric pollutants. The symbiotic microbial association may be epiphytic (i.e., attached to the surface of plant organs) and/or endophytic (i.e., occurring within plant tissues). It is estimated that the phyllosphere area is up to 4×10^8 km^2 on earth and contains up to 10^{26} bacterial cells (Kembel et al., 2014). Phyllosphere bacterial communities consist of Proteobacteria, such as *Methylobacterium, Sphingomonas, Beijerinckia, Azotobacter*, and *Klebsiella*, and Cyanobacteria such as *Nostoc, Scytonema,* and *Stigonema* (Vacher et al., 2016; Wei et al., 2017). Major fungi in the phyllosphere are Ascomycota, such as *Aureobasidium, Cladosporium,* and *Taphrina* (Kembel and Mueller 2014; Wei et al., 2017), and Basidiomycetous yeasts of genera *Cryptococcus* and *Sporobolomyces* (Cordier et al., 2012; Ottesen et al., 2013). Different plant species contain various microbial communities. The composition of the communities differs with changes in leaf chemistry, leaf morphology, plant growth status, mortality, climatic conditions, season of the year, urban or rural area, and geographical locations (Redford and Fierer, 2009; Jumpponen and Jones, 2010; Finkel et al., 2011; Rastogi et al., 2012; Kembel and Mueller, 2014).

18.4.2 Phytoremediation mechanism for removal of air pollutants

Particles are removed from the atmosphere and come in direct contact with the plant surface by either dry or wet deposition. Metal-enriched particles deposited on the leaf surface may enter the internal tissues through cuticle and stomata (Schreck et al., 2012). Once a pollutant enters the internal tissues of the plant it may undergo

Table 18.6 Air pollutants and their phytoremediation.

Pollutant	Plant species	References
Particulate matter	Neem (*Azadirachta indica*), Gulmohar (*Delonix regia*), Silk cotton (*Bombax ceiba*), Pipal (*Ficus religiosa*), Indian laburnum (*Cassia fistula* and *Cassia siamea*), Indian lilac (*Lagerstroemia indica*), Pagoda tree (*Plumeria rubra* and *Plumeria alba*), Java plum (*Syzygium cumini*)	Pokhriyal and Subba Rao (1986)
	Mango (*Mangifera indica*), Pongamia (*Derris indica*), Umbrella trees (*Thespesia populnea*)	Shetye and Chaphekar (1989)
	Stone pine (*Pinus pinea*), Cornel (*Cornus mas*), Maple (*Acer pseudoplatanus*)	Terzaghi et al. (2013)
	Ficus bengalensis, *Mangifera indica*, *Azadirachta indica*, *Hibiscus rosa-sinensis* (shrub)	Mate and Deshmukh (2015)
	Shrubs: *Pinus mugo*, *Spireae* sp., *Stephanandra incisa*, *Taxus baccata*, *Taxus media*, *Acer campestre*, *Hydrangea arborescens*, *Sorbaria sorbifolia*	Wang et al. (2015)
	Catalpa speciosa, *Broussonetia papyrifera*, *Ulmus pumila*	Chen et al. (2017)
CO	**Trees:** *Acer saccharum*, *Acer saccharinum*, *Gleditsia triacanthos*, *Pinus resinosa*, *Populus nigra*, *Fraxinus pennsylvanica* **Shrubs:** *Syringa vulgaris*, *Hydrangea* sp., **Ornamental indoor plants:** *Ficus variegata* and *Phenix roebelenii*	Bidwell and Bebee (1974)

Continued

Table 18.6 Air pollutants and their phytoremediation.—*cont'd*

Pollutant	Plant species	References
NO$_2$	**Trees:** *Magnolia kobus,* *Eucalyptus viminalis,* *Eucalyptus grandis,* *Eucalyptus globulus,* *Populus nigra,* *Populus* sp., *R. pseudoacacia,* *S. japonicum,* *Prunus cerasoides* **Herbaceous species:** *Nicotiana tabacum* *Erechtites hieracifolia*	Morikawa et al. (1998)
	Robinia pseudoacacia, Sophora japonica, *Populus nigra, Prunus lannesiana*	Takahashi et al. (2005)
Formaldehyde	*Glycine max,* *Chlorophytum comosum* L.	Giese et al. (1994)
	Nerium indicum	Kondo et al. (1995)
	Epipremnum aureum, *Ficus benjamina*	Schmitz et al. (2000)
	Osmunda japonica, Selaginella tamariscina, *Davallia mariesii, Polypodium formosanum*	Wei et al. (2017)
Benzene, Toluene and Xylene (BTX)	*Spinacia oleracea*	Ugrekhelidze et al. (1997)
	Zamioculcas zamiifolia	Sriprapat and Thiravetyan (2013)
Toluene	*Spathiphyllum* *Dracaena deremensis*	Orwell et al. (2006)
	Herbs: *H. helix,* *Philodendron* spp., *Schefflera elegantissima,* *Sansevieria* spp. **Trees:** *Pittosporum tobira* *Pinus densiflora*	Kim et al. (2011)
Benzene	*Dracaena deremensis,* *Dracaena marginata,* *Schefflera actinophylla,* *Howea forsteriana,* *Spathiphyllum floribundum*	Orwell et al. (2004)

Table 18.6 Air pollutants and their phytoremediation.—*cont'd*

Pollutant	Plant species	References
Benzene and Toluene	*Chrysanthemum morifolium, Calathea rotundifolia*	Yang and Liu (2011)
Ethylbenzene and xylene	*D. deremensis, D. marginata, Spathiphyllum*	Wolverton et al. (1989); Wood et al. (2006)
Xylene	*Dieffenbachia seguine* var. *seguine Nephrolepis exaltata Schott*	Wolverton and Wolverton (1996)
	Zamioculcas zamiifolia Aglaonema commutatum, Philodendron martianum, Sansevieria hyacinthoides, Aglaonema rotundum, Fittonia albivenis	Sriprapat et al. (2013)

any of the following processes, i.e., it may be sequestered inside the plant, detoxified, or metabolized to yield CO_2 and H_2O (Agarwal et al., 2018). The process of phytoremediation comprises several processes that differ substantially based on the nature of the pollutant and the physiology and biochemistry of the plant. Major processes of phytoremediation are phytostabilization, rhizodegradation, phytofiltration, phytoextraction, and phytovolatilization. Phytostabilization involves the sequestration of pollutants at first for further metabolism but without degrading them. Once taken up by the plant, the sequestered pollutants may be degraded in the rhizosphere through plant—microbe interactions by the process of rhizodegradation (Yang and Liu, 2011). The screening of airborne PMs by the surface of a leaf is called phytofiltration (Singh and Verma, 2007). Phytoextraction is the process of removing contaminants from soil and translocating them from root to shoots. Phytovolatilization involves the release of volatile pollutants into the air through the stomata of the leaf.

18.4.3 **Remediation of particulate matter/aerosol**

PM is the most dominant air pollutant, especially in urban areas. PM is a combination of solid and liquid particles suspended in the air. According to their mass and composition PM are divided into two groups: coarse particles mostly larger than 2.5 μm in aerodynamic diameter, and fine particles mostly smaller than 2.5 μm in aerodynamic diameter ($PM_{2.5}$) (World Health Organization, 2003).

Large amounts of PM can be removed from the air by the process of adsorption in the aboveground parts of plants or by getting embedded in the waxy cuticle that does not get removed easily due to rain or wind (Beckett et al., 2000; Gawronski et al., 2017). A significant correlation has been found between leaf wax contents and the amount of PM on leaves (Popek et al., 2013). PM embedded in cuticular wax, particularly lipophilic organic compounds (PAHs, PCBs, PBDEs, dioxins, and the charged PM) can gradually diffuse into the leaf and may get displaced and deposited in the cell wall or the vacuole (Gawronski et al., 2017).

Many studies confirm that phytofiltration is a promising mechanism to control pollution by PM_{10} especially in urban areas (Nowak et al., 2014; Agarwal et al., 2018). A thick green belt around the industrial and urban areas can efficiently reduce the spread of air pollutants to the surrounding areas and can reduce dust fall to a large extent (Banerjee and Srivastava 2010; Singh and Verma 2007). The extent of filtration done is different for different plant species and mainly depends on factors such as leaf characteristics, leaf shape, stomata, presence, and density of trichomes, leaf hair density, and the structure and composition of cuticular wax (Sæbø et al., 2012; Weyens et al., 2015; Wei et al., 2017). Studies confirm that needle-shaped leaves accumulated more $PM_{2.5}$ compared with broad leaves (Terzaghi et al., 2013; Chen et al., 2017). Chen et al. (2017) found that small individual leaf areas, abundant wax layers, profuse hairs, and high leaf trichome density showed an increase in $PM_{2.5}$ accumulation and both parameters were positively correlated. Trees have proven to be effective in controlling PM in urban areas along the roadsides. Studies indicate that trees, such as oaks and cedars, can efficiently reduce 50% of PM_{10} and $PM_{2.5}$ concentrations, growing at a distance of 25 m from the roadside, while tall prairie grass can reduce their concentrations by 35% (Cowherd et al., 2006).

Heavy metals may also be found in association with PM and are generally released with the combustion of fossil fuels or mining (Gawronski et al., 2017). Some metals may be highly toxic and nonessential for plants (Cd, As) while some may be necessary in trace amounts but get toxic at higher concentrations (Zn, Cu, Mn). Studies have shown that some elements as Pb, Ni, Cr, V, and Fe are not easily taken up from soils and usually do not translocate to the aerial plant parts, so their presence in leaves and twigs indicates that they get absorbed from the air through their aerial parts (Megremi, 2010; Ny and Lee, 2011; Sæbo et al., 2015; Gawronski et al., 2017). Some plants have a strong defensive mechanism against some harmful species of metals such as toxic forms of metal(loid)s such as As–V and Cr-VI are reduced to less toxic forms, As-III and Cr-III, similarly methylation of As, Cr, and Se may also take place with their subsequent release in the air (Gawronski et al., 2017). Winter melon (*Benincasa hispida*) is an example of a plant that can effectively take up di (2-ethylhexyl) phthalate (DEHP) from polluted air through its leaves, shoots, and flowers (Wu et al., 2013).

18.4.4 **Remediation of gaseous pollutants**

Some common gaseous air pollutants comprise SO_2, CO_2, CO, NOx, and O_3. Major entry points of the gaseous pollutants in plants are stomata and waxy cuticle though at a low level (Gawronski et al., 2017).

18.4.4.1 Carbon monoxide

CO is a toxic gas for humans, and it readily combines with hemoglobin in the blood to form carboxyhemoglobin, which is fatal. However, plants can tolerate CO and they can metabolize it by oxidation to CO_2 or may reduce and incorporate it into amino acids (Gawronski et al., 2017). Different plant species have different abilities to uptake CO.

18.4.4.2 Carbon dioxide

Though CO_2 is a major greenhouse gas leading to global warming, it contributes to the synthesis of plant biomass and sustains the trophic levels of a food chain. So CO_2 can be used to enhance the rate of phytoremediation. Diao et al. (2013) studied the impact of CO_2 fertilization on mung bean (*Vigna radiate*) and maize (*Zea mays*) and found an enhance in their phytoremediation potential denoted by an increase in aboveground dry mass, improved rhizosphere microenvironment, and making plant tolerant to DEHP stress. However, indoor CO_2 pollution can be controlled by phytoremediation. A study was conducted by Tarran et al. (2007) in urban offices, it was revealed that three or more potted plants could reduce about 10% CO_2 concentrations in an air-conditioned building, and a 25% reduction in a non-air-conditioned building.

18.4.4.3 Sulfur oxides

Sulfur oxides are emitted by fossil fuel combustion, industrial activities, etc., and are responsible for acid rain. SO_2 is taken up by plants through stomata. Excessive amounts of SO_2 may be toxic to plants as it gets hydrated and oxidized to sulfite and sulfate, and these are responsible for the inhibition of photosynthesis at high concentrations. It has been observed that plants can assimilate SO_2 and H_2S into reduced sulfur forms such as cysteine and sulfates, but SO_2 concentrations ranging from 131 to 1310 µg/m could be injurious to plants (Wei et al., 2017).

18.4.4.4 Nitrogen oxides

Vehicular pollution is the major source of nitrogen oxides in urban areas. NO_2 enters the plant through the stomata, cuticle, and water film on leaf surfaces. Studies indicate that entry through the stomata is more common and is almost negligible through the cuticle (Geßler et al., 2000). NO_2 gets metabolized in plants to nitrate, nitrite, and organic form such as amino acids with the help of nitrate assimilation pathway, and the process is aided by enzymes such as *nitrate reductase, nitrite reductase, or glutamine synthetase* (Yang and Liu, 2011). Morikawa et al. (2003) in a study exposed plants for 10 weeks to 282 $\mu g/m^{-3}$ NO_2 (which is similar to NO_2 level in heavily polluted urban air) and found that biomass, total leaf area, the contents of carbon (C), nitrogen (N), sulfur (S), phosphorus (P), potassium (K), calcium (Ca), and magnesium (Mg) as well as free amino acid contents and crude proteins doubled, and thus, they concluded that plants could not only utilize NO_2 as a fertilizer but also reduce its concentrations in the atmosphere.

18.4.4.5 Ozone

O_3 also enters the plant mainly through stomata and to some extent by deposition on the cuticle. O_3 readily dissolves in water and releases reactive oxygen species i.e., ROS (O_2^-, H_2O_2, and OH radical) upon reacting with apoplastic structures and plasma membranes (Wei et al., 2017). O_3 or ROS can disturb cell membrane integrity and cause phytotoxicity. Different plant species have different tolerance levels for O_3. In areas with high levels of O_3 concentration, plants develop mechanisms to minimize their toxicity and reduce their concentration from the atmosphere. Some plants emit reactive compounds such as monoterpenes that react with ozone and thus remove it from the air (Di Carlo et al., 2004). Leaves of certain plants such as tobacco secret semivolatile organic compounds such as diterpenoids from their trichomes that can capture and deplete O_3 at the leaf surface and prevent its entry into plants through the stomata (Jud et al., 2016). Plants intolerant to O_3 show activation of their antioxidant systems, ascorbate-glutathione cycle, and antioxidant enzymes upon absorption of O_3 by their leaves to ease the oxidative burden ensued from O_3 exposure (Gawronski et al., 2017). The use of synthetic antioxidants has proved to be effective in helping plants to fight O_3 stress. For instance, European ash (*F. excelsior*) showed a very effective protection mechanism against O_3 injury upon treating with synthetic antioxidants, i.e., ethylene urea (Paoletti et al., 2011).

18.4.4.6 Volatile organic compounds

VOCs include those organic compounds that have a low boiling point and high vapor pressure at room temperature, and thus, they can easily vaporize in the air. They may be naturally occurring and known as biogenic VOCs or may be anthropogenic in origin. VOCs at high levels may cause hematotoxicity and neurotoxic effect, some may cause leukemia, and other compounds such as benzene may cause cancer (Wallace, 2001). They are also involved in various photochemical reactions and lead to the generation of O_3 and secondary aerosols. Some plants release biogenic VOC from their stomata as a part of a defense mechanism to protect themselves from pathogens, herbivores, and other stresses (Gawronski et al., 2017: Agarwal et al., 2018). These compounds include isoprene, terpenes, monoterpenes, sesquiterpenes, and C-10 to C-15 alkanes (Gawronski et al., 2017; Wei et al., 2017).

Anthropogenic VOCs comprise formaldehyde, polycyclic aromatic hydrocarbons (PAHs), and benzene, toluene, and xylenes (BTX) and are responsible for the pollution of indoor spaces as these are found in adhesives, resins, paints, carpets, wallpapers, electronic equipment, tobacco smoke, etc. Plants mostly take up VOC through their stomata or cuticle and may degrade, store or excrete them (Weyens et al., 2015). Studies confirm that benzene, toluene, ethylbenzene, and xylene can be adsorbed by the wax layer of the leaf (Cruz et al., 2014). The mechanism for VOC uptake by the aboveground plant parts is reliant on the characteristics of the VOCs, for example, VOC that are hydrophilic (e.g., formaldehyde) cannot diffuse simply through the cuticle that consists of lipids; however, a lipophilic VOC (e.g., benzene) is more expected to penetrate through the cuticle (Cruz et al., 2014). In an experiment conducted by Tani and Hewitt (2009) it was detected that for eight aldehydes and five ketones, the uptake rate for pollutants was 30−100 times higher compared with what theoretically could be absorbed by the aqueous phase inside the leaves, thus indicating metabolism or translocation of the pollutants inside the plant. Some xenobiotic VOCs are detoxified through metabolism by the use of *oxidoreductase* or *hydrolases*, bioconjugation with sugars, amino acids, organic acids, or peptides, and subsequent removal from the cytoplasm for storage in the vacuole (Edwards et al., 2011). Studies confirm that some plants may detoxify VOCs by converting them into harmless forms for example formaldehyde undergoes oxidation and C1 metabolism in the tissues of *Glycine max\1*, while *Epipremnum aureum* and *Ficus benjamina* can transform formaldehyde into CO_2 and incorporate it into the plant material via the Calvin cycle (Giese et al., 1994; Schmitz et al., 2000). If VOCs cannot be degraded by plants, then their storage or excretion becomes important. At a very high concentration, stored VOC may be toxic to plants. Microorganisms existing in the rhizosphere of potted plants are involved in the absorption and metabolism and thus removal of VOCs from indoor air (Orwell et al., 2006; Wood et al., 2006; Irga et al., 2013).

18.4.4.7 Benzene, toluene, and xylenes
BTXs include benzene, toluene, and three xylene isomers (ortho−, meta−, and para−), which are found in gasoline and have low water solubility. They are considered harmful pollutants due to their acute toxicity and genotoxicity. BTXs are also absorbed through the stomata of the leaf and are transformed to phenol or pyrocatechol, and later to muconic acid and fumaric acid (Ugrekhelidze et al., 1997). The wax of certain plants such as *Sansevieria trifasciata* and *S. hyacinthoides* is rich in hexadecanoic acid and plays very significant role in toluene absorption (Sriprapat et al., 2013). Collins et al. (2000) observed that apple and blackberry leaves and fruits of apple, blackberry, and cucumber accumulate benzene.

18.4.4.8 Polycyclic aromatic hydrocarbons and phenols
PAHs are hydrocarbons that have only carbon and hydrogen with multiple aromatic rings. Phenols consist of a hydroxyl group attached directly to an aromatic hydrocarbon group. Leaves of some plants can absorb gaseous PAHs through their stomata

and cuticular lamellae whereas PAHs that get adsorbed on PM mainly deposit on leaves and are subsequently desorbed from the particles (De Nicola et al., 2017). The PAHs captured and held by leaves, may get adsorbed or absorbed in the leaf tissues (Desalme et al., 2013). Huang et al. (2018) did a comparative study to find the capacity of PAH accumulation by pine, oak, and moss. They observed that moss had a greater tendency to trap high molecular weight PAHs (such as Dibenzo[a,h]anthracene, Benzo[a]pyrene) than oak leaves and pine needles. Due to the presence of many resin channels that link the inside the leaf with the outside, pine needles have a strong tendency to accumulate gaseous PAHs; however, pine needles have a lower tendency to accumulate high molecular weight PAHs that may be attributed to the relatively high lipid content. On the other hand, they observed that oak leaves exhibited a wide range of PAHs including low molecular weight, medium molecular weight PAHs, and a small fraction of high molecular weight PAHs, which may be due to higher values of stomatal density and the occurrence of trichomes in oak leaves compared with pine needles and mosses.

18.4.5 Phytoremediation of indoor and outdoor pollutants

Air pollution has now become a global problem and all spaces whether indoor or outdoor are affected. Phytoremediation has proved to be a very economical and efficient technology that usually does not have any side effects, but very careful analysis is required before selecting any plant species for phytoremediation.

In the case of outdoor phytoremediation, forest covers act as an effective barrier and absorb a significant amount of air pollutants. On relatively smaller scales parks and gardens also act as barriers. The selection of plants for outdoor phytoremediation is very crucial. Plants selected for phytoremediation must have the following characteristics (Wei et al., 2017; Agarwal et al., 2018):

(1) Large leaf area and rough stem;
(2) Broad-leaved and always with many hairs, trichomes, stomata, and waxy cuticle;
(3) Tall and straight with a spreading canopy and medium growth rate;
(4) Ecological compatible and indigenous; use of exotic species must be avoided;
(5) Low water requirement and little agricultural input;
(6) High pollutant-absorption potential;
(7) Tolerant to synergistic effects of pollutants;
(8) Climatic suitability;
(9) Aesthetic values such as conspicuous, attractive foliage and flowers;
(10) Easily multiplied;
(11) Ability to face abiotic stresses such as heat, cold, and drought;
(12) Ability to withstand biotic stresses.

Leaf morphology, such as the presence of a waxy layer, the shape of the leaf, simple or compound leaf, trichomes, etc., greatly affects the phytoremediation potential. Studies show evergreen, conifer plants that have needles covered with profuse waxy layer act as exceptional pollutant accumulators throughout the year (Agarwal et al., 2018). Trees such as Tamarind (*Tamarindus indica*) having smaller compound leaves and are more effective in capturing dust compared with trees with larger simple leaves. Trees, shrubs, herbs, and foliage plants and their combinations may prove to give good results. The combination of plant species, age, and the proper space for the air to move into the filter is helpful for promising results (Gawronski et al., 2017).

The presence of high buildings near the road leads to a canyon effect in which pollutant dispersion is hindered. If trees are planted in the canyon, they may bring down the concentration of NO_2 and PM by 40% and 60%, respectively (Pough et al., 2012). Although in some cases, the effect may be damaging because planting trees in urban street canyons, result in high $PM_{2.5}$ concentrations at street level as trees cause a reduction in the wind speed and ventilation of streets, thus reducing dispersion of the pollutants. Hence, canopy or phytobarrier density, the position of the vegetation, leaf area index, rate of wind speed were found to be significant indicators of PM accumulation in urban street canyons (Jin et al., 2014). A phytobarrier must be sufficiently porous to allow easy airflow through it and sufficiently dense to keep flowing air close to the canopy of the vegetation (Janhall, 2015). For example, Popek et al. (2015), found that more than 50% reduction in PM concentration was observed in a 50 m wide green belt, with a significant decrease in fine and coarse PM, and a slight decline in the ultrafine fraction.

Line sources of air pollution such as highways, streets, roads with excessive traffic, and vehicular pollution may also be controlled by phytoremediation, which is otherwise difficult to control through other technologies. Takahashi et al. (2005) identified four plant species viz., *R. pseudoacacia, Sophora japonica, Populus nigra,* and *Prunus lannesiana* that may control NO_2 pollution in the areas close to the roads with heavy traffic. A layer of herbs and shrubs act as a potential deposition area for air pollutants because they can be planted close to the roads and dividers, do not block the view or traffic lights, and make it possible to remove pollutants by mowing and cutting periodically; their shallow roots can take up deposited pollutants from the top layer of the soil (Gawronski et al., 2017). Examples of herbaceous plants that may be successfully used in phytoremediation include *Achillea millefolium, Polygonum aviculare, Berteroa incana, Flaveria trinervia,* and *Aster gymnocephalus* (Weber et al., 2014). Morikawa et al. (1998) identified *Magnolia kobus, Eucalyptus globulus, Eucalyptus grandis, P. nigra, Populus* sp., *Prunus cerasoides, Nicotiana tabacum,* and *Erechtites hieracifolia* as efficient species that can uptake and assimilate NO_2 after studying 217 herbaceous and woody plant species. Climbers can prove to be effective accumulators of PM when grown on the buildings as green walls and in areas where land for growing plants is inadequate or with dense habitation (Ottele et al., 2010). *Parthenocissus tricuspidata, Parthenocissus quinquefolia, Hedera helix,* and *Polygonum aubertii* are examples of some climbers that can effectively be used in the phytoremediation of PM (Borowski et al., 2009).

There is a great difference between the indoor and outdoor environments, and hence, the selection of plant species is crucial in the phytoremediation of indoor spaces. Indoor plants have been used for quite a long time for decoration and to improve indoor air quality. Using plants as biological filters is advantageous because of their low cost, easy execution, trouble-free maintenance, efficient uptake of pollutants by some species, and being nature-friendly technology (Gawronski et al., 2017). However, plants selected for indoor plantation should not be a source of aeroallergens such as pollens, VOC, microbial spores from degrading plant parts that may adversely affect human health (Agarwal et al., 2018).

Cruz et al. (2014) reviewed the use of various species of potted plants to remove VOCs from indoor air and listed the species. Enzyme-mediated degradation of form-aldehyde by the enzyme *formaldehyde dehydrogenase* has been studied and reported by Xu et al. (2011) in three plant species i.e., *Chlorophytum comosum, Aloe vera*, and *Epipremnum aureum* of these *Chlorophytum* showed better results. Of 86 plant species studied for eliminating formaldehyde by Kim et al. (2010), *Osmunda japonica, Selaginella tamariscina, Davallia mariesii, Polypodium formosanum, Psidium guajava, Lavandula* spp., *Pteris dispar, Pteris multifida,* and *Pelargonium* spp., were found to have high removal capacity.

Benzene may be removed by plants such as *Crassula portulacae, Hydrangea macrophylla, Cymbidium "golden elf," Syngonium podophyllum, S. trifasciata, C. comosum, Euphorbia milii, Dracaena sanderiana, H. helix,* and *Clitoria ternatea* (Liu et al., 2007; Sriprapat and Thiravetyan, 2016). Ornamental plants such as *Hoya carnosa, Hemigraphis alternata,* and *Asparagus densiflorus* are reported to have the capacity to remove VOCs such as benzene, toluene, octane, TCE, and a-pinene, while *Fittonia argyroneura,* can remediate benzene, toluene, and TCE removal, and *Ficus benjamina* can remove octane and a-pinene (Yang et al., 2009). Toluene may be effectively removed by *Schefflera elegantissima, Philoden-dron* spp. "sunlight," and *H. helix* (Kim et al., 2011). *Zamioculcas zamiifolia* has shown an excellent capability to remove xylene (Sriprapat et al., 2013), and to some extent, ethylbenzene also (Tobata et al., 2016). These plants have been used as active green walls in many studies and found to have good potential for the removal of air pollutants. Active green walls have mechanical systems to capture polluted indoor air and force it to pass through the planted porous filter matrix to allow increased contact between the pollutants such as VOCs, PMs, and plant-microorganism association (Irga, 2018). Thomas et al. (2015) developed a mathe-matical model that takes into account the amounts of plant material, building air volume, VOC concentrations, and air exchange.

18.5 Phytoremediation of water pollutants

Water is an essential natural resource on the earth. Life cannot be assumed without water. About 71% of the earth's surface is covered by water. Approximately 97.2% (320 million cubic miles) of the earth's water is found in the oceans. Oceanic water is too salty and not useful for drinking, growing crops, and industrial uses except cooling. The remaining 3% of the earth's water is fresh. But 2.5% of fresh water is locked up in polar icecaps, atmosphere, and soil. This water lies too far under the earth's surface and similarly unavailable. So only 0.5% of the earth's water is available as fresh water and only 0.01% of the fresh water is available for human use. Water pollution is one of the major risks to public health particularly in developing and underdeveloped countries due to poor management and monitoring of water (Mwegoha, 2008; Azizullah et al., 2011).

The majority of the world population is facing the problem of water scarcity. The main reasons for water contamination and scarcity is rapid uncontrolled population growth and urbanization, industrialization, agricultural activities, discharge of geothermal waters (Aguliar, 2009; Rahman et al., 2011; Renuka et al., 2013; Goncalves et al., 2017), and uncontrolled exhaustion of natural resources (Calzadilla, 2011; Connell, 2018). Water contamination by heavy metals is also a substantial problem for us. Heavy meals are difficult to degrade and can accumulate in the organism and the food chain and produce a potential threat to living beings and cause ecological disturbances (Akpor and Muchie, 2010). Heavy metal pollution in ground water causes a serious threat to human health and the aquatic ecosystem (Ali et al., 2020). These pollutants are noxious and toxic for the aquatic ecosystem and human health (Mendoza et al., 2015; Carstea et al., 2016; Ahmed et al., 2017). Many developing countries are facing the problem of potable water. According to WHO, UNICEF (2000) more than 2.2 million people die every year due to drinking polluted water. The availability of safe and fresh water is an utmost challenge for the government due to ground water is gets polluted in the developed and developing world (World Bank, 1998).

Reclamation of wastewater has been the only option left to meet the increasing demand for water in growing industrial and agricultural sectors (Tee et al., 2016). Phytostabilization, phytostimulation, phytovolatilization, phytoextraction, and phytotransformation processes are utilized for remediation of contaminated soil and rhizofiltration and vegetative caps for contaminated waters in the phytoremediation process (Warrier, 2012).

18.5.1 Aquatic plants and phytoremediation

We can utilize the aquatic ecosystem for the heavy metal-contaminated area. It is a cost-effective phytoremediation technique that uses aquatic plants to act as a natural absorber for contaminants and heavy metals (Pratas et al., 2014). Treatment of heavy metal contamination with aquatic plants is the most capable and profitable method (Ali et al., 2013; Guittonny-Philippe et al., 2015).

Aquatic plants are categorized as free-floating aquatic plants, submerged aquatic plants, and emerging aquatic plants. These plants have extensive root systems, which helps them and makes them the best option for the accumulation of heavy metal contaminants in their roots and shoots (Mays et al., 2001; Stoltz and Greger, 2002). Some species of aquatic plants are extensively used for the treatment of wastewater throughout the world (Mesa et al., 2015; Gorito et al., 2017). The success of this technique is depended on the uptake and accumulation capacity of the aquatic plant species (Fritiof and Greger, 2003; Galal et al., 2018). Table 18.7 presents the removal of various contaminants by aquatic plant species.

Table 18.7 Aquatic plants and their application in removal of water pollutants.

	Characteristics	Examples

1.	**Free-floating aquatic plants**
	Characteristics
	• Found in waters with moderate current velocity.
	• High temperature favored the growth.
	• Grow in phosphorus and nitrogen rich water.
	• Heavy metals removed by following two mechanism-
	  (1) Active transport of heavy metals in these plants occurs by the roots, and then transferred to other parts of the body.
	  (2) Passive transport occurs by direct contact of the plant body with pollution medium. In this mechanism, heavy metals mainly accumulate in upper parts of the plant.
	• These plants can effectively be used for the phytoremediation of several heavy metals such as Pb, Hg, Cu, Cr, Ni, Zn, Fe, Mn, As, Cd, Al (Muthusaravanan et al., 2018; Maine et al., 2001; Olguin et al., 2002)
		Examples
		Water hyacinth (*Eichhornia crassipes*)
		Water ferns (*Salvinia minima*)
		Duckweed (*Lemna minor*)
		Duckweed (*Spirodela intermedia*)
		Water lettuce (*Pistia stratiotes*)
		Water cress (*Nasturtium officinale*)

2.	**Submerged Aquatic Plants** • These plants absorb metals mainly by leaves through passive transport. • Metal uptake into plants is due to presence of polygalacturonic acid (cell wall), and pectin (cuticle). • They can accumulate and remove Zn, Cr, Fe, Cu, Cd, Ni, Hg and Pb heavy metals from water and sediments (Branković et al., 2012; El-Khatib et al., 2014; Borisova et al., 2014; Peng, 2008)	Parrot feather (*Myriophyllum spicatum*) Hornwort (*Ceratophylla demersum*) Pond weed (*Potamogeton crispus*) American pondweed (*Potamogeton pectinatus*) Water mint (*Mentha aquatica*) Tape grass (*Vallisneria spiralis*)
3.	**Emergent Aquatic Plants** • These plants are usually found on submerged soil where the water table is half meter below the soil. • Heavy metal accumulations vary in different plant species. • These plants can bioconcentrate most metals in roots from water and sediments and in some emergent plants heavy metals distributed in aerial parts as well. • These plants can effectively be used for the phytoremediation of several heavy metals such as Cd, Fe, Pb, Cr, Zn, Ni, Cu (Sasmaz et al., 2008; Kutty and Al-Mahaqeri, 2016; Rudin et al., 2017)	Smooth cordgrass (*Spartina alterniflora*) Common reed (*Phragmites australis*) Common cattail (*Tyoha latifolia*), Bulrush (*Scirpus* spp.), Common reed (*Phragmites australis*) Smartweed (*Polygonum hydropiperoides*)

18.6 Advantages of phytoremediation

The most serious problem for the environment is the pollution of air, water, and soil. Several physicochemical methods are available to remove pollutants, but after treatment, these methods also produce unwanted toxic products. One of the promising technologies to solve environmental problems is phytoremediation technology, which is an energy-efficient, eco-friendly method of remediating low to moderate levels of pollutants. It can be used in combination with other physicochemical remedial methods to provide excellent results. The cost of phytoremediation is half the price of some alternative methods. It provides a permanent in situ remediation rather than translocating the problem. The key factor of this technology is the identification of suitable plant species that tolerate contaminants and can degrade/remove these contaminants. To date, several plant species have been reported to have the potential to remove various air and water pollutants. But the application of phytoremediation technology has not been widely and commercially used thus far. Growing concern over human health will soon lead to incorporation of phytoremediation technology to combat air and water pollution at various levels. By selecting suitable plant species, it is possible to reduce pollution in heavily polluted areas.

References

Abhilash, P., Jamil, S., Singh, N., 2009. Transgenic plants for enhanced biodegradation and phytoremediation of organic xenobiotics. Biotechnol. Adv. 27, 474–488.

Agarwal, P., Sarkar, M., Chakraborty, B., Banerjee, T., 2018. Phytoremediation of air pollutants: prospects and challenges. In: Pandey, V.C., Bauddh, K. (Eds.), Phytomanagement of Polluted Sites, Market Opportunities in Sustainable Phytoremediation. Elsevier, pp. 221–241. https://doi.org/10.1016/B978-0-12-813912-7.00007-7.

Aguliar, M.J., 2009. Olive oil mill wastewater for soil nitrogen and carbon conservation. J. Environ. Manag. 90, 2845–2848.

Ahmed, M.B., Zhou, J.L., Ngo, H.H., Guo, W., Thomaidis, N.S., Xu, J., 2017. Progress in the biological and chemical treatment technologies for emerging contaminant removal from wastewater: a critical review. J. Hazard Mater. 323, 274–298.

Akpor, O.B., Muchie, M., 2010. Remediation of heavy metals in drinking water and wastewater treatment systems: processes and applications. Int. J. Phys. Sci. 5 (12), 1807–1817.

Ali, H., Khan, E., Sajad, M.A., 2013. Phytoremediation of heavy metals—concepts and applications. Chemosphere 91, 869–881.

Ali, S., Abbas, Z., Rizwan, M., Zaheer, I.E., Yavas, I., Ünay, A., Abdel-DAIM, M.M., Bin-Jumah, M., Hasanuzzaman, M., Kalderis, D., 2020. Application of floating aquatic plants in phytoremediation of heavy metals polluted water: a review. Sustainability 2020 (12), 1–33. https://doi.org/10.3390/su12051927, 1927.

Aronsson, P., Perttu, K., 1994. Willow Vegetation Filters for Municipal Wastewaters and Sludges: A Biological Purification System. Swedish University of Agricultural Science, Uppsala, p. 230.

Arslan, M., Imran, A., Khan, Q.M., Afzal, M., 2017. Plant—bacteria partnerships for the remediation of persistent organic pollutants. Environ. Sci. Pollut. Res. 24, 4322—4336.

ATSDR, 2004. Toxicological Profile for Copper. CAS#: 7440-50. https://www.atsdr.cdc.gov/toxprofiles/tp.asp?id=206&tid=37.

Avudainayagam, S., Megharaj, M., Owens, G., Kookana, R.S., Chittleborough, D., Naidu, R., 2003. Chemistry of chromium in soils with emphasis on tannery waste sites. Rev. Environ. Contam. Toxicol. 178, 53—91.

Azizullah, A., Khattak, M.N., Richter, P., Hader, D.P., 2011. Water pollution in Pakistan and its impact on public health- A review. Environ. Int. 37 (02), 479—497.

Baker, A.J.M., 1981. Accumulators and excluders—strategies in the response of plants to heavy metals. Plant Nutr. 3, 643—654.

Baker, A.J.M., Reeves, R.D., Hajar, A.S.M., 1994. Heavy metal accumulation and tolerance in British populations of the metallophyte *Thlaspi caerulescens*. J. & C. Presl (Brassicaceae). New Phytologist 127, 61—68.

Banerjee, T., Srivastava, R.K., 2010. Estimation of the current status of floral biodiversity at surroundings of Integrated Industrial Estate-Pantnagar, India. Int. J. Environ. Res. 4 (1), 41—48.

Banerjee, T., Kumar, M., Mall, R.K., Singh, R.S., 2017. Airing "clean air" in clean India mission. Environ. Sci. Pollut. Res. 24, 6399—6413.

Bañuelos, G.S., Ajwa, H.A., Mackey, L.L., Wu, C., Cook, S., Akohoue, S., 1997. Evaluation of different plant species used for phytoremediation of high soil selenium. J. Environ. Qual. 26, 639—646.

Beckett, K.P., Freer-Smith, P.H., Taylor, G., 1998. Urban Woodlands: their role in reducing the effects of particulate pollution. Environ. Pollut. 99, 347—360.

Beckett, K.P., Freer-Smith, P.H., Taylor, G., 2000. Particulate pollution capture by urban trees: effect of species and windspeed. Global Change Biol. 6, 995—1003.

Benavides, L.C.L., Pinilla, L.A.C., Serrezuela, R.R., Serrezuela, W.F.R., 2018. Extraction in laboratory of heavy metals through rhizofiltration using the plant *Zea mays* (maize). Int. J. Appl. Environ. Sci. 13, 9—26.

Bernhoft, A.R., 2013. Cadmium toxicity and treatment. Sci. World J. 2013, 7. https://doi.org/10.1155/2013/394652. Article ID 394652.

Bidwell, R.G.S., Bebee, G.P., 1974. Carbon monoxide fixation by plant. Can. J. Bot. 174 (8), 1841—1847.

Bieby, V.T., Siti Rozaimah, S., Hassan, B., Mushrifah, I., Nurina, A., Muhammad, M., 2011. A review on heavy metals (as, Pb, and Hg) uptake by plants through phytoremediation. Int. J. Chem. Eng. 31. https://doi.org/10.1155/2011/939161. Article ID 939161.

Bitther, O.P., Pilon-Smits, E.A.H., Meagher, R.B., Doty, S., 2012. Biotechnological approaches for phytoremediation. In: Arie Altman, A., Hasegawa, P.M. (Eds.), Plant Biotechnology and Agriculture. Academic Press, Oxford, UK, pp. 309—328.

Bittsanszkya, A., Kömives, T., Gullner, G., Gyulai, G., Kiss, J., Heszky, L., Radimszky, L., Rennenberg, H., 2005. Ability of transgenic poplars with elevated glutathione content to tolerate zinc^{2+} stress. Environ. Int. 31, 251—254.

Bollag, J.M., Mertz, T., Otjen, L., 1994. Role of microorganisms in soil remediation. In: Anderson, T.A., Coats, J.R. (Eds.), Bioremediation through Rhizosphere Technology. ACS Symposium Series 563.American Chemical Society. Maple Press, York, PA.

Bondada, B.R., Ma, L.Q., 2003. Tolerance of heavy metals in vascular plants: arsenic hyperaccumulation by Chinese, brake fern (*Pterzs vzttata* L.). In: Chandra, S., Srivastava, M. (Eds.), Pteridology in the New Millennium, pp. 397—420.

Borisova, G., Chukina, N., Maleva, M., Prasad, M., 2014. Ceratophyllum demersum L. and Potamogeton alpinus Balb. From Iset'river, Ural region, Russia differ in adaptive strategies to heavy metals exposure—a comparative study. Int. J. Phytoremediat 16, 621—633.

Borowski, J., Loboda, T., Pietkiewicz, S., 2009. Photosynthetic rates and water efficiencies in three climber species grown in different exposures a turban and suburban sites. Dendrobiology 62, 55—61.

Bose, S., Vedamati, J., Rai, V., Ramanathan, A.L., 2008. Metal uptake and transport by *Tyaha angustata* L. grown on metal-contaminated waste amended soil: an implication of phytoremediation. Geoderma 145, 136—142.

Branković, S., Pavlović-Muratspahić, D., Topuzović, M., Glišić, R., Milivojević, J., Đekić, V., 2012. Metals concentration and accumulation in several aquatic macrophytes. Biotechnol. Biotechnol 26, 2731—2736. Equip.

Bringel, F., Couee, I., 2015. Pivotal roles of phyllosphere microorganisms at the interface between plant functioning and atmospheric trace gas dynamics. Front. Microbiol. 6, 486. https://doi.org/10.3389/fmicb.2015.00486.

Bücker-Neto, L., et al., 2017. Interactions between plant hormones and heavy metals responses. Genet. Mol. Biol. 40 (11), 373—386. https://doi.org/10.1590/1678-4685-gmb-2016-0087.

Burges, A., Alkorta, I., Epelde, L., Garbisu, C., 2018. From phytoremediation of soil contaminants to phytomanagement of ecosystem services in metal contaminated sites. Int. J. Phytoremediation 20, 384—397.

Burken, J.G., Schnoor, J.L., 1997. Uptake and metabolism of atrazine by poplar trees. Environ. Sci. Technol. 31, 1399—1406.

Burken, J.G., Schnoor, J.L., 1998. Distribution and Volatilisation of Organic Compounds.

Calzadilla, A., Rehdanz, K., Tol, R.S., 2011. Water scarcity and the impact of improved irrigation management: a computable general equilibrium analysis. Agric. Econ. 42, 305—323.

Carstea, E.M., Bridgeman, J., Baker, A., Reynolds, D.M., 2016. Fluorescence spectroscopy for wastewater monitoring: a review. Water Res. 95, 205—219.

Cempel, M., Nikel, G., 2005. Nickel: a review of its sources and environmental toxicology. Polish J. Environ. Stud. 15 (3), 375—382 (2006).

Chandra, R., Kumar, V., Tripathi, S., Sharma, P., 2018. Heavy metal phytoextraction potential of native weeds and grasses from endocrine-disrupting chemicals rich complex distillery sludge and their histological observations during *in-situ* phytoremediation. Ecol. Eng. 111, 143—156.

Chatterje, S., 2015. Chromium toxicity and its health hazards. Int. J. Adv. Res. 3 (7), 167—172.

Che, D., Meagher, R.B., Heaton, A.C., Lima, A., Rugh, C.L., Merkle, S.A., 2003. Expression of mercuric ion reductase in Eastern cottonwood (*Populus deltoides*) confers mercuric ion reduction and resistance. Plant Biotechnol. J. 1, 311—319.

Chen, L., Liu, C., Zhang, L., Zou, R., Zhang, Z., 2017. Variation in tree species ability to capture and retain airborne fine particulate matter ($PM_{2.5}$). Sci. Rep. 7 (3206). https://doi.org/10.1038/s41598-017-03360-1.

Collins, C.D., Bell, J.N.B., Crews, C., 2000. Benzene accumulation in horticultural crops. Chemosphere 40, 109—114.

Connell, D.W., 2018. Pollution in Tropical Aquatic Systems. CRC Press, Boca Raton, FL, USA.

Cordier, T., Robin, C., Capdevielle, X., Fabreguettes, O., Desprez-Loustau, M.L., Vacher, C., 2012. The composition of phyllosphere fungal assemblages of European beech (*Fagus sylvatica*) varies significantly along an elevation gradient. New Phytol. 196, 510−519. https://doi.org/10.1111/j.1469-8137.2012.04284.x.

Cowherd, C., Muleski, G., Gebhart, D., 2006. Development of an emission reduction term for near-source depletion. In: Proceeding of 15th International Emission Inventory Conference: Reinventing Inventories d New Ideas in New Orleans. 15 May 2006, New Orleans, LA.

Cristaldi, A., Conti, G.O., Jho, E.H., Zuccarello, P., Grasso, A., Copat, C., Ferrante, M., 2017. Phytoremediation of contaminated soils by heavy metals and PAHs. A brief review. Environ. Technol. Innov. 8, 309−326.

Cruz, M.D., Christensen, J.H., Thomsen, J.D., Muller, R., 2014. Can ornamental potted plants remove volatile organic compounds from indoor air? A review. Environ. Sci. Pollut. Res. Int. 21, 13909−13928. https://doi.org/10.1007/s11356-014-3240-x.

Cunningham, S.D., Berti, W.R., Huang, J.W., 1995. Phytoremediation of contaminated soils. Trends Biotechnol. 13, 393−397.

Cunningham, S.D., Ow, D.W., 1996. Promises and prospects of phytoremediation. Plant Physiol 110, 715.

De Nicola, F., Graña, E.C., Mahía, P.L., Lorenzo, M.S., Rodríguez, P.D., Retuerto, R., et al., 2017. Evergreen or deciduous trees for capturing PAHs from ambient air? A case study. Environ. Pollut. 221, 276−284.

Demirezen, D., Aksoy, A., 2004. Accumulation of heavy metals in *Typha angustifolia* (L.) and *Potamogeton pectinatus* (L.) living in sultan marsh (Kayseri, Turkey). Chemosphere 56, 685−696.

Desalme, D., Binet, P., Chiapusio, G., 2013. Challenges in tracing the fate and effects of atmospheric polycyclic aromatic hydrocarbon deposition in vascular plants. Environ. Sci. Technol. 47, 3967−3981.

Di Carlo, P., Brune, W.H., Martinez, M., Harder, H., Lesher, R., Ren, X., 2004. Missing OH reactivity in a forest: evidence for unknown reactive biogenic VOCs. Science 304, 722−725. https://doi.org/10.1126/science.1094392.

Diao, X.J., Wang, S.G., Mu, N., 2013. Effects of enhanced CO_2 fertilization on phytoremediation of DEHP-polluted soil. Yingyong Shengtai Xuebao 24 (3), 839−846.

Dushenkov, V., Kumar, P., Motto, H., Raskin, I., 1995. Rhizofiltration − the use of plants to remove heavy − metals from aqueous streams. Environ. Sci. Technol. 29, 1239−1245.

Edwards, R., Dixon, D.P., Cummins, I., Brazier-Hicks, M., Skipsey, M., 2011. New perspectives on the metabolism and detoxification of synthetic compounds in plants. In: Schrode, P., Collins, C.D. (Eds.), Organic Xenobiotics and Plants. Springer, New York, NY, pp. 125−148.

El-Khatib, A., Hegazy, A., Abo-El-Kassem, A.M., 2014. Bioaccumulation potential and physiological responses of aquatic macrophytes to Pb pollution. Int. J. Phytoremediation 16, 29−45.

Erakhrumen, A., Agbontalor, A., 2007. Review Phytoremediation: an environmentally sound technology for pollution prevention, control and remediation in developing countries. Educ. Res. Rev. 2 (7), 151−156.

Erakhrumen, A.A., 2017. Phytoremediation: an environmentally sound technology for pollution prevention, control and remediation in developing countries. Educ. Res. Rev. 2, 151−156.

Eswara, R.O., Kesavulu, M.M., Devarajan, S.K., 2019. Pilot study on phytoremediation of contaminated soils with different plant species. J. Hazardous Toxic Radioactive Waste 23 (4).

Finkel, O.M., Burch, A.Y., Lindow, S.E., Post, A.F., Belkin, S., 2011. Geographical location determines the population structure in phyllosphere microbial communities of a salt-excreting desert tree. Appl. Environ. Microbiol. 77, 7647−7655. https://doi.org/10.1128/AEM.05565-11.

Fritiof, Å., Greger, M., 2003. Aquatic and terrestrial plant species with potential to remove heavy metals from stormwater. Int. J. Phytoremediation 5, 211−224.

Galal, T.M., Eid, E.M., Dakhil, M.A., Hassan, L.M., 2018. Bioaccumulation and rhizofiltration potential of *Pistia stratiotes* L. for mitigating water pollution in the Egyptian wetlands. Int. J. Phytoremediation 20, 440−447.

Gardea-Torresdey, J.L., 2003. Phytoremediation: where does it stand and where will it go? Environ. Prog. 22 (1), A2−A3.

Gawronski, S.W., Gawronska, H., Lomnickix, S., Sæbo, A., Vangronsweldk, J., 2017. Plants in air phytoremediation. Adv. Bot. Res. 83, 319−346.

Geßler, A., Rienks, M., Rennenberg, H., 2000. NH_3 and NO_2 fluxes between beech trees and the atmosphere e correlation with climatic and physiological parameters. New Phytol. 147, 539−560.

Giese, M., Bauerdoranth, U., Langebartels, C., Sandermann, H., 1994. Detoxification of form-aldehyde by the spider plant (*Chlorophytum comosum* L.) and by soybean (*Glycine max* L.) cell-suspension cultures. Plant Physiol. 104, 1301−1309.

Glimmerveen, I., 1996. Heavy Metals and Trees. Institute of Chartered Foresters, Edinburgh, p. 206.

Goncalves, A.L., Pires, J.C., Simões, M., 2017. A review on the use of microalgal consortia for waste water treatment. Algal Res. 24, 403−415.

Gorito, A.M., Ribeiro, A.R., Almeida, C.M.R., Silva, A.M., 2017. A review on the application of constructed wetlands for the removal of priority substances and contaminants of emerging concern listed in recently launched EU legislation. Environ. Pollut. 227, 428−443.

Guittonny-Philippe, A., Petit, M.-E., Masotti, V., Monnier, Y., Malleret, L., Coulomb, B., Lafont-Schwob, I., 2015. Selection of wild macrophytes for use in constructed wetlands for phytoremediation of contaminant mixtures. J. Environ. Manag. 147, 108−123.

GWRTAC, 1997. "Remediation of Metals-Contaminated Soils and Groundwater," Tech. Rep. TE-97-01, GWRTAC. GWRTAC-E Series, Pittsburgh, Pa, USA.

Harvey, P.J., Campella, B.F., Castro, P.M.L., Harms, H., Litchtfouse, E., Schaffner, A.R., et al., 2002. Phytoremediation of polyaromatic hydrocarbons, anilines and phenols. Environ. Sci. Pollut. Res. Int. 9 (1), 29−47.

Henry, J.R., 2000. In an Overview of Phytoremediation of Lead and Mercury. NNEMS Report, Washington, D.C., pp. 3−9

Huang, H., Yu, N., Wang, L., Gupta, D.K., He, Z., Wang, K., Zhu, Z., Yan, X., Li, T., Yang, X.E., 2011. The phytoremediation potential of bioenergy crop *Ricinus communis* for DDTs and cadmium co-contaminated soil. Bioresour. Technol. 102, 11034−11038.

Huang, S., Dai, C., Zhou, Y., Peng, H., Yi, K., Qin, P., et al., 2018. Comparisons of three plant species in accumulating polycyclic aromatic hydrocarbons (PAHs) from the atmosphere: a review. Environ. Sci. Pollut. Control Ser. 25 (17), 16548−16566. https://doi.org/10.1007/s11356-018-2167-z.

Hutton, M., 1983. Sources of cadmium in the environment. Ecotoxicol. Environ. Saf. 7 (1), 9−24. https://doi.org/10.1016/0147-6513(83)90044-1.

Iqbal, L.M., He, Z.-li, Stoffella, P.J., Yang, X.-e, 2008. Phytoremediation of heavy metal polluted soils and water: progresses and perspectives. J. Zhejiang Univ. - Sci. B 9 (3), 210−220.

Irga, P.J., Pettit, T.J., Torpy, F.R., 2018. The phytoremediation of indoor air pollution: a review on the technology development from the potted plant through to functional green wall biofilters. Rev. Environ. Sci. Biotechnol. 17 (2), 395−415. https://doi.org/10.1007/s11157-018-9465-2.

Irga, P.J., Torpy, F.R., Burchett, M.D., 2013. Can hydroculture be used to enhance the performance of indoor plants for the removal of air pollutants? Atmos. Environ. 77, 267−271.

Jadia, C.D., Fulekar, M., 2009. Phytoremediation of heavy metals: recent techniques. Afr. J. Biotechnol. 8, 921−928.

Jamil, S., Abhilash, P.C., Singh, N., Sharma, P.N., 2009. *Jatropha curcas*: a potential crop for phytoremediation of coal fly ash. J. Hazard Mater. 172, 269−275.

Janhall, S., 2015. Review on urban vegetation and particle air pollution - deposition and dispersion. Atmos. Environ. 105, 130−137.

Jin, S., Guo, J., Wheeler, S., Kan, L., Che, S., 2014. Evaluation of impacts of tree on $PM_{2.5}$ dispersion in urban streets. Atmos. Environ. 99, 277−287.

Johannes, G., Scheidig, F., Grosse-Siestrup, C., Esche, V., Brandenburg, P., Reich, A., Groneberg, D.A., 2006. The toxicity of cadmium and resulting hazards for human health. J. Occup. Med. Toxicol. 1 (22). https://doi.org/10.1186/1745-6673-1-22.

Jud, W., Fischer, L., Canaval, E., Wohlfahrt, G., Tissier, A., Hansel, A., 2016. Plant surface reactions: an opportunistic ozone defence mechanism impacting atmospheric chemistry. Atmos. Chem. Phys. 16, 277−292. https://doi.org/10.5194/acp-16-277-2016.

Jumpponen, A., Jones, K.L., 2010. Seasonally dynamic fungal communities in the *Quercus macrocarp* phyllosphere differ between urban and nonurban environments. New Phytol. 186, 496−513. https://doi.org/10.1111/j.1469-8137.2010.03197.x.

Kabata-Pendias, A., Pendias, H., 1991. Trace Elements in Soils and Plants, second ed. CRC Press, Boca Raton, FL.

Kembel, S.W., Mueller, R.C., 2014. Plant traits and taxonomy drive host associations in tropical phyllosphere fungal communities. Botany 92, 303−311. https://doi.org/10.1139/cjb-2013-0194.

Kembel, S.W., O'connor, T.K., Arnold, H.K., Hubbell, S.P., Wright, S.J., Green, J.L., 2014. Relationships between phyllosphere bacterial communities and plant functional traits in a neotropical forest. Proc. Natl. Acad. Sci. U.S.A. 111, 13715−13720. https://doi.org/10.1073/pnas.1216057111.

Kim, K.J., Jeong, M.I.I., Lee, D.W., Song, J.S., Kim, H.D., Yoo, E.H., Jeong, S.J., Han, S.W., 2010. Variation in formaldehyde removal efficiency among indoor plant species. Hortscience 35 (10), 1489−1495.

Kim, K.J., Yoo, E.H., Jeong, M.I., Song, J.S., Lee, S.Y., Keys, S.J., 2011. Changes in the phytoremediation potential of indoor plants with exposure to toluene. Hortscience 46 (12), 1646−1649.

Kondo, T., Hasegawa, K., Uchida, R., Onishi, M., Mizukami, A., Omasa, K., 1995. Absorption of formaldehyde by oleander (*Nerium indicum*). Environ. Sci. Technol. 29, 2901−2903.

Krämer, U., 2010. Metal hyperaccumulation in plants. Annu. Rev. Plant Biol. 61, 517−534.

Kumar, M., Singh, R.S., Banerjee, T., 2015. Associating airborne particulates and human health: exploring possibilities. Environ. Int. 84, 201–202.

Kumar, P., Dushenkov, V., Motto, H., Rasakin, I., 1995. Phytoextraction: the use of plants to remove heavy metals from soils. Environ. Sci. Technol. 29, 1232–1238.

Kutty, A.A., Al-Mahaqeri, S.A., 2016. An investigation of the levels and distribution of selected heavy metals in sediments and plant species within the vicinity of ex-iron mine in Bukit Besi. J. Chem 2096147, 2016.

Lasat, M.M., 2000. Phytoextraction of metals from contaminated soil: a review of plant, soil, metal interaction and assessment of pertinent agronomic issues. J. Hazardous Substance Res. 2, 1–25.

Last, F.T., 1955. Seasonal incidence of *Sporobolomyces* on cereal leaves. Trans. Br. Mycol. Soc. 38, 221–239. https://doi.org/10.1016/S0007-1536(55)80069-1.

Lee, A.N., Reynolds, C.M., 2004. Phytodegradation of organic compounds. Curr. Opin. Biotechnol. 15, 225–230.

Liu, Y.J., Mu, Y.J., Zhu, Y.G., Ding, H., Arens, N.C., 2007. Which ornamental plant species effectively remove benzene from indoor air? Atmos. Environ. 41 (3), 650–654.

Luqman, M., Tahir, M.B., Tanvir, A., Atiq, M., Zakaria, M., Hussan, Y., Yaseen, M., 2013. Phytoremediation of polluted water by trees: a review. Afr. J. Agric. Res. 8 (17), 1591–1595. https://doi.org/10.5897/AJAR11.1111. ISSN 1991-637X ©2013 Academic Journals.

Mahmood, T., Malik, S.A., Hussain, Z., Qamar, I., Mateen, H.A., 2007. A Review of Phytoremediation Technology for Contaminated Soil and Water. ESDev. CIIT, Abbottabad, Pakistan.

Maine, M.A., Duarte, M.V., Suñé, N.L., 2001. Cadmium uptake by floating macrophytes. Water Res 35, 2629–2634.

Mate, A.R., Deshmukh, R.R., 2015. To control effects of air pollution using roadside trees. Int. J. Innov. Res. Sci. Eng. Technol. 4 (11), 11167–11171.

Mays, P., Edwards, G., 2001. Comparison of heavy metal accumulation in a natural wetland and constructed wetlands receiving acid mine drainage. Ecol. Eng. 16, 487–500.

McCutcheon, S.C., Schnoor, J.L., 2004. Phytoremediation: transformation and control of contaminants. Environ. Sci. Pollut. Res. 11, 40. https://doi.org/10.1007/BF02980279.

McDonald, A.G., Bealey, W.J., Fowler, D., Dragosits, U., Skiba, U., Smith, R.I., et al., 2007. Quantifying the effect of urban tree planting on concentrations and depositions of PM10 in two UK conurbations. Atmos. Environ. 41 (38), 8455–8467.

Meers, E., van Slycken, S., Adriaensen, K., Ruttens, A., Vangronsveld, J., Du Laing, G., Witters, N., Thewys, T., Tack, F.M., 2010. The use of bio-energy crops (*Zea mays*) for "phytoattenuation" of heavy metals on moderately contaminated soils: a field experiment. Chemosphere 78, 35–41.

Megremi, L., 2010. Distribution and bioavailability of Cr in central Europe, EOB Greece. Cent. Eur. J. Geosci. 2 (2), 103–123.

Mendoza, R.E., García, I.V., de Cabo, L., Weigandt, C.F., de Iorio, A.F., 2015. The interaction of heavy metals and nutrients present in soil and native plants with arbuscular mycorrhizae on the riverside in the Matanza-Riachuelo River Basin (Argentina). Sci. Total Environ. 505, 555–564.

Mesa, J., Mateos-Naranjo, E., Caviedes, M., Redondo-Gómez, S., Pajuelo, E., Rodríguez-Llorente, I., 2015. Scouting contaminated estuaries: heavy metal resistant and plant growth promoting rhizobacteria in the native metal rhizoaccumulator *Spartina maritima*. Mar. Pollut. Bull. 90, 150–159.

Miao, Y., Xi-yuan, X., Xu-feng, M., Zhao-hui, G., Feng-yong, W., 2012. Effect of amendments on growth and metal uptake of giant reed (*Arundo donax* L.) grown on soil contaminated by arsenic, cadmium and lead. Trans. Nonferrous Metals Soc. China 22, 1462−1469.

Mirza, N., Qaisar, M., Shah, M.M., Pervez, A., Sultan, S., 2014. Plants as useful vectors to reduce environmental toxic arsenic content. Sci. World J. 2014, 11. https://doi.org/10.1155/2014/921581. Article ID 921581.

Monsant, A.C., Kappen, P., Wang, Y.D., Pigram, P.J., Baker, A.J.M., Tang, C.X., 2011. *In-vivo* speciation of zinc in *Noccaea caerulescens* in response to nitrogen form and zinc exposure. Plant Soil 348, 167−183.

Morikawa, H., Higaki, A., Nohno, M., Takahashi, M., Kamada, M., Nakata, M., et al., 1998. More than a 600- fold variation in nitrogen dioxide assimilation among 217 plant taxa. Plant Cell Environ. 21, 180−190.

Morikawa, H., Takahashi, M., Kawamura, Y., 2003. Metabolism and genetics of atmospheric nitrogen dioxide control using pollutant-philic plants. In: McCutcheon, S.C., Schnoor, J.L. (Eds.), Phytoremediation: Transformation and Control of Contaminants. John Wiley, Hoboken, NJ, pp. 765−786.

Muthusaravanan, S., Sivarajasekar, N., Vivek, J., Paramasivan, T., Naushad, M., Prakashmaran, J., Al-Duaij, O.K., 2018. Phytoremediation of heavy metals: mechanisms, methods and enhancements. Environ. Chem. Lett 16, 1339−1359.

Mwegoha, W.J.S., 2008. The use of phytoremediation technology for abatement soil and groundwater pollution in Tanzania: opportunities and challenges. J. Sustain. Dev. Afr. 10 (01), 140−156.

Negri, C.M., Hinchman, R.R., 1996. Plants that remove contaminants from the environment. Lab. Med. 27 (1), 36−40.

Nowak, D.J., Hirabayashi, S., Bodine, A., Greenfield, E., 2014. Tree and forest effects on air quality and human health in the United States. Environ. Pollut. 193, 119−129.

Nriagu, J.O., 1994. Arsenic in the Environment: Cycling and Characterization. Wiley, New York, USA, p. 430.

Ny, M.T., Lee, B.K., 2011. Size distribution of airborne particulate matter and associated elements in an urban area an industrial cities in Korea. Aerosol Air Qual. Res. 11, 643−653.

Olguín, E., Hernández, E., Ramos, I., 2002. The effect of both different light conditions and the pH value on the capacity of *Salvinia minima* Baker for removing cadmium, lead and chromium. Acta Biotechnol. 22, 121−131.

Orwell, R.L., Wood, R.A., Burchett, M.D., Tarran, J., Torpy, F., 2006. The potted-plant microcosm substantially reduces indoor air VOC pollution: II. Laboratory study. Water Air Soil Pollut. 177, 59−80.

Orwell, R.L., Wood, R.L., Tarran, J., Torpy, F., Burchett, M.D., 2004. Removal of benzene by the indoor plant/substrate microcosm and implications for air quality. Water Air Soil Pollut. 157, 193−207.

Ottele, M., von Bohemen, H.D., Fraaij, A.L.A., 2010. Quantifying the deposition of particulate matter on climber vegetation on living walls. Ecol. Eng. 36, 154−163.

Ottesen, A.R., Gonzalez, P.A., White, J.R., Pettengill, J.B., Li, C., Allard, S., et al., 2013. Baseline survey of the anatomical microbial ecology of an important food plant: *Solanum lycopersicum* (tomato). BMC Microbiol. 13 (14). https://doi.org/10.1186/1471-2180-13-114.

Paoletti, E., Manning, W.J., Ferrara, A.M., Tagliaferro, F., 2011. Soil drench of ethylenediurea (EDU) protects sensitive trees from ozone injury. iFor. Biogeosci. For. 4 (2), 66−68.

Papazogloua, G., Karantounias, G.A., Vemmos, S.N., Bouranis, D.L., 2005. Photosynthesis and growth responses of giant reed to the heavy metals Cd and Ni. J. Environ. Int. 31 (1), 243—249.

Peng, K., Luo, C., Lou, L., Li, X., Shen, Z., 2008. Bioaccumulation of heavy metals by the aquatic plants Potamogeton pectinatus L. and Potamogeton malaianus Miq and their potential use for contamination indicators and in wastewater treatment. Sci. Total Environ 392, 22—29.

Peuke, A.D., Rennenberg, H., 2005. Phytoremediation: molecular biology, requirements for application, environmental protection, public attention and feasibility. Eur. Mol. Biol. Org. 6, 497—501.

Pokhriyal, T.C., Subba Rao, B.K., 1986. Role of forests in mitigating air pollution. Indian For. 112, 573—582.

Popek, R., Gawronska, H., Gawronski, S.W., 2015. The level of particular matter on foliage depends on the distance from the source of emission. Int. J. Phytoremediation 17, 1262—1268.

Popek, R., Gawronska, H., Wrochna, M., Gawronski, S.W., Sæbo, A., 2013. Particulate matter on foliage of 13 woody species: deposition on surfaces and phytostabilisation in waxes—a 3-year study. Int. J. Phytoremediation 15, 245—256.

Pough, T., MacKenzie, A.R., Whyatt, J.D., Hevitt, C.N., 2012. Effectiveness of green infrastructure for improvement of air quality in urban street canyons. Environ. Sci. Technol. 46, 7692—7699.

Pradhan, S.P., Conrad, J.R., Paterek, J.R., Srivastava, V.J., 1998. Potential of phytoremediation treatment of PAHs in soil at MGP sites. J. Soil Contam. 7, 467—480.

Prasad, M.N.V., 2004. Heavy metal stress in plants: from biomolecules to ecosystem, 2nd ed. Narosa Publishing House, Dyryaganj, New Delhi.

Pratas, J., Paulo, C., Favas, P.J., Venkatachalam, P., 2014. Potential of aquatic plants for phytofiltration of uranium-contaminated waters in laboratory conditions. Ecol. Eng. 69, 170—176.

Pulford, I., Watson, C., 2003. Phytoremediation of heavy metal-contaminated land by trees-a review. Environ. Int. 29, 529—540.

Rahman, M.A., Hasegawa, H., 2011. Aquatic arsenic: phytoremediation using floating macrophytes. Chemosphere 83, 633—646.

Rastogi, G., Sbodio, A., Tech, J.J., Suslow, T.V., Coaker, G.L., Leveau, J.H., 2012. Leaf microbiota in an agroecosystem: spatiotemporal variation in bacterial community composition on field-grown lettuce. ISME J. 6, 1812—1822. https://doi.org/10.1038/ismej.2012.32.

Ratnaike, R.N., 2003. Acute and chronic arsenic toxicity. Postgrad. Med. J. 79, 391—396.

Redford, A.J., Fierer, N., 2009. Bacterial succession on the leaf surface: a novel system for studying succesional dynamics. Microb. Ecol. 58, 189—198. https://doi.org/10.1007/s00248-009-9495-y.

Reeves, R.D., Baker, A.J., Jaré, T., Erskine, P.D., Echevarria, G., van der Ent, A., 2018. A global database for plants that hyperaccumulate metal and metalloid trace elements. New Phytol. 218, 407—411.

Renuka, N., Sood, A., Ratha, S.S.K., Prasanna, R., Ahluwalia, A.S., 2013. Evaluation of microalgal consortia for treatment of primary treated sewage effluent and biomass production. J. Appl. Phycol. 25, 1529—1537.

Roy, S., Labelle, S., Mehta, P., et al., 2005. Phytoremediation of heavy metal and PAH-contaminated brownfield sites. Plant Soil 272 (1—2), 277—290.

Rudin, S.M., Murray, D.W., Whitfeld, T.J., 2017. Retrospective analysis of heavy metal contamination in Rhode Island based on old and new herbarium specimens. Appl. Plant Sci 5, 1600108.

Ruttens, A., Boulet, J., Weyens, N., Smeets, K., Adriaensen, K., Meers, E., van Slycken, S., Tack, F., Meiresonne, L., Thewys, T., et al., 2011. Short rotation coppice culture of willows and poplars as energy crops on metal contaminated agricultural soils. Int. J. Phytoremediation 13, 194–207.

Sæbo, A., Popek, R., Nawrot, B., Hanslin, H.M., Gawronska, H., Gawronski, S.W., 2012. Plant species differences in particulate matter accumulation on leaf surfaces. Sci. Total Environ. 427 (428), 347–354. https://doi.org/10.1016/j.scitotenv.2012.03.084.

Sandhu, A., Halverson, L.J., Beattie, G.A., 2007. Bacterial degradation of airborne phenol in the phyllosphere. Environ. Microbiol. 9, 383–392. https://doi.org/10.1111/j.14622920.2006.01149.x.

Santiago, M., Bolan, N.S., 2010. Phytoremediation of arsenic contaminated soil and water. In: Proceedings of the 19th World Congress of Soil Science. Soil Solutions for a Changing World, Brisbane, Australia, August 2010.

Sarwar, N., Imran, M., Shaheen, M.R., Ishaque, W., Kamran, M.A., Matloob, A., Hussain, S., 2017. Phytoremediation strategies for soils contaminated with heavy metals: modifications and future perspectives. Chemosphere 171, 710–721.

Sasmaz, A., Obek, E., Hasar, H., 2008. The accumulation of heavy metals in Typha latifolia L. grown in a stream carrying secondary effluent. Ecol. Eng. 33, 278–284.

Schmitz, H., Hilgers, U., Weidner, M., 2000. Assimilation and metabolism of formaldehyde by leaves appear unlikely to be of value for indoor air purification. New Phytol. 147, 307–315.

Schreck, E., Foucault, Y., Sarret, G., Sobanska, S., Cécillon, L., Castrec-Rouelle, M., et al., 2012. Metal and metalloid foliar uptake by various plant species exposed to atmospheric industrial fallout: mechanisms involved for lead. Sci. Total Environ. 427 (428), 253–262.

Sela, M., Fritz, E., Hutterman, A., Tel-Or, E., 1990. Physiol. Plantarum 79, 547.

Sela, M., Garty, J., Tel-Or, E., 1989. New Phytol. 112, 7.

Shammas, N.K., 2009. Remediation of metal finishing brown field sites, as chapter 14. In: Wang, L.K., Chen, J.P., Hung, Y., Shammas, N.K. (Eds.), Heavy Metals in the Environment. CRC Press, Boca Raton.

Shetye, R.P., Chaphekar, S.B., 1989. Some estimation on dust fall in the city of Bombay using plants. Prog. Ecol. 4, 61–70.

Singh, S., Verma, A., 2007. Phytoremediation of air pollutants: a review. In: Singh, S.N., Tripathi, R.D. (Eds.), Environmental Bioremediation Technologies. Springer, Berlin, Heidelberg, pp. 293–314. https://doi.org/10.1007/978-3-540-34793-4_13.

Smith, R.A.H., Bradshaw, A.D., 1972. Stabilization of toxic mine wastes by the use of tolerant plant populations. Trans. Inst. Mininig Metallurgy Sec. A 81, 230–237.

Sokol, R.Z., Berman, N., 1991. The effect of age of exposure on lead-induced testicular toxicity. Toxicology 69, 269–278.

Sreelal, G., Jayanthi, R., 2017. Review on phytoremediation technology for removal of soil contaminant. Indian J. Sci. Res. 14, 127–130.

Sriprapat, W., Thiravetyan, P., 2013. Phytoremediation of BTEX from indoor air by *Zamioculcas zamiifolia*. Water Air Soil Pollut. 224 (3). https://doi.org/10.1007/s11270-013-1482-8.

Sriprapat, W., Boraphech, P., Thiravetyan, P., 2013. Factor affecting xylene contaminated air removal by the ornamental plants *Zamioculcas zamiifolia*. Environ. Sci. Pollut. Res. 21, 2603–2610. https://doi.org/10.1007/s11356-013-2175-y.

Sriprapat, W., Thiravetyan, P., 2016. Efficacy of ornamental plants for benzene removal from contaminated air and water: effect of plants associated bacteria. Int. Biodeterior. Biodegrad. https://doi.org/10.1016/j.ibiod.2016.03.001.

Stoltz, E., Greger, M., 2002. Accumulation properties of As, Cd, Cu, Pb and Zn by four wetland plant species growing on submerged mine tailings. Environ. Exp. Bot. 47, 271–280.

Stratford, H.K., William, T.H., Leon, A., 1984. Effects of heavy metals on water hyacinths (*Eichhornia crassipes*). Aquat. Toxicol. 5 (2), 117–128.

Takahashi, M., Higaki, A., Nohno, M., Kamada, M., Okamura, Y., Matsui, K., et al., 2005. Differential assimilation of nitrogen dioxide by 70 taxa of roadside trees a tan urban pollution level. Chemosphere 61 (5), 633–639.

Tani, A., Hewitt, C.N., 2009. Uptake of aldehydes and ketones at typical indoor concentrations by houseplants. Environ. Sci. Technol. 43, 8338–8343.

Tarran, J., Torpy, F., Burchett, M.D., 2007. Use of living pot-plants to cleanse indoor air. Research Review. In: Proceedings of Sixth International Conference on Indoor Air Quality, Ventilation & Energy Conservation in Buildings — Sustainable Built Environment, 3. Sendai, Japan, pp. 249–256.

Tee, P.F., Abdullah, M.O., Tan, I.A.W., Rashid, N.K.A., Amin, M.A.M., Nolasco-Hipolito, C., Bujang, K., 2016. Review on hybrid energy systems for wastewater treatment and bio-energy production. Renew. Sustain. Energy Rev. 54, 235–246.

Terzaghi, E., Wild, E., Zacchello, G., Cerabolini, E.L., Jones, K.V., Di Guardo, A., 2013. Forest filter effect: role of leaves in capturing/releasing air particulate matter and its associated PAHS. Atm. Environ 74, 378–384. https://doi.org/10.1016/j.atmosenv.2013.04.013.

Tewes, L.J., Stolpe, C., Kerim, A., Krämer, U., Müller, C., 2018. Metal hyperaccumulation in the *Brassicaceae* species *Arabidopsis halleri* reduces camalexin induction after fungal pathogen attack. Environ. Exp. Bot. 153, 120–126.

Thomas, C.K., Kim, K.J., Kays, S.J., 2015. Phytoremdiation of indoor air. Hortscience 50 (5), 765–768.

Tobata, M., Vangnai, A.S., Thiravetyan, P., 2016. Removal of ethylbenzene from contaminated air by *Zamioculcas zamiifolia* and microorganisms associated on *Z. zamiifolia* leaves. Water Air Soil Pollut. 227 (4), 115. https://doi.org/10.1007/s11270-016-2817-z.

Ugrekhelidze, D., Korte, F., Kvesitadze, G., 1997. Uptake and transformation of benzene and toluene by plant leaves. Ecotoxicol. Environ. Saf. 37, 24–29. https://doi.org/10.1006/eesa.1996.1512.

USEPA (United State Environmental Protection Agency), 2000. Introduction to Phytoremediation. National Risk Management Research Laboratory. EPA/600/R-99/107, 2000. http://www.clu-in.org/download/remed/introphyto.pdf.

USEPA (United States Environmental Protection Agency), 2017. https://www.epa.gov/mercury/basic-information-about-mercury.

USEPA (United States Environmental Protection Agency), 2009. EPA. Drinking Water Contaminants, Washington, DC, USA.

Vacher, C., Hampe, A., Porte, A.J., Sauer, U., Compant, S., Morris, C.E., 2016. The phyllosphere: microbial jungle at the plant-climate interface. Annu. Rev. Ecol. Evol. Syst. 47, 1–24. https://doi.org/10.1146/annurev-ecolsys-121415-032238.

Vajpayee, P., Tripathi, R.D., Rai, U.N., Ali, M.B., Singh, S.N., 2000. Chemosphere 41, 1075.

Varma, A., Abbott, L., Werner, D., Hampp, R., 2008. Plant Surface Microbiology. Springer, Berlin.

Volk, T.A., Abrahamson, L.P., Nowak, C.A., Smart, L.B., Tharakan, P.J., White, E.H., 2006. The development of short-rotation willow in the northeastern United States for bioenergy and bioproducts, agroforestry and phytoremediation. Biomass Bioenergy 30, 715–727.

Wallacc, L., 2001. Human cxposurc to volatilc organic pollutants: implications for indoor air studies. Annu. Rev. Energy Environ. 26, 269–301.

Wang, H., Shi, H., Wang, Y., 2015. Effects of weather time and pollution level on the amount of particulate matter deposited on leaves of *Ligustrum lucidum*. Sci. World J. https://doi.org/10.1155/2015/935942.

Wang, L.K., Chen, J.P., Hung, Y., Shammas, N.K., 2009. Heavy Metals in the Environment. CRC Press, Boca Raton.

Wani, A.L., Ara, A., Usmani, J.A., 2015. Lead toxicity: a review. Interdiscipl. Toxicol. 8 (2), 55–64. https://doi.org/10.1515/intox-2015-0009.

Warrier, R.R., 2012. Phytoremediation for environmental clean up. Forestry Bull. 12 (2), 2012.

Warrier, R.R., Saroja, S., 2008. Impact of depolluted effluents on the growth and productivity of selected crops. Ecol. Environ. Conserv. 6 (2), 251–253.

Weber, F., Kowarik, I., Saumel, I., 2014. Herbaceous plants as filters: immobilization of particulates along urban street corridors. Environ. Pollut. 186, 234–240.

Wei, X., Lyu, S., Yu, Y., Wang, Z., Liu, II., Pan, D., et al., 2017. Phylloremediation of air pollutants: exploiting the potential of plant leaves and leaf-associated microbes. Front. Plant Sci. 81 (318). https://doi.org/10.3389/fpls.2017.01318.

Weis, J.S., Weis, P., 2004. Metal uptake, transport and release by wetland plants: implications for phytoremediation and restoration review. Environ. Int. 30, 685–700.

Weyens, N., Thijs, S., Popek, R., Witters, N., Przybysz, A., Espenshade, J., et al., 2015. The role of plant-microbe interactions and their exploitation for phytoremediation of air pollutants. Int. J. Mol. Sci. 16, 25576–25604.

WHO, 2014. Mortality and burden of disease from ambient air pollution. Global Health Observ. Data. http://www.who.int/gho/phe/outdoor_air_pollution/burden_text/en/. (Accessed 6 August 2020).

WHO, UNICEF, 2000. Global Water Supply and Sanitation Assessment 2000 Report. World Health Organization and United Nations Children's Fund, USA.

Wolverton, B.C., Wolverton, J.D., 1996. Interior plants their influence on airborne microbes inside energy-efficient buildings. J. Miss. Acad. Sci. 41, 99–105.

Wolverton, B.C., Johnson, A., Bounds, K., 1989. Interior Landscape Plants for Indoor Air Pollution Abatement. Final Report, NASA. John C. Stennis Space Center, Hancock County, MS.

Wood, R.A., Burchett, M.D., Alquezar, R., Orwell, R.L., Tarran, J., Torpy, F., 2006. The potted-plant microcosm substantially reduces indoor air VOC pollution: I. Office field-study. Water Air Soil Pollut 175, 163–180. https://doi.org/10.1007/s11270-006-9124-z.

World Bank, 1998. World Resources 1998–99, Oxford. Oxford University Press, New York.

World Health Organization. Regional Office for Europe, 2003. Health Aspects of Air Pollution with Particulate Matter, Ozone and Nitrogen Dioxide: Report on a WHO Working Group, Bonn, Germany 13–15 January 2003. WHO Regional Office for Europe, Copenhagen.

Wu, Z., Zhang, X., Wu, X., Shen, G., Du, Q., Mo, C., 2013. Uptake of di(2-ethylhexyl) phthalate (DEHP) by plants *Benincasa hispida* and its use for lowering DEHP content in intercropped vegetables. J. Agric. Food Chem. 6 (22), 5220−5225.

Xu, Z., Wang, L., Hou, H., 2011. Formaldehyde removal by potted plant-soil system. J. Hazard Mater. 192 (1), 314−318.

Yang, D.S., Pennisi, S.V., Son, K.C., Kays, S.J., 2009. Screening indoor plants for volatile organic pollutants removal efficiency. Hortscience 44 (5), 1377−1381.

Yang, H., Liu, Y., 2011. Phytoremediation on air pollution. In: Khallaf, M. (Ed.), The Impact of Air Pollution on Health, Economy, Environment and Agricultural Sources. https://doi.org/10.5772/19942. Available from: https://www.intechopen.com/books/the-impact-of-air-pollution-on-health-economy-environment-and-agricultural-sources/phytoremediation-on-air-pollution.

Zacchini, M., Pietrini, F., Mugnozza, G.S., Iori, V., Pietrosanti, L., Massacci, A., 2009. Metal tolerance, accumulation and translocation in poplar and willow clones treated with cadmium in hydroponics. Water Air Soil Pollut. 197, 23−34.

Zeng, P., Guo, Z., Cao, X., Xiao, X., Liu, Y., Shi, L., 2018. Phytostabilization potential of ornamental plants grown in soil contaminated with cadmium. Int. J. Phytoremediation 20, 311−320.

Zhang, S., Li, Q., Lü, Y., Zhang, X., Liang, W., 2013. Contributions of soil biota to C sequestration varied with aggregate fractions under different tillage systems. Soil Biol. Biochem. 62 (2013), 147−156.

Zhu, Y., Zayed, A., Qian, J., De Souza, M., Terry, N., 1999. Phytoaccumulation of trace elements by wetland plants: II. Water hyacinth. J. Environ. Qual. 28, 339−344.

Aquatic plant remediation to control pollution

19

M. Muthukumaran

PG and Research Department of Botany, Ramakrishna Mission Vivekananda College (Autonomous), Affiliated to the University of Madras, Chennai, Tamil Nadu, India

Chapter outline

Biological Approaches to Controlling Pollutants. https://doi.org/10.1016/B978-0-12-824316-9.00004-5

19.1 Introduction

19.1.1 Pollution

Contamination is a troublesome change in the physical, chemical, or natural attributes of air, land, and water squanders or weakens our raw materials assets. Present-day industrial developments have expanded contamination levels beyond the self-cleaning limits of nature. Lately, a significant issue has been the danger to human life from dynamic disintegration of the earth. The rapid development of industrialization, urbanization, agrarian changes, and a vital coming-of-age about the abuse of normal assets to satisfy human needs and wants have contributed much in upsetting the biological equilibrium on which the conditions of our lives depend (Anton and Mathe-Gaspar, 2005). Contamination is the entry of hurtful materials into the earth. These hurtful materials are called contamination. The most significant wellspring of heavy metal (HM) contamination is industrial activity. The principal chemical components delivered into nature because of these exercises may be As, Cd, Cr, Cu, Zn, Ni, Pb, and Hg. Industrial and untreated household wastewater contains pesticides, oils, colors, phenol, cyanides, poisonous organics, phosphorous, suspended solids, and HMs (Pakdel and Peighambardoust, 2018). Commercial activities, for example, metal handling, mining, geothermal vitality plants, car, paper, pesticide fabricating, tanning, and plating, are considered the causes of worldwide HM pollution. The expulsion of HMs from the wastewater is troublesome because they exist in various concoction structures. Most metals are not biodegradable, and they can, without much of a stretch, go through various trophic levels to constantly amass in the biota (Peligro et al., 2016; Raval et al., 2016). The impacts of some HMs, for example, arsenic, lead, mercury, cadmium, chromium, aluminum, and iron, on nature and living creatures, primarily people (Jaishankar et al., 2014).

Phytoremediation is by all accounts a less problematic, conservative, and naturally stable tidy-up innovation. The decision about the proper plant to use is the most critical element in phytoremediation. The cycles related to phytoremediation include cleanup or control methods for remediation of dangerous waste destinations. The various types of phytoremediation are characterized, and their applications are examined. The sorts of polluted media and contaminants appropriate for phytoremediation are summarized for lake and reservoir execution (Jeppesen et al., 2009). Bacterially helped phytoremediation is given here to both natural and metallic contaminants to provide an understanding of how these microscopic organisms help phytoremediation so that future field studies may be encouraged (Glick, 2010). The ongoing turns of events and specialized pertinence of different medicines for the expulsion of HMs from industrial wastewater (Barakat, 2011). An assortment of choices is possible when considering phytoremediation, including wild plant–microorganism affiliations or specific planting and advanced methods (DalCorso et al., 2019).

19.1.2 Pollutant contaminants in aquatic ecosystem

Water contamination by HM particles is an overall ecological issue. HM contamination from industrialization and urbanization is a worldwide issue (Hong et al., 2011). The various contaminants of aquatic environments include (1) inorganic pollutants, (2) organic pollutants, and (3) radionuclide contaminants (Plate 19.1). Aquatic plants and built wetlands have been structured and utilized for the treatment of a wide scope of inorganic pollutants and mine seepage, saltwater, and evacuation of radionuclides.

19.1.2.1 Inorganic pollutants

These incorporate nitrate, phosphate, perchlorate, cyanide, and so on; minor components basic to plants when present in abundance viz., B, Cu, Fe, Mn, Mo, and Zn; minor components basic for creature nourishment when present in overabundance,

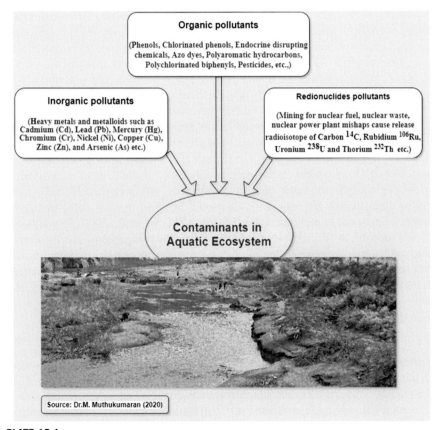

PLATE 19.1

Sources of pollutants in the aquatic ecosystem.

for example, As, Co, Fe, Mn, Zn, Cr, F, Ni, Se, Sn, and V, and the most mephitic minor components such as Cd, Hg, and Pb that are not needed by any creatures. Minor components basic for human nourishment are indistinguishable from creature sustenance except for As and V. HMs among these poisonous substances can undoubtedly be amassed in the general condition (An et al., 2015). HMs are known to be normally happening mixes; however, anthropogenic exercises present them in extraordinary amounts in various natural settings. This causes nature's life-encourage capacity to be diminished as human, creature, and plant well-being become compromised. This happens because of bioaccumulation in the evolved ways of life because of the nondegradable condition of HMs. Remediation of heavy metals requires uncommon regard for secure soil quality, air quality, water quality, human well-being, creature well-being, and all of these in combination (Olusegun et al., 2012).

19.1.2.2 Organic pollutants

Various aquatic plants function admirably for the remediation of natural pollutants. Silt tainted with organics can be cleaned up with plant catalysts (e.g., dehalogenase, laccase, peroxidase, nitrilase, and nitrate reductase). Compounds discharged from plant roots into the rhizosphere can debase the natural particles. On account of PAHs, there is proof for both uptake and digestion by plants (McCutcheon and Schnoor, 2003). In any case, the uptake of huge atoms by plant cells is troublesome when relying on tight "channels" in the structure of the cell divider framework, particularly when they are lipophilic. Oxygenation is a significant introductory method of assault, and this progression serves to expand water dissolvability and gives a chance to formation by means of a glycosidic bond arrangement. Cytochrome P450, peroxygenases, and peroxidases are engaged with plant oxidation of xenobiotics. Other protein classes such as glutathione S-transferases, carboxylesterases, O-glucosyltransferases, malonyltransferases, N-glucosyltransferases, and N-malonyltransferases are related to xenobiotic digestion in plant cells, transport of intermediates, and compartmentation measures (Macek et al., 2000). In addition, the plant roots fill in as a territory for biodegrading microorganisms, and these microorganisms flourish much better and debase organics more quickly in the rhizosphere of explicit plant species. Remediation of water polluted with chlorinated alkanes and other natural synthetics has appeared with aquatic plants. Phytotransformation of perchlorate utilizing parrot-plume (*Myriophyllum aquaticum*) was depicted by Susarla et al. (1999).

19.1.2.3 Radionuclide contamination

Significant wellsprings of radioactive sullying and major radionuclides are delivered with long-term effects. Atomic weapon creation and testing (arrival of basically ^{14}C, ^{106}Ru, 137Cs, ^{90}Sr, and ^{95}Zr) and atomic force creation: (1) During the mining activity, the principle radionuclide released is ^{222}Rn; nature of the mining and processing locales is debased through the scattering of ^{238}U (and descendants—for example, ^{226}Ra, ^{210}Pb, ^{210}Po, and ^{232}Th). (2) During the operational stage, limited quantities of radionuclides are regularly delivered, basically ^{14}C. (3) atomic mishaps can include just a little neighborhood sullying (mixed drink of ^{37}Cs, ^{90}Sr, ^{131}I, ^{210}Po, ^{95}Zr, ^{144}Ce) and (4) normal wellsprings of tainting. Others, for example, zircon and uncommon earth, the grouping of ^{238}U and ^{232}Th might be impressively raised. The transition of a component from aquatic condition to plant frequently alludes to plant take-up or expulsion (Sheppard and Motycka, 1997; Hattink et al., 2000; Pelec et al., 2002). The expulsion of the radionuclide from water with the reaped aspect of the plant (in Bq/zone or volume) is the result of the focus in the plant (plant, in Bq/kg) and the yield of the collected biomass (kg per unit territory/volume of water). Thus, the exchange factor is a significant boundary for deciding the capability of phytoextraction of aquatic macrophytes. *Alternanthera philoxeroides* is one exemplary model that was broadly utilized for the expulsion of radionuclides (Prasad, 2001).

19.1.3 **Phytotechnologies**

Ecological insurance systems including plants are classified as "phytotechnologies." Phytotechnologies utilize plants to remediate, settle, or control tainted or dirtied locales. Phytoremediation is one of the methodologies of phytotechnologies (COST activity 837, 2003). Aquatic plants are spoken to by an assortment of macrophytes, including algal species that happen in different living spaces. They are significant in supplement cycling, control of water quality, residue adjustment, and arrangement of natural surroundings for aquatic living beings. Aquatic macrophytes have fundamental hugeness in the observing of metals in aquatic environments (e.g., *Lemna minor, Azolla pinnata*, and *Eichhornia crassipes*). The utilization of aquatic macrophytes in water quality appraisal has been a typical practice for biomonitoring (Mohan and Hosetti, 1999). Pteridophytes (ferns) could introduce new transformations to the activity of stress factors because of the wide geological spread and the decent variety of the territories where they vegetate, variations that could permit them, in addition to other things hyper-amassing metals (Drăghiceanu et al., 2014).

These species are valuable in contaminated water treatment through phytoremediation or bioremediation innovations. Among the different aquatic plant species, *Azolla, Eichhornia, Lemna, Potamogeton, Spirodela, Wolffia*, and *Wolffiella* have been accounted for as phytoremediators and furthermore, they are exceptionally productive in decreasing aquatic defilement through bioaccumulation of contaminants in their biochemical oxygen request (BOD) tissues. Among the different aquatic species, water hyacinth (*Eichhornia*) is exceptionally safe and can endure the harmfulness of heavy metals, phenols, formaldehyde, formic acids, acidic acids, and oxalic acids even in their high fixations. In like manner, some different types of the family Lemnaceae are proficient at lessening the level of BOD, synthetic oxygen request (COD), just as the effect of HMs, and different ionic types of nitrogen and phosphorus. Here in this audit, we are giving forward-thinking data with respect to the use of these aquatic plants for the bioremediation of debased waters. The survey is fundamentally centered around the particular capacities of aquatic plants and as a significant apparatus in phytotechnologies in the administration of contaminants in aquatic conditions (Ansari et al., 2020).

19.2 Materials and methods

The accompanying strategies were utilized for the investigation of aquatic plants, phytoplankton, biomass creation, and different physic-compound boundaries, including particles and heavy metals of aquatic examples, squander water, and effluents. The relating natural boundaries are distributed (Tadros et al., 2016). Phytoplankton tests were gathered from the surface with a Ruttner water-sampler and protected in 4% killing formaldehyde arrangement left 4 days for setting by Ütermohl (1958). IDs followed Van Heurck (1896) and refreshed the nonexclusive names utilizing Anagnostidis and Komárek (1988). Beginning macroalgae biomass was estimated as the normal of 10 examples weighing from new algal example to dry load of macroalgae biomass at 105°C for 4 h (Horwitz, 2000). The expulsion abundance of Mg^{2+} particles was from the Azolla by Cohen-Shoel et al. (2002), and the metal particle examination arrangement was by Sanyahumbi et al. (1998). Broken down oxygen was resolved by the old-style Winkler's strategy changed by Grasshoff (1976). Disintegrated inorganic nitrogen mixes Clamor (NH_4/N, NO_2/N, and NO_3/N), responsive phosphate (PO_4/P), and receptive silicate (SiO_4) were resolved by Grasshoff (1976). All nitrogen and phosphorous (TN and TP) were resolved in unfiltered water tests following the strategy portrayed by Koroleff (1977) and changed by Valderrama (1981). Chlorophyll in the surface water was removed with 90% $(CH_3)_2CO$ and estimated spectrophotometrically utilizing the SCORE UNESCO condition given by Jeffrey and Humphrey (1975). The eutrophication list was determined by Tomotoshi (1972). The location of arsenic species was cultivated with an ELAN DRC-e ICP-MS (PerkinElmer SCIEX) at Tokyo College of Drug store and Life Science (Kaimidate et al., 2000). Absolute arsenic fixations were likewise estimated without speciation, utilizing ordinary nebulization into the ICP-MS at Ehime College (Sano et al., 2005). The bioadequacy of bacterial strains to decrease contamination was dissected without cell concentrate and communicated as a percentage of pollutant expulsion by Sharma et al. (2014). In light of the development of radiance, the most encouraging strain was biochemically portrayed (Cappucino and Shermann, 2009). A debasement study was completed after the convention devised by Rajeswari et al. (2013). Nitrate and phosphate were resolved by standard strategies (González-Camejo et al., 2018). Aquatic/squander water/effluent tests were examined for the decrease in contamination markers viz. shading, pH, BOD, COD, chloride, nitrate, phosphate, and heavy metals (APHA, 2000).

19.3 Results and discussion

19.3.1 Phytoremediation technology

Phytoremediation is viewed as a successful, tastefully satisfying, practical, and natural innovation for the remediation of conceivably harmful metals from the earth. Plants in phytoremediation amass contaminants through their underlying foundations and afterward move these contaminants in the over-the-ground part of their BOD (Ashraf et al., 2018; Sharma et al., 2015). The thought of utilizing metal gatherer plants for the expulsion of HMs and a few different contaminants in phytoremediation was first presented in 1983, yet this thought has just been embedded throughout the previous 300 years (Blaylock, 2008). Phytoremediation is referred to by various names, for example, agroremediation, green remediation, vegetative remediation, green innovation, and botanoremediation (Sarwar et al., 2017; Kushwaha et al., 2018). Aquatic plants are basic in the treatment of aquatic environment/organic wastewater treatment frameworks because they can be utilized for phytoremediation through rhizofiltration, phytoextraction, phytovolatilization, phytodegradation, or phytotransformation methods (Table 19.1). The destruction of pollutants relies on the span of introduction, centralization of toxins, ecological elements (pH, temperature), and plant attributes (species, root framework, and so on.). In any case, it should be noted that various types of aquatic plants have been used in the phytoremediation cycle of wastewater with impressive results (Akinbile et al., 2016).

Aquatic macrophytes are spoken to by an assortment of algal and macrophytic species in various territories. Individuals from Cyperaceae, Potamogetonaceae, Ranunculaceae, Typhaceae, Haloragaceae, Hydrocharitaceae, Najadaceae, Juncaceae, Pontederiaceae, Zosterophyllaceae, and Lemnaceae, for the most part, speak to aquatic plants. These plants are either developing, lowered, or free coasting. Some nonvascular plants, similar to full-scale green growth, are rootless and equipped for developing with their thalli in the water. Aquatic macrophytes are critical segments of a marine environment for essential profitability and supplement cycling (Prasad, 2007). Aquatic plants perform lively jobs remediating HMs from contaminated sites without the need for work by other hyperaccumulator plants. The use of aquatic plants both in bioaccumulation (with living plant biomass) and biosorption (with dead plant biomass) should be possible effectively for the annihilation of HMs (Ali et al., 2020). Throughout the long term, aquatic plants have increased their notoriety tremendously as a result of their ability to tidy up sullied destinations all through the world. Aquatic plants consistently build up a broad arrangement of roots that causes them and makes them an ideal alternative for the aggregation of contaminants in their foundations and shoots. The development and development of aquatic plants are tedious, which may confine the developing interest in phytoremediation. By the by, this weakness is counteracted by the number of favorable effects this innovation has for the treatment of wastewater (Fritioff and Greger, 2003; Syukor et al., 2014). An interdisciplinary innovation can profit by various methodologies that preowned aquatic plants are appropriate for wastewater treatment because they have a colossal limit of engrossing supplements and eliminates HMs from wastewater and subsequently cut the contamination load down. This audit demonstrated that aquatic plants, for example, *Pistia*, Duckweed, water hyacinth, and *Hydrilla,* can have remediating impacts on lead expulsion from wastewater (Singh et al., 2012).

Water quality was assessed by phytoplankton thickness and dependent on inorganic supplement information (phosphate—PO_4, nitrate—NO_3, nitrite—NO_2, smelling salts—NH_3), the aquatic framework was partitioned into oligotrophic, mesotrophic, and eutrophic water types. The aquatic biological system is a savvy

Table 19.1 Indicated the kinds of phytoremediation, capacities, plant species that eliminate the pollutants (Uqab et al., 2016; Wani et al., 2017).

Phytoremediation process	Functions	Pollutants	Sources	Plant species
Phytoextraction	Eliminates metal contaminations that collect in plants. Eliminates organics from soil by gathering them in plant parts	Cd, Cu, Pb, Zn, Cr, Ni petroleum, hydrocarbons, and radionuclides of uranium and thorium series	Soil and aquatic parts	Eichhornia crassipes, Nymphaea violacea, Viola baoshanensis, Rumex crispus, Sedum alfredii, etc.
Phytodegradation	Plants and related microorganisms degrade organic pollutants	DDT, explosives, and nitrates	Aquatic parts	Ceratophyllum demersum, Elodea Canadensis, and Pueraria sp., etc.
Phytostabilization	Utilizations plants to lessen the bioavailability of pollutants in the environment	Cu, Cd, Cr, Fe, Ni, Pb, Zn	Soil parts	Anthyllis vulneraria and Festuca arvernensis, Hydrocotyle umbellata, etc.
Phytotransformation	Plants take up and corrupt organic compounds	Xenobiotic constituents	Soil parts	Carex gracilis, Canna flaccida, Salvinia rotundifolia
Phytovolatilization	The progression in which usually organic compounds are released from the aboveground sectors of the plant into the atmosphere	Retention of some metalloids or heavy metals	Aquatic and terrestrial plant biomass	Phragmites australis, Pteris vittata
Rhizofiltration	Retains essential metals from water and waste streams	As, Cd, Cu, Pb, Zn	Aquatic parts	Nymphaea alba, Trifolium pratense

and ingenious tidy-up procedure for the phytoremediation of an enormous defiled region. As an extra to different phytoremediation systems and as a component of a push to make this innovation stronger, various researchers have started to investigate the chance of utilizing different soil microorganisms along with plants. These microorganisms incorporate biodegradative microbes, plant development advancing microscopic organisms, and microbes that encourage phytoremediation by different methods (Glick, 2010). The reason for this investigation was to acquire the expulsion effectiveness esteems from glucose, ammonium, and formaldehyde dependent on poisonousness units. The nitrification cycle is applied utilizing nitrifying microscopic organisms and the cycle of plants utilizing water hyacinth (Paris and Mangkoedihardjo, 2020). Phytoremediation is a viable and reasonable arrangement used to remediate harmful toxins from aquatic environments. The different aquatic plants that can possibly eliminate heavy metals from wastewater have been studied (Soni and Kaur, 2018, Plate 19.2).

19.3.2 Characteristics of phytoremediation of aquatic plants

Plants ought to have the accompanying qualities that make phytoremediation an eco-maintainable innovation: a local and snappy development rate, high biomass yield, the take-up of a substantial volume of HMs, the capacity to move metals in over-the-ground portions of the plant, and a system to endure metal poisonousness (Ali et al., 2013; Arslan et al., 2017; Burges et al., 2018; Cunningham and Ow, 1996). Different elements such as pH, sun-oriented radiation, supplement accessibility, and saltiness incredibly impact the phytoremediation potential and development of the plant (Reeves et al., 2018; Tewes et al., 2018). *Salvinia* spp. display limits with regards to eliminating contaminants, for example, HMs, inorganic supplements, explosives from wastewaters. Properties, for example, including high profitability, high sorption limit, and high metal evacuation potential, set up *Salvinia* as an aquatic plant with enormous potential for use in phytoremediation (Dhir, 2009, Table 19.2).

19.3.3 Mechanism of phytoremediation

Phytoremediation follows various instruments, for example, phytoextraction, phytostabilization, phytovolatilization, and rhizofiltration during the take-up or aggregation of HMs in the plant (Rahman and Hasegawa, 2011; Sarwar et al., 2017). The various instruments associated with the phytoremediation cycle are quickly portrayed beneath (Plates 19.3 and 19.4).

19.3.3.1 Phytoextraction

Phytoextraction is likewise called phytoaccumulation, and it includes the take-up of heavy metal in the plant roots and afterward their movement into an over-the-ground-level bit of the plantlike shoots, and so on. When the phytoextraction is done, the plant can be reaped and consumed for picking up vitality and recuperating/reusing metal whenever required from the debris (Erakhrumen, 2017; Chandra

Nature of some aquatic plants

Eg: Nelumbo nucifera, Nymphaea alba, Hydrilla verticillata, Chara vulgaris, Myriophyllum verticellatum, Salvinia molesta, Pistia stratiotes, Eichornia crassipes, Marsilea minuta and *Azolla filiculoides* etc.

Source: Dr.M. Muthukumaran (2020)

PLATE 19.2

Some aquatic plants in the natural ecosystem.

et al., 2018). Now and then, phytoremediation and phytoextraction are utilized equivalently, which is a misguided judgment; phytoextraction is a cleanup innovation, while phytoremediation is the name of an idea (Prasad et al., 2005). Phytoextraction is an appropriate phytoremediation strategy for the remediation of HMs from wastewater, dregs, and soil (Ali et al., 2013; Kocon and Jurga, 2017).

Table 19.2 Aquatic micro/macrophytes for phytotechnologies to treated natural, inorganic, radionuclides toxins, industrial effluents, corrosive mine seepage, saltwater, the guideline of water, and expulsion of different physic-compound boundaries including HMs (Ali et al., 2020; COST action 859, 2005; Farahdiba et al., 2020; Jasrotia et al., 2017; Muthukumaran, 2009; Muthukumaran et al., 2012; Muthukumaran and Sivasubramaian, 2017; Prasad, 2007).

Plant name	Plant group	Common name	Bioremediation/ Phycoremediation/ Phytoremediation functions
Alisma subcordatum	Angiosperms (Alismataceae)	Water-plantain	Uptake of explosives
Amphora turgida	Algae (Bacillariophyceae)	Diatom	Involved significant correction of pH, BOD, COD, removal of color, TDS, nitrate, phosphate, heavy metals, etc., around 60%–80% in the industrial effluents
Anthoceros sp.	Bryophyte (Anthocerotaceae)	Hornworts	Removal of pesticides
Apium graveolens	Angiosperms (Apiaceae)	Celery	Removes sulfonated anthraquinones in textile wastewater
Azolla filiculoides	Pteridophyte (Salviniaceae)	Water fern	Metal hyperaccumulation
Bacopa monnieri	Angiosperms (Plantaginaceae)	Water hyssop	Metal accumulation
Canna flaccida	Angiosperms (Cannaceae)	Canna plant	Heavy metals removal in constructed wetland
Carex gracilis	Angiosperms (Cyperaceae)	Acute sedge	Degrades trinitrotoluene (TNT)
Carex pendula	Angiosperms (Cyperaceae)	Pendulous sedge	Heavy metals removal in constructed wetland
Chara sp.	Algae (Characeae)	Chara alga	To effectively remove U and other radionuclides in the mining effluents
Ceratophyllum demersum	Angiosperms (Ceratophyllaceae)	Coontail	Degradation of organics
Cladium jamaicense	Angiosperms (Cyperaceae)	Sawgrass	Brine concentration
Chlorella vulgaris	Algae (Chlorophyceae)	Chlorella alga	Involved significant correction of pH, BOD, COD, removal of color, TDS, nitrate, phosphate, heavy metals, etc., around 60%–80% in the industrial effluents

Continued

Table 19.2 Aquatic micro/macrophytes for phytotechnologies to treated natural, inorganic, radionuclides toxins, industrial effluents, corrosive mine seepage, saltwater, the guideline of water, and expulsion of different physic-compound boundaries including HMs (Ali et al., 2020; COST action 859, 2005; Farahdiba et al., 2020; Jasrotia et al., 2017; Muthukumaran, 2009; Muthukumaran et al., 2012; Muthukumaran and Sivasubramaian, 2017; Prasad, 2007).—*cont'd*

Plant name	Plant group	Common name	Bioremediation/ Phycoremediation/ Phytoremediation functions
Chlorococcum humicola	Algae (Chlorophyceae)	Chlorococcum alga	Involved significant correction of pH, BOD, COD, removal of color, TDS, nitrate, phosphate, heavy metals, etc., around 60%–80% in the industrial effluents
Chroococcus turgidus	Algae (Cyanophyceae)	Chroococcus alga	Involved significant correction of pH, BOD, COD, removal of color, TDS, nitrate, phosphate, heavy metals, etc., around 60%–80% in the industrial effluents
Cladophora glomerata	Algae (Chlorophyceae)	Cladophora alga	To reduce nitrate and phosphate
Cladophora sp.	Algae (Chlorophyceae)	Cladophora alga	To reduce arsenic by 40%–50%
Chlorodesmis sp	Algae (Chlorophyceae)	Chlorodesmis alga	To reduce arsenic by 40%–50%
Eichhornia crassipes	Angiosperms (Pontederiaceae)	Water hyacinth	Metal accumulation and biosorption
Eleocharis obtusa	Angiosperms (Cyperaceae)	Blunt spike	Transformation of TNT
Eleocharis tuberosa	Angiosperms (Cyperaceae)	Water chestnut	Transformation of TNT
Elodea canadensis	Angiosperms (Hydrocharitaceae)	Pondweed	Phytofiltration of stormwater and removal of zinc
Eriophorum angustifolium	Angiosperms (Cyperaceae)	Cottonsedge	Phytostabilization of metal-rich mine tailings
Eriophorum scheuchzeri	Angiosperms (Cyperaceae)	White cottongrass	Phytostabilization of metal-rich mine tailings
Glyceria fluitans	Angiosperms (Poaceae)	Floating sweet grass	Phytostabilization of mine tailings, treatment of acid mine drainage
Hydrilla verticillata	Angiosperms (Hydrocharitaceae)	Water thyme	TNT transformation and metals accumulation

Table 19.2 Aquatic micro/macrophytes for phytotechnologies to treated natural, inorganic, radionuclides toxins, industrial effluents, corrosive mine seepage, saltwater, the guideline of water, and expulsion of different physic-compound boundaries including HMs (Ali et al., 2020; COST action 859, 2005; Farahdiba et al., 2020; Jasrotia et al., 2017; Muthukumaran, 2009; Muthukumaran et al., 2012; Muthukumaran and Sivasubramaian, 2017; Prasad, 2007).—cont'd

Plant name	Plant group	Common name	Bioremediation/ Phycoremediation/ Phytoremediation functions
Hydrocotyle umbellata	Angiosperms (Araliaceae)	Dollarweed	Biosorption of toxic metals
Ipomea aquatica	Angiosperms (Convolvulaceae)	Water spinach	Metal accumulation
Iris pseudacorus	Angiosperms (Iridaceae)	Swamp/yellow iris	Methyl bromide and TNT transformation
Juncus articulatus	Angiosperms (Juncaceae)	Jointleaf rush	Phytostabilization of mine tailings
Juncus glaucus	Angiosperms (Juncaceae)	Rushes	Degrades TNT
Lemna minor	Angiosperms (Araceae)	Duckweed	Concentrates technetium-99
Miscanthus floridulus	Angiosperms (Poaceae)	Silvergrass	Heavy metals removal in constructed wetland
Miscanthus sacchariflorus	Angiosperms (Poaceae)	Amur silvergrass	Heavy metals removal in constructed wetland
Mougeotia sp.	Algae (Chlorophyceae)	Mougeotia alga	To effectively remove U and other radionuclides in mining effluents
Myriophyllum aquaticum	Angiosperms (Haloragaceae)	Parrot feather	Explosives sensitivity to and transformation, halocarbon metabolism, halogenated organics transformation, hormesis, organophosphorus degradation, perchlorate degradation
Myriophyllum spicatum	Angiosperms (Haloragaceae)	Milfoil	TNT monitoring and transformation
Nelumbo nucifera		Indian lotus	TNT transformation
Nitella sp.	Algae (Characeae)	Nitella alga	To effectively remove U and other radionuclides in the mining effluents
Nymphaea violacea	Angiosperms (Nymphaeaceae)	Waterlily	Uranium and thorium series

Continued

Table 19.2 Aquatic micro/macrophytes for phytotechnologies to treated natural, inorganic, radionuclides toxins, industrial effluents, corrosive mine seepage, saltwater, the guideline of water, and expulsion of different physic-compound boundaries including HMs (Ali et al., 2020; COST action 859, 2005; Farahdiba et al., 2020; Jasrotia et al., 2017; Muthukumaran, 2009; Muthukumaran et al., 2012; Muthukumaran and Sivasubramaian, 2017; Prasad, 2007).—*cont'd*

Plant name	Plant group	Common name	Bioremediation/ Phycoremediation/ Phytoremediation functions
Nymphaea odorata	Angiosperms (Nymphaeaceae)	Fragrant water lily	TNT transformation
Pistia stratiotes	Angiosperms (Araceae)	Water lettuce	Metal accumulation
Phragmites australis	Angiosperms (Poaceae)	Common reed	Methyl iodide volatilization, integral component in wetlands, and treatment of dairy wastes in constructed wetland
Phormidium papyraceum	Algae (Cyanophyceae)	Phormidium algae	The removal of nitrate, phosphate, and heavy metals
Polygonum punctatum	Angiosperms (Polygonaceae)	Dotted smartweed	Radiocesium (^{137}Cs)
Potamogeton nodosus	Angiosperms (Potamogetonaceae)	Pondweed	Explosives degradation
Potamogeton natans	Angiosperms (Potamogetonaceae)	Floating pondweed	Phytofiltration of (1) stormwater and removal of zinc
Potamogeton pectinatus	Angiosperms (Potamogetonaceae)	Sago pondweed	Transformation of explosives, uptake of metals
Potamogeton pusillus	Angiosperms (Potamogetonaceae)	Small pondweed	Used in free-surface wetlands
Rumex hydrolapathum	Angiosperms (Polygonaceae)	Water dock	Removes sulfonated anthraquinones in textile wastewater
Sagittaria latifolia	Angiosperms (Alismataceae)	Arrowhead	Explosives degradation and radiocesium (^{137}Cs)
Salicornia virginica	Angiosperms (Amaranthaceae)	Perennial glasswort	Perchlorate tolerance, brine concentrator
Salvinia molesta	Angiosperms (Salviniaceae)	Kariba weed	Metals accumulation
Salvinia rotundifolia	Angiosperms (Salviniaceae)	Floating moss	TNT transformation
Scirpus validus	Angiosperms (Cyperaceae)	Bulrush	Brine concentration

Table 19.2 Aquatic micro/macrophytes for phytotechnologies to treated natural, inorganic, radionuclides toxins, industrial effluents, corrosive mine seepage, saltwater, the guideline of water, and expulsion of different physic-compound boundaries including HMs (Ali et al., 2020; COST action 859, 2005; Farahdiba et al., 2020; Jasrotia et al., 2017; Muthukumaran, 2009; Muthukumaran et al., 2012; Muthukumaran and Sivasubramaian, 2017; Prasad, 2007).—*cont'd*

Plant name	Plant group	Common name	Bioremediation/ Phycoremediation/ Phytoremediation functions
Spartina alterniflora	Angiosperms (Poaceae)	Cordgrass	Saltwater, brine concentration
Spirodela oligorrhiza	Angiosperms (Araceae)	Giant duckweed	Organic degradation and metals accumulation
Spirodela oligorrhiza	Angiosperms (Araceae)	Giant duckweed	Metal accumulation
Spirogyra sp.	Algae (Chlorophyceae)	Spirogyra alga	Removal of pesticides
Spirulina platensis	Algae (Cyanophyceae)	Spirulina alga	Removal of nitrate, phosphate, and heavy metals
Sporobolus virginicus	Angiosperms (Poaceae)	Coastal dropseed	Brine concentration
Tamarix spp.	Angiosperms (Tamaricaceae)	Salt cedar	Hydraulic control of arsenic
Trifolium pratense	Angiosperms (Fabaceae)	Red clover	Rhizodegradation
Typha angustifolia	Angiosperms (Typhaceae)	Cattail	Degradation of explosives
Typha latifolia	Angiosperms (Typhaceae)	Cattail	Degrades TNT, biosorption and perchlorate degradation
Ulothrix sp.	Algae (Chlorophyceae)	Ulothrix alga	To effectively remove U and other radionuclides in the mining effluents
Vallisneria americana	Angiosperms (Hydrocharitaceae)	Tape grass	TCE transformation and metals accumulation
Vallisneria spiralis	Angiosperms (Hydrocharitaceae)	Eelgrass	Metal hyperaccumulation
Wolffia globosa	Angiosperms (Araceae)	Asian watermeal	Biosorbents of inorganic and organic pollutants and metals accumulation
Zizania aquatica	Angiosperms (Poaceae)	Wild rice	Uptake of ^{129}I

BOD, *biochemical oxygen demand;* COD, *chemical oxygen demand;* TCE, *trichloroethylene;* TDS, *total dissolved solids;* TNT, *trinitrotoluene.*

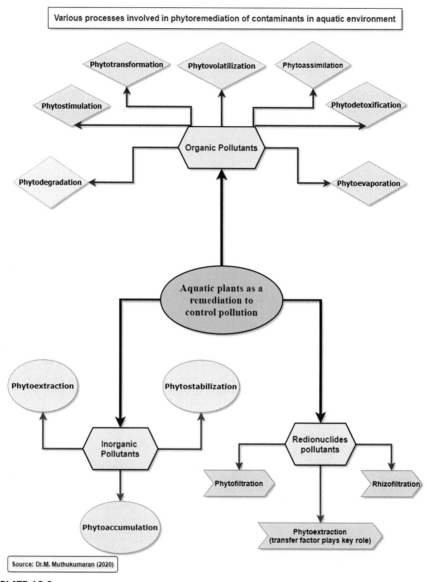

Various processes involved in phytoremediation of contaminants in aquatic environment

Source: Dr.M. Muthukumaran (2020)

PLATE 19.3

Various processes involved in phytoremediation of contaminants in aquatic environments.

19.3.3.2 Phytostabilization

Phytostabilization includes the utilization of the plant to confine the development of contaminants in the dirt. The term phytostabilization is otherwise called setup deactivation. Remediation of soil, muck, and residue can be viably done by utilizing this innovation. It does not meddle with the indigenous habitat and is a much more

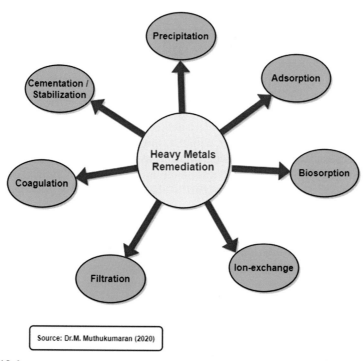

PLATE 19.4

Mechanisms for the phytoremediation of heavy metals.

secure elective alternative (Cundy et al., 2013; Najeeb et al., 2017). In phytostabilization, plants restrain or go about as an obstruction for the permeation of water inside the
dirt. At the point, when we must continue rebuilding our surface water, groundwater,
and soil quality, this innovation is most appropriate because it stops the development
of contaminants (Jadia and Fulekar, 2009; Labidi et al., 2017). Phytostabilization is
successful for an enormous site, which is vigorously influenced by the contaminants
(Mahdavian et al., 2017). Phytostabilization is just an overseeing approach for inacti-
vating/immobilizing the possibly destructive contaminants. It is anything but a perpetual goal, on the grounds that solitary the development of metals is confined, yet they
keep on remaining in the dirt (Zeng et al., 2018).

19.3.3.3 Rhizofiltration

Rhizofiltration includes the utilization of the plant to retain/adsorb the contaminants, bringing about confined development of these contaminants in underground
water (Abhilash et al., 2009; Benavides et al., 2018). Roots have a tremendous

impact on rhizofiltration. Factor, for example, changing pH in the rhizosphere and root exudates helps the precipitation of HMs on the outside of the roots. When the plant has absorbed all the contaminants, they can, without much of a stretch, be collected and arranged (Zhu et al., 1999). Plants for rhizofiltration ought to have the capacity; to create a far-reaching root framework, gather high groupings of heavy metals, be anything but difficult to deal with, and have low upkeep cost (Raskin and Ensley, 2000; Kushwaha et al., 2018). Both aquatic and earthbound plants with long sinewy root frameworks can be utilized in rhizofiltration (Raskin and Ensley, 2000). Rhizofiltration is profitably utilized for dealing with and treatment of farming runoff, industrial release, radioactive contaminant, and metals (Galal et al., 2018). HMs that are generally held in the dirt, for example, cadmium, lead, chromium, nickel, zinc, and copper, can be remediated enough through rhizofiltration (Sreelal and Jayanthi, 2017). Another innovation, called phytoremediation, utilizes green plants to dewater, eliminate inorganic contaminants as HMs and radionuclides, and debase natural contaminants as plants retain supplements with their underlying foundations. Evapotranspiration is misused to diminish the volume of fluid Free-living microorganisms of the rhizosphere that encompass the root framework take an interest in the debasement of manufactured natural contaminants (Negri and Hinchman, 1996).

19.3.3.4 Phytovolatilization

Phytovolatilization is the cycle wherein a plant changes toxins into a different unpredictable nature, and afterward, their progressive delivery into the general condition with the assistance of the plant's stomata (Ghosh and Singh, 2005; Leguizamo et al., 2017). Plant species such as canola and Indian mustard are valuable for the phytovolatilization of selenium. Mercury and selenium are the best contaminants for remediation by phytovolatilization (Karami and Shamsuddin, 2010). Perhaps the best-preferred position of phytovolatilization is that it does not need any extra administration once the ranch is finished. Different advantages are limiting soil disintegration, no aggravation to the dirt, pathetic reaping, and the removal of plant biomass (Cristaldi et al., 2017). Microbes present in the rhizosphere additionally help in the biotransformation of the contaminant, which inevitably supports the pace of phytovolatilization. It includes retention of some metalloids or HMs from the dirt/aquatic, passing them on to the leaves, and changing them into unpredictable mixes in the stomata. As has been effectively volatilized in the leaves of *Pteris vittata*—As mixes (Sakakibara et al., 2007).

19.3.3.5 Phytodegradation

Because of the instruments of the plant, a few items are corrupted by the digestion measures and are normally natural; however, there are some inorganic intensifies that can likewise be debased, among them trinitrotoluene (dynamite), dichlorodiphenyltrichloroethane (DDT), trichloroethylene, hexachlorobenzene, and polychlorinated biphenyls (PCBs) (Serrat, 2014).

19.3.3.6 Phytotransformation

The contaminant item is biotransformed—that is, changed over—to the less destructive substance, an element that is extremely helpful for the remediation of natural mixes, and this cycle can be perceived as the utilization of the contamination in less harmful structures by the plant itself (Tadros et al., 2016).

19.3.4 The potential roles of aquatic plants in remediation of polluted water resources

Aquatic plants have important jobs and possibilities in the remediation of contamination in wastewater. It has likewise investigated ongoing examination chip away at the proficiency of *Salvinia molesta* and *Pistia stratiotes* plants in wastewater remediation and distinguished zones for additional investigations as we find stoichiometric homeostatic list and asset beat impacts investigations of these plants is fundamental in wastewater phytoremediation measures (Mustafa and Hayder, 2020). The aquatic plants, this bioaccumulative impact can be misused, pointing to a biotechnological and bioengineering application to eliminate metals, called phytoremediation, utilizing drifting aquatic macrophytes, which have high potential because of their properties holding contaminants. Results obtained were definitive for variations in *Eichhornia crassipes* and *Salvinia auriculata* as better phytoremediation specialists individually, while *Lemna minor* and *Pistia stratiotes* fit better in biomonitoring, which has protection from specific groupings of metals when identified with Album, Hg, Zn, Ni, and Pb (de Souza and Silva, 2019).

The lowered aquatic macrophytes have extremely flimsy fingernail skin and in this way, promptly take up metals from water through the whole surface. Subsequently, the incorporated measures of bioavailable metals in water and residue can be shown somewhat by utilizing macrophytes. Aquatic macrophytes, with their capacity to endure unfavorable conditions and high colonization rate, are outstanding instruments for phytoremediation. Further, they redistribute metals from dregs to water, and lastly, they take up in the plant tissues and henceforth look after flow. Benthic established macrophytes (both lowered and developing) assume a significant job in metal bioavailability from residue through rhizosphere trades and other transporter chelates. This normally encourages metal take-up by other gliding and emanant structures. The aquatic macrophytes promptly take up metals in their decreased structure from dregs and oxidize them in the plant tissues making them stationary and thus bioconcentrating them to a serious degree (Prasad, 2007). There is capacity by four aquatic macrophytes, *Eichhornia crassipes, Ludwigia stolonifera, Echinochloa stagnina,* and *Phragmites australis,* to gather Disc, Ni, and Pb and their utilization for demonstrating and phytoremediating these metals in polluted wetlands (Eid et al., 2020). The *Eleocharis acicularis* is handily developed and controlled in situations with bountiful water, hyperaccumulates Cu, Zn, As, Pb, Album, Cs, and Hg, demonstrating the incredible potential for use in the phytoremediation of water conditions tainted by the HMs (Sakakibara, 2016). Fundamental appraisals of phytoremediation potential propose that the plant *Pityrogramma calomelanos* may eliminate around 2% of the dirt arsenic load every year. With due thought to the sort of arsenic mixes present in the greenery and their water dissolvability, the alternative of arranging high arsenic plants adrift is raised for conversation (Francesconi et al., 2002). The results are shown in Table 19.3 and Plate 19.5.

Table 19.3 Collection capability of different aquatic plants for biomonitoring of harmful minor components in a wide scope of harmfulness bioassays (Ahmad et al., 2011; Ali et al., 2020; Jasrotia et al., 2017; Muthukumaran, 2009; Prasad, 2007).

Aquatic plant	Common name	Metals/Metalloids
Azolla filiculoides	Water fern	Cr, Ni, Zn, Fe, Cu, Pb
Bacopa monnieri	Water hyssop	Hg, Cr, Cu, Cd
Chara sp.	Chara alga	U and other radionuclides
Chlorococcum humicola	Chlorococcum alga	Cr, Cu, Ni
Ceratophyllum demersum	Coontail	Cd, Cu, Cr, Pb, Hg, Fe, Mn. Zn, Ni, Co, and radionuclides
Cladophora sp.	Cladophora alga	As
Eichhornia crassipes	Water hyacinth	As, Cd, Co, Cr Cu, Al, Ni, Pb, Zn, Hg, P, Pt, Pd, Os, Ru, Ir, Rh
Hydrilla verticillata	Water thyme	Hg, Fe, Ni, Pb
Lemna minor	Duck weed	As, Mn, Pb, Ba, B, Cd, Cu, Cr, Ni, Se, Zn, Fe
Mentha aquatica	Water mint	Pb, Cd, Fe, Cu
Myriophyllum spicatum	Milfoil	Cd, Cu, Zn, Pb, Ni, Cr
Nitella sp.	Nitella alga	U and other radionuclides
Pistia stratiotes	Water lettuce	Cr, Zn, Fe, Cu, Mn
Salvinia minima	water spangles	As, Ni, Cr, Cd
Salvinia herzogii	Water fern	Cd, Cr
Spirodela intermedia	Duckweed	Fe, Zn, Mn, Cu, Cr, Pb
Nasturtium officinale	Watercress	Cr, Ni, Zn, Cu,
Nymphaea alba	White water Lily	Ni, Cr, Co, Zn, Mn, Pb, Cd, Cu, Hg, Fe
Mentha aquatica	Water mint	Cd, Zn, Cu, Fe, Hg
Myriophyllum spicatum	Parrot feathers	Pb, Cd, Fe, Cu
Pistia stratiotes	Water lettuce	Cu, Al, Cr, P, Hg
Phormidium papyraceum	Blue-green algae	Fe, Cr, Zn, Pb, As, Hg, Cd, and Cu

Table 19.3 Collection capability of different aquatic plants for biomonitoring of harmful minor components in a wide scope of harmfulness bioassays (Ahmad et al., 2011; Ali et al., 2020; Jasrotia et al., 2017; Muthukumaran, 2009; Prasad, 2007).—cont'd

Aquatic plant	Oommon name	Metals/Metalloids
Phragmites australis	Common reed	Fe, Cu, Cd, Pb, Zn
Polygonum hydropiperoides	Smartweed	Cu, Pb, Zn
Potamogeton crispus	Pondweed	Cu, Fe, Ni, Zn, and Mn
Potamogeton pectinatus	American pondweed	Cd, Pb, Cu, Zn
Ranunculus aquatilis	Water-crowfoot	Mn, Pb, Cd, Fe, Pb
Salvinia molesta	Kariba weed	Mn, Pb
Scapania uliginosa	Liverworts	B, Ba, Cd, Co, Cr, Cu, li, Mn, Mo, Ni, Pb, Sr, V, Zn
Spartina alterniflora	Cordgrass	Cu. Cr, Zn, Ni, Mn, Cd, Pb, As
Spirulina platensis	Blue-green algae	Fe, Cr, Zn, Pb, As, Hg, Cd, and Cu
Vallisneria americana	Tape grass	Cd, Cr, Cu, Ni, Pb, Zn
Typha latifolia	Common cattail	Zn, Mn, Ni, Fe, Pb, Cu
Wolffia globosa	Asian watermeal	Cd, Cr

19.3.4.1 Bioremediation

Bioremediation implies utilizing natural operators to clean conditions. HM contamination being the center everywhere on over the world needs quick consideration with the goal that our debasing surroundings will be remediated. Phytoremediation is eco-friendly and has indicated promising outcomes for contaminants such as HMs. The essential components in phytoremediation are plants, whether earthbound or aquatic, that assume a key job for remediation of HM-influenced situations. Phytoremediation has additionally been an answer for different developing issues (Wani et al., 2017). Numerous aquatic plants (rising, lowered, or free streaming) have been applied widely as of late and generally directed utilizing tank-farming or field try by developed wetlands. Results from writing looked into have commonly settled the viability in remediating natural contaminations and HMs by aquatic plants, albeit HMs have been widely concentrated than natural pollutants. Most ordinarily utilized plants incorporate duckweed (*L. minor*), water hyacinth (*Eichhornia crassipes*), and water lettuce (*P. stratiotes*) because of their omnipresent nature, obtrusive system, irregular conceptive limit, bioaccumulation possibilities, and versatility in contaminated condition (Beniah Obinna and Ebere, 2019).

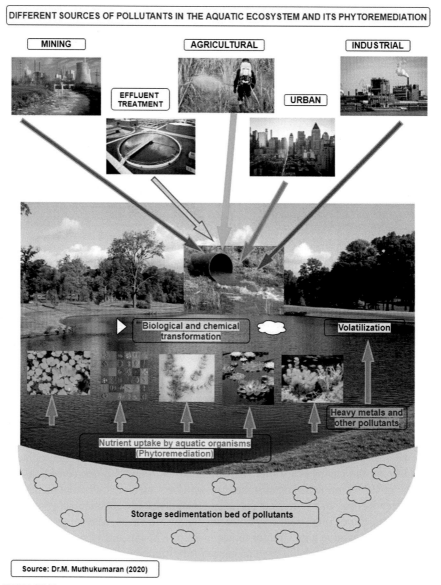

Source: Dr.M. Muthukumaran (2020)

PLATE 19.5

Different sources of pollutants in the aquatic ecosystem and its phytoremediation.

19.3.4.2 Phycoremediation—an emerging technology

Phycoremediation is characterized as the utilization of green growth to eliminate contaminations from the earth or to deliver them innocuous (Dresback et al., 2001). Olguin (2003) characterizes phycoremediation in a more extensive sense as the utilization of macroalgae or microalgae for the evacuation or biotransformation of pollutants, including supplements and xenobiotics from wastewater and CO_2 from squandering air. The small-scale green growth, *Chlorococcum humicola, Desmococcus olivaceus, Chlorella vulgaris, Chroococcus turgidus,* and *Amphora turgida* could treat different industrial effluents (Phycoremediation), for example, textile dyeing, soft drink, chemical, detergent, oil-drilling, leather processing, alginate, and petrochemical industrial effluents. The above green growth was included in the substantial remedy of pH, BOD, COD, the expulsion of shading, total dissolved solids (TDS), nitrate, phosphate, and so forth., around 60%−80% in (Muthukumaran, 2009; Muthukumaran et al., 2012; Muthukumaran and Sivasubramaian, 2017).

The impacts of high nitrate focus on the energy of cell development during nitrate and phosphate evacuation by a macroalga *Cladophora glomerata*. The algal development and nitrate expulsion from media containing starting nitrate convergences of 5 to 400 mg/L were observed in bunch development, while control media has no extra nitrate (Farahdiba et al., 2020). It was likewise discovered that during a maintenance time of 10 days under surrounding temperature conditions, *Cladophora* sp. could cut down arsenic fixation from 6 to 0.1 mg/L, *Chlorodesmis* sp. had the option to decrease arsenic by 40%−50%, whereas the water hyacinth could lessen arsenic by just 20%. *Cladophora* sp. is accordingly appropriate for cotreatment of sewage and arsenic-improved saltwater in an algal lake, making some maintenance memories of 10 days. The distinguished plant species gives a straightforward and practical technique for application in country territories influenced by arsenic issues (Jasrotia et al., 2017). A relationship coefficient of organic elements with some ecological conditions is given and examined. The successive appearance of red tides created by the dinoflagellates, *Alexandrium minutum,* and *Prorocentrum triestinum* reflected high estimations of assorted phytoplankton variety, just as changes in the prevailing species from one to the next. This recommends the water nature of the EH might be recuperating (Zaghloul et al., 2020).

19.3.4.3 *Phytoremediation of industrial effluents*
Phytoremediation is, along these lines, on a basic level, a low natural effect bioremediation method as contrasted and other physical, industrial, or organic procedures (Warrier, 2012). The release of industrial waste into soil and water implies a more basic danger to human well-being, living creatures, and different assets (Francová et al., 2017). Phytoremediation, alongside recently created designing and organic systems, has encouraged the effective expulsion of HMs from industrial wastewater through both phytostabilization and phytoextraction (Cheraghi et al., 2011). Twelve aquatic plants were tried for their phytoremediation capacity for various HMs starting from the industrial wastewater in the Swabi region, Pakistan. Results exhibited that these aquatic plants essentially eliminated heavy metals from industrial wastewater with fantastic expulsion efficiencies: Album (90%), Cr (89%), Fe (74.1%), Pb (half), Cu (48.3%), and Ni (40.9%), separately (Khan et al., 2009). Southern cattail (*Typha domingensis*) demonstrated the greatest aggregation of zinc, aluminum, iron,

and lead, particularly in roots instead of leaves from the industrial wastewater lake. Rhizofiltration was found as the predominant system that clarified the phytoremediation capability of *Typha domingensis* (Hegazy et al., 2011). Promising aquatic plants indicated better quality for the phytoremediation of the industrial effluents than different plants. Aquatic macrophytes *Marsilea quadrifolia, Hydrilla verticillata,* and *Ipomea aquatica* showed much better collection potential and transcription factor (TF) esteem for HMs (Fe, Cr, Zn, Pb, As, Hg, Cd, and Cu) from the industrial effluents when contrasted with the earthly plants *Sesbania cannabina, Eclipta alba* and algal species (*Phormidium papyraceum, Spirulina platensis*) (Ahmad et al., 2011). A decent variety of aquatic plants has the accompanying favorable circumstances for remediation: high expulsion effectiveness, better living space, and appropriation versatility (Gerhardt et al., 2017; Chandra et al., 2018). *Typha domingensis* is such a predominant aquatic plant-animal type having a high resilience to a harmful situation and capable gathering of HMs. Maine et al. (2017) additionally announced that *T. domingensis* indicated much better endurance and evacuation effectiveness for iron, zinc, nickel, and chromium delivered from industrial wastewater of metallurgy plant over other higher differentiated aquatic plants. Such a sort of aquatic plant can be utilized for an enormous scope to examine the drawn-out evacuation execution. The water hyacinth (*Eichhornia crassipes*) and two green growth (*Chlorodesmis* sp. what's more, *Cladophora* sp.) found close to arsenic-improved water bodies were utilized to decide their resistance toward arsenic and their viability to take up arsenic accordingly decreasing natural contamination in arsenic-advanced wastewater of various fixations. *Cladophora* sp. was found to get by up to an arsenic centralization of 6 mg/L, though water hyacinth and *Chlorodesmis* sp. could get by up to arsenic groupings of 2 and 4 mg/L, separately (Jasrotia et al., 2017).

19.3.4.4 Phytoremediation of metropolitan wastewaters

Metropolitan wastewater is a huge danger for the aquatic condition as it is a primary driver of HM contamination. Zn, Cu, Ni, Pb, and Hg are conceivably more toxic metals, and they may cause incessant and intense well-being impacts, bio-accumulation, and phytotoxicity (Alhadrami et al., 2016; Mânzatu et al., 2015; Wacławek et al., 2017). The use of aquatic plants for the expulsion of heavy metals from city wastewater, sewage water, spillage zones, and other contaminated destinations has become a typical practice and test procedure (Akpor and Muchie, 2010; Carolin et al., 2017). Aquatic plants can be utilized as bio-aggregators as they can collect high groupings of HMs in their biomass (Bonanno et al., 2017; Bravo et al., 2017). Root and shoot tissues of Typha domingensis demonstrated the greatest gathering of Zn, Cd, Ni, Fe, and Mn during the initial 48 h of study, planted in pots loaded up with city wastewater (Mojiri, 2012). *L. gibba* was read for the gathering of arsenic, boron, and uranium from the metropolitan wastewater as an elective expulsion technique. Results uncovered that U, As, and B were quickly consumed by the plant during the initial 2 days of a 7-day exploratory examination (Sasmaz and Obek,

2009). Two established macrophytes *Typha angustifolia* and *Phragmites australis* eliminated 14%−85% of heavy metals, for example, zinc, lead, arsenic, nickel, iron, copper, aluminum, and magnesium, from metropolitan wastewater in an aquaculture study (Pedescoll et al., 2015). So in addition, aquatic plants *Typha latifolia* and *Phragmites australis* indicated fantastic expulsion effectiveness of heavy metals from the city wastewater. Both these aquatic plants demonstrated higher evacuation rate for aluminum (96%), copper (91%), lead (88%) and zinc (85%) and somewhat less expulsion rate for iron (44%), boron (40%) and cobalt (31%) (Morari et al., 2015).

19.3.5 Other preferences for aquatic plants

The investigation of phytoremediation uncovers that the aquatic macrophytes have a favorable position over different plants in the remediation of HMs (Galal et al., 2018; Daud et al., 2018; Bravo et al., 2017). Their boundless accessibility, quick development rate, high biomass, cost-adequacy, and resilience to poisonous toxins make them the most appropriate and accessible phytoremediation plants. Refinements framework utilizing these aquatic plants have increased more consideration overall in light of their ability to collect and eliminate a tenacious natural toxin from water bodies (Leguizamo et al., 2017; Daud et al., 2018). Inclusion of fitting phytoremediation innovation needs intervallic gathering of the plant biomass so as to acclimatize and appropriate heavy metals and supplements from water bodies. The transformation of biomass into restrictive material is a noteworthy factor in advancing this method for the treatment of contaminants. An aquatic plant's biomass can last on be utilized as creature feed, helpful in the creation of biogas and fertilizer as detailed in numerous examinations (Shahid et al., 2018). Biosorption alongside bioaccumulation *of Lemna minor* biomass was assessed and demonstrated its conceivable use as creature feed (Daud et al., 2018; Olguin et al., 2003). Aquatic plants have starch, cellulose, and hemicelluloses that look like sugar. Starch hydrolysis of this fermentable sugar brings about the creation of lactic corrosive, ethanol, and other significant items. In this way, the sugar present in aquatic plants is another promising component supporting their job in the eco-practical condition. Aquatic plants, i.e., *Pistia stratiotes* and *Eichhornia crassipes*, have been accounted for to create sugar during their enzymatic hydrolysis measure (Mishima et al., 2006). Free-skimming aquatic plants (*Azolla* spp., *Wolffia* spp., *Spirodela* sp. furthermore, Duckweeds) can be utilized as a food hotspot for water-winged animals. They additionally give a safe house to creepy crawly hatchlings and little mollusks. Fishes additionally utilize the mats of these plants as spread and utilize their shade for proliferation (Villamagna, 2009). Aquatic plants can be effectively used to improve the hydroponics for fishponds. End of nitrogenous misuse of aquatic plants is additionally an additional advantage for their application in hydroponics, i.e., *Canna generalis* L., *Typha angustifolia*. *Echinodorus cordifolius* and *Cyperus involucratus* eliminated smelling salts, nitrate, and nitrite effectively (Nakphet et al., 2017).

19.4 Conclusion

Water contamination is one of the greatest ecological concerns, and it is obvious that phytoremediation is giving a superior answer for handle this issue. sHMs are taken up by plants utilizing phytoremediation innovation, which is by all accounts a prosperous method to remediate HM-polluted condition. It has a few favorable circumstances contrasted and other normally utilized traditional advances. A few components must be considered so as to achieve elite remediation results. The most significant factor is a reasonable plant species that can be utilized to take up the contaminant. Extensive communication, transport, and chelator exercises direct the capacity and collection of HMs by the aquatic macrophytes, including algae. The use of phytoremediation is by all accounts a less troublesome, practical, and naturally stable tidy-up innovation. Utilization of aquatic plants both in bioaccumulation (with living plant biomass) and biosorption (with dead plant biomass) should be possible effectively for the destruction of HMs. Indeed, even the phytoremediation strategy is by all accounts a standout among the other options, but it additionally has a few constraints. A thorough investigation should be directed toward limiting this constraint so as to viably apply this procedure. The advantages of utilizing aquatic plants to treat contaminants are enormous on the grounds that this innovation treats the contaminants yet is practical and outwardly satisfying as well as favorable for managing entire biological systems.

Acknowledgment

Author thanks to the Secretary, Principal and HOD of Botany, Ramakrishna Mission Vivekananda College (Autonomous), Chennai — 600004, Tamil Nadu, India for providing the facilities to complete the research work.

References

Abhilash, P., Jamil, S., Singh, N., 2009. Transgenic plants for enhanced biodegradation and phytoremediation of organic xenobiotics. Biotechnol. Adv. 27, 474–488.

Ahmad, A., Ghufran, R., Zularisam, A., 2011. Phytosequestration of metals in selected plants growing on a contaminated Okhla industrial areas, Okhla, New Delhi, India. Water Air Soil Pollut. 217, 255–266.

Akinbile, C.O., Ogunrinde, T.A., Che Bt Man, H., Aziz, H.A., 2016. Phytoremediation of domestic wastewaters in free water surface constructed wetlands using *Azolla pinnata*. Int. J. Phytoremediation 18 (1), 54–61. https://doi.org/10.1080/15226514.2015.1058330.

Akpor, O., Muchie, M., 2010. Remediation of heavy metals in drinking water and wastewater treatment systems: processes and applications. Int. J. Phys. Sci. 5, 1807–1817.

Alhadrami, H.A., Mbadugha, L., Paton, G.I., 2016. Hazard and risk assessment of human exposure to toxic metals using in vitro digestion assay. Chem. Speciat. Bioavailab. 28, 78–87.

Ali, H., Khan, E., Sajad, M.A., 2013. Phytoremediation of heavy metals—concepts and applications. Chemosphere 91, 869—881.

Ali, S., Abbas, Z., Rizwan, M., Zaheer, I.E., Yavas, I., Ünay, A., Abdel-Daim, M.M., Bin-Jumah, M., Hasanuzzaman, M., Kalderis, D., 2020. Application of floating aquatic plants in phytoremediation of heavy metals polluted water: a review. Sustainability 12 (5), 1—33. https://doi.org/10.3390/su12051927.

An, B., Lee, C.-G., Song, M.-K., Ryu, J.-C., Lee, S., Park, S.-J., Lee, S.H., 2015. Applicability and toxicity evaluation of an adsorbent based on jujube for the removal of toxic heavy metals. React. Funct. Polym. 93, 138—147.

Anagnostidis, A., Komárek, J., 1988. Industrial approach to the classification system of cyanobacteria-3 oscillatoriales. Archiv für Hydrobiol./Algolog. Stud. 50 (53), 327—472.

Ansari, A.A., Naeem, M., Gill, S.S., AlZuaibr, F.M., 2020. Phytoremediation of contaminated waters: an eco-friendly technology based on aquatic macrophytes application. Egypt. J. Aquatic Res. 46 (4), 371—376. https://doi.org/10.1016/j.ejar.2020.03.002.

Anton, A., Mathe-Gaspar, G., 2005. Factors affecting heavy metal uptake in plant selection for phytoremediation. Z. Naturforsch. 60 (3—4), 244—246.

APHA, 2000. Standard Methods for Examination of Waters and Wastewater, twentieth ed. American Public Health Association, Washington American Water Works Association (AWWA), Water Environment Federation (WEF), Washington DC USA.

Arslan, M., Imran, A., Khan, Q.M., Afzal, M., 2017. Plant—bacteria partnerships for the remediation of persistent organic pollutants. Environ. Sci. Pollut. Res. 24, 4322—4336.

Ashraf, S., Afzal, M., Naveed, M., Shahid, M., Zahir, Z.A., 2018. Endophytic bacteria enhance remediation of tannery effluent in constructed wetlands vegetated with *Leptochloa fusca*. Int. J. Phytoremediation 20, 121—128.

Barakat, M.A., 2011. New trends in removing heavy metals from industrial wastewater. Arab. J. Chem. 4 (4), 361—377. https://doi.org/10.1016/j.arabjc.2010.07.019.

Benavides, L.C.L., Pinilla, L.A.C., Serrezuela, R.R., Serrezuela, W.F.R., 2018. Extraction in laboratory of heavy metals through rhizofiltration using the plant *zea mays* (maize). Int. J. Appl. Environ. Sci. 13, 9—26.

Beniah Obinna, I., Ebere, E.C., September 2019. A review: water pollution by heavy metal and organic pollutants: brief review of sources, effects and progress on remediation with aquatic plants. Anal. Methods Environ. Chem. J. 5—38. https://doi.org/10.24200/amecj.v2.i03.66.

Blaylock, M., 2008. Phytoremediation of Contaminated Soil and Water: Field Demonstration of Phytoremediation of Lead Contaminated Soils. Lewis Publishers, Boca Raton, FL, USA.

Bonanno, G., Borg, J.A., Di Martino, V., 2017. Levels of heavy metals in wetland and marine vascular plants and their biomonitoring potential: a comparative assessment. Sci. Total Environ. 576, 796—806.

Bravo, S., Amorós, J., Pérez-de-los-Reyes, C., García, F., Moreno, M., Sánchez-Ormeño, M., Higueras, P., 2017. Influence of the soil pH in the uptake and bioaccumulation of heavy metals (Fe, Zn, Cu, Pb and Mn) and other elements (Ca, K, Al, Sr and Ba) in vine leaves, Castilla-La Mancha (Spain). J. Geochem. Explor. 174, 79—83.

Burges, A., Alkorta, I., Epelde, L., Garbisu, C., 2018. From phytoremediation of soil contaminants to phytomanagement of ecosystem services in metal contaminated sites. Int. J. Phytoremediation 20, 384—397.

Cappucino, J.G., Shermann, N., 2009. Microbiology: A Laboratory Manual, sixth ed. Pearson Education. Benjamin Cummings, San Francisco.

Carolin, C.F., Kumar, P.S., Saravanan, A., Joshiba, G.J., Naushad, M., 2017. Efficient techniques for the removal of toxic heavy metals from aquatic environment: a review. J. Environ. Chem. Eng. 5, 2782–2799.

Chandra, R., Kumar, V., Tripathi, S., Sharma, P., 2018. Heavy metal phytoextraction potential of native weeds and grasses from endocrine-disrupting chemicals rich complex distillery sludge and their histological observations during in-situ phytoremediation. Ecol. Eng. 111, 143–156.

Cheraghi, M., Lorestani, B., Khorasani, N., Yousefi, N., Karami, M., 2011. Findings on the phytoextraction and phytostabilization of soils contaminated with heavy metals. Biol. Trace Elem. Res. 144, 1133–1141.

Cohen-Shoel, N., Ilzycer, D., Gilath, I., 2002. The involvement of pectin in Sr (II) biosorption by *Azolla*. Water Air Soil Pollut. 135, 195–205.

COST Action 837 View, 2003. In: Vanec, T., Schwitzguebel, J.P. (Eds.). UOCHB AVCR, Prague, pp 41.

COST Action 859, 2005. In: Schwitzguebel, J.-P. (Ed.), Phytotechnologies to Promote Sustainable Land Use and Improve Food Safety.

Cristaldi, A., Conti, G.O., Jho, E.H., Zuccarello, P., Grasso, A., Copat, C., Ferrante, M., 2017. Phytoremediation of contaminated soils by heavy metals and PAHs. A brief review. Environ. Technol. Innov. 8, 309–326.

Cundy, A.B., Bardos, R., Church, A., Puschenreiter, M., Friesl-Hanl, W., Müller, I., Vangronsveld, J., 2013. Developing principles of sustainability and stakeholder engagement for "gentle" remediation approaches: the European context. J. Environ. Manag. 129, 283–291.

Cunningham, S.D., Ow, D.W., 1996. Promises and prospects of phytoremediation. Plant Physiol. 110, 715.

DalCorso, G., Fasani, E., Manara, A., Visioli, G., Furini, A., 2019. Heavy metal pollution: state of the art and innovation in phytoremediation. Int. J. Mol. Sci. 20 (14) https://doi.org/10.3390/ijms20143412.

Daud, M., Ali, S., Abbas, Z., Zaheer, I.E., Riaz, M.A., Malik, A., Zhu, S.J., 2018. Potential of duckweed (*Lemna minor*) for the phytoremediation of landfill leachate. J. Chem. 1–9.

de Souza, C.B., Silva, G.R., 2019. Phytoremediation of effluents contaminated with heavy metals by floating aquatic macrophytes species. Biotechnol. Bioeng. IntechOpen (Chapter), 1–8. https://doi.org/10.5772/intechopen.83645.

Dhir, B., 2009. Salvinia: an aquatic fern with potential use in phytoremediation. Environ. Int. J. Sci. Tech. 7112 (4), 23–27.

Drăghiceanu, O.A., Dobrescu, C.M., Soare, L.C., 2014. Applications of pteridophytes in phytoremediation. Curr. Trends Nat. Sci. 3 (6), 68–73.

Dresback, K., Ghoshal, D., Goyal, A., 2001. Phycoremediation of trichloroethylene (TCE). Physiol. Mol. Biol. Plants 7 (2), 117–123.

Eid, E.M., Galal, T.M., Sewelam, N.A., Talha, N.I., Abdallah, S.M., 2020. Phytoremediation of heavy metals by four aquatic macrophytes and their potential use as contamination indicators: a comparative assessment. Environ. Sci. Pollut. Res. 27, 12138–12151. https://doi.org/10.1007/s11356-020-07839-9.

Erakhrumen, A.A., 2017. Phytoremediation: an environmentally sound technology for pollution prevention, control and remediation in developing countries. Educ. Res. Rev. 2, 151–156.

Farahdiba, A.U., Hidayah, E.N., Win Myint, A.Y., 2020. Growth and removal of nitrogen and phosphorus by a macroalgae *Cladophora glomerata* under different nitrate concentrations. Nat. Environ. Pollut. Technol. 19 (02), 809–813. https://doi.org/10.46488/nept.2020.v19i02.038.

Francesconi, K., Visoottiviseth, P., Sridokchan, W., Goessler, W., 2002. Arsenic species in an arsenic hyperaccumulating fern, Pityrogramma calomelanos: a potential phytoremediator of arsenic-contaminated soils. Sci. Total Environ. 284 (1−3), 27−35. https://doi.org/10.1016/S0048-9697(01)00854-3.

Francová, A., Chrastný, V., Šillerová, H., Vítková, M., Kocourková, J., Komárek, M., 2017. Evaluating the suitability of different environmental samples for tracing atmospheric pollution in industrial areas. Environ. Pollut. 220, 286−297.

Fritioff, Å., Greger, M., 2003. Aquatic and terrestrial plant species with potential to remove heavy metals from stormwater. Int. J. Phytoremediation 5, 211−224.

Galal, T.M., Eid, E.M., Dakhil, M.A., Hassan, L.M., 2018. Bioaccumulation and rhizofiltration potential of *Pistia stratiotes* L. for mitigating water pollution in the Egyptian wetlands. Int. J. Phytoremediation 20, 440−447.

Gerhardt, K.E., Gerwing, P.D., Greenberg, B.M., 2017. Opinion: taking phytoremediation from proven technology to accepted practice. Plant Sci. 256, 170−185.

Ghosh, M., Singh, S., 2005. A review on phytoremediation of heavy metals and utilization of it's by products. Asian J. Energy Environ. 6, 18.

Glick, B.R., 2010. Using soil bacteria to facilitate phytoremediation. Biotechnol. Adv. 28 (3), 367−374. https://doi.org/10.1016/j.biotechadv.2010.02.001.

González-Camejo, J., Barat, R., Pachés, M., Murgui, M., Seco, A., Ferrer, J., 2018. Wastewater nutrient removal in a mixed microalgae-bacteria culture: effect of light and temperature on the microalgae-bacteria competition. Environ. Technol. 39 (4), 503−515.

Grasshoff, K., 1976. Methods of Seawater Analysis. Verlag Chemie, Weinkeim and New York, 317 pp.

Hattink, J., Goeij, J.J.M., Wolterbeek, H.T., 2000. Uptake kinetics of 99Tc in common duckweed. Environ. Exp. Bot. 44 (1), 9−22.

Hegazy, A., Abdel-Ghani, N., El-Chaghaby, G., 2011. Phytoremediation of industrial wastewater potentiality by *Typha domingensis*. Int. J. Environ. Sci. Technol. 8, 639 648.

Hong, K.S., Lee, H.M., Bae, J.S., Ha, M.G., Jin, J.S., 2011. Removal of heavy metal ions by using calcium carbonate extracted from starfish treated by protease and amylase. J. Anal. Sci. Technol. 2 (2), 75−82.

Horwitz, W., 2000. Official Methods of Analysis of AOAC. Association of Official Analytical Chemists, Washington, DC.

Jadia, C.D., Fulekar, M., 2009. Phytoremediation of heavy metals: recent techniques. Afr. J. Biotechnol. 8, 921−928.

Jaishankar, M., Tseten, T., Anbalagan, N., Mathew, B.B., Beeregowda, K.N., 2014. Toxicity, mechanism and health effects of some heavy metals. Interdiscipl. Toxicol. 7 (2), 60−72. https://doi.org/10.2478/intox-2014-0009.

Jasrotia, S., Kansal, A., Mehra, A., 2017. Performance of aquatic plant species for phytoremediation of arsenic-contaminated water. Appl. Water Sci. 7 (2), 889−896. https://doi.org/10.1007/s13201-015-0300-4.

Jeffrey, S.W., Humphrey, G.F., 1975. New spectrophotometric equations for determining chlorophylls a, b, c1 and c2 in higher plants, algae and natural phytoplankton. Biochem. Physiol. Pflanz. (BPP) 167, 191−194.

Jeppesen, E., Kronvang, B., Meerhoff, M., Søndergaard, M., Hansen, K.M., Andersen, H.E., Lauridsen, T.L., Beklioglu, M., Özen, A., Olesen, J.E., 2009. Climate change effects on runoff, catchment phosphorus loading and lake ecological state, and potential adaptations. J. Envir. Qual. 38, 1930−1941.

Kaimidate, Y., Yamada, M., Furusho, Y., Kuroiwa, T., Kaise, T., Fujiwara, K., 2000. A rapid method for the detection of arsenic compounds in biological samples with semimicro column and an inductively coupled plasma mass spectrometer. Biomed. Res. Trace Elem. 11, 445−446.

Karami, A., Shamsuddin, Z.H., 2010. Phytoremediation of heavy metals with several efficiency enhancer methods. Afr. J. Biotechnol. 9, 3689−3698.

Khan, S., Ahmad, I., Shah, M.T., Rehman, S., Khaliq, A., 2009. Use of constructed wetland for the removal of heavy metals from industrial wastewater. J. Environ. Manag. 90, 3451−3457.

Kocoń, A., Jurga, B., 2017. The evaluation of growth and phytoextraction potential of Miscanthus x giganteus and Sida hermaphrodita on soil contaminated simultaneously with Cd, Cu, Ni, Pb, and Zn. Environ. Sci. Pollut. Res. 24, 4990−5000.

Koroleff, F., 1977. Simultaneous persulphate oxidation of phosphorus and nitrogen compounds in water. In: Grassof, K. (Ed.), Report of Baltic Intericalibration Workshop, Annex, Interim Commission for the Protection of the Environment of the Baltic Sea.

Kushwaha, A., Hans, N., Kumar, S., Rani, R., 2018. A critical review on speciation, mobilization and toxicity of lead in soil-microbe-plant system and bioremediation strategies. Ecotoxicol. Environ. Saf. 147, 1035−1045.

Labidi, S., Firmin, S., Verdin, A., Bidar, G., Laruelle, F., Douay, F., Sahraoui, A.L.H., 2017. Nature of fly ash amendments differently influences oxidative stress alleviation in four forest tree species and metal trace element phytostabilization in aged contaminated soil: a long-term field experiment. Ecotoxicol. Environ. Saf. 138, 190−198.

Leguizamo, M.A.O., Gómez, W.D.F., Sarmiento, M.C.G., 2017. Native herbaceous plant species with potential use in phytoremediation of heavy metals, spotlight on wetlands—a review. Chemosphere 168, 1230−1247.

Macek, T., Mackova, M., Kas, J., 2000. Exploitation of plants for the removal of organics in environmental remediation. Biotechnol. Adv. 18, 23−34.

Mahdavian, K., Ghaderian, S.M., Torkzadeh-Mahani, M., 2017. Accumulation and phytoremediation of Pb, Zn, and Ag by plants growing on Koshk lead−zinc mining area, Iran. J. Soils Sediments 17, 1310−1320.

Maine, M., Hadad, H., Sánchez, G., Di Luca, G., Mufarrege, M., Caffaratti, S., Pedro, M., 2017. Long-term performance of two free-water surface wetlands for metallurgical effluent treatment. Ecol. Eng. 98, 372−377.

Mânzatu, C., Nagy, B., Ceccarini, A., Iannelli, R., Giannarelli, S., Majdik, C., 2015. Laboratory tests for the phytoextraction of heavy metals from polluted harbor sediments using aquatic plants. Mar. Pollut. Bull. 101, 605−611.

McCutcheon, S.C., Schnoor, J.L. (Eds.), 2003. Phytoremediation − Transformation and Control of Contaminants. Wiley Interscience, p. 985.

Mishima, D., Tateda, M., Ike, M., Fujita, M., 2006. Comparative study on chemical pretreatments to accelerate enzymatic hydrolysis of aquatic macrophyte biomass used in water purification processes. Bioresour. Technol. 97, 2166−2172.

Mohan, B.S., Hosetti, B.B., 1999. Aquatic plants for toxicity assessment. Environ. Res. 81, 259−274.

Mojiri, A., 2012. Phytoremediation of heavy metals from municipal wastewater by *Typha domingensis*. Afr. J. Microbiol. Res. 6, 643−647.

Morari, F., Dal Ferro, N., Cocco, E., 2015. Municipal wastewater treatment with *Phragmites australis* L. and *Typha latifolia* L. for irrigation reuse. Boron and heavy metals. Water Air Soil Pollut. 226, 56.

Muthukumaran, M., Thirupathi, P., Chinnu, K., Sivasubramanian, V., 2012. Phycoremediation efficiency and biomass production by micro alga *Desmococcus olivaceus* (Persoon et

Acharius) J.R. Laundon treated on chrome-sludge from an electroplating industry-A open raceway pond study. Int. J. Curr. Sci. 52–62. Special Issue.

Muthukumaran, M., 2009. Studies on the Phycoremediation of Industrial Effluents and Utilization of Algal Biomass (Ph.D. thesis). University of Madras.

Muthukumaran, M., Sivasubramanian, V., 2017. Microalgae cultivation for biofuels: cost, energy balance, environmental impacts and future perspectives. In: Sarup Singh, R., Pandey, A., Gnansounou, E. (Eds.), Biofuels: Production and Future Perspectives. To be published by Taylor & Francis Group, CRC Press, USA, ISBN 9781498723596, pp. 363–411 (Chapter 15).

Mustafa, H.M., Hayder, G., 2020. Recent studies on applications of aquatic weed plants in phytoremediation of wastewater: a review article. Ain Shams Eng. J. 12 (1), 355–365. https://doi.org/10.1016/j.asej.2020.05.009.

Najeeb, U., Ahmad, W., Zia, M.H., Zaffar, M., Zhou, W., 2017. Enhancing the lead phytostabilization in wetland plant *Juncus effusus* L. through somaclonal manipulation and EDTA enrichment. Arab. J. Chem. 10, 3310–3317.

Nakphet, S., Ritchie, R.J., Kiriratnikom, S., 2017. Aquatic plants for bioremediation in red hybrid tilapia (*Oreochromis niloticus & Oreochromis mossambicus*) recirculating aquaculture. Aquacult. Int. 25, 619–633.

Negri, M.C., Hinchman, R.R., 1996. Plants that remove contaminants from the environment. Lab. Med. 27 (1), 36–40. https://doi.org/10.1093/labmed/27.1.36.

Olguin, E., Rodriguez, D., Sanchez, G., Hernandez, E., Ramirez, M., 2003. Productivity, protein content and nutrient removal from anaerobic effluents of coffee wastewater in *Salvinia min*ima ponds, under subtropical conditions. Acta Biotechnol. 23, 259–270.

Olguin, E.J., 2003. Phycoremediation: key issues for cost-effective nutrient removal process. Biotechnol. Adv. 22, 81–91.

Olusegun, A., Makun, H.A., Ogara, I.M., Edema, M., Idahor, K.O., Oluwabamiwo, B.F., Eshiett, M.E., 2012. We Are IntechOpen, the World'S Leading Publisher of Open Access Books Built by Scientists, for Scientists TOP 1 %. Intech, p. 38. https://doi.org/10.1016/j.colsurfa.2011.12.014 i(tourism).

Pakdel, P.M., Peighambardoust, S.J., 2018. A review on acrylic based hydrogels and their applications in wastewater treatment. J. Environ. Manag. 217, 123–143.

Paris, D.N., Mangkoedihardjo, S., 2020. Detoxification of glucose, ammonium and formaldehyde using nitrification and plant processes. Nat. Environ. Pollut. Technol. 19 (1), 385–388.

Pedescoll, A., Sidrach-Cardona, R., Hijosa-Valsero, M., Bécares, E., 2015. Design parameters affecting metals removal in horizontal constructed wetlands for domestic wastewater treatment. Ecol. Eng. 80, 92–99.

Peles, J.D., Smith, M.H., Brisbin, I.L., 2002. Ecological half-life of 137Cs in plants associated with a contaminated stream. J. Environ. Radioact. 59 (2), 169–178.

Peligro, F.R., Pavlovic, I., Rojas, R., Barriga, C., 2016. Removal of heavy metals from simulated wastewater by in situ formation of layered double hydroxides. Chem. Eng. J. 306, 1035–1040.

Prasad, M.N.V., 2001. Bioremediation potential of amaranthaceae. In: Leeson, A., Foote, E.A., Banks, M.K., Magar, V.S. (Eds.), Phytoremediation, Wetlands, and Sediments, Proc 6th Int in Situ and On-Site Bioremediation Symposium, vol. 6. Battelle Press, Columbus, OH, pp. 165–172 (5).

Prasad, M.N.V., Greger, M., Aravind, P., 2005. Biogeochemical cycling of trace elements by aquatic and wetland plants: relevance to phytoremediation. Trace Elem. Environ. Biogeochem. Biotechnol. Bioremediat. 1, 451–474.

Prasad, M.N.V., 2007. Aquatic Plants for Phytotechnology. Environmental Bioremediation Technologies, pp. 259–274. https://doi.org/10.1007/978-3-540-34793-4_11.

Rahman, M.A., Hasegawa, H., 2011. Aquatic arsenic: phytoremediation using floating macrophytes. Chemosphere 83, 633–646.

Rajeswari, K., Subhaskumar, R., Vijayaraman, K., 2013. Decoloriza- tion and degradation of textile dyes by Stenotrophomonas maltophila RSV-2. Int. J. Environ. Bioremed. & Biodegr. 1, 60–65.

Raskin, I., Ensley, B.D., 2000. Phytoremediation of Toxic Metals. John Wiley and Sons, Hoboken, NJ, USA.

Raval, N.P., Shah, P.U., Shah, N.K., 2016. Adsorptive removal of nickel (II) ions from aqueous environment: a review. J. Environ. Manag. 179, 1–20.

Reeves, R.D., Baker, A.J., Jaffré, T., Erskine, P.D., Echevarria, G., van der Ent, A., 2018. A global database for plants that hyperaccumulate metal and metalloid trace elements. New Phytol. 218, 407–411.

Sakakibara, M., June 2016. Phytoremediation of toxic elements-polluted water and soils by aquatic macrophyte *Eleocharis acicularis*. AIP Conf. Proc. 1744, 1–6. https://doi.org/10.1063/1.4953512.

Sakakibara, M., Watanabe, A., Sano, S., Inoue, M., Kaise, T., January 2007. Phytoextraction and phytovolatili-zation of arsenic from as-contaminated soils by *Pteris vittata*. In: Association for Environmental Health and Sciences - 22nd Annual International Conference on Contaminated Soils, Sediments and Water, vol. 12, pp. 258–263.

Sano, S., Sakakibara, S., Watanabe, A., 2005. Quantitative analyses of arsenic in plants by instrumental neutron activation analyses and inductively coupled plasma-mass spectrometry. Mem. Fac. Sci. Ehime Univ. 11, 1–6.

Sanyahumbi, D., Duncan, J.R., Zhao, M., 1998. Removal of lead from solution by the non-viable biomass of the water fern *Azolla filiculoides*. Biotechnol. Lett. 20 (8), 745–747.

Sarwar, N., Imran, M., Shaheen, M.R., Ishaque, W., Kamran, M.A., Matloob, A., Hussain, S., 2017. Phytoremediation strategies for soils contaminated with heavy metals: modifications and future perspectives. Chemosphere 171, 710–721.

Sasmaz, A., Obek, E., 2009. The accumulation of arsenic, uranium, and boron in *Lemna gibba* L. exposed to secondary effluents. Ecol. Eng. 35, 1564–1567.

Serrat, B.M., 2014. Metodologia para Seleção de Técnica de Fitorremediação em Áreas Contaminadas Methodology for selection of Phytoremediation technique in brownfields RESUMO. Revista Brasileira de Ciências Ambientais — Número 31, 97–104. http://abes-dn.org.br/publicacoes/rbciamb/PDFs/31-12_Materia_10_artigos395.pdf.

Sheppard, S.C., Motycka, M., 1997. Is the akagare phenomenon important to iodine uptake by wild rice (*Zizania aquatica*)? J. Environ. Radioact. 37, 339–353.

Singh, D., Tiwari, A., Gupta, R., 2012. Phytoremediation of lead from wastewater using aquatic plants. J. Agric. Technol. 8 (81), 1–11.

Shahid, M.J., Arslan, M., Ali, S., Siddique, M., Afzal, M., 2018. Floating wetlands: a sustainable tool for wastewater treatment. Clean 46, 1800120.

Sharma, N., Saxena, S., Fatima, M., Iram, B., Datta, A., Gupta, S., 2014. Microcosm analysis of untreated textile effluent for cod reduction by autochthonous bacteria. Int. J. Curr. Res. Chem. & Pharm. Sci. 1, 15–23.

Sharma, S., Singh, B., Manchanda, V., 2015. Phytoremediation: role of terrestrial plants and aquatic macrophytes in the remediation of radionuclides and heavy metal contaminated soil and water. Environ. Sci. Pollut. Res. 22, 946–962.

Sreelal, G., Jayanthi, R., 2017. Review on phytoremediation technology for removal of soil contaminant. Indian J. Sci. Res. 14, 127–130.

Susarla, S., Bacchus, T.S., Wolfe, N.L., McCutcheon, C.S., 1999. Phytotransformation of perchlorate using parrot feather. Soil & Groundw. Cleanup 2, 20–23.

Syukor, A.A., Zularisam, A., Ideris, Z., Ismid, M.M., Nakmal, H., Sulaiman, S., Nasrullah, M., 2014. Performance of phytogreen zone for BOD5 and SS removal for refurbishment conventional oxidation pond in an integrated phytogreen system. World Acad. Sci. Eng. Technol. 8, 159.

Soni, V., Kaur, P., 2018. Efficacy of aquatic plants for removal of heavy metals from wastewater. Int. J. Life Sci. Sci. Res. 4 (1), 1527–1531. https://doi.org/10.21276/ijlssr.2018.4.1.1.

Tadros, H.R.Z., Ibrahim, G.H., Zokm, G. M. El, Okbah, M.A., 2016. Multivariate analysis to investigate the organization of physicochemical parameters in the Eastern Harbour Alexandria. Egypt. Int. J. Contemp. Appl. Sci. 3 (3). www.ijcas.net.

Tewes, L.J., Stolpe, C., Kerim, A., Krämer, U., Müller, C., 2018. Metal hyperaccumulation in the Brassicaceae species *Arabidopsis halleri* reduces camalexin induction after fungal pathogen attack. Environ. Exp. Bot. 153, 120–126.

Tomotoshi, O., 1972. In IOC/Westpac. Workshop. Penang.Malaysia, 26–29 Nov., 1991, Report, No.79, Annex III-P167.

Uqab, B., Mudasir, S., Sheikh, A.Q., Nazir, R., 2016. Bioremediation: a management tool. J. Biorem. Biodegrad. 7, 331.

Ütermohl, H., 1958. Zur vervollkommnung der quantitativen phytoplanktonmethodik (On improvements in quantitative phytoplankton methods). Mitteilungen: internationale Vereinigung Theoretische. Angewandte Limnologie 9, 1–38.

Valderrama, J.C., 1981. The simultaneous analysis of total nitrogen and total phosphorus in natural waters. Mar. Chem. 10, 109–122.

Van Heurck, M., 1896. Treatise on the Diatomaceae. Wesley, London, United Kingdom.

Villamagna, A.M., 2009. Ecological Effects of Water Hyacinth (*Eichhornia crassipes*) on Lake Chapala, Mexico (Ph.D. thesis). Virginia Polytechnic Institute and State University, Blacksburg, Virginia.

Wacławek, S., Lutze, H.V., Grübel, K., Padil, V.V., Černík, M., Dionysiou, D.D., 2017. Chemistry of persulfates in water and wastewater treatment: a review. Chem. Eng. J. 330, 44–62.

Wani, R.A., Ganai, B.A., Shah, M.A., Uqab, B., 2017. Heavy metal uptake potential of aquatic plants through phytoremediation technique - a review. J. Biorem. Biodegrad. 08 (04) https://doi.org/10.4172/2155-6199.1000404.

Warrier, R.R., 2012. Phytoremediation for environmental clean up. Forest. Bull. 12 (2), 1–7.

Zaghloul, F.A.E.R., Khairy, H.M., Hussein, N.R., 2020. Assessment of phytoplankton community structure and water quality in the Eastern Harbor of Alexandria, Egypt. Egypt. J. Aquatic Res. 46 (2), 145–151. https://doi.org/10.1016/j.ejar.2019.11.008.

Zeng, P., Guo, Z., Cao, X., Xiao, X., Liu, Y., Shi, L., 2018. Phytostabilization potential of ornamental plants grown in soil contaminated with cadmium. Int. J. Phytoremediation 20, 311–320.

Zhu, Y., Zayed, A., Qian, J., De Souza, M., Terry, N., 1999. Phytoaccumulation of trace elements by wetland plants: II. Water hyacinth. J. Environ. Qual. 28, 339–344.

Biofilm in remediation of pollutants

20

Tanushri Chatterji[1], Sunil Kumar[2]

[1]*Department of Microbiology, Babu Banarasi Das College of Dental Sciences, Babu Banarasi Das University, Uttar Pradesh, Lucknow, India;* [2]*Faculty of Biosciences, Institute of Biosciences and Technology, Shri Ramswaroop Memorial University, Barabanki, Uttar Pradesh, India*

Chapter outline

Biological Approaches to Controlling Pollutants. https://doi.org/10.1016/B978-0-12-824316-9.00019-7

20.1 Introduction

Biofilms are defined as the assemblage of surface-associated microbial cells adhered to an extracellular polymeric substance matrix. A special feature of these microbial cells is that they are irreversibly associated (not removed by gentle rinsing) with a surface (Donlan 2002). This was first observed by Van Leeuwenhoek on tooth surfaces and named microbial biofilm by Heukelekian and Heller (1940). Biofilm was well explained using high-resolution photomicroscopy at much higher magnifications (Jones et al., 1969). Its external surface and matrix material are composed of polysaccharides, which was confirmed using a specific polysaccharide-stain called Ruthenium red and coupling this with osmium tetroxide fixative (Characklis 1973).

The organisms associated with the formation of biofilms differ from their planktonic (freely suspended) counterparts in terms of the genes transcribed. The development of biofilm occurs on a wide variety of surfaces, including living tissues, indwelling medical devices, industrial or portable water system piping, or natural aquatic systems. However, the nature of the biofilms varies from place to place, which can be illustrated by scanning electron micrographs. Using scanning and transmission electron microscopy, the role of biofilms on trickling filters in a wastewater treatment plant was also examined. The water system of biofilm is extraordinarily complex, containing corrosion products, clay material, freshwater diatoms, and filamentous bacteria. It was observed to be composed of a variety of organisms based on cell morphology (Donlan 2002).

20.2 Characteristic features of biofilm

Primarily, the environmental factors that contribute to the formation of biofilm are pH, moisture, nutrient levels, ionic strength, oxygen content, and temperature. As site characteristics, site of interaction temperature (water temperature) also plays a vital role, and hence, the seasonal effect on bacterial attachment and biofilm formation in different systems is marked significantly (Fera et al., 1989; Donlan et al., 1994). Elevation in the concentration of several cations (sodium, calcium, lanthanum, ferric iron) and nutrient concentration also regulates the attachment of microbial cells on the matrix (Fletcher, 1988; Donlan, 2002).

The structural frame forming biofilms is composed of exopolysaccharides. In several bacteria, N-acyl-homoserine lactones (AHLs) of quorum-sensing (QS) signaling molecules are involved in biofilm formation. Multiple biological and cellular factors are also involved in the formation of biofilms. The former involves the role of enzymes for metabolizing the contaminants. Selected bacterial cell characteristics—flagella and swarming motility, extracellular polymer substances (EPSs), and production of AHL QS signaling molecules—also contribute to the formation of biofilm. These characteristic features were determined and their relative importance in biofilm formation was analyzed by path analysis (Meng-Ying et al., 2009; Abatenh et al., 2017).

20.3 **Bioremediation**

Bioremediation is defined as an environmentally friendly process for the removal of harmful pollutants from soil, water, and air using microbes (Alexander and Loehr, 1992). Bioremediation is encouraged by consortia of microbes, supported by supplying optimum levels of nutrients and other chemicals essential for their metabolism to degrade/detoxify substances hazardous to the environment and living things. The metabolic reactions are enzyme mediated. Enzymes actively involved are oxidoreductases, hydrolases, lyases, transferases, isomerases, and ligases. Multiple mechanisms and pathways are associated with biodegradation in a wide range of compounds (Abatenh et al., 2017). The process of bioremediation is preferred, as environmental destruction is reduced, and consequently, no or minimal damage of land or wildlife is experienced (Vidali, 2001; Mitra and Mukhopadhyay, 2016).

The other advantages are reduction in noise and dust during treatment as well as avoidance of harsh chemicals. The process of bioremediation is economical when applied to large areas compared to conventional decontamination methods. It implements the usage of varied microorganisms for degradation or treatment of xenobiotic compounds, aromatic hydrocarbons, volatile organic compounds, pesticides, herbicides, heavy metals, radionuclides, crude oil, jet fuels, petroleum products, and explosives (Gaur et al., 2014). Nowadays, bioremediation contributes significantly in the hazardous waste industry as an era of development in the field of bioremediation across the world (Day, 1993; Mitra and Mukhopadhyay, 2016).

Microorganisms such as bacteria and fungi are easily modified genetically, and hence, they can be used for pollution removal and makes them suitable for bioremediation (Cerniglia, 1997; Balaji et al., 2014; Mishra and Malik, 2014). These microbes adapt to the existing environment and degrade organic compounds or waste pollutants via diverse catabolic pathways (Bouwer and Zehnder, 1993; Bruins et al., 2000). The microbes involved in the process of bioremediation are usually extremophiles, which can resist acidic or heavy metal-contaminated or radioactive environments (Mitra and Mukhopadhyay, 2016). Conversion of these hazardous wastes into less toxic components such as carbon dioxide and water is carried out by the enzymatic action of microbes (Das and Dash, 2014). The process of metabolic pathways requires the transfer of electrons from electron donors to electron acceptors. The electron donors act as food for microbes, which is usually limited in a noncontaminated site. The pollutants release organic electron donors that stimulate the microorganisms to compete for available acceptors to restore the balance of the system. Microbe-dependent degradation is either in aerobic or in anaerobic mode. In aerobic degradation, microbes utilize oxygen as the final electron acceptor to convert organic and inorganic pollutants and produce carbon dioxide and water, whereas in anaerobic degradation, oxygen is not present or limited; hence, the microbes use other electron acceptors such as nitrate, iron, manganese, sulfate, etc. to break down organic compounds often into carbon dioxide and methane. Though the process of aerobic microbial degradation is rapid than anaerobic degradation; however, it depends on the pollutants, both electron donor and acceptor may be supplied to promote the degradation process. Sometimes, many redox reactions also immobilize trace elements found in the polluted sites. Variation in oxidation potential of metals is responsible for regulating the change in toxicity or solubility as evident in the case of uranium and chromium (Kumar et al., 2007; Gao and Francis, 2008; Joutey et al., 2015). In the case of heavy metals, the conversion of sulfate to sulfide changes solubility and facilitates immobilization and removal of sulfate from wastewater by sulfur-reducing bacteria (Beyenal et al., 2004).

20.4 Mechanism of action of biofilms in bioremediation

Degradation of several pollutants is carried out either by collective microbial species forming biofilms (Horemans et al., 2013; Gieg et al., 2014). Microbes supporting bioremediation adapt themselves to the harsh environment and compete with nutrients and oxygen for their survival. The use of biofilms is efficient for bioremediation as biofilms absorb, immobilize and degrade various environmental pollutants (Mitra and Mukhopadhyay, 2016).

Microbial species responsible for the formation of biofilms expressed with genes are distinct to free-floating planktonic cells. Their genetic makeup promotes them to the variable local concentration of nutrients and oxygen within the biofilm matrix. Consequently, genetic expression facilitates the degradation of divergent pollutants by various metabolic pathways. One of the significant microbial characteristic features for biofilm formation is chemotaxis and flagellar-dependent motility (Pratt and Kolter, 1999). The properties enable them in swimming, swarming, twitching motility, chemotaxis, QS in presence of xenobiotics commonly present in soil and water assist microbes to coordinate movement toward pollutant and improved biodegradation (Lacal et al., 2013).

A significant feature of biofilm-forming bacteria encased in an EPS matrix contributes to bioremediation (Flemming and Wingender,2001, 2010; More et al., 2014). The EPS is composed of a secreted form of polysaccharides, proteins, lipids, nucleic acids, humic substances, and water (Branda et al., 2005) and the ratio of these components varies in species and is dependent on growth factors. The formation of biofilm and production of EPS depends on environmental conditions. It also undergoes variation both in structure and content (Kreft and Wimpenny, 2001; Miqueleto et al., 2010; Jung, Choi et al., 2013). The biofilm structure observes as filamentous and mushroom-like shapes in fast-moving and static water, respectively (Edwards et al., 2000; Reysenbach and Cady, 2001). The matrix of the biofilm allows the microbes to overcome environmental stress, shear stress, acid stress, antimicrobial agents, UV damage, desiccation, predation, biocides, solvents, high concentration of toxic chemicals and pollutants (Davey and O'toole, 2000; Mah and O'Toole, 2001). In contrast, free-floating planktonic cells decontaminate environmental pollutants by metabolic activity but such cells are not stationary and not adapted to persist under mechanical and environmental stress (Sutherland, 2001).

The 3D structure of EPS with reduced oxygen concentrations toward the center attracts aerobes and anaerobes, heterotrophs near nitrifiers, and sulfate reducers with sulfate oxidizers, which promotes the rapid degradation of varied pollutants (Field et al., 1995). EPS matrix of cyanobacteria removes heavy metals from the aqueous phase and acts as a biosorbent (De Philippis et al., 2007; Micheletti et al., 2008; De Philippis et al., 2011). In some cases, EPS also act as surfactants and serves in the solubilization of hydrophobic or other refractory substrates (Iwabuchi et al., 2002). In the biofilm community, the exchange of nutrients and removal of by-products is encouraged by the presence of EPS. Pollutants such as heavy metals and organic compounds are decontaminated by extracellular enzymes within EPS of biofilms (Flemming and Wingender, 2010). Metals and metalloids get trapped on EPS due to their charge and this property enables them to form complexes with heavy metals and organic contaminants and subsequently remove the pollutants (Pal and Paul, 2008; Li and Yu, 2014). The metals that bind firmly with EPS are lead, copper, manganese, magnesium, zinc, cadmium, iron, and nickel (Ferris et al., 1989). Shortage of available nutrients promotes EPS synthesis, which is followed by absorption of metals and pollutants from the environment. EPS of biofilms comprising phosphorous-accumulating microorganisms act as storage and facilitates removal and recovery of phosphorous from wastewater (Yuan et al., 2012; Zhang et al., 2013).

The steps involved in the mechanism of action of biofilms in biofilms are as follows (Batoni et al., 2016) (Fig. 20.1):

(1) Presence of slow-growing or nongrowing cells (persisters)—activity on metabolically inactive cells.
(2) Maturation with the production of abundant EPS—inhibition of matrix synthesis or accumulation.
(3) Regulatory signals mediate biofilm formation or dispersal—interference with biofilm regulatory pathways.
(4) Adhesion—possibility to be immobilized on a surface and inhibit adhesion phase.
   ❖ Possibility to be released in a controlled manner at the pollutant site.
(5) Activity in different environmental niches.
   ❖ Multimodal mechanism of action (heavy metal, hydrocarbons, etc.)
(6) Fast mechanism of action to perform degradation in harsh environmental conditions.
(7) Action on cells released by biofilm.

20.5 **Role of microbes in bioremediation**

Usually, biofilms are formed of diverse mixed microbial species that may further promote enhanced biofilm formation and have the potential to withstand environmental stress (Burmølle et al., 2006). The existence of divergence in species helps

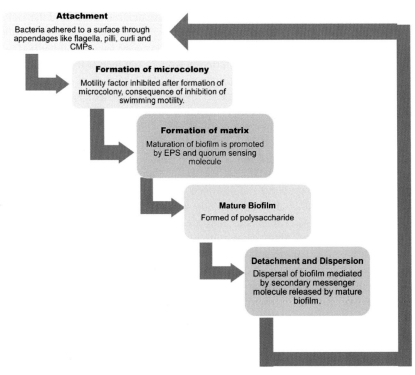

FIGURE 20.1

Formation and mechanism of action of biofilm.

in multiplication within biofilms in close proximity and promotes interaction among its members. Biofilms are traced in natural as well as engineered systems. Biofilms are found naturally in soil, aquatic plants, sediments, covering rocks in streams and plants, lakes, rivers, and wetlands. Natural biofilms are capable of degrading and removing pollutants from soil and river.

Bacteria, protozoa, fungi, and algae form biofilms on water surfaces. The following are examples of the activity of biofilms for removing pollutants (Bengtsson and Øvreås, 2010; Kartal et al., 2010):

Nitzschia-biofilms by algae in organic sediments can aid recovery of sediments by the production of oxygen, which in turn facilitates the biodegradation of aerobic bacteria present in the biofilm.

Phototrophic bacteria and algae- contribute to aquatic systems. Microbial aggregation forms microbial mats followed by the formation of sediments on the surface under extreme conditions to survive predation from grazers and stress. Flocs, fragile structures are formed during bloom periods.

Planctomycetes—biofilms of seaweed in marine waters to remove nitrogen from wastewater because of its capacity to perform anammox reactions in which ammonia is anaerobically oxidized to dinitrogen.

20.6 Types of bioremediation

The process of bioremediation is categorized into two types based on the location of pollutant treatment (Jørgensen et al., 2010; Vogt and Richnow, 2014), are as follows:

In situ bioremediation—the waste sample is treated in the original location. The process of bioremediation is more considerable, as it minimizes transportation costs and site disruption.

Ex situ bioremediation—the waste sample is treated typically off-site.

In one of the bioremediation processes, the microbes are optimized physically and chemically (such as pH, aeration, moisture) and supplemented with nutrients, to achieve rapid and effective degradation. Another approach for effective and efficient degradation by modifying pathways is to use a particular microbial strain or genetically modified microbe (Hedlund and Staley, 2001; Nakajima-Kambe et al., 2009; Singh et al., 2011).

20.7 Approaches for use of biofilms based remediation (in situ)

In situ bioremediation process involves natural attenuation based on natural processes without the intervention of engineered steps.

20.7.1 Biostimulation or natural attenuation

In this approach, microbial strains of the pollutants are degraded, transformed, immobilized, and detoxified without human intervention (Saylern et al., 1995). Though the procedures involve the addition of extra nutrients into the process, such as carbon and phosphorus compounds and air to improve oxygen availability, enhance growth, and degrade pollutants rapidly. Among the nutrition nitrogen, phosphorous and carbon are the basic sources required for the development of life (Abatenh et al., 2017). The process is known as **biostimulation** (Jørgensen et al., 2010; Vogt and Richnow, 2014). The method of natural attenuation is considerable, while the level of contaminant is relative. It is commonly practiced in the remediation of petroleum hydrocarbon sites.

20.7.2 Bioaugmentation or bioenhancement

In a few cases, specific competent microbes are introduced at the pollutant site to carry out degradation (Tyagi et al., 2011). Though the consortia of these microbes may themselves reside at the contaminant site, they also may require supplemental nutrition to perform their action of degradation. Alternately, these particular microbial strains can be extracted, isolated, stored under laboratory conditions, and could be added to the pollutant site whenever required for bioremediation purposes. The biomarkers *gfp* and *luc* are used to assess the efficacy of inoculated microbes (Jansson et al., 2000). The process of bioaugmentation can be enhanced by incorporating conventional genetic engineering techniques or by methods to increase nutrient concentrations or the persistence of microbes, or by air venting and biostimulation methods (Gentry et al., 2004). In air venting, air is pumped into the site of contamination present below the soil surface to enrich the aerobic microbial community and establish biofilms. Bioaugmentation is advantageous in remediation for petroleum hydrocarbon oil and to enhance nitrification performance (Abeysinghe et al., 2002).

20.7.3 Approaches for biofilm-based remediation (ex situ)

In large-scale pollutant sites for degradation of heavy metals, hydrocarbons, and industrial and municipal wastewater, the ex situ approach is a method of choice. In this strategy, bioreactors-based remediation is performed for sorption and biochemical conversion of pollutants (Boon et al., 2002). They are usually applicable for commercially cleaning up pollutants from the contaminant site (Bryers, 1993; Qureshi et al., 2005). Bioreactor dependent bioremediation is considered over the conventional method because of the following reasons:

❖ High concentration and retention of biomass for extended periods.
❖ Elevated metabolic activity.
❖ Enhanced process flow rates.
❖ Highly tolerant to harsh pollutants.
❖ Huge mass transfer area.
❖ Increased volumetric biodegradation capacity.
❖ Coexistence of anoxic and aerobic metabolic activity and less interruption in the bioreactor.

Bioreactor-based remediation depends on the availability of free-floating microbes in suspension. This restricts the production of adequate biomass or biomass to that which is not retained long enough for efficient volumetric conversion. This results from the gradual growth of microbes or when diluted feed streams are used in bioreactors. On the other hand, a disadvantage of bioreactor-dependent remediation is that it requires a support medium adhesion microbes and the development of biofilms (Mitra and Mukhopadhyay, 2016).

Various bioreactors used in the process of bioremediation are batch, continuous stirred tank, trickle-bed, airlift reactors, upflow anaerobic sludge blanket, fluidized-bed, expanded granular sludge blanket, biofilm airlift suspension batch reactors (Mitra and Mukhopadhyay, 2016).

20.7.4 Fixed-bed reactor

It is also known as a **packed bed reactor**. It requires solid support media, packed tightly for colonization of the biofilms. This media promotes the interaction of the biofilm mass and liquid. In this type of bioreactor, strains of mercury-resistant biofilms are successfully carried out. One of the conventional and special types of fixed-bed bioreactor is the trickle-bed biofilm bioreactor (TBR). It is commonly used for the treatment of wastewater. The support media in TBR is composed of plastic, rock, ceramics, and other materials. The working principle of TBR is that the contaminated water flows down from the top through the distribution system over the biofilm surface of the media. The pollutants present in the wastewater undergoes metabolic activities for degradation as it diffuses through the biofilms. Oxygen supply is maintained through upward or downward diffusion through the water. The nutrient reduction is directly proportional to reduced productivity (Mitra and Mukhopadhyay, 2016).

20.7.5 Fluidized-bed reactor

It is a column-based bioreactor coated with beads and polluted water is gradually pumped upward. During the treatment of contaminated water, biofilm beads remain suspended rather than the media. Suspension of solids is maintained by flow or gas at a certain velocity. Fluidization facilitates biofilm growth on large surface areas and produces large quantities of biomass. Aeration is maintained through an oxygenator or passed through the bottom of the reactor. This bioreactor is used for the treatment of polluted streams possessing organic and inorganic compounds (Shieh and Keenan, 1986; Denac and Dunn, 1988; Kumar and Saravanan, 2009).

20.7.6 Rotating biological contactors

Bioremediation is performed by reducing chemical oxygen demand and biochemical oxygen demand and includes nitrification and denitrification process (Costley and Wallis, 2001; Eker and Kargi 2008, 2010). A bioreactor is composed of a rotating cylinder or biodiscs on which the thin biofilms of aerobic microbes are grown. In effluent, the discs are partly submerged, and gradual rotation of biofilm microbes are alternately exposed to the effluent. The mechanism of action works as air passes on the disc, which promotes degradation by the biofilms. Clarifiers are used to remove the extra biomass from rotating biological contactor (RBC) media. RBCs are preferred because of their economical functioning and because they require less land or space. Pollutant removal is affected by the parameters such as the rotational speed of the discs, disc submergence, and hydraulic retention time. RBC-dependent bioremediation is used for polluted water heavy metals, degradation of dyes, volatile organic compounds, and polycyclic aromatic hydrocarbons (PAHs) (Costley and Wallis, 2001; Abraham et al., 2003; Jeswani and Mukherji, 2012; Sarayu and Sandhya, 2012).

20.7.7 **Membrane biofilm reactor**

The bioreactor is designed with permeable membranes to the attached biofilms formed on the membrane exterior. The membranes also promote the growth of the biofilm. The working principle to perform degradation is based on pressurized air or oxygen via gas is delivered to the permeable membranes. Biofilms composed of autotrophic bacteria are preferably degraded by the use of a hydrogen-based membrane biofilm reactor (MBfR). In the process of degradation, bacteria oxidize hydrogen and use electron donors to various pollutants such as chlorate and nitrate. The process avoids the formation of a bubble. Additionally, high transfer of oxygen prevents stripping of volatile organic compounds and greenhouse gases and foaming when a surfactant is used. In the treatment of nitrate and pesticides containing polluted water, a methane-fed MBfR is used. It is used for the treatment of high oxygen demanding wastewater (Nerenberg and Rittmann, 2004; Rittmann, 2006; Modin et al., 2008).

20.7.8 **Sequential aerobic-anaerobic two-stage biofilm reactor**

It is an airlift internal loop biofilm bioreactor that helps degrade polychlorinated hydrocarbons. Aerobic and anoxic biofilms undergo nitrification and denitrification. Removal of nitrate is possible due to denitrification of combining autotrophic and heterotrophic bacteria in an intensified biofilm-electrode reactor. Biofilm reactors using sulfate-reducing bacteria entrap or precipitate metals such as copper and zinc at the interface of biofilms (White and Gadd, 1998, 2000; Smith and Gadd, 2000).

20.8 **Types of pollutants remediated by biofilms**

Nowadays, biofilm-based bioremediation is highly preferred for the removal of pollutants such as organic pollutants, oil spills, heavy metals pesticides, and xenobiotics. It degrades heavy metals such as cadmium, copper, uranium, and chromium (Diels et al., 1999; Valls and de Lorenzo, 2002). Combination of phosphatase enzyme in presence of biofilm matrix enhances metal precipitation both for aerobic bacteria and for anaerobic conditions (Macaskie et al., 1997). It is also observed that the addition of carbon sources to polluted water hinders or lowers the flow of pollutants away from the contaminated site and minimizing its dispersion.

20.8.1 **Persistent organic pollutants**

Persistent organic pollutants (POPs) are found in air, water, and sediments. They have long half-lives due to hydrophobicity and proved to be toxic if they enter the food chain (Bonefeld et al., 2006). A few POPs are PAHs, polychlorinated biphenyls, and polychlorinated ethenes (Mitra and Mukhopadhyay, 2016). From the environment, many bacterial biofilms are isolated that participate in biodegradation by degrading POPs. In some cases, they are engineered accordingly to perform degradation of POP PAHs (Johnsen and Karlson, 2004; Mitra and Mukhopadhyay, 2016).

20.8.1.1 Petroleum industry

Water pollution due to the petroleum industry has reached an alarming state and is causing toxicity to marine life. An oil spill has hydrocarbons that degrade the soil microbes (Song and Bartha, 1990). Examples of bacteria exhibiting speedy growth utilized for decontamination of hydrocarbons decontamination are *Pseudomonas, Arthrobacter, Rhodococcus, Bacillus, Alcanivorax, and Cycloclasticus* spp. of gammaproteobacteria (Harayama et al., 2004; Atlas and Hazen, 2011; Ron and Rosenberg, 2014). *Bacillus subtilis* and *Acinetobacter radioresistens* with a surfactant-producing strain efficiently participate in the degradation of pollutants in marine environments (Mnif et al., 2015). Hydrocarbon degrading bacteria use uric acid from water-insoluble fertilizers as their nitrogen source (Koren et al., 2003).

20.8.2 Heavy metals

Bioremediation is applicable for the degradation of heavy metals, and metals such as copper, zinc, nickel, cadmium, and cobalt have been remediated using various biofilm reactors. In mines for scavenging metals from metal-polluted water into precipitates of metal sulfides are by biofilm-forming sulfate-reducing bacteria (Smith and Gadd, 2000; Muyzer and Stams, 2008). Few bacteria have the potential of forming electroactive films or electrochemically active biofilms which directly exchange electrons with a conductive solid surface (Erable et al., 2010). These types of biofilms are actively involved in bioremediation and are utilized for the treatment of heavy metals from polluted groundwater and soil (Li et al., 2008; Cong et al., 2013).

Various other microbes are involved in the process of biodegradation of pollutants are summarized in Table 20.1 (Abatenh et al., 2017).

Few advantages and disadvantages of biofilm-dependent bioremediation are as follows (Mitra and Mukhopadhyay, 2016; Abatenh et al., 2017):

20.9 Advantages of biofilm-based bioremediation

(1) Biofilm-based remediation for cleanup of environmental pollutants.
(2) Environmentally friendly, cost-effective, and requires less effort.
(3) In sludge plants, municipal wastewater treatment using floc activity.
(4) Process transforms hazardous compounds into harmless products, thus destroying pollutants.
(5) Nonintrusive, potentially allowing for continued site use and sustainability.
(6) Requires less effort for implementation.
(7) Effective way of remediating natural ecosystem from many contaminants and offers environmentally friendly options.

Table 20.1 List of pollutants and microorganisms involved in bioremediation.

Pollutants	Microorganisms
Hydrocarbons	
Monocyclic aromatic hydrocarbons, benzene, toluene, ethylbenzene and xylene, phenol compounds	*Penicillium chrysogenum, P. putida*
Petrol and diesel polycyclic aromatic hydrocarbons, toluene	*P. alcaligenes, P. mendocina, P. putida, P. veronii, Achromobacter, Flavobacterium,* and *Acinetobacter*
Biphenyl and triphenylmethane	*Phanerochaete chrysosporium*
Hydrocarbon	*Aspergillus niger, A. fumigatus, Fusarium solani,* and *Penicillium funiculosum, Tyromyces palustris, Gloeophyllum trabeum, Trametes versicolor*
PAHs, methylnaphthalenes, and dibenzofurans	*Coprinellus radians*
Phenol	*Alcaligenes odorans, Bacillus subtilis, Corynebacterium propinquum, Pseudomonas aeruginosa*
Phenanthrene, benzopyrene	*Candida viswanathii*
Naphthalene	Cyanobacteria, green algae and diatoms, and *Bacillus licheniformis*
Aromatic hydrocarbons	*Acinetobacter* sp., *Pseudomonas* sp., *Ralstonia* sp., and *Microbacterium* sp.
Striatum pyrene, anthracene, 9-methyl anthracene, dibenzothiophene, lignin peroxidase	*Gloeophyllum striatum*
Oils	
Oil	*Fusarium* sp., A. odorans, B. subtilis, C. propinquum, P. aeruginosa
Diesel oil	*Bacillus cereus A, P. aeruginosa, P. putida, Arthrobacter* sp., and *Bacillus* sp.
Crude oil	*A. niger, Candida glabrata, Candida krusei,* and *Saccharomyces cerevisiae, B. brevis, P. aeruginosa KH6, B. licheniformis,* and *Bacillus sphaericus*
Diesel oil, crude oil	*Pseudomonas cepacia, B. cereus, Bacillus coagulans, Citrobacter koseri,* and *Serratia ficaria*

Table 20.1 List of pollutants and microorganisms involved in bioremediation.—*cont'd*

Pollutants	Microorganisms
Dyes	
Oil-based based paints	*B. subtilis* strain NAP1, NAP2, NAP4
Industrial dyes	*Myrothecium roridum* IM 6482, *Pycnoporus sanguineous, Phanerochaete chrysosporium,* and *Trametes trogii, Penicillium ochrochloron*
Textile azo dyes	*Micrococcus luteus, Listeria denitrificans,* and *Nocardia atlantica*
Textile dye (Remazol black B), sulfonated di-azo dye, reactive red HE8B, RNB dye	*Bacillus* spp. ETL-2012, *P. aeruginosa,* and *Bacillus pumilus* HKG212
Azo dye effluents	*Exiguobacterium indicum, Exiguobacterium aurantiacum, B. cereus,* and *Acinetobacter baumanii*
Vat dyes, textile effluents	*Bacillus firmus, Bacillus macerans, Staphylococcus aureus,* and *Klebsiella oxytoca*
Heavy metals	
Heavy metals, lead, mercury, and nickel	*Saccharomyces cerevisiae*
Heavy metals	*Cunninghamella elegans*
Fe^{2+}, Zn^{2+}, Pb^{2+}, Mn^{2+}, and Cu^2	*Pseudomonas fluorescens* and *P. aeruginosa*
Cobalt, copper, chromium, and lead	*Lysinibacillus sphaericus* CBAM5
Fe	*Microbacterium profundi* strain Shh49T
Cadmium	*Aspergillus versicolor, A. fumigatus, Paecilomyces* sp., *Paecilomyces* sp., *Trichoderma* sp., *Microsporum* sp., and *Cladosporium* sp.
Fe (III), U (VI)	*Geobacter* spp.
Cadmium	*Bacillus safensis* (JX126862) strain (PB-5 and RSA-4)
U, Cu, Ni, Cr	*P.* aeruginosa, *Aeromonas* sp.
Pb, Cr, Cd	*Aerococcus* sp., *Rhodopseudomonas palustris*
Pesticides	
Endosulfan	*Bacillus, Staphylococcus*
Chlorpyrifos	*Enterobacter*
Ridomil MZ 68 MG, Fitoraz WP 76, Decis 2.5 EC, malathion	*Pseudomonas putida, Acinetobacter* sp., *Arthrobacter* sp.
Chlorpyrifos and methyl parathion	*Acinetobacter* sp., *Pseudomonas* sp., *Enterobacter* sp., and *Photobacterium* sp.

20.10 Disadvantages of biofilm-based bioremediation

(1) Restricted to biodegradable compounds.
(2) Products of biodegradation possibly more persistent or toxic than the parent compound.
(3) Troublesome to extrapolate from bench and pilot-scale studies to full-scale pollutant sites.
(4) Often takes longer than other treatment options, such as excavation and removal of soil or incineration.

20.11 Conclusion

Nowadays, pollution is a major concern. To overcome the critical issues, bioremediation is quite effective at eliminating pollutants from contaminated sites. Bioremediation using microbes plays a vital role in the degradation of pollutants with the release of no or minimal toxic or hazardous compounds. In this practice, novel approaches are constantly being tested and implemented in the process of bioremediation. However, the emerging trend of biofilm-dependent remediation is robust and reliable, although implementation on a massive scale remains challenging. Thus, in future work to improve biofilm-mediated remediation, we will need an in-depth understanding of the biofilm formation processes, specific mechanisms, and diverse microbial actions for better outcome.

References

Abatenh, E., Gizaw, B., et al., 2017. The role of microorganisms in bioremediation- A review. Open J. Environ. Biol. 2 (1), 038–046.
Abeysinghe, D.H., De Silva, D.G., et al., 2002. The effectiveness of bioaugmentation in nitrifying systems stressed by a washout condition and cold temperature. Water Environ. Res. 74 (2), 187–199.
Abraham, T.E., Senan, R.C., et al., 2003. Bioremediation of textile azo dyes by an aerobic bacterial consortium using a rotating biological contactor. Biotechnol. Prog. 19 (4), 1372–1376.
Alexander, M., Loehr, R.C., 1992. Bioremediation review. Science 258 (5084), 874.
Atlas, R.M., Hazen, T.C., 2011. Oil biodegradation and bioremediation: a tale of the two worst spills in U.S. history. Environ. Sci. Technol. 45 (16), 6709–6715.
Balaji, V., Arulazhagan, P., et al., 2014. Enzymatic bioremediation of polyaromatic hydrocarbons by fungal consortia enriched from petroleum contaminated soil and oil seeds. J. Environ. Biol. 35 (3), 521–529.
Batoni, G., Maisetta, G., et al., 2016. Antimicrobial peptides and their interaction with biofilms of medically relevant bacteria. Biochim. Biophys. Acta 1858 (5), 1044–1060.
Bengtsson, M.M., Øvreås, L., 2010. Planctomycetes dominate biofilms on surfaces of the kelp Laminaria hyperborea. BMC Microbiol. 10, 261.

Beyenal, H., Sani, R.K., et al., 2004. Uranium immobilization by sulfate-reducing biofilms. Environ. Sci. Technol. 38 (7), 2067−2074.

Bonefeld-Jorgensen, E.C., Hjelmborg, P.S., et al., 2006. Xenoestrogenic activity in blood of European and Inuit populations. Environ. Health 5 (12).

Boon, N., De Gelder, L., et al., 2002. Bioaugmenting bioreactors for the continuous removal of 3-chloroaniline by a slow release approach. Environ. Sci. Technol. 36, 4698−4704.

Bouwer, E.J., Zehnder, A.J., 1993. Bioremediation of organic compounds–putting microbial metabolism to work. Trends Biotechnol.: Cell Press 11 (8), 360−367.

Branda, S.S., Vik, S., et al., 2005. Biofilms: the matrix revisited. Trends Microbiol.: Cell Press 13 (1), 20−26.

Bruins, M.R., Kapil, S., et al., 2000. Microbial resistance to metals in the environment. Ecotoxicol. Environ. Saf. 45 (3), 198−207.

Bryers, J.D., 1993. Bacterial biofilms. Curr. Opin. Biotechnol. 4 (2), 197−204.

Burmølle, M., Webb, J.S., et al., 2006. Enhanced biofilm formation and increased resistance to antimicrobial agents and bacterial invasion are caused by synergistic interactions in multi-species biofilms. Appl. Environ. Microbiol. 72 (6), 3916−3923.

Cerniglia, C.E., 1997. Fungal metabolism of polycyclic aromatic hydrocarbons: past, present and future applications in bioremediation. J. Ind. Microbiol. Biotechnol. 19 (5−6), 324−333.

Characklis, W.G., 1973. Attached microbial growths-II. Frictional resistance due to microbial slimes. Water Res. 7, 1249−1258.

Cong, Y., Xu, Q., et al., 2013. Efficient electrochemically active biofilm denitrification and bacteria consortium analysis. Bioresour. Technol. 132, 24−27.

Costley, S.C., Wallis, F.M., 2001. Bioremediation of heavy metals in a synthetic wastewater using a rotating biological contactor. Water Res. 35, 3715−3723.

Das, S., Dash, H.R. (Eds.), 2014. Microbial Bioremediation: A Potential Tool for Restoration of Contaminated Areas. Microbial Biodegradation and Bioremediation. Elsevier.

Davey, M.E., O'toole, G.A., 2000. Microbial biofilms: from ecology to molecular genetics. Microbiol. Mol. Biol. Rev. 64 (4), 847−867.

Day, S.M., 1993. US environmental regulations and policies–their impact on the commercial development of bioremediation. Trends Biotechnol.: Cell Press 11 (8), 324−328.

De Philippis, R., Colica, G., et al., 2011. Exopolysaccharide-producing cyanobacteria in heavy metal removal from water: molecular basis and practical applicability of the biosorption process. Appl. Microbiol. Biotechnol. 92 (4), 697−708.

De Philippis, R., Paperi, R., et al., 2007. Heavy metal sorption by released polysaccharides and whole cultures of two exopolysaccharide-producing cyanobacteria. Biodegradation 18 (2), 181−187.

Denac, M., Dunn, I.J., 1988. Packed- and fluidized-bed biofilm reactor performance for anaerobic wastewater treatment. Biotechnol. Bioeng. 32, 159−173.

Diels, L., De Smet, M., et al., 1999. Heavy metals bioremediation of soil. Mol. Biotechnol. 12, 149−158.

Donlan, R.M., 2002. Biofilms: microbial life on surfaces. Emerg. Infect. Dis. 8 (9), 881−890.

Donlan, R.M., Pipes, W.O., et al., 1994. Biofilm formation on cast iron substrata in water distribution systems. Water Res. 28, 1497−1503.

Edwards, K.J., Bond, P.L., et al., 2000. An archaeal iron-oxidizing extreme acidophile important in acid mine drainage. Science 287 (5459), 1796−1799.

Eker, S., Kargi, F., 2008. Biological treatment of 2,4-dichlorophenol containing synthetic wastewater using a rotating brush biofilm reactor. Bioresour. Technol. 99 (7), 2319−2325.

Eker, S., Kargi, F., 2010. COD, para-chlorophenol and toxicity removal from synthetic wastewater using rotating tubes biofilm reactor (RTBR). Bioresour. Technol. 101 (23), 9020–9024.

Erable, B., Duteanu, N.M., et al., 2010. Application of electro-active biofilms. Biofouling 26 (1), 57–71.

Fera, P., Siebel, M.A., et al., 1989. Seasonal variations in bacterial colonization of stainless steel, aluminum, and polycarbonate surfaces in a seawater flow system. Biofouling 1, 251–261.

Ferris, F.G., Schultze, S., et al., 1989. Metal interactions with microbial biofilms in acidic and neutral pH environments. Appl. Environ. Microbiol. 55 (5), 1249–1257.

Field, J.A., Stams, A.J., et al., 1995. Enhanced biodegradation of aromatic pollutants in cocultures of anaerobic and aerobic bacterial consortia. Antonie Van Leeuwenhoek 67 (1), 47–77.

Flemming, H.C., Wingender, J., 2001. Relevance of microbial extracellular polymeric substances (EPSs)–Part I: structural and ecological aspects. Water Sci. Technol. 43 (6), 1–8.

Flemming, H.C., Wingender, J., 2010. The biofilm matrix. Nat. Rev. Microbiol. 8, 623–633.

Fletcher, M., 1988. In: Houghton, D.R., Smith, R.N., Eggins, H.O.W. (Eds.), The Applications of Interference Reflection Microscopy to the Study of Bacterial Adhesion to Solid Surfaces. Springer, Dordrecht.

Gao, W., Francis, A.J., 2008. Reduction of uranium(VI) to uranium(IV) by clostridia. Appl. Environ. Microbiol. 74 (14), 4580–4584.

Gaur, N., Flora, G., et al., 2014. A review with recent advancements on bioremediation-based abolition of heavy metals. Environ. Sci. Processes Impacts 16 (2), 180–193.

Gentry, T., Rensing, C., et al., 2004. New approaches for bioaugmentation as a remediation Technology. Crit. Rev. Environ. Sci. Technol. 34, 447–494.

Gieg, L.M., Fowler, S.J., et al., 2014. Syntrophic biodegradation of hydrocarbon contaminants. Curr. Opin. Biotechnol. 27, 21–29.

Harayama, S., Kasai, Y., et al., 2004. Microbial communities in oil-contaminated seawater. Curr. Opin. Biotechnol. 15 (3), 205–214.

Hedlund, B.P., Staley, J.T., 2001. Vibrio cyclotrophicus sp. nov., a polycyclic aromatic hydrocarbon (PAH)-degrading marine bacterium. Int. J. Syst. Evol. Microbiol. 51 (Pt 1), 61–66.

Heukelekian, H., Heller, A., 1940. Relation between food concentration and surface for bacterial growth. J. Bacteriol. 40 (4), 547–558.

Horemans, B., Breugelmans, P., et al., 2013. Environmental dissolved organic matter governs biofilm formation and subsequent linuron degradation activity of a linuron-degrading bacterial consortium. Appl. Environ. Microbiol. 79, 4534–4542.

Iwabuchi, N., Sunairi, M.U., Urai, M., et al., 2002. Extracellular polysaccharides of Rhodococcus rhodochrous S-2 stimulate the degradation of aromatic components in crude oil by indigenous marine bacteria. Appl. Environ. Microbiol. 68 (5), 2337–2343.

Jansson, J.K., Björklöf, K., et al., 2000. Biomarkers for monitoring efficacy of bioremediation by microbial inoculants. Environ. Pollut. 107 (2), 217–223.

Jeswani, H., Mukherji, S., 2012. Degradation of phenolics, nitrogen-heterocyclics and polynuclear aromatic hydrocarbons in a rotating biological contactor. Bioresour. Technol. 111, 12–20.

Johnsen, A.R., Karlson, U., 2004. Evaluation of bacterial strategies to promote the bioavailability of polycyclic aromatic hydrocarbons. Appl. Microbiol. Biotechnol. 63 (4), 452–459.

Jones, H.C., Roth, I.L., et al., 1969. Electron microscopic study of a slime layer. J. Bacteriol. 99 (1), 316—325.

Jørgensen, K.S., Salminen, J.M., et al., 2010. Monitored natural attenuation. Methods Mol. Biol. 599, 217—233.

Joutey, N.T., Sayel, H., et al., 2015. Mechanisms of hexavalent chromium resistance and removal by microorganisms. Rev. Environ. Contam. Toxicol. 233, 45—69.

Jung, J.H., Choi, N.Y., et al., 2013. Biofilm formation and exopolysaccharide (EPS) production by Cronobacter sakazakii depending on environmental conditions. Food Microbiol. 34 (1), 70—80.

Kartal, B., Kuenen, J.G., et al., 2010. Engineering. Sewage treatment with anammox. Science 328 (5979), 702—703.

Koren, O., Knezevic, V., et al., 2003. Petroleum pollution bioremediation using water-insoluble uric acid as the nitrogen source. Appl. Environ. Microbiol. 69, 6337—6339.

Kreft, J.U., Wimpenny, J.W., 2001. Effect of EPS on biofilm structure and function as revealed by an individual-based model of biofilm growth. Water Sci. Technol. 46 (6), 135—141.

Kumar, R., Singh, S., et al., 2007. Bioremediation of radionuclides: emerging technologies. OMICS 11 (33), 295—304.

Kumar, T.A., Saravanan, S., 2009. Treatability studies of textile wastewater on an aerobic fluidized bed biofilm reactor (FABR): a case study. Water Sci. Technol. 59 (9), 1817—1821.

Lacal, J., Reyes-Darias, J.A., et al., 2013. Tactic responses to pollutants and their potential to increase biodegradation efficiency. J. Appl. Microbiol. 114 (4), 923—933.

Li, W.W., Yu, H.Q., 2014. Insight into the roles of microbial extracellular polymer substances in metal biosorption. Bioresour. Technol. 160, 15—23.

Li, Z., Zhang, X., et al., 2008. Electricity production during the treatment of real electroplating wastewater containing Cr^{6+} using microbial fuel cell. Process Biochem. 43 (12), 1352—1358.

Macaskie, L.E., Yong, P., et al., 1997. Bioremediation of uranium-bearing wastewater: biochemical and chemical factors influencing bioprocess application. Biotechnol. Bioeng. 53 (1), 100—109.

Mah, T.F., O'Toole, G.A., 2001. Mechanisms of biofilm resistance to antimicrobial agents. Trends Microbiol.: Cell Press 9 (1), 34—39.

Meng-Ying, L., Ji, Z., et al., 2009. Evaluation of biological characteristics of bacteria contributing to biofilm formation. Pedosphere 19 (5), 554—561.

Micheletti, E., Colica, G., et al., 2008. Selectivity in the heavy metal removal by exopolysaccharide-producing cyanobacteria. J. Appl. Microbiol. 105 (1), 88—94.

Miqueleto, A.P., Dolosic, C.C., et al., 2010. Influence of carbon sources and C/N ratio on EPS production in anaerobic sequencing batch biofilm reactors for wastewater treatment. Bioresour. Technol. 101 (4), 1324—1330.

Mishra, A., Malik, A., 2014. Novel fungal consortium for bioremediation of metals and dyes from mixed waste stream. Bioresour. Technol. 171, 217—226.

Mitra, A., Mukhopadhyay, S., 2016. Biofilm mediated decontamination of pollutants from the environment. AIMS Bioeng. 3 (1), 44—59.

Mnif, I., Mnif, S., et al., 2015. Biodegradation of diesel oil by a novel microbial consortium: comparison between co-inoculation with biosurfactant-producing strain and exogenously added biosurfactants. Environ. Sci. Pollut. Control Ser. 22 (19), 14852—14861.

Modin, O., Fukushi, K., et al., 2008. Simultaneous removal of nitrate and pesticides from groundwater using a methane-fed membrane biofilm reactor. Water Sci. Technol. 58 (6), 1273—1279.

More, T.T., Yadav, J.S., et al., 2014. Extracellular polymeric substances of bacteria and their potential environmental applications. J. Environ. Manag. 1 (144), 1−25.

Muyzer, G., Stams, A.J., 2008. The ecology and biotechnology of sulphate-reducing bacteria. Nat. Rev. Microbiol. 6 (6), 441−454.

Nakajima-Kambe, T., Ichihashi, F., et al., 2009. Degradation of aliphatic−aromatic copolyesters by bacteria that can degrade aliphatic polyesters. Polym. Degrad. Stabil. 94 (11), 1901−1905.

Nerenberg, R., Rittmann, B.E., 2004. Hydrogen-based, hollow-fiber membrane biofilm reactor for reduction of perchlorate and other oxidized contaminants. Water Sci. Technol. 49 (11−12), 223−230.

Pal, A., Paul, A.K., 2008. Microbial extracellular polymeric substances: central elements in heavy metal bioremediation. Indian J. Med. Microbiol. 48 (1), 49−64.

Pratt, L.A., Kolter, R., 1999. Genetic analyses of bacterial biofilm formation. Curr. Opin. Microbiol. 2 (6), 598−603.

Qureshi, N., Annous, B.A., et al., 2005. Biofilm reactors for industrial bioconversion processes: employing potential of enhanced reaction rates. Microb. Cell Fact. 4, 24.

Reysenbach, A.L., Cady, S.L., 2001. Microbiology of ancient and modern hydrothermal systems. Trends Microbiol.: Cell Press 9 (2), 79−86.

Rittmann, B.E., 2006. The membrane biofilm reactor: the natural partnership of membranes and biofilm. Water Sci. Technol. 53 (3), 219−225.

Ron, E.Z., Rosenberg, E., 2014. Enhanced bioremediation of oil spills in the sea. Curr. Opin. Biotechnol. 27, 191−194.

Sarayu, K., Sandhya, S., 2012. Rotating biological contactor reactor with biofilm promoting mats for treatment of benzene and xylene containing wastewater. Biotechnol. Appl. Biochem. 168 (7), 1928−1937.

Sayler, G.S., Layton, A., et al., 1995. Molecular site assessment and process monitoring in bioremediation and natural attenuation. off. Appl. Biochem. Biotechnol. 54 (1−3), 277−290.

Shieh, W.K., Keenan, J.D., 1986. Fluidized bed biofilm reactor for wastewater treatment. In: Bioproducts. Springer, Berlin, Heidelberg.

Singh, J.S., Abhilash, P.C., et al., 2011. Genetically engineered bacteria: an emerging tool for environmental remediation and future research perspectives. Gene 480 (1−2), 1−9.

Smith, W.L., Gadd, G.M., 2000. Reduction and precipitation of chromate by mixed culture sulphate-reducing bacterial biofilms. J. Appl. Microbiol. 88 (6), 983−991.

Song, H.G., Bartha, R., 1990. Effects of jet fuel spills on the microbial community of soil. Appl. Environ. Microbiol. 56 (3), 646−651.

Sutherland, I.W., 2001. The biofilm matrix–an immobilized but dynamic microbial environment. Trends Microbiol.: Cell Press 9 (5), 222−227.

Tyagi, M., da Fonseca, M.M., et al., 2011. Bioaugmentation and biostimulation strategies to improve the effectiveness of bioremediation processes. Biodegradation 22 (2), 231−241.

Valls, M., de Lorenzo, V., 2002. Exploiting the genetic and biochemical capacities of bacteria for the remediation of heavy metal pollution. Fed. Eur. Microbiol. Soc. Microbiol. Rev. 26 (4), 327−338.

Vidali, M., 2001. Bioremediation. an overview. Pure Appl. Chem. 73 (7), 1163−1172.

Vogt, C., Richnow, H.H., 2014. Bioremediation via in situ microbial degradation of organic pollutants. Adv. Biochem. Eng. Biotechnol. 142, 123−146.

White, C., Gadd, G.M., 1998. Accumulation and effects of cadmium on sulphate-reducing bacterial biofilms. Microbiology 144 (5), 1407−1415.

White, C., Gadd, G.M., 2000. Copper accumulation by sulfate-reducing bacterial biofilms. Fed. Eur. Microbiol. Soc. Microbiol. Lett. 183 (2), 313–318.

Yuan, Z., Pratt, S., et al., 2012. Phosphorus recovery from wastewater through microbial processes. Curr. Opin. Biotechnol. 23 (6), 878–883.

Zhang, H.L., Fang, W., et al., 2013. Phosphorus removal in an enhanced biological phosphorus removal process: roles of extracellular polymeric substances. Environ. Sci. Technol. 47 (20), 11482–11489.

Index

Note: 'Page numbers followed by "f" indicate figures and "t" indicates tables.'